Blue and Green

Urban and Industrial Environments
Series editor: Robert Gottlieb, Henry R. Luce Professor of Urban and Environmental Policy, Occidental College

For a complete list of books published in this series, please see the back of the book.

Blue and Green

The Drive for Justice at America's Port

Scott L. Cummings

The MIT Press
Cambridge, Massachusetts
London, England

© 2018 Massachusetts Institute of Technology

All rights reserved. No part of this book may be reproduced in any form by any electronic or mechanical means (including photocopying, recording, or information storage and retrieval) without permission in writing from the publisher.

This book was set in ITC Stone Serif Std by Toppan Best-set Premedia Limited.

Library of Congress Cataloging-in-Publication Data

Names: Cummings, Scott L., 1969-, author.
Title: Blue and green : the drive for justice at America's port / Scott L. Cummings.
Description: Cambridge, MA : The MIT Press, 2018. | Series: Urban and industrial environments | Includes bibliographical references and index.
Identifiers: LCCN 2017014397| ISBN 9780262036986 (hardcover : alk. paper) | ISBN 9780262534314 (pbk. : alk. paper)
Subjects: LCSH: Harbors--Law and legislation--California. | Truck drivers--Legal status, laws, etc.--California--Los Angeles. | Labor laws and legislation--California--Los Angeles. | Harbors--Law and legislation--California--Long Beach. | Harbors--Law and legislation--California--Los Angeles. | Harbors--Environmental aspects--California--Los Angeles. | Harbors--Environmental aspects--California--Long Beach. | Long Beach Harbor (Calif.) | Los Angeles Harbor (Calif.) | Immigrants--Law and legislation--California--Los Angeles.
Classification: LCC KFC30.5.D4 C86 2017 | DDC 343.79409/67--dc23 LC record available at https://lccn.loc.gov/2017014397

To Mom and Dad—
For nurturing and sustaining the drive for justice in me.

Contents

Preface ix
List of Acronyms xiii

1 The Movement for Economic and Environmental Justice at America's Port 1
2 Law in the Development of the Port 17
3 The Port as a Unit of Legal Analysis 51
4 Resisting the Port: Activism in Separate Spheres 59
5 Reforming the Port: The Campaign for Clean Trucks 103
6 Reshaping the Industry: The Challenge to Misclassification 185
7 Assessing the Movement: Impact and Implications 293

Notes 355
Index 499
Series List 525

Preface

This book is a testament to following a road wherever it may lead you—even when you do not know in advance precisely where that point will be. I did not sit down with a plan to write a book about the activists and lawyers whose fight for economic and environmental justice at America's largest port complex has profoundly changed the lives of some of the most marginalized workers and community members in my home of Los Angeles. But once I began learning about their struggle, their courage in the face of long odds, their commitment to the idea that our society could and should do better, and their resilience in the face of setbacks, I was unable to stop writing a story that for me came to transcend the labor and environmental movements. It was a story of solidarity and sacrifice—of choosing to stand for something and living it out—that I wanted to tell in full so that others after me could be similarly moved and inspired.

Telling this story would not have been possible without the support of a number of special people. I am particularly grateful to Richard Abel, who not only read through the entire manuscript but also nudged me to publish it as a book. As always, his mentorship and guidance have been essential.

I am forever grateful to Catherine Fisk for inviting me to participate in the *UC Irvine Law Review*'s symposium on Re-imagining Labor Law—for which I wrote an initial article version of this project entitled "Preemptive Strike: Law in the Campaign for Clean Trucks"—and for hanging with me as it ballooned in size and scope. There were other incredible people who were instrumental at various points along the way, particularly Tony Alfieri, Sameer Ashar, Devon Carbado, Rashmi Dyal-Chand, Ingrid Eagly, Sean Hecht, Victor Narro, Doug NeJaime, Katherine Stone, and Noah Zatz, all of whom provided comments and other help at critical junctures. While visiting at Harvard Law School, I received thoughtful feedback from Elizabeth Bartholet, Jody Freeman, Lani Guinier, Michael Klarman, Ben Sachs, Lucie White, and David Wilkins. Jon Zerolnick at the Los Angeles Alliance for a

New Economy was crucial for helping me understand the campaign and giving me access to the materials I needed to do it justice.

Bob Gottlieb, professor at Occidental College and series editor for the MIT Press's Urban and Industrial Environments series, provided excellent comments on an earlier version of the manuscript, and gave invaluable support in my effort to turn it into a book and navigate the publication process. Bob has been an inspiration to those of us who aspire to combine academics and activism, and it is a privilege to be part of a series associated with him.

I am deeply indebted to my own institution for giving me the time and resources I needed to complete the project. I am particularly grateful to former Dean Rachel Moran and Vice Dean Laura Gómez for encouragement and assistance at critical stages and to my current Dean Jennifer Mnookin for giving me the time to complete the final manuscript. My work benefited tremendously from the generous and enduring support of the UCLA Institute for Research on Labor and Employment and its leadership, particularly Chris Tilly and Abel Valenzuela. I also relied on the stalwart assistance of a dream team of research assistants over several years—including Nareg Essaghoolian, Michael Fenne, Tala Oszkay, Doug Smith, and Alyssa Titche. Lindsay Cutler, in particular, deserves special recognition for her extraordinary commitment to helping me get the manuscript in final shape for publication. Without her, I would have certainly lost track of all the details of the editorial process. I am thankful for the incredible efforts of the students who helped edit parts of the earlier article, "Preemptive Strike," especially Amy Bowles, Anne Conley, Catriona Lavery, Margaux Poueymirou, Marco Pulido, Joseph Roth, and Thomas Worger. In addition, the UCLA School of Law library staff—Rachel Green, Jodi Kruger, and Lynn McClelland—were invaluable and infinitely patient. Elyse Meyers' special friendship and generosity were indispensable to getting me over the final hurdles, talking through problems and providing excellent editorial support on the final chapter. Frank Lopez, manager of Publications and Graphic Design, gave freely of his time to enhance the book's design. Finally, I have been privileged to work with UCLA's greatest faculty assistant, Jamie Libonate, who has helped in too many ways to enumerate.

I benefited from the insights of extremely helpful and engaged colleagues in different workshops, including: the Labor Law Group meeting at Lake Arrowhead in 2012; the UC Irvine Law Review Conference, Re-imagining Labor Law in 2013; the Law & Society Association Annual Meeting in 2014; the International Legal Ethics Conference VI in London, 2014; the Harvard Law School Faculty Workshop in 2014; the ClassCrits VIII conference at the

University of Tennessee in 2015; the Northeastern Law School conference on The Puzzle of the Urban Core in 2015; and the AALS Annual Meeting panel on Local Laboratories of Workplace Regulation in 2016.

Last but not least, I have the good fortune of having the most wonderful family and I could not have written this book without their incredible, multigenerational support. I am particularly grateful for the tenacious devotion of my parents, who connect me to my past; my wife, who anchors me in the present; and my daughters, who give me hope for the future.

List of Acronyms

AFL	American Federation of Labor
AQMD	Air quality management district
ARB	Air Resources Board
ATA	American Trucking Associations
BNSF	Burlington Northern Santa Fe Railroad
CAAP	Clean Air Action Plan
CARB	California Air Resources Board
CEQA	California Environmental Quality Act
CIO	Congress of Industrial Organizations
CLUE	Clergy and Laity United for Economic Justice
COSCO	China Ocean Shipping Company
CWA	Communications Workers of America
DLSE	Department of Labor Standards Enforcement
DOL	Department of Labor
EIR	Environmental Impact Report
EIS	Environmental Impact Statement
FAAAA	Federal Aviation Administration Authorization Act
FMC	Federal Maritime Commission
IBT	Intermodal Bridge Transport
ICC	Interstate Commerce Commission
ICTF	Intermodal Container Transfer Facility
ILA	International Longshoremen's Association
ILWU	International Longshore and Warehouse Union
LAANE	Los Angeles Alliance for a New Economy
LAFCO	Local Agency Formation Commission
LBACA	Long Beach Alliance for Children with Asthma
LMC	Licensed Motor Carrier
MATES	Multiple Air Toxics Exposure Study
MWIU	Marine Workers Industrial Union

NELP	National Employment Law Project
NEPA	National Environmental Policy Act
NLRA	National Labor Relations Act
NLRB	National Labor Relations Board
PAGA	Private Attorney General Act
PMA	Pacific Maritime Association
PMSA	Pacific Merchant Shipping Association
SCAQMD	South Coast Air Quality Management District
SCIG	Southern California International Gateway
SEIU	Service Employees International Union
TEU	Twenty-foot equivalent unit
TMA	Transport Maritime Association
TRO	Temporary restraining order
TTSI	Total Transportation Services, Inc.
UCLA	University of California Los Angeles
ULP	Unfair labor practice
UNITE HERE	Union of Needletrades, Industrial, and Textile Employees and Hotel Employees and Restaurant Employees union
UP	Union Pacific Railroad
USC	University of Southern California
WRTU	Waterfront Rail Truckers Union

1 The Movement for Economic and Environmental Justice at America's Port

Overview

This book is about the struggle over the future of work and the environment on the edge of the global economy. It traces the history of conflict in an industry that is not widely known, but sits at the epicenter of the global supply chain: short-haul trucking responsible for moving the mass of imports from enormous cargo ships to warehouses and retailers around the country. The book's specific focus is on the nation's largest and most important port complex—America's port—which straddles the border of Los Angeles and Long Beach, California, and includes the ports in both cities.[1] Here, for nearly two decades, labor and environmental groups—bound together in a pivotal "blue-green" alliance—carried forward a monumental campaign to transform working conditions for drivers and environmental conditions for communities. At bottom, the book tells a story of the unceasing resolve of courageous people working together to make lives better for some of the most marginalized members of society: immigrant truck drivers barely scraping by as they deliver goods to be sold by the richest and most powerful companies in the world and residents of neighborhoods whose poverty consigns them to inhale the noxious residue of global trade. The central question is: How does law serve as a tool in their struggle?

To answer this question, the book probes deeply into the history and contemporary results of the fight to reform the trucking industry at America's port. Doing so spotlights the transformational role played by lawyers and activists from social movements historically at odds, who came together around a complex problem—a dysfunctional trucking industry producing negative economic and environmental externalities—to define and implement an ambitious and comprehensive solution. Using the port struggle as a lens, the book explores how movements deploy law to advance challenges to corporate power and how opponents mobilize law, in turn, to limit achievements in one domain—while inspiring new pathways of reform in others.[2] Centered on a close analysis of advocacy and organizing in Los Angeles, the port story draws attention to a broader political phenomenon: the creation of social movement coalitions using interconnected legal and political strategies to reshape local regulation in the pursuit of progressive policy reform. Tracing the long arc of this cross-sectoral movement activism, the book reveals how the fight for economic and environmental change never ends, but rather takes different shapes and forms, finding ways to circumvent barriers and relentlessly forge new legal and political paths on the long road to justice.

The book advances these goals by documenting and analyzing the multifaceted struggle to reshape port trucking in Los Angeles and Long Beach, which began in the mid-2000s and continues to this day. The struggle unfolded in three key phases: the first organized around passing a new local policy that would jointly address the economic and environmental harms of port trucking (called the "Campaign for Clean Trucks"); the second focused on defending that policy from industry-led attack in court; and the third directed toward litigation and organizing designed to create the conditions for unionizing port truck drivers (called the "Justice for Port Drivers campaign").[3] As the book shows, these efforts emerged as a conflict over air quality, but developed as a fight over the conditions of short-haul truck drivers, whose precarious economic status as independent contractors contributed to poorly maintained trucks that were a key cause of air pollution. The mobilization itself thus became an all too rare moment of labor-environmental interest convergence and an opportunity to rebuild a frayed alliance. The labor movement also viewed the port struggle as a chance to test a new strategy of investing in campaigns around regionally sticky industries—those connected to the local economy in ways that protected them against outsourcing—in order to advance a more ambitious project of citywide economic change.

The Campaign for Clean Trucks, the central policy-making phase, rested on an innovative legal hook: the ports, as publicly owned and operated entities, had the power to define the terms of entry for trucking companies through concession agreements—essentially private contracts permitting trucks onto port property. The campaign therefore hinged on how these agreements treated drivers and what types of standards the agreements set. Labor and environmental activists formed a coalition with residents of low-income communities adjacent to the ports. Together, they sought to pass a local law that would convert truck drivers into employees of their companies (thus enabling unionization), while requiring the gradual upgrading of trucks to reduce pollution. Lawyers working with the coalition crafted that law to minimize ex ante legal risk and ultimately defended—with mixed results—the law against a federal preemption challenge. In the aftermath of that challenge, the labor side of the coalition revised its goals and strategy, launching the new Justice for Port Drivers campaign based on a novel litigation-and-organizing strategy to fight driver misclassification, which has made significant progress toward changing driver status and promoting unionization industry wide.

The book offers a comprehensive account of these efforts to explore how law operates both as a powerful tool and a formidable barrier in social movement challenges to environmental degradation and low-wage work. Its goal is to use the port struggle to explore how distinct social movements—environmental and labor—come together in *coalition* around an issue of joint interest—port trucking—and how they mobilize law to create policy reform, defend against opposition, and adapt to unforeseen challenges. Doing so spotlights the role of *lawyers* as collaborators in—and sometimes architects of—innovative strategies to change law and corporate practice to benefit those with less power. It also focuses attention on the role of *local government* as an alternative forum for progressive law making in a political context where national-level opportunities are limited. From this vantage point, the book seeks to deepen empirical and theoretical understanding of local legal mobilization as a social movement strategy, while addressing a prominent policy issue with national implications: how to regulate labor and environmental standards at U.S. ports, which continue to grow in scale and importance as the main conduit to the global economy. Especially as ports rapidly expand on the eastern seaboard in connection with the reopening of the Panama Canal, lessons from the struggle at Los Angeles and Long Beach promise to shape how activists and policy makers craft regulations going forward that strike the proper balance between promoting strong economic and environmental standards, and sustaining robust global trade. Indeed, the outcome of the struggle has already reverberated through other ports—particularly those in New York and New Jersey, as well as Oakland—which have attempted to craft their own Clean Truck Programs that avoid the challenges faced by their Southern California counterpart.

The broader implications of this study are significant for the future of work in the United States. As recent job growth has occurred primarily in occupational categories outside the scope of traditional full-time permanent work (i.e., temporary employees and independent contractors), the American work force as a whole is starting to look more and more like the port trucking industry. Risk is being shifted from companies to workers throughout the economy. Understanding what that means for the workers involved, how they might resist these changes, and how regulation may be adapted to support their resistance are critical questions for the larger movement for economic justice. And, as the port struggle also highlights, it is a movement that cannot be won by organized labor alone, but requires sustainable alliances to create a sustainable economy.

Law and Social Movement Coalitions at the Local Level

Although this book is empirical in its fundamental approach, it is motivated by a set of theoretical questions about the role of law and lawyers in progressive social movements, and the relation between those movements and local government law in contemporary efforts to improve conditions for low-wage workers and low-income communities. As such, it seeks to intervene in important scholarly conversations, while providing lessons of a deeply practical nature for lawyers and activists searching for models of viable economic and environmental reform. From this perspective, a distinguishing feature of the book is its innovative theoretical focus on the relationship between *law* and social movement *coalitions* operating *locally* to change policy and transform conditions for the most marginalized members of society. As such, it brings together related strands of research that have previously been disconnected: on legal mobilization, social movement coalitions, and local workplace and environmental regulation.

From a theoretical perspective, this project is centrally interested in the complex relationship between law and social movements. After the civil rights era, scholars in law and social science painted a skeptical picture of the power of law and lawyers to promote fundamental social change.[4] Lawyers were viewed with suspicion, too quick to turn to court and too removed from grassroots activism and constituent accountability.[5] Courts were seen as places where complex and controversial social problems were converted into technical legal arcana, where collective struggles were reframed as individual grievances, and where movement momentum languished as the gears of the justice system moved at a glacial pace.[6] Overall, the nationally oriented legal and political strategy that defined the civil rights struggle was viewed by many critical scholars as ultimately inadequate to the task of producing enduring democratic change. And as the political terrain shifted to the right, opportunities for national-level reform for progressive causes faded from view.

It is against this backdrop that scholars have sought to rethink the relationship between law and social movements with an eye toward recapturing productive linkages and contesting critical accounts. Drawing on the idea of law as a resource for social change that can be leveraged inside and outside of court, Michael McCann has outlined the productive uses of legal mobilization in movement building: leveraging the symbolic dimensions of law in support of "developing an agenda," "generating mass involvement," promoting "rights consciousness and political identity," and helping "provide reform activists a variety of more refined tools—procedures, standards, practices—useful in the struggle to win *effective* policy agreements and implementation from those negotiations."[7] Building upon the legal mobilization concept, social science scholars have explored how law shapes the opportunity structure for contentious politics and helps to frame movement claims,[8] how legal mobilization impacts movement building and policy development,[9] and how movement lawyers use legal tactics in a *"skeptical, politically sophisticated manner"* to advance movement efforts to build power and create sustainable change.[10] In the legal academy, scholars have incorporated social movements into legal theory as the prime engines of progressive legal reform, developing new accounts of popular constitutionalism and movement lawyering in which movements change law from the "bottom up," thereby responding to concerns about activist courts overriding the majority will and activist lawyers dominating marginalized clients.[11]

Within social movement theory, researchers have drawn increasing attention to the role of coalitions in representing movement constituencies and leading movement campaigns. This work has emphasized the tradeoffs of coalition politics: on the one hand expanding organizational resources and clout, while permitting an efficient division of labor; on the other hand, reinforcing professionalization and an emphasis on insider political strategies over disruption. On the positive side, Suzanne Staggenborg has argued that coalitions may play a productive role in mounting challenges by consolidating resources for collective activities (like lobbying) and fostering specialization allowing individual social movement organizations to "conserve resources for tactics other than those engaged in by coalitions."[12] She has also emphasized that coalitions may sustain a movement when environmental conditions turn adverse by promoting collaboration and reducing costs.[13] On top of these advantages, coalitions may offer potential opportunities for distinct social movements to coalesce around issues of mutual concern—which may lead to more effective mobilization while also increasing the potential for collaboration over time.

Understanding this potential directs attention to the contours of collaboration between two of the country's most important and dynamic social movements—around labor and the environment, the "blue" and the "green"—which traditionally have been on opposite sides of many policy disputes with labor activists emphasizing development and growth, while environmentalists have focused on sustainability. Within both movements, there have been deep historic tensions between mainstream white movement leadership and the concerns of people of color around issues of environmental justice and workplace equity (e.g., the exclusion of black workers from many of the benefits of New Deal labor law and organized labor's early opposition to immigration reform), which have contributed to legacies of distrust and antagonism. How the labor and environmental movements might overcome the past to build a viable blue-green alliance for the future is a critical question with significant theoretical and practical implications.

Finally, legal and urban studies scholars have begun to look at the city generally and at local government law in particular as a tool and a target of social movement campaigns, especially around issues that find a sympathetic reception from local policy makers.[14] The labor and environmental movements, both confronting strong national-level opposition, have turned to progressive state and local governments to advance policy reform and promote enforcement. In progressive cities around the country, labor has scored important recent victories, most notably the $15-per-hour minimum wage in Los Angeles, San Francisco, Seattle, and elsewhere, while the environmental movement (in the post-Trump era) has successfully enlisted cities to implement emission reduction commitments imperiled by the United States withdrawal from the Paris climate accord. These victories have been built over decades of strategic investment in local politics and careful planning about how to use the levers of local government authority to win new workplace regulations that improve labor conditions and create the possibility for building greater union density, and new environmental regulations that promote green building initiatives and reduce noxious land uses. Overall, efforts by progressive social movements to advance local regulatory reform have thus depended on political and legal opportunities that leverage the unique powers of local government, while avoiding the constraints of federal preemption.[15]

This book brings together these different scholarly conversations as an analytical framework for studying the formation, execution, and impact of the blue-green challenge to the trucking industry at America's port. It analyzes social movement political and legal strategy through the lens of the *campaign*—an interconnected set of deliberate strategic and tactical choices oriented toward a predefined goal. Using the campaign as the central unit of analysis, the book's narrative focus is on how movement lawyers and activists interact in representing the labor-environmental coalition at the ports of Los Angeles and Long Beach, framing the nature of the coalition's economic and environmental challenge around a dysfunctional trucking market in which drivers are misclassified as independent contractors, and devising and revising a solution to end that misclassification. Core issues of leadership accountability to the movement's constituency, the form and impact of legal and political mobilization, the role of litigation, and the legacy of the campaign are judged in relation to the structural conditions surrounding the campaign and the subjective perceptions and choices of the key actors involved. Toward this end, the book focuses on three clusters of research questions.

The first explores the role of law in shaping cross-sectoral social movement coalitions. How do cross-sectoral movement coalitions form and what role do law and lawyers play in that formative process? Within such coalitions, how does the political and legal power of key organizational actors affect the way that lawyers and activists understand the nature of representation and their accountability to different interests within the affected constituencies?

The second set of questions revolves around the role of lawyering and legal mobilization in movements for local reform. How do movement lawyers interact with activists and other stakeholders to define campaign goals and strategy, and what are the points of conflict and commonality? How do movement leaders understand the utility of law and who decides how it is mobilized? How do lawyers and activists navigate different institutional environments—the ports as regulatory agencies, city governments, courts, the media, and the streets—over different phases of the campaign to create and protect opportunities for campaign success? How does litigation relate to other types of movement advocacy, what nonlitigation roles do lawyers play, and how do campaigns seeking to enhance economic and environmental standards respond to industry countermobilization?

The final set of questions is oriented around the city as a site of progressive law making. How do cross-sectoral movements craft policy reform to jointly advance labor and environmental campaign goals, particularly, unionization and the reduction of air pollution? What factors ultimately influence the capacity to achieve policy goals, implement them over time, and build the power of the movements involved? What legal and political constraints do local governments confront in enacting and defending progressive policy reform, and how do those constraints operate differently in relation to labor and environmental objectives?

Themes and Plan of the Book

The fight to transform port trucking in Los Angeles and Long Beach is, at bottom, a story about the shifting geography of legal power and how movements and countermovements seek to use legal tools at different levels of regulatory authority within the structure of federalism to advance their ends.[16] In such struggles, weaker movements always start off at a legal and political disadvantage and thus face a more difficult challenge: requiring greater ingenuity, solidarity, luck, and perseverance. Less powerful movements look for opportunities to mobilize for change within the crevices that the legal and political systems afford. In this process, law is omnipresent: it creates a set of baseline rights that benefit movement constituents at one level, it undermines these rights through deregulatory changes at another, and it presents opportunities for and constraints on movements seeking to change conditions at still another.[17]

In Los Angeles and Long Beach, the ports were initially built through assertions of local power to advance regional industrialization in the context of strong federal regulation and limited global trade. Port authorities were given autonomy to promote industrialization and succeeded in facilitating manufacturing-led regional growth, while the rise of the federal regulatory state in the New Deal era empowered labor to share in its benefits. Beginning in the 1970s, globalization disempowered local governments, which lost their manufacturing base, and federal deregulation disempowered port drivers, who lost their union representation and their fundamental status as employees. Federal labor and transportation law, created in part to benefit workers, became a hindrance to them by decentralizing industry control and affirmatively preventing collective worker action by truck drivers—who were recast as independent contractors prohibited by antitrust and labor law from organizing. In the 1990s, local governments reasserted control over the ports to appropriate their revenue, but externalized the costs of growth on local communities.

In the Campaign for Clean Trucks, these communities fought back by turning to still-potent federal and state environmental law. Organized labor joined with environmentalists to seize the opportunity to reshape the entire port trucking industry—harnessing the city as a market participant to create local law that reduced diesel pollution and enabled labor organizing, while seeking to avoid federal preemption. The labor–environmental coalition at the helm of the campaign thus attempted to reregulate a sector of the globalized logistics industry—port trucking—tethered to the regional economy by *keeping law local*: changing the rules of the ports to facilitate driver unionization, while simultaneously addressing the environmental and community impacts of port growth. The trucking industry responded by taking law out of the local arena, using litigation to bring the policy outcome of the campaign back under the very federal regulatory regime the coalition sought to avoid. Yet while the industry succeeded in nullifying labor's attempt to change driver status through local ordinance, it trained the spotlight on a deeper problem: that the drivers themselves were misclassified employees under state and federal law. And thus the labor movement pivoted away from local law reform, and back to court, to jumpstart an innovative union organizing effort that promised to transform the application of labor law to the "gig" economy.

It is against this backdrop that the book examines how the labor and environmental movements' challenge to port trucking operated and what it achieved. The analysis proceeds from the perspective of movement actors advancing the campaigns, and draws upon interviews, public records review, and a systematic analysis of archived movement documents and legal materials.[18] The arc of the story builds from separate and uncoordinated activism by different movement actors around the negative impact of port operation and expansion on local communities, to a moment of interest convergence and coalition building that resulted in the passage of clean truck ordinances at both ports, to an industry-led legal challenge that succeeded in carving apart what the coalition had done, to an emergent labor-led campaign to advance a new theory of misclassification.

Focusing on campaign formulation and execution, the book explores why labor lawyers and activists came to focus on the ports as a target, how they joined a coalition with environmentalists and environmental justice advocates, and what factors influenced decisions about campaign objectives and the mix of tactics to achieve them. Thematically, it emphasizes the role of law in building the coalition and defining its goals, shaping coalition strategy and tactics, and affecting outcomes and ongoing struggles.

In terms of coalition formation and goal definition, the book shows how law shaped the way movement actors understood labor and environmental issues at the ports and how to address them. There were top-down and bottom-up movement processes at work. The study traces how these forces were joined around a mutual analysis of convergent legal interests. From the top down, organized labor had developed a sophisticated legal analysis of the trucking industry, identifying its independent-contractor structure as the main impediment to unionization. Labor strategists identified the ports as a target of legal opportunity: they were embedded in the region and the Port of Los Angeles was under the authority of a local government friendly to labor interests. From the bottom up, community groups, increasingly in concert with environmentalists, understood the local impacts of port expansion as a problem of regulatory capture: port governance was controlled by logistic industry firms and their governmental allies, who excluded communities from meaningful participation. Community groups also identified trucking as a key cause of local pollution and congestion, and focused on participation in port governance as the path to change. The Campaign for Clean Trucks was explicitly designed to harmonize labor, environmental, and community interests by crafting a master legal solution to the intertwined problems of deunionization and pollution. The drivers' independent-contractor status was defined as the causal link: forcing low-paid, mostly immigrant drivers to operate as owners simultaneously decreased their pay and increased pollution since they could not afford to upgrade and maintain their trucks. Changing the drivers' legal status was the campaign's linchpin.

The Los Angeles Alliance for a New Economy—known as LAANE—organized the Coalition for Clean and Safe Ports with key labor, community, and environmental allies, particularly the International Brotherhood of Teamsters (Teamsters) and the Natural Resources Defense Council (NRDC). Together, they redesigned local port law to advance mutual interests: "greening" the port, while improving economic conditions for drivers. In 2008, after two years of impressive organizing, the coalition won a Clean Truck Program in Los Angeles that amended port rules to permit trucking companies to enter port property only if they converted their trucks, from dirty old diesel-fuel to modern low-emission vehicles, and their drivers, from independent contractors to employees. From the campaign's perspective, using the city's legal power to force trucking companies to internalize the costs of employment and maintenance would create a sustainable foundation for clean trucking over time and permit driver unionization.

Achieving this policy reform required more than just a well-designed plan. It required local politics to line up in the coalition's favor and coalition actors to execute their plan at a high level—drawing attention to the complex interplay of opportunity and resources in enabling the legal campaign and moving it forward. Since trucking deregulation in the 1980s, the Teamsters had long sought ways to organize port truck drivers. The confluence of a prolabor Los Angeles mayor and a decisive environmental legal challenge by NRDC to block port expansion created the possibility of achieving labor's goals. The 2005 launch of Change to Win—a federation of unions, including the Teamsters, which broke away from the AFL-CIO to promote aggressive organizing—and its association with successful local labor organizations, like LAANE, provided the resources to make it happen. The formation of a coalition of labor, environmental, and community groups brought the political muscle necessary to move local officials to produce change. In this way, top-down labor planning intersected with bottom-up resistance to port activities at a moment of political opportunity to create a powerful coalition with the political leverage to make law, which the coalition succeeded in doing.

The book also focuses on how law shaped campaign strategy and tactics. In a system of weak federal labor regulation, organized labor relied on environmental law—with the crucial legal power to delay port growth—as the initial lever to create space for policy change. From there, labor turned to the local government, where it had built political power, to advance local policy reform that would facilitate achievement of employee status for port truck drivers. However, for political and legal reasons, that reform was framed in terms of environmental justice: employee status was necessary to green the port and reduce impacts on local communities. Politically, this facilitated forming the coalition and persuading local officials. Legally, employee status was linked to the goal of reducing port emissions and thus avoiding further environmental litigation challenge—which was deemed critical for the port, as a market participant, to ensure orderly and efficient operations. In this way, federal preemption law shaped how movement lawyers, in particular, understood the possibilities for regulatory change at the local level—and how that understanding was translated into policy reform. The book thus highlights *how federal preemption was a primary battleground on which the contest over the labor and environmental movements' local strategy played out*. Movement lawyers mobilized law in the administrative and legislative process to support readings of preemption doctrine in a context of jurisprudential uncertainty in order to minimize preemption risk and validate the Clean Truck Program. Although they succeeded in getting the law passed, the lawyers ultimately were not entirely successful in defending the law from a preemption attack by the trucking industry in a case that ultimately made its way to the United States Supreme Court.

Finally, the book explores how law shaped the nature of movement opposition, outcomes, and ongoing struggles. In its conception and execution, the Campaign for Clean Trucks was highly effective. Powerful alliances were built, a sophisticated policy was crafted that achieved labor and environmental goals, opposition was thwarted, and legislative passage secured. Yet the policy was only partially implemented. The Clean Truck Program's labor centerpiece—the provision requiring port truck drivers to become employees to enable unionization—was enjoined and invalidated by the industry's preemption challenge under the Federal Aviation Administration Authorization Act. Yet, the industry lawsuit did not challenge the entire Clean Truck Program and left standing the conversion of the port trucking fleet to low-emission vehicles. As a result, what remained of the Clean Truck Program advanced environmental interests by mandating clean trucks, but undercut labor interests by withdrawing the legal basis to organize drivers. In so doing, the drivers themselves suffered a setback: with the employee provision undermined in federal court, drivers faced the burden of purchasing and maintaining clean trucks without the economic benefits promised by employee status.

Yet from that setback, there has been a labor resurgence. As trucking companies saddled drivers with expensive clean truck leases and onerous driving conditions, the labor movement retooled its strategy and reframed the problem in terms of *misclassification*, in which companies designated drivers as independent contractors as a formal legal matter but treated them as employees. The resulting Justice for Port Drivers campaign has focused on an innovative mix of litigation and organizing to do what was previously deemed impossible: unionize port truck drivers at firms formally operating under an independent-contractor model (the vast majority serving the Los Angeles-Long Beach port complex). How that has happened and what it means for the future of labor organizing in connection with the growing number of nominal contract workers in the American economy are questions at the heart of the port trucking story.

To tell that story, chapter 2 traces the development of the Los Angeles and Long Beach ports, providing an in-depth analysis of how law shaped port evolution from export-oriented engines of regional prosperity to import hubs of the global economy that externalized economic and environmental impacts on the most disadvantaged local communities.

Chapter 3 then presents a fuller picture of the legal framework governing port operations and the relationship between the ports and surrounding communities. Doing so provides context for the struggle by outlining the legal space within which reform was possible. Specifically, this chapter describes how the port, as a city entity, exercises local power within a framework of state and federal regulation. Understanding the port as a regulatory actor with the power to affect environmental and labor standards requires appreciating three aspects of the legal framework: (1) the port's local power as a city entity to make law through its internal governance structure, (2) nonlocal government regimes, particularly environmental law, that could be leveraged to force the port to act, and (3) the scope of local port authority in relation to federal preemption doctrine.

Having laid this foundation, chapters 4, 5, and 6 detail the origin, execution, and outcome of the struggle to reform port trucking. Chapter 4 focuses on the struggle's origin by weaving together three strains of activism: community-based resistance to port expansion, the labor movement's efforts to organize drivers, and the environmental movement's legal challenge to port air pollution, which set in motion the policy process that focused movement leaders and policy makers on the potential to restructure port trucking. Chapter 5 traces the process of coalition formation and strategic execution in relation to the Campaign for Clean Trucks, illuminating how law was mobilized through insider and outsider strategies to achieve final passage of the Clean Truck Program in Los Angeles (and enactment of a related, but weaker, program in Long Beach). This chapter then analyzes the industry litigation challenging the Los Angeles program, which resulted in an appellate court ruling invalidating the crucial employee provision and a Supreme Court decision striking down other aspects of the concession plan. Chapter 6 focuses on the improbable aftermath of the litigation, showing how the labor movement—despite nearly abandoning the port struggle—managed to convert initial ad hoc legal challenges to driver misclassification into a novel and ultimately quite powerful strategy, advanced in the Justice for Port Drivers campaign, to use misclassification as an affirmative basis for striking trucking companies and unionizing drivers.

In conclusion, chapter 7 explores theoretical and practical lessons of the movement for economic and environmental justice at the ports. Reflecting on how law shaped strategy and outcomes, the chapter shows how movement leaders identified legal problems at the heart of drayage trucking and designed solutions to address them, while continuously revising strategy and leveraging favorable institutional conditions to address labor's fundamental legal weakness: the independent-contractor status of drivers.

Turning to outcomes, chapter 7 evaluates the tradeoffs of using alternative legal frameworks—in this case, environmental and local government law—as proxies for advancing the economic rights of marginalized low-wage workers. This move is necessitated by weak labor law, but to be successful it must thread a difficult needle. As the Campaign for Clean Trucks shows, these proxy battles aim to achieve industry restructuring that indirectly facilitates unionization by converting independent contractors to employees who may organize unions under U.S. labor law. This is a powerful tool that has been used successfully in other contexts.[19] But it also raises challenges. Joining labor policy to alternative regulatory frameworks—like environmental law—risks the alternative regulatory claim being validated to the detriment of the labor claim. Thus, if the argument for a Clean Truck Program is centrally about reducing port pollution and avoiding environmental lawsuits (rather than unionization), then the environmental objective could be read by policy makers or courts as trumping the labor one.

The joinder of labor and environmental claims, which strengthens the coalition, also makes it vulnerable to industry countermobilization that seeks to "divide and conquer." As the book shows, industry litigation on preemption grounds succeeded in splitting apart and reallocating the gains from a policy campaign built upon mutual interest—resulting in environmental victory but labor defeat. However, as the Justice for Port Drivers misclassification strategy illustrates, the resilience of social movements rests in part on their ability to adapt law to new constellations of power—even when, for organized labor, that means charting a separate path in pursuit of labor peace.

Building on the discussion of industry countermobilization, the last chapter further explores the dynamic relation between law and social movement coalitions, illustrating how law serves to bring coalitions together, while simultaneously pulling them apart. In the port struggle, the preexisting legal regime endowed the labor and environmental movements with different tools and shaped distinctive goals; there were moments when these goals converged around a common legal and political analysis, but also powerful incentives to pursue divergent outcomes, which industry opponents exploited.

Turning to implications for legal mobilization and local law making, the chapter analyzes the challenge that federal preemption poses to movement efforts to nest prolabor and proenvironmental regulation in local government law—and the particular difficulties that presents for crafting and advocating local reform.[20] Ultimately, such local reform strategies are never entirely local. Rather, they are framed by the overhang of federal law—and not just environmental and labor law, as it turns out, but other federal regulatory structures as well—that both shapes legal strategy and pulls the movement back into the federal system to either defend or circumvent challenges to carefully crafted local policy.

This insight highlights the role of lawyering and legal mobilization in local movements and how they are influenced by the context in which campaigns play out. Focusing on this context, chapter 7 examines innovative campaign efforts to link litigation to organizing and policy development, revealing how, despite structural disadvantages, labor law may be strengthened through strategic lawyering and robust regulatory enforcement. The chapter also illuminates underappreciated roles lawyers have played over the long arc of port struggle, particularly their behind-the-scenes mobilization of legal doctrine, both to validate the legality of prospective policy reform, like the Clean Truck Program, and to gain legal authorization for collective action, like misclassification strikes.

As this discussion underscores, planning around preemption is a key part of the lawyering process in local policy campaigns, but one fraught with uncertainty, since predicting judicial outcomes is such an inexact science. How lawyers manage this uncertainty—and how they understand and relate to their social movement clients in the process—draws attention to the contested nature of representation in bottom-up social movement campaigns. The immediate consequence of the Campaign for Clean Trucks for the drivers themselves—burdened with the costs of leasing and maintaining expensive new low-emission trucks, but without employee status and potential union benefits—raises questions about who should make risky movement decisions and what role those affected by the decisions should play. Within legal scholarship, the classic accountability concern is with top-down lawyers making choices that are inconsistent with constituent interests. Yet here movement lawyers effectively served as outside counsel to the organizations driving policy development. It was movement leaders, and not movement lawyers, making the major campaign policy calls—and deciding to take the risk to pursue the Clean Truck Program, even though there was a chance that truck drivers might bear the brunt of clean truck conversion without reaping the benefit. Whether that was the right choice—and whether the right people made it—are key questions.

The chapter concludes by reflecting on the meaning of success and failure in the context of social movement challenges to entrenched power. It emphasizes the ongoing, cyclical nature of conflict and the role of expertise, tenacity, and courage in enabling and sustaining struggle. Despite repeated obstacles and reversals at America's port, the movement for economic and environmental justice continues—and continues to win victories—through a savvy combination of employment litigation and union organizing. Thus, although port truck drivers may have suffered an important setback, the movement marches forward with new hope for reshaping the port trucking industry. This march forward has provided lessons that have informed success in other low-wage worker campaigns in Los Angeles, particularly around waste hauling. And it has laid the groundwork for larger national efforts to fuse the labor and environmental movements to build mutually reinforcing, sustainable regulatory reform in U.S. ports and beyond.

2 Law in the Development of the Port

The Port of Los Angeles sits in the San Pedro Bay, directly adjacent to the Port of Long Beach to the east. The bay itself is tucked under the Palos Verdes Peninsula, which juts out prominently south of Santa Monica Bay. The ports are located in distinct municipalities, subjecting them to different rules and political pressures—and making them competitors for cargo business. However, as a functional matter, they form an integrated unit: sharing the same land mass, benefitting from the same infrastructure, and connecting to a unified transportation system of roads and rail.[1] Individually, the ports of Los Angeles and Long Beach are the first and second largest, respectively, in the United States; together, they constitute America's port, one of the largest port complexes in the world.[2]

The geography of the ports, both physical and man-made, is a central feature in the struggle over their impact and control. This geography—and the inequality it demarcates—has been shaped by interlocking international, federal, state, and local legal decisions. These decisions have facilitated the ports' development as an engine of regional economic growth—and a gateway to globalization—while concentrating its most harmful externalities in some of the region's lowest-income communities. As law has contributed to the growth of the port complex, it has also distributed the costs and benefits of that growth unequally—enabling some communities to escape the worst impacts, while appropriating others in the project of expanding global trade. This project has resulted in transportation and land use decisions that have contributed to segregation and environmental degradation in surrounding communities, while also creating winners and losers among workers.

The history of the ports' legal development can be roughly broken into three phases. In the first, from the mid-1800s through the 1920s, law was used to appropriate the harbor—created by and beholden to outside capital—for the project of city building. In Los Angeles and Long Beach, the ports were wrested from private ownership, constituted as public entities, and given broad powers as independent agencies to build the infrastructure necessary to promote economic development. This process was led by local business elites advancing a vision of the ports as key to regional industrialization. While both ports fueled this growth, they increasingly began to compete, establishing an interport dynamic that would shape future development.

The second phase, from the Great Depression to the 1970s, was marked by the rise of the regulatory state and a working compromise among business elites, labor, and local communities to share the benefits of port growth. In this period, the ports were harnessed to fuel industrialization and facilitate U.S. exports, building the Los Angeles region as a manufacturing stronghold, led by aerospace and auto production. Strong federal regulation of transportation and labor produced stable industry patterns and powerful unions, which were able to negotiate their share of the peace dividend. Trade barriers permitted internal manufacturing development. Port expansion occurred, but had yet to achieve a scale that impaired surrounding communities, whose residents reaped economic benefits of jobs and local investments (though their communities began to suffer from the effects of intensifying oil extraction).

In the third phase, beginning in the 1970s, this arrangement unraveled in the face of globalization and the decline of the regulatory state. Against the backdrop of free trade, the ports became conduits of globalization, powered by the rise of intermodalism, which connected the U.S. market to East Asian exporters and fueled prodigious growth. Federal deregulation and weakened labor laws contributed to industry reorganization that empowered shippers and negatively impacted the least powerful workers in the logistics supply chain—namely, port truck drivers whose downgrading to independent contractors undermined their economic position. Globalization also disempowered city governments, which saw the development benefits of export-led growth dwindle, and with it, the job and tax benefits of local manufacturing. Los Angeles and Long Beach responded by reasserting greater legal control over the ports in order to shore up faltering city budgets and harness port growth to power development of the regional service economy. This move allied cities with the project of continuous port expansion—since it was through expansion that jobs were created and local revenues grew. Although the ports continued to compete, they also made joint investments in transportation and logistics infrastructure to maintain their comparative advantage over other locations.

Law in the Development of the Port

Increasingly, this growth came at a cost to regional air quality, which was polluted by the diesel-fuel-driven port transportation network. The ports' negative externalities fell with greatest force on adjacent low-income communities, which were made to absorb the most significant environmental and land use impacts. Disempowered by a legal system in which local elites worked with global capital to expand port capacity, these communities—in collaboration with a resurgent labor movement—sought to gain greater input into port governance in order to adapt local control to their own ends. To do so, they turned back to a tool from the regulatory state—environmental law—to leverage changes to local policy that would better align port growth with community and labor interests. That is where the Campaign for Clean Trucks began.

Local Power: Annexation and Autonomy

The creation of the Port of Los Angeles was shaped by the clash of competing economic ambitions for the region.[3] As local business elites used law to ultimately wrest control of the port from outside capital around the turn of the twentieth century, they built the foundations of a transportation infrastructure within the city's jurisdictional boundaries that connected shipping, rail, and roads. In this process, local elites used legal strategies to annex the harbor property—facilitating dramatic growth in city territory and population—and created a municipal governance structure that conferred broad legal authority on the Port of Los Angeles to pursue dynamic expansion plans. The goal was to take port control away from outside entrepreneurs in order to build the economic foundations for manufacturing-driven city growth.

First used as an outpost to supply Spanish missionaries in the late 1700s, the San Pedro harbor became an important trading center after Mexico took control in 1822, and American land acquisition and commercial activity expanded.[4] The U.S. annexation of California after the Mexican-American War, followed by the Gold Rush, brought American settlers streaming west.[5] As the volume of passenger and commercial shipping increased, the need for transportation infrastructure grew. Investment was spurred by the struggle for control over lucrative regional trade. Rancheros gained early advantage, developing the first San Pedro-to-Los Angeles stagecoach shipping route after the war,[6] but the balance quickly shifted to new entrepreneurs.

Delaware transplant Phineas Banning entered the market and swiftly established a stagecoach business between the harbor and Los Angeles that extended on to Salt Lake and Fort Yuma.[7] When competition for the San Pedro route became too fierce, Banning bypassed it altogether, transferring his shipping business to land he purchased north of the harbor, which he named Wilmington.[8] While Banning's quest for market dominance was disrupted by the Civil War, his postwar strategy sought to monopolize trade to Los Angeles through the construction of a rail line from the harbor.[9] Banning thus entered politics, where he used his influence as a state senator to gain passage of a bill authorizing the Los Angeles & San Pedro Railroad charter, and then won a hard-fought local ballot initiative authorizing municipal bond financing.[10] Construction of the line, which ran along Alameda Street, was completed in 1869, marking the creation of the Alameda transportation corridor.[11] With the rail line in place, Banning then turned to improving the port itself, which was too narrow and shallow for large sea vessels. He persuaded the federal government to add two jetties and then dredged the channel to a serviceable depth—thereby facilitating a nearly hundredfold increase in total port commerce between the late 1860s and 1886.[12]

The Los Angeles & San Pedro Railroad did not stay within Banning's control for long. Indeed, its initial construction was motivated in part by the desire to connect Los Angeles to the approaching transcontinental railroad, which local elites believed could only be secured by offering its owner—Leland Stanford's powerful Southern Pacific Railroad—ready-made rail access to the harbor.[13] The Southern Pacific, anchored in San Francisco, threatened to bypass Los Angeles without a generous public subsidy,[14] which included acquiring Banning's rail line.[15] After fierce lobbying, Congress passed a law directing the transcontinental railroad to run through Los Angeles.[16] Yet the terms of any deal between Los Angeles and the Southern Pacific were yet to be worked out and ultimately subject to local voter approval. Determined to bring transcontinental service to Los Angeles, Banning—along with other Angeleno businessmen who formed the "Committee of Thirty"—reluctantly agreed to support a $600,000 subsidy to the Southern Pacific, which included a controlling share in Banning's Los Angeles & San Pedro Railroad.[17] A bitter election contest ensued, but the Southern Pacific subsidy was passed handily by county voters in 1872.[18] Four years later, the construction of the Southern Pacific line to Los Angeles was completed[19]—officially connecting the city, and the San Pedro harbor, to the national market. The population of Los Angeles at the time was approximately 10,000.[20]

Growth occurred rapidly, but the position of the San Pedro harbor as the gateway to Los Angeles was still uncertain. Local boosterism helped attract a wave of new immigrants, who the Southern Pacific eagerly transported west.[21] Yet, with Los Angeles firmly within its grasp, "The Octopus" (as the Southern Pacific was called) squeezed local shippers subject to its virtual monopoly.[22] Saddled with debt and eager to protect its investment in the San Francisco port, the Southern Pacific raised rates on Los Angeles shippers and refused to build out the San Pedro harbor.[23] Competitors sought to challenge the Southern Pacific with rival rail lines and ports—provoking harsh reprisals by the railroad. In the 1870s, the Southern Pacific crushed a plan to build a new railroad and port in Santa Monica; and as Banning's efforts to dredge the San Pedro harbor began to pay off in the 1880s, the Southern Pacific rerouted its own line from Los Angeles to the west side of the main channel in San Pedro, thereby circumventing Wilmington—and effectively putting the Wilmington port out of business.[24]

Still, the competition was unrelenting: with one rival building a port in Redondo Beach and another laying new rail tracks through East Wilmington, the Southern Pacific made a dramatic play to defeat both threats by abandoning San Pedro altogether.[25] With local businessmen consumed by the threat of losing regional shipping to San Diego's superior natural harbor, the Los Angeles Chamber of Commerce (known as the Chamber), led by *Los Angeles Times* owner Harrison Gray Otis, pressed Congress to fund construction of an artificial deep-water harbor at San Pedro.[26] As Congress vacillated, the Southern Pacific, under the control of Collis Huntington, surprised local leaders by opposing the selection of San Pedro as the harbor site, instead endorsing Santa Monica, where it had quietly made significant waterfront investments.[27] The Chamber, which resented the Southern Pacific's power and opposed its tourism-oriented vision for Los Angeles development,[28] seized the chance to have a decisive confrontation with the railroad. With Congress unwilling to choose sides without the backing of the California congressional delegation, the Chamber lobbied Senator Stephen White (Otis's personal lawyer),[29] who championed San Pedro and vilified the already-unpopular Southern Pacific.[30] After a bitter political struggle, the California delegation eventually coalesced around White's leadership, defeating a proposed $3 million appropriation for Santa Monica. The delegation secured massive federal funding for the Army Corp of Engineers to build a breakwater in the San Pedro Bay, which commenced in 1899 (and was completed in 1912)—finally securing San Pedro's place as the port of entry to the Los Angeles region.[31]

The "free harbor" movement, however, was not a complete success. The city of Los Angeles lacked legal control over the harbor, which lay sixteen miles to the south of downtown, within San Pedro and Wilmington.[32] And despite its failed Santa Monica gambit, the Southern Pacific still monopolized port operations in the San Pedro harbor through its ownership of the waterfront.[33] Municipal control was therefore necessary to build the port and ultimately break the Southern Pacific monopoly. The U.S. acquisition of the Panama Canal in 1904 and its impending completion heightened the sense of urgency among local businessmen eager to solidify Los Angeles's place as the major western port city.[34]

The stakes became higher still with the machinations of adjacent Long Beach—on the east side of San Pedro harbor—which had been incorporated in 1888 and steadily grew with the arrival of rail connection.[35] As the free harbor fight solidified plans to dredge and improve the west side of the harbor, Long Beach expansionists sought to exploit the commercial prospects of the east side. After securing federal funds for inner harbor dredging in 1903, business leaders urged Long Beach to annex Terminal Island, the massive landmass running the width of the San Pedro Bay that separated the outer harbor from an inner channel connected to a tributary of the Los Angeles River.[36] Although this effort failed, an annexation battle ensued, with Los Angeles unsuccessfully trying to annex Long Beach as the latter acquired more land up to the Wilmington border. In 1909, Long Beach won an election to acquire the eastern half of Terminal Island.[37] In the face of this incursion, Los Angeles moved to exert greater control over the western part of the harbor.

Doing so required a series of legal maneuvers. Because state law only allowed the consolidation of contiguous cities, Los Angeles first had to extend the reach of its jurisdictional border down to the port, which it did in 1906 by annexing the unincorporated "shoestring district"[38]—a one-mile-wide strip of land from Los Angeles's southern border due south to San Pedro.[39] From there, the Los Angeles City Council took the symbolic step of creating the Los Angeles Board of Harbor Commissioners (also called the harbor commission) in 1907,[40] as it turned to the more formidable task of actually acquiring the harbor itself by annexing San Pedro and Wilmington, whose skeptical residents had to be convinced to vote for consolidation.[41]

Before annexation could be formally considered, state law had to be amended to authorize the consolidation of charter cities (those, like Los Angeles, which had chosen home rule by ratifying their own city constitution) and noncharter cities (those, like San Pedro and Wilmington, which had not opted for charter status and were thus governed under the state's general law). The consolidation law was duly amended in 1908, after spirited lobbying by local business elites—and over the Southern Pacific's objection.[42] In the electoral campaign for consolidation, Los Angeles used its most powerful form of persuasion: the promise of its vast resources. Realizing that they lacked the funds to significantly improve the port, which powered the local economies, San Pedro and Wilmington residents acceded to the annexation plan, in exchange for Los Angeles committing $10 million for harbor improvement, agreeing to build a truck highway from the harbor to downtown, and promising additional infrastructure investments.[43] San Pedro and Wilmington formally voted in favor of consolidation with Los Angeles in 1909,[44] within days of Long Beach's Terminal Island annexation.[45] Doing so married the interests of Los Angeles's white business elites with those of the predominantly white ethnic working-class residents of Wilmington and San Pedro—seamen and fishermen from Croatia, who had settled beginning a century earlier, joined by immigrants from Italy, Portugal, Norway, Ireland, and Greece around the turn of the twentieth century.

Consolidation did not fully settle the matter since ownership of much of the waterfront property remained in dispute. The city of Los Angeles challenged title of the Southern Pacific and other purported landowners under the antiquated State Admissions Act, which assigned ownership of navigable waters to the state.[46] Los Angeles brought a series of lawsuits to perfect its title, which was settled once and for all by the 1911 passage of the state Tidelands Trust Act. The act made Los Angeles trustee of the tidelines—the land under the normal ebb and flow of the tide, as well as submerged land and navigable waterways—that constituted the harbor.[47] Now firmly located on city land, the Port of Los Angeles—an independent municipal department governed by an appointed board of harbor commissioners—was officially born.

The history of the Los Angeles port as an instrument of private enterprise appropriated to municipal control influenced its subsequent role in regional growth. After consolidation, the port remained semiautonomous, but its mission was shaped by local business elites who sought to build its power in order to facilitate Los Angeles's growth as an export-led manufacturing economy. To accomplish this, the port was placed under the power of a proprietary department (established in the model of the city's formidable Department of Water and Power) and governed by the harbor commission.[48] Under the 1913 Los Angeles charter amendment, a harbor commission of three members appointed by the mayor and approved by the City Council, was given "possession and control ... of the entire water front of the city."[49] The commission's power included broad authority to manage and lease port property, hire personnel, and pass rules of operation, as well as the right to set rates (subject to City Council approval), collect revenue, and issue bonds (subject to voter approval).[50] Although technically independent, the harbor commission in its early phase relied on support from local business elites to win greater authority and control. In collaboration with the Chamber, the commission secured a series of charter amendments that enlarged its bureaucratic authority, expanding its size to five commissioners, and giving it greater power over budget, personnel, policy making, and contracting in ways that further diminished mayoral and City Council control.[51] Because it kept shipping rates low to promote trade, the port required public financing for major improvements and leveraged local business support to win approval of more than $30 million in municipal bond funds by 1932.[52] However, as Progressive Era citizen resistance to public subsidies grew, the harbor commission eventually was forced to abandon bond referenda and become self-financing through shipping fees and tariffs.[53] Thus dependent on revenue from shipper and carrier use to fund operations and improvements, the commission became increasingly focused on the satisfaction of its main customers: import-exporters, ocean carriers, railroads, and trucking companies.[54] Yet—still insulated from intense competition—the harbor commission was at this point able to strike bargains that fueled Los Angeles's rapid growth.

Powered by the real estate boom in the late 1880s (and undeterred by the subsequent bust),[55] Los Angeles's population grew tenfold to 100,000 in 1900 and then more than tripled to 320,000 in 1910; by 1930, the city's population had surpassed one million.[56] During this time, port commerce shifted from imports to a more balanced two-way flow, as the discovery of oil and the beginnings of Los Angeles's industrialization significantly increased export traffic. In the period before World War I, immigrants in search of the California dream fueled a strong demand for building construction and, as a result, lumber imports dominated port trade, driving an elevenfold increase in total port commerce from 1900 to 1917.[57] After the war, oil production skyrocketed with a series of major oil strikes around Long Beach (beginning in 1921) and oil exports—which had been growing in the prewar period—increased dramatically, facilitated by the opening of the Panama Canal that same year, which permitted oil to be shipped immediately for refining on the East Coast.[58] From 1921 to 1922, port commerce doubled to over ten million net tons.[59] As a Chamber-led push to promote Los Angeles industrialization won some early success—with Ford and major tire companies opening regional plants to take advantage of Los Angeles's shipping facilities—exports surged and port commerce grew further, reaching nearly thirty million net tons by 1930 and establishing the Port of Los Angeles as the largest on the West Coast.[60]

Law in the Development of the Port

The Port of Los Angeles's growing regional dominance occurred alongside the upstart ambitions of neighboring Long Beach. As the Port of Los Angeles began to take shape in the early 1900s, local developers purchased harbor property and began dredging to create a rival deep-water port in Long Beach.[61] The arrival of new industry—most notably the Craig Shipyard and then the Southern California Edison power plant—added momentum to the harbor project, which remained in the hands of private developers even after the official creation of the Port of Long Beach in 1911.[62] Despite a series of city-backed bond measures to support harbor development,[63] World War I and major flooding reinforced the perception that the harbor's private owners were unable to undertake improvements at the necessary scale to build and maintain a world-class port. As a result, the city of Long Beach finally acquired ownership of its port in 1916, promptly issuing bonds for further upgrades.[64]

The 1921 Signal Hill oil strike radically changed the fortunes of Long Beach, newly awash in "black gold" and able to finance the massive improvements necessary to create a world-class port.[65] That year, the city passed a new charter, establishing a harbor department, with authority to manage the city-owned harbor asset.[66] A $5 million bond issue in 1924 financed a breakwater that transformed the port into a deep-water rival to its Los Angeles neighbor,[67] separated by an invisible jurisdictional line, but otherwise integrated into a massive port complex. On the Long Beach side, a series of ballot initiatives through the early 1930s gave the harbor department proprietary status along the model of Los Angeles, with a harbor commission that possessed similar independent powers.[68] Los Angeles, appreciating the threat, attempted again to consolidate authority by creating a unified port district, but Long Beach rejected the overture.[69] A wealthy city with larger aspirations, Long Beach preferred to challenge Los Angeles head-on, quadrupling its port tonnage to four million between 1926 and 1930.[70] Though still far below the Port of Los Angeles in overall volume, the Port of Long Beach had established itself as a serious competitor, causing each city to ratchet up investment to secure its share of trade.

Federal Power: Industrialization in the Shadow of Regulation

Although the Great Depression slowed growth dramatically at both ports, wartime mobilization and the postwar prosperity that flowed from U.S. economic dominance once again transformed the ports—and their relationship with the communities connected to them. The regulatory state that emerged from the Depression set the template for postwar growth. Wartime industrial investment fueled a postwar manufacturing boom, particularly in Southern California, where wartime manufacturers of aircraft and ships were retooled for the peacetime economy. Import tariffs reduced foreign competition and encouraged export-driven industrialization, in which the ports became key distribution centers. Federal regulation of transportation permitted the ports to negotiate favorable terms with shippers and carriers, which they could then reinvest in infrastructure development. Transportation regulation, coupled with newly minted federal labor laws, also gave unions power to negotiate a favorable share of growth proceeds for port workers. Those workers, particularly truck drivers, benefitted from the postwar regime, while local communities—increasingly under stress from oil production—had not yet incurred the blight of rapid port expansion. It was a fragile stability that rested on federally regulated industrial prosperity.

Trade was significantly interrupted by the Depression—which decreased port revenues and forced greater reliance on federal assistance for harbor improvements[71]—and World War II. These events nonetheless drew attention to two aspects of port development that would prove crucial in the postwar period. One was oil production, which despite decreased demand remained a mainstay of harbor exports during the 1930s and, in Long Beach, generated revenues that financed ongoing harbor improvements. The 1932 discovery of the Wilmington Oil Field under the harbor (the third largest oil field in the United States) spurred increasing oil extraction and refining activities in the harbor area, while also contributing to harbor subsidence on the Long Beach side.[72] By the mid-1930s, Wilmington and Long Beach were marked by the relentless rise and fall of oil pumps, and significant areas had been conveyed to oil companies, whose operations often abutted the houses, schools, and stores that residents used.[73] At the beginning of World War II, 75 percent of all cargo shipped through the Port of Los Angeles was oil.[74] The cataclysm of World War II appropriated the ports for the national interest, while simultaneously laying the foundation for an even greater postwar role for the ports in building regional industrialization. Petroleum was needed in huge amounts to support the war effort and drilling intensified around the ports, reaching 17,000 barrels a day in 1943.[75] Wartime brought naval bases to the strategically valuable San Pedro harbor, which became the central conduit for the transportation of military personnel and the distribution of locally manufactured aircraft and ships to Allied forces in the Pacific.[76]

Manufacturing production spurred by the war became the basis for regional economic growth after the war's end. Both ports invested substantially in postwar repurposing to convert facilities back to civilian uses and

Law in the Development of the Port

to build for increasing trade afforded by the peacetime dividend.[77] Under pressure to be financially self-sufficient, the ports promoted new local development,[78] while also cultivating global connections, sending trade missions to Asia and Europe.[79] The pressure on Long Beach, in particular, to increase port revenues grew more intense in the 1950s, when it lost control over its lucrative oil revenue after the state amended the Tidelands Trust Act in 1951 to allocate 50 percent of oil revenues to the state for purposes unrelated to the harbor.[80] Faced with a dwindling oil subsidy, the Port of Long Beach launched an aggressive pricing strategy to lure shipping away from Los Angeles, which allowed Long Beach to quadruple its port tonnage in the 1960s, causing it to nearly equal its rival Los Angeles's total by 1971.[81]

The rise of manufacturing powered postwar economic growth in the Los Angeles region and the ports grew in relation to regional prosperity. Ports and their workers shared in some of the benefits of growth under a set of federal laws that regulated transportation and labor, giving ports and unions negotiating strength to extract benefits. Port transportation was tightly controlled by an interlocking federal regulatory system governing carriers: the ocean steamship companies, railroad lines, and trucking firms that moved cargo. This system gave ports more authority to set rates, while consolidating the trucking industry in ways that facilitated unionization. Part of this structure predated the New Deal. The Interstate Commerce Act of 1887 governed interstate railroad companies and established the Interstate Commerce Commission (ICC) to police unfair competition by mandating reasonable shipping rates.[82] The Shipping Act of 1916 similarly regulated ocean carriers,[83] establishing the Shipping Board (which became the Federal Maritime Commission) to police anticompetitive practices by setting uniform price schedules and exempting port-to-port rate agreements from antitrust law.[84] The Motor Carrier Act, which regulated trucking, was passed in 1935.[85] The act set routes, regulated rates, and limited market entry to firms able to secure a certificate of "convenience and necessity" from the federal government.[86]

Taken together, this regulatory system had two important effects that benefited the ports and organized labor. First, it strengthened port negotiating power relative to shippers and carriers. Companies that wanted to ship goods had to contract with ocean carriers to transport cargo along authorized routes from port to port, and then separately contract with inland carriers (rail or trucking) to haul cargo to and from the ports.[87] Because federal agencies controlled shipping rates and routes, shippers were not able to negotiate single "through rates" to move their cargo from door to door on a single bill of lading.[88] Fixed carrier pricing meant that shippers saw little financial advantage to rerouting, which gave ports greater bargaining power to negotiate higher fees for access.[89] These fees supported further port expansion.

Federal regulation also shaped labor relations for port workers. For these workers, the Great Depression exacerbated what had long been the painful reality of substandard and often inhumane working conditions.[90] Harbor railroads had been built using low-paid and sometimes forced labor, while maritime workers on ships and their longshore counterparts, who loaded and unloaded cargo on the docks, labored in dangerous settings and often for little pay.[91] The labor militancy of the 1930s—culminating in the 1935 passage of the National Labor Relations Act (NLRA), which established employee collective bargaining rights—began to challenge these conditions.

For longshoremen on the West Coast, the 1930s were a pivotal decade, marked by "major institutional gains, sustained bursts of self-activity, and expanding consciousness."[92] Labor militancy at the ports was radical and often violent: influenced by burgeoning Communism, fueled by "desire to transform the world by fundamentally reshaping the patterns of authority and organization in the realm of work," and motivated by a "determination to cross traditional craft union barriers in order to build solidarity with other workers."[93] The roots of this militancy and aspiring solidarity ran deep. The International Longshoremen's Association (ILA) was formed in 1892, receiving a charter from the craft-based American Federation of Labor (AFL) in 1896 and quickly joining forces with the incipient Teamsters in San Francisco to beat back the Chamber of Commerce's effort to destroy the city's growing union movement.[94] Pre-war solidarity across craft sectors was also evident in the decision by some longshoremen in San Francisco to leave the ILA to join with the powerful (though deeply white supremacist) sailor's union toward the goal of uniting all West Coast unions engaged in maritime trade.[95]

The ILA's ranks swelled in the first part of the 1900s. San Pedro's longshoremen organized a union around the turn of the century, adding to a growing number along the coast that became affiliated with the ILA's Pacific Coast District. With membership growing, the ILA tested its power in a bloody 1916 strike focused on San Francisco, where most of the West Coast's maritime trade was centered. This effort was met with forceful employer pushback culminating in a bitter 1919 strike that ended with a significant loss for the longshoremen, who were forced to accept a handpicked employer-backed, gangster-organized "Blue Book" union (so named because of its blue dues book, distinguishing it from Communist red), which was awarded a sweetheart contract establishing a closed shop (requiring all longshoremen to pay dues to the Blue Book union).[96] Discontent and conflict quickly spread to San Pedro, where employers fought to maintain open shop policies (in which employees were not required to join duly elected unions as a condition of employment), sending in strikebreakers and police to quash a 1923 walkout.[97]

By the early 1930s, these simmering disputes began to erupt in acts of open defiance by longshoremen. Depression-era conditions were the fuel that sparked protest. At the San Pedro ports, longshoremen had to wait nine months in the union hall for a job and those "who were unable or unwilling to find a 'flop' at a waterfront charity were forced to sleep under viaducts, in lumberyards, or in some other spot that provided a measure of protection from the elements and the San Pedro police."[98] In 1933, dockworkers in San Pedro received an average weekly wage of $10.45, with one federal investigator stating that half of the longshoremen received public relief.[99] The rise of the Blue Book union sucked the life from the ILA, now challenged by the Communist-backed Marine Workers Industrial Union (MWIU), which gained adherents through the early 1930s as class struggle built to a crescendo. By 1933, a radical splinter group from the MWIU (which decided to devote its energy to organizing seamen), led by Harry Bridges in San Francisco, gained power within the ILA and succeeded in winning a vote at the coastwide ILA convention in 1934 calling for what would come to be known as the "Big Strike."[100]

This strike began on May 9, 1934 with a walkout by West Coast longshoremen demanding uniform wage rates and union-controlled hiring halls.[101] Ship owners refused, brought in strikebreakers, and enlisted local police in cracking down on the protesters.[102] After an eighty-three day strike, during which two ILA members were killed in violent clashes memorialized as "Bloody Thursday,"[103] the federal government intervened and pressured both sides to accept an arbitration that ultimately established the first industry-wide collective bargaining unit, created jointly controlled hiring halls, limited working hours, and raised longshoremen's base pay.[104]

Though successful on these terms, the strike set in motion a pattern of intra-union conflict that would shape relations between unions at the port, undermining the dream of a federation of all waterfront unions—"One Big Union"—championed by militant leaders like Bridges.[105] While the Big Strike began with solidarity between longshoremen and the Teamsters—with the Teamsters' San Francisco local voting "not to transport merchandise to and from the docks"[106]—and rank-and-file Teamsters continued to support the longshoremen throughout the protracted battle,[107] tensions between the ILA and Teamsters leadership emerged as the strike wore on. Early on, the San Francisco Teamsters' president, Mike Casey, fought to limit his union's involvement in the waterfront boycott (he refused to back a prohibition, passed by the members, on Teamsters handling cargo that had been smuggled out of the docks).[108] As the strike wore on, Casey was viewed by rank-and-file ILA members as the embodiment of AFL conservatism: too cozy with employers and too quick to support a deal on terms favorable to them.[109] In this vein, Casey voted against a motion (which still passed) by the AFL's General Strike Committee for an all-union walk off;[110] when the Teamsters' sympathy strike officially ended, Casey rejected a proposal to refuse hauling cargo to and from the San Francisco port, instead ordering his local back to work "without reservation."[111] After the strike ended, the Teamsters rejected an invitation to join a coastwide labor federation led by Bridges.[112]

That federation ultimately failed for other reasons, fractured by internal dissension among longshoremen themselves. The rise of the Congress of Industrial Organizations (CIO), created in 1935 with the goal of organizing the unorganized masses of workers, sparked intense jurisdictional conflict with the AFL, which began expelling CIO affiliates and forcing local unions to take sides in the "fratricidal warfare."[113] Bridges, whose commitment to inter-union solidarity meshed with the CIO's advocacy for a mass trade union movement opposed to discrimination, called for longshoremen to side with the CIO.[114] This provoked a deep fissure that resulted in the ILA's Pacific Coast District breaking away in 1937 to form the CIO-affiliated International Longshore and Warehouse Union (ILWU), which rose under Bridges' leadership to become a powerful force across West Coast ports in the postwar era.[115] Known for its radical politics and militant tactics (it was famous for calling "quickie strikes"), the ILWU "won some of the most restrictive work rules of any industry,"[116] although San Pedro was "less responsive to radical politics" by virtue of its relative isolation[117]—with "rednecks" in the ILWU local there leading "the fight against the ILWU's no-discrimination policy" through "continuing harassment of blacks."[118] Indeed, from World War II through the 1970s, white ethnic control of the ILWU in San Pedro led to refusals to hire workers of color, protests when they were hired, denials of promotions and harassment, and the creation of a hiring hall system in which ILWU members were required to sponsor new workers to join the union, which "tended to create ethnic (white) homogeneity among members"—a practice that did not end until "a series of affirmative action lawsuits ... [forced the union] to open up membership to women and people of color."[119]

As its rift with the longshoremen during the Big Strike augured, the Teamsters' growing success organizing port truck drivers, who hauled cargo to and from the ocean steamships,[120] created an ambiguous and contested jurisdictional boundary that would shape ongoing relations between the Teamsters and the ILWU. During the first part of the twentieth century, the Teamsters had made little progress against the strong open-shop forces in Los Angeles.[121] The turning point occurred in an audacious 1937 campaign that used the threat of the then-still-legal secondary boycott to force Los Angeles trucking companies to recognize the union and negotiate a contract.[122] The 1935 Motor Carrier Act prevented a carrier from shipping cargo outside its region if another carrier refused to connect.[123] Because the San Francisco trucking firms had already unionized, the Teamsters used the pressure of their refusal to accept Southern California hauls to force Los Angeles firms to unionize upon risk of losing access to the lucrative Bay Area market.[124] This, combined with the newly formed ILWU's refusal to cross the Teamsters picket lines at the port, succeeded in unionizing the largest—and most antiunion—regional carrier, Pacific Freight Lines, and to subsequently win an agreement that unionized the regional trucking industry.[125] Building on the foundation of this agreement, the Teamsters became one of the most successful unions in the state (and also the nation), achieving dramatic union growth that helped increase trucker wages and benefits through the 1960s.[126]

The labor movement's legacy from this period was decidedly mixed. Although the ILWU and Teamsters independently rose to become dominant players at the ports, winning significant benefits and influence for their members, jurisdictional conflict undercut the promise of inter-union solidarity. Although truckers and longshoremen had benefited from sporadic coordination in the past, their narrow self-interests often clashed, with truckers loath to lose opportunities to haul due to work stoppages by longshoremen, while longshoremen feared the power of truckers to reject cargo and effectively shut down the loading and unloading process upon which their well-being depended. Moreover, there was always the overlay of protectionism. The ILWU resisted Teamsters incursion onto the docks, just as the Teamsters opposed ILWU extension outward into the trucking sector. And, in both unions, persistent racism and nativism maintained white economic privilege at the cost of building bridges with the region's increasingly diverse working population.

Yet during the boom period of postwar prosperity, such internecine conflict and exclusionism did little to dampen organized labor's overall optimism. The Teamsters' success contributed to the broader rise of the postwar labor movement in Los Angeles, which at its height in the mid-1950s had more than 35 percent of nonagricultural private-sector workers under union contract.[127] Much of the increase in union density was attributable to the growth of manufacturing, particularly in the aerospace industry.[128] This growth depended on the ports to facilitate exports to the expanding global marketplace. During this time, the interests of the ports, local business, and organized labor aligned around the project of port expansion.

This alignment, which lasted from World War II to the 1970s, marked a transitional moment. As the growth engine of local trade shifted from city building to globalization, and the federal regulatory regime governing transportation and labor relations crumbled, the ports and some workers—specifically truckers—lost power. Globalization, deregulation, and new transport technologies shifted power to corporate shippers (i.e., large importing manufacturers and retailers, like Walmart) and global shipping firms (i.e., owners of ocean carriers and related intermodal services, like Maersk), which were increasingly able to set terms with the ports and other carriers. As the ports grew to meet demands for expanded facilities to accommodate booming global trade, the ports' relations with workers and local communities once again were recast—with new tensions emerging.

Global Power: Free Trade, Deregulation, and the Logistics Revolution

Globalization would lift port activity to new heights and also fundamentally change its nature. As the volume of global trade through the ports began to expand dramatically in the 1970s, it also changed in composition from a balanced export-import flow to an import-dominated stream.[129] This transformation profoundly altered the role of the ports: from building the local economy to facilitating the global one.

The result was a ratchet effect in growth. Rapidly expanding global trade, deregulation, and more powerful shippers weakened port negotiating strength, as shippers of goods could drive a harder bargain by threatening to direct cargo to different West Coast ports. To maintain their advantage, the ports had to outcompete rivals—and each other—at the level of infrastructure and service. This required massive new investments, typically publicly financed, in port facilities and transportation networks. As port infrastructure was developed, it became more attractive for shippers; as more cargo flowed through the ports, the transportation infrastructure had to be expanded to accommodate the increased volume; as infrastructure was built out, the harbor attracted even more shipping in an iterative cycle. Competition between Los Angeles and Long Beach contributed to this growth pressure, which was no longer consistent with the interests of labor unions and surrounding communities. Indeed, the emergence of Los Angeles and Long Beach as the global point of entry to the United States—America's port—depended on industry restructuring, which undermined the labor bargain struck in the postwar period, and infrastructure expansion, which encroached on the ports' low-income community neighbors. As a result, the ports' integration into the global market imposed significant local externalities and generated intense local friction—provoking political efforts to reign in port autonomy and ultimately igniting community and labor mobilization against port expansion.

During this period, global trade—and port growth—was facilitated by technological and legal changes that served to reinforce one another. Beginning in the 1950s, transportation innovations promoted growth by making it more efficient and cost effective to move production farther away from the point of sale. The key advance was the advent of containerization and, from that, the rise of intermodalism, which dramatically reduced the cost of moving goods from one form (or modality) of transportation to another.[130] "Containerization" was the term given to the creation of shipping containers in standardized sizes (typically eight by six by twenty feet, often called twenty-foot equivalent units, or TEUs) that could be locked in place on different types of transport systems—steamships, trains, and trucks—and could also be stacked on top of each other for maximum shipping volume.[131] This allowed goods to be packed in containers at the point of origin and then shipped unaltered via an interconnected transport system comprised of different modalities to the destination (hence "intermodalism"). Costly and time-consuming loading and unloading of cargo under the break-bulk system—in which pallets of cargo would be transported by crane and loaded by hand—were thereby eliminated.[132] As a result, goods production could be increasingly remote from the point of sale and transportation could be made more mechanized and efficient.[133] This appealed to shippers, which sought to reduce labor costs by outsourcing production to countries with lower labor standards, and carriers, which could begin to create standardized equipment and envision door-to-door service. In the process of containerization, the immediate power of the ILWU was weakened as the number of cargo processing jobs was reduced.

Achieving the long-term efficiencies of containerization, however, required substantial short-term capital investment to create the necessary port facilities. With interport competition constraining how much the ports could exact through user fees (charged to shippers and carriers), the

Figure 2.1
Container ship entering the Port of Los Angeles. Photo by the author.

resources for infrastructure development came from public subsidies, as the cities competed to maintain their share of trade. In 1960, Los Angeles launched a $50 million project, financed by municipal bonds, to build new berths and terminals and to upgrade other facilities.[134] Long Beach used its declining oil revenues to follow suit.[135] Although Long Beach lost its oil revenues in 1965, it turned to municipal bond financing to fund further improvements in 1970, including the creation of new container terminals and a freight station. Los Angeles also issued more bonds to finance the expansion of its container terminal.[136]

Despite parity in infrastructure investment, the port rivalry began to tilt in Long Beach's favor as Los Angeles's too-shallow harbor impeded entry of the large "post-Panamax" containerships that moved containers from Asia (so named because they were too big to fit through the Panama Canal).[137] These ships were the logical extension of containerization, which incentivized ocean carriers to build bigger ships to haul more containers per trip, thus reducing the number of trips (and their associated costs), while increasing revenue per trip.[138] In the Port of Long Beach, oil-extraction-induced subsidence had the benefit of naturally deepening the harbor and permitting the docking of post-Panamax vessels.[139] As Los Angeles began to lag behind, local political officials lobbied for federal financial assistance for additional dredging, which it won in 1981; the project was completed two years later.[140] By this point, the Port of Long Beach had surpassed its Los Angeles counterpart in total cargo, although Los Angeles remained more profitable because of higher fees and rents.[141] Containers constituted an increasing share of port cargo: the proportion of cargo shipped via containers through West Coast ports grew from roughly 15 percent in 1970 to 30 percent by 1980.[142] By 2000, containers comprised fully two-thirds of West Coast port traffic, and approximately 70 percent of those containers came through the ports of Los Angeles and Long Beach.[143]

Containerization promoted, and was also a product of, rapidly expanding global trade routed through the ports. Despite the global recession in the mid-1970s, port traffic continued to grow geometrically, increasing due to more manufactured imports from the emerging markets of the Pacific Rim.[144] Still critical to regional economic activity, with one estimate suggesting that more than two hundred thousand jobs depended on maritime trade,[145] the ports became increasingly geared toward facilitating imports,[146] and routing them to delivery points deep within the national economy—and oftentimes beyond to Europe. This transformation of the ports into central nodes in the global supply chain was authorized and promoted by interrelated legal change.

The decline of trade barriers permitted the rise of Asian imports. Trade liberalization through multilateral agreements, particularly the General Agreement on Tariffs and Trade, and bilateral agreements with trading partners, significantly reduced the cost of imports to the United States and thus helped fuel the growth of export-driven economies, particularly China and the so-called East Asian Tigers.[147] As manufactured goods could be produced more cheaply in foreign countries with lower labor standards, production was outsourced and U.S. trade shifted toward imports. Whereas in 1970 the United States still had nearly a $4 billion overall trade surplus, by 1976, it had turned into a deficit; exports of manufactured goods fell behind imports beginning in 1983.[148] The postwar industrialization enabled by trade barriers gave way to deindustrialization associated with free trade. As Asia came to dominate the import market, the strategically positioned San Pedro ports reinvented themselves, becoming the gateway of this new trading regime.[149]

That this occurred was not preordained by geographic advantage. The northwest ports of Seattle and Tacoma were closer to East Asia.[150] However, the Los Angeles and Long Beach ports offered shippers superior infrastructure and service—an advantage that had to be maintained.[151] Because of their political autonomy, both ports were able to move quickly—and in coordination—to build facilities for container traffic and upgrade the transportation infrastructure required to move it.[152] Intermodalism became even more important with the development of "landbridge," by which containers shipped into the ports were loaded onto trains for further transport across the United States.[153] Landbridge was quicker and more cost effective for large-volume exporters from China, which could load up post-Panamax ships and bypass the Panama Canal for East Coast and transatlantic shipments.[154] But landbridge's efficiency depended on robust intermodalism—an integrated transportation system that relied on legal deregulation to permit ocean carriers, railroads, and trucking firms to enter into rate-setting agreements that allowed door-to-door service.[155]

Deeper integration began to take shape in the 1970s, when the federal system of transportation regulation that had enabled the ports to set favorable rates was dismantled. Railroad deregulation came first, followed by the 1980 Motor Carrier Act, which deregulated trucking. The act dramatically changed the trucking industry: making it easier for new companies to enter the market, deregulating routes, and reducing industry authority to set general rates,[156] which permitted discriminatory pricing (through, for example, high-volume discounts).[157] In 1984, Congress passed the U.S. Shipping Act, which deregulated ocean transport,[158] allowing rates and routes to be set by individual companies.[159] In addition, the Shipping Act permitted ocean carriers to contract directly with trucking and rail carriers to set door-to-door rates,[160] thus authorizing them to establish "single through rates on intermodal shipments" without incurring antitrust liability.[161]

Taken together, deregulation completed the legal transformation necessary to achieve intermodalism. By authorizing intermodal contracts, shippers were able to negotiate through rates directly with ocean carriers, which contracted with trucking and rail carriers to provide door-to-door service on one bill of lading[162]—without regulatory barriers or antitrust exposure.[163] Because standardized rates were no longer required, ocean carriers could negotiate directly with individual rail and trucking carriers for the best prices to reduce overall shipping costs.[164] Because ocean carriers dealt in such high container volume, they could exert downward price pressure on rail and trucking companies, which were forced to compete among themselves (and authorized to do so by deregulation) in order to be part of intermodal contracts. In addition, the ability to set door-to-door rates gave shippers greater power vis-à-vis the ports. By threatening to run their intermodal routes through other ports, shippers could negotiate more favorable port access fees and demand improvements to facilitate intermodal connections.

Law in the Development of the Port

To maintain their dominance over container traffic, the Los Angeles and Long Beach ports were forced to respond to these changes. This required building the infrastructure needed to permit efficient container transport from ocean steamship vessels to rail, which became the critical mode of transportation in the landbridge system.[165] Containers coming off steamships were placed on rail cars in two locations: some were moved directly from steamships to railcars at on-dock rail facilities, while others were transported to off-dock rail yards by short-haul, or "drayage," truckers.[166] On-dock loading required interconnected rail lines and loading facilities at the port. The Harbor Belt Line Railroad, unifying the tangle of separately owned railways in the port complex, was completed by the Los Angeles harbor commission and railroad companies in the 1930s.[167] This system was augmented and loading facilities expanded to permit intermodalism.[168]

Off-dock shipping required new investments to create massive areas where trucks could congregate to transfer their loads to rail cars that could then easily connect to transcontinental lines. To achieve this, the San Pedro ports coordinated their first major joint project, in concert with the Southern Pacific Railroad (later Union Pacific, or UP): construction of a $50 million Intermodal Container Transfer Facility (ICTF), completed in 1986, to allow mass movement of containers from ships to off-dock rail lines operated by UP.[169] The other major off-dock rail connection was located near downtown Los Angeles, with four major intermodal rail yards: three operated by UP (the East LA Yard in Commerce, the LA Trailer and Intermodal Container Facility just east of the Los Angeles River, and the City of Industry Yard) and the fourth (Hobart, just west of the Interstate 710 freeway in Commerce) operated by Burlington Northern Santa Fe (BNSF).[170] All of these yards were serviced by drayage trucks.[171]

The 150-acre ICTF was built five miles from the ports in Carson (on the northern border of Wilmington), at the terminus of State Highway 103 (called the Terminal Island Freeway),[172] and adjacent to the interchange of two major freeways (Interstate 405 and Interstate 710).[173] The ICTF was designed to alleviate truck impacts at the ports by routing traffic to a massive facility with ample parking and faster loading service.[174] Yet in its attempt to reduce port congestion, the ICTF introduced a new source of gridlock into the community: increasing drayage truck traffic on the freeways and surface streets coming to and from the ports. This increase in truck traffic highlighted a counterintuitive problem. Although landbridge placed railroads at the center of trade distribution, drayage trucking was essential to move containers from ocean vessels to off-dock rail lines—and also to move cargo from the ports to local warehouses.[175] Containerization thus increased demand for trucking in proportion to rail, placing strain on the region's overtaxed freeway system, while overtaking local roadways in port communities and in the communities around the downtown intermodal yards.

The symbols of Southern California mobility—the freeways—were supposed to alleviate the burdens of local traffic. But freeway expansion in Los Angeles and Long Beach occurred without containerization in mind—and ultimately could not handle the ever-increasing volume of truck traffic necessary to serve the ports. Freeway development was fueled by postwar suburbanization that created the vast car-dependent metropolis. The design of the freeways was, however, done with the ports in mind. Construction began in the 1950s, spurred by federal investments and local pressures. The primary route into the Port of Los Angeles was built through northern San Pedro via the Harbor Freeway (Interstate 110), running due south from downtown Los Angeles. That freeway, funded by a state gas tax and federal interstate highway money,[176] was built between 1952 and 1970, extending piece by piece from Pasadena, south of downtown Los Angeles, and then bisecting African-American communities in the south-central part of the city.[177] What is now known as the Long Beach Freeway (Interstate 710) forms the eastern border of Wilmington. With federal money, Interstate 710 was built from 1954 to 1975, designed to connect Long Beach to Pasadena, bypassing downtown Los Angeles; however, it was only extended just past the Interstate 10 freeway in Alhambra as the proposed link to Pasadena was thwarted by community opposition.[178] These freeways became the main conduits for the increasing volume of heavy-duty drayage trucks pulled to the harbor by free trade.

Increasingly linked through a dense intermodal transportation system, container shipments through Los Angeles and Long Beach surged, growing from 9 million tons to 122 million tons between 1970 and 1994.[179] In 1986, the San Pedro port complex passed the Port of New York and New Jersey as the largest in the United States; the next decade, container volume doubled.[180] Auto imports, particularly from Asia, powered this growth—with more than 150,000 autos coming into the Port of Los Angeles in the first five months of 1985—creating jobs in the regional auto-processing industry, but also producing severe space constraints in the port itself.[181] In 1990, nudged forward by frequent trade missions of public officials,[182] the Port of Los Angeles surpassed New York as the nation's busiest by volume.[183]

Once again, this growth—achieved by creating better intermodal connections—placed new pressures on existing infrastructure. Specifically, enhanced links to rail transport began to overtax the rail system itself. More—and more efficient—rail connection was thus needed to avoid another bottleneck. By the 1980s, the rail system—operated primarily by the two major railroads, UP and BNSF—was a complex web viewed by industry observers as impeding the movement of ports-related goods by forcing rail cars to travel old branch lines, pass through numerous crossings, and share track with other freight and passenger trains. This slowed on-dock rail loaded directly at the terminals and off-dock rail loaded at the ICTF.[184] In response, Los Angeles and Long Beach created a joint powers authority in 1985 authorizing the development of the Alameda Corridor project, a twenty-mile high-speed elevated line from both ports connecting to the transcontinental railroad.[185] The Alameda Corridor rail, running through Wilmington (then north through Carson, Compton, Lynwood, Watts, South Gate, Huntington Park, and Vernon), was completed in 2002 with $2.4 billion in federal, state, and local financing.[186] Carrying roughly fifteen thousand trains a year,[187] it consolidated track to more efficiently link on-dock rail to eastern destinations,[188] while creating better connections to the ICTF for off-dock transfer.[189]

The Alameda Corridor was one of several large-scale megaprojects coordinated between the ports to deal with massive projected increases in port traffic.[190] In the mid-1980s, both ports adopted the "2020 Plan" to upgrade and integrate maritime trade and land transport systems to deal with an anticipated 250 percent increase in tonnage.[191] A 1998 study predicted that, with appropriate infrastructure investments, cargo at the ports would double by 2020, making Los Angeles the "trading center of the world."[192] Although Long Beach eventually withdrew from formal coordination, both ports nonetheless completed nearly $4 billion in joint investments by 2000, with Long Beach focusing on land acquisition and redevelopment, and Los Angeles on dredging and the creation of new terminals and rail lines.[193] These investments correlated with growth. From 1990 to 2000, total TEUs increased by 130 percent at the Los Angeles port and by 188 percent at Long Beach.[194] By 2005, the Los Angeles-Long Beach port complex was the fifth largest in the world, with a combined fourteen million TEUs of traffic.[195] Three-quarters of trade into the Los Angeles Customs District were imports and most of those (85 percent in 2005) were from Asia (with nearly half from China).[196] Despite the ports' joint investment, growth produced new challenges. Port space for tenants remained a concern.[197] Although the ports coordinated on megaprojects, they continued to compete on price and service to attract more tenants and cargo.[198] As figure 2.2 shows, during the 1990s, both ports saw containerized cargo increase at roughly the same rate, with one port surging ahead and then the other. In 1995, Long Beach

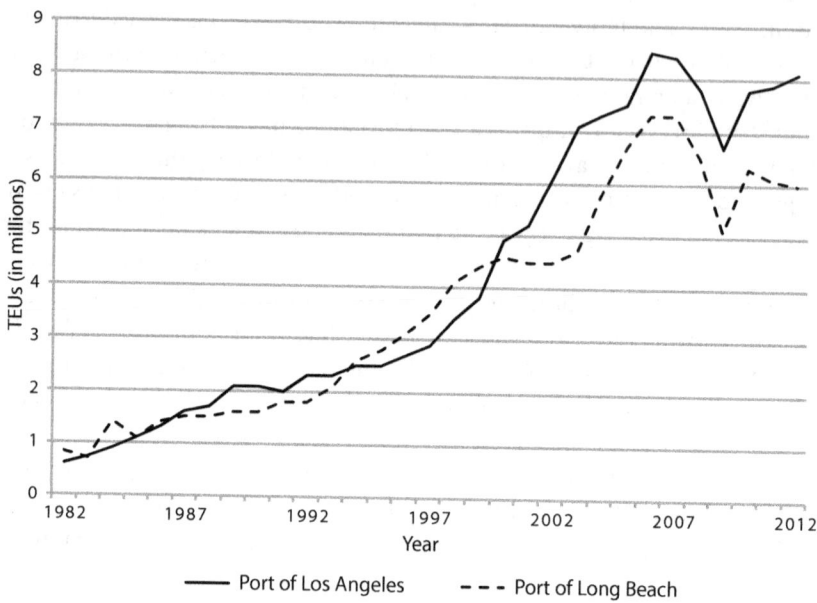

Figure 2.2
Annual container trade in TEUs. Graph based on TEU data provided by the ports. Port of Los Angeles: "TEU Statistics (Container Counts)," www.portoflosangeles.org/maritime/stats.asp; Port of Long Beach: "Yearly TEUs," www.polb.com/economics/stats/yearly_teus.asp (both accessed Nov. 20, 2014). Data for Long Beach from 1982 to 1994 are based on information from the Los Angeles Almanac, "Ocean-Going Cargo Container Traffic Los Angeles and Long Beach Harbors (TEUs)," www.laalmanac.com/transport/tr46.htm (accessed Nov. 20, 2014).

surpassed Los Angeles as the nation's biggest port.[199] Scoring a major coup by enticing Maersk, the world's largest container carrier,[200] to move from Long Beach,[201] Los Angeles eked back ahead in 2000, handling 4.9 million TEUs to Long Beach's 4.6 million.[202]

Even with new infrastructure, however, the ports were tested by the surge in overall volume.[203] The most prominent challenge came in 2004, when more than one hundred ships were diverted to other ports because of an unanticipated increase in ship traffic that could not be handled by existing longshore, rail, and trucking systems.[204] To prevent future bottlenecks, more longshoremen were hired and the acute stress was alleviated.[205] Yet the episode underscored the transformation of labor relations in the context of intense port competition.[206]

The advent of containerization had initially threatened longshore jobs.[207] No longer needed for the difficult and time-consuming loading process, longshoremen were redeployed for container transport, which involved movement via overhead crane and attachment to trucks and trains.[208] The immediate consequence of containerization was to reduce the demand for longshore work since container transport, which relied on mechanization, required fewer labor hours.[209] This provoked labor unrest and the ILWU struck both ports in 1971.[210] The strike lasted roughly four months and caused a cargo reduction of two million tons at Los Angeles.[211] With the power to choke port traffic, longshore workers demonstrated that even though they were fewer in number, they remained a force to be reckoned with.[212] And the ILWU leveraged that force to negotiate agreements with maritime employers that protected its control over the explosion of container processing work. In this way, the ILWU was able to turn the threat of containerization into a long-term strategy to insulate longshoremen from downsizing and maintain their privileged position as the gatekeepers of global trade.

That strategy, however, brought to the surface long-simmering tensions between Bridges' ILWU and the Teamsters, which bristled at the ILWU's effort to wrest exclusive control over containers.[213] As Bridges advanced an ill-fated proposal to reunite the ILWU and ILA, the Teamsters picketed the Los Angeles Container Company as "a public warning that if it gives longshoremen the right to load and unload most containers, then the Teamsters will shut down harbor operations throughout the West Coast."[214] This jurisdictional dispute precipitated discussions between Bridges and Teamsters president Frank Fitzsimmons over the possibility of a "broad alliance, if not outright merger,"[215] which resulted in a preliminary "letter of intent" to merge.[216] In a sign of solidarity during the ILWU's 1972 West Coast strike, the Teamsters went so far as to coordinate a joint picket of non-union trucks crossing into the United States from Mexico, carrying cargo that had been diverted from U.S. ports to avert the strike.[217] However, the merger proposal, which would have created a separate longshore division of the Teamsters,[218] was defeated by the ILWU membership, which remained bullish about the containerized future and thus reluctant to cede control and benefits to another union.

As container trade grew rapidly following deregulation, demand for dockworkers began to grow again and their bargaining position strengthened as the ports became crucial nodes in the import chain. Just as intermodalism forced the ports to invest in infrastructure and service to keep shipping lines satisfied, it also made the ports invest in labor peace, since even a minor disruption could send shippers elsewhere.[219] This significantly bolstered the longshoremen's bargaining position relative to the ports, permitting them to grow their membership, increase wages, and build the ILWU's organizational strength.[220]

The longshoremen's new position reflected a broader power shift. As a gateway to the regional market, the ports in the industrial era were empowered by federal regulation to charge higher fees and align port growth with local interests. However, deregulation and intermodalism changed this equation,[221] rendering the ports a pass-through to the global market. Particularly as shippers could divert cargo to different ports, they gained more bargaining power to drive down rates and demand port amenities that permitted larger volume. The ports were forced to continuously invest in new infrastructure to maintain their advantage. This investment no longer fostered local industrial development as it had in the postwar period. To the contrary, Los Angeles and Long Beach found themselves increasingly under fiscal strain because of deindustrialization, which was now itself intrinsically linked to the ports. The political autonomy the ports had acquired to build the regional economy became an increasing liability, as port revenues were used to benefit the ports' global shipping clientele by continuously upgrading the intermodal system.[222]

No longer reaping a return on local industrial development, Los Angeles and Long Beach sought to assert greater control over the ports in an effort to claim more local fiscal benefits. By the early 1990s, the Los Angeles harbor commission's vaunted independence still existed, but had been reined in by charter amendments that gave the City Council greater oversight authority: imposing limits on significant contracts, requiring council approval for certain types of leases, making it easier to terminate key personnel, and ultimately giving the council authority to approve important commission decisions.[223] Long Beach underwent similar changes to limit harbor commission power.[224] These changes made it easier for local politicians to check port spending and adapt port activity to city agendas shaped by declining revenues.[225]

During this time, Los Angeles and Long Beach—suffering from manufacturer outsourcing and the end of Cold War-driven defense production—sought a share of port resources to infill dwindling city taxes.[226] In 1992, state lawmakers permitted the two cities to divert some port discretionary reserves to replace property tax taken by the state to fund its own budget shortfall.[227] Once this temporary provision expired, the cities sought to use their greater power over the harbor commissions to extract revenue by charging the ports more for city services (like police and fire

Law in the Development of the Port

protection).[228] When this practice was challenged under the Tidelands Trust Act,[229] cities changed course by using port funds to build tax-revenue-generating harbor projects, like the Long Beach Aquarium, under an expanded definition of "public benefit."[230] With the ports no longer fueling local industrialization, city governments looked to them to play a new regional role: creating logistics industry jobs and spurring retail growth foundational to the ascendant service-based economy. While this strategy sought to address local fiscal needs, it exacerbated the impact of port expansion on local communities as infrastructure megaprojects like the ICTF and Alameda Corridor rail resulted in increased congestion and pollution. By linking municipal finances to port growth, cities committed themselves to a development program with increasingly serious local consequences.

Local Impact: Community, Labor, and the Environment

The ports' local impact is a function of their dual identity: at once "an integral function in the globalization of production," the ports are also "one of the most localized and embedded industries of all."[231] As such, they are special kinds of agglomeration economies, where cargo distribution facilities—steamships, rail, trucks, and support services—all cluster.[232] It is precisely this clustering that creates externalities, both positive and negative, for local communities. These externalities stem from two types of organizational relationships within the global supply chain.[233] One is *inter*organizational: the relationship between different economic actors linked across the chain—from shippers to ocean carriers to dock workers to railways and trucks to warehouses.[234] From a geographic point of view, there has to be space appropriated to permit "transshipment": the transfer of cargo, especially containerized cargo, from one transport mode to the next. Over time, with port growth, that space becomes more built out, putting more pressure on surrounding communities and increasing environmental risk. The second type of relationship is *intra*organizational: the formal division of labor within specific firms, like trucking.[235] In firms connected to port logistics, there are different models of providing services—through employees and independent contractors—that are authorized by distinct legal standards. These intraorganizational relationships have significant implications for workers in two key areas of intermodal logistics: terminal operations and drayage trucking.[236] This section examines how these inter- and intraorganizational relationships have contributed to the creation of environmental justice problems in low-income communities with weak political power and the degradation of labor standards in industries with weak legal protection.

The communities of San Pedro and Wilmington (shown in figure 2.3) have been shaped in inverse relationship to port development—suffering economically as the ports, and the global economy they serve, have thrived. The impact can be measured in relation to neighboring communities. On the north face of the San Pedro Peninsula, rising dramatically above the coastline, is a group of cities collectively known as Palos Verdes—an area noted for stunning bluffs, excellent public schools, and some of the most coveted real estate in the Los Angeles area (as the city names—such as Palos Verdes Estates and Rolling Hills Estates—suggest). These cities, formally within Los Angeles County, are all legally separate, with small populations (Rolling Hills Estates has only 8,000 residents), some of which are enclosed in gated communities, and all of which enjoy high levels of services and amenities—a function of their legal distinction from the city of Los Angeles, with which they do not share their tax base.[237]

As the peninsula slopes southeastward, toward the ports, its change in municipal jurisdiction is marked by dramatically different socioeconomic conditions. Although separated by only a few miles, San Pedro is distinguished from Palos Verdes in crucial respects. It encompasses the Port of Los Angeles, which lies on the eastern edge of the community, abutting the harbor's main channel, and also includes the western part of Terminal Island. San Pedro is therefore a key point of access to the port, which drayage trucks traverse from Interstate 110 in order to cross into Terminal Island or to travel down to facilities along the main channel.[238] On-dock rail lines also run along San Pedro's eastern edge as they snake their way to the Alameda Corridor exchange. The neighborhood's northeastern border also abuts the ConocoPhillips Oil Refinery in Wilmington, which creates a cluster of environmental hazards in that corner. Because it is within Los Angeles city proper, San Pedro receives a lower level of services than Palos Verdes does, symbolized by the chasm between the quality of the public school systems.[239] There are other markers of socioeconomic divide. Palos Verdes is nearly four-fifths white, highly educated (nearly 60 percent holding a college degree), older on average (median age of fifty), and more affluent (median income of approximately $130,000) than San Pedro, whose residents—two-fifths Latino and a quarter immigrants—are relatively younger (median age of thirty-four), less educated (roughly one-quarter are college educated), and less well-off (median household income of $57,000).[240]

Law in the Development of the Port 45

Figure 2.3
The Ports of Los Angeles and Long Beach and surrounding communities. Map created by the author using Esri ArcGIS.

Just before it terminates in San Pedro, the Harbor Freeway cuts along the western border of Wilmington, which is bounded by Interstate 710 on the east. Wilmington's northern border is defined by the ConocoPhillips Los Angeles Refinery (in addition to the ConocoPhillips Oil Refinery to the west and Tesoro Los Angeles to the east), thus encircling it with environmental hazards. The strip of land on its southern border is part of the port's functioning inner channel, lined with terminals (including the massive TraPac Container Terminal) and crossed by streets and rail lines, separated from the residential part of the city by Harry Bridges Boulevard (named after the pioneering founder of the ILWU[241]) and above that, in the western part of the community, the recently developed Wilmington Waterfront Park. Trucks access this part of the port from the freeways on both sides, as well as the surface streets, which are often traveled by trucks connecting between the docks and the ICTF. In this way, Wilmington, even more so than San Pedro, exists as an adjunct to the port transportation system. As a result, the community itself is more disadvantaged, with a higher level of segregation and lower socioeconomic indicators than San Pedro. Nearly 90 percent of Wilmington residents are Latino and almost half are immigrants; the community has a median household size of four and a median income of $40,000; and only 5 percent of residents have a college degree.[242] It is, to a greater degree than its neighbor, a dramatically different place than the white ethnic enclave that defined the port community a century ago—doubly disempowered by race and class in ways that translate into economic neglect and political disenfranchisement.

As with land use, the labor impacts of port development also vary.[243] The market for landside workers at the port is highly segmented in ways that reflect legal differentiation—and build on historical divisions within the labor movement. As that history demonstrates, all port workers, from longshoremen to truckers, are theoretically in a position to choke distribution along the supply chain. But the legal power to take advantage of that position differs, cementing the labor hierarchy that emerged during the postwar conflict between the ILWU and Teamsters. Because longshoremen are employed by port firms (ocean carriers and terminal operators), they are legally empowered to organize and exempted from antitrust law. It is the combination of their legal and market position that gives them significant bargaining power, which they have been able to use—as the Harry Bridges-led strikes of the 1970s demonstrated—to unionize and negotiate relatively high wages and benefits.[244] From a market perspective, their leverage rests not just in complete shutdown, but also in delay. Given the "just-in-time" nature of global distribution, slowdowns pose a significant threat to shippers, who prefer to buy labor peace to ensure logistical efficiency.[245] The risk of capital flight (transferring cargo operations to other ports) is minimized because of the massive up-front investments required to facilitate transport, which enhances longshoremen's bargaining power.[246]

The longshoremen have used that power to win global labor agreements covering dock workers at multiple ports in a region—preventing shippers and carriers from rerouting cargo to reduce labor costs. In the 1990s, the ILWU negotiated a contract with the Pacific Maritime Association (PMA)—the West Coast employer trade group—that increased wages by 9 percent, expanded the union's jurisdiction into harbor trucking, and rejected the creation of a computerized job-dispatch system that longshoremen believed would take job assignment power away from the union hiring hall.[247] In 2002, alleging work slowdowns, the PMA locked out the longshoremen in an effort to decrease their clout,[248] producing a six-week backlog and causing President George W. Bush to invoke the Taft-Hartley Act to reopen the ports.[249] As shippers began to reroute cargo to the East Coast, the PMA backed down, agreeing to a six-year contract—"the most lucrative in the union's seventy-year history"—that increased hourly base wages to $30, substantially increased pension benefits, and provided strong employment security protections.[250] As the episode reinforced, longshoremen had become the ports' "labor aristocracy."[251]

Drayage truck drivers, in contrast, had sunk to the bottom of the labor hierarchy.[252] Although also in a position to choke supply, their formal status as independent contractors—a consequence of deregulation—meant that they could not organize and therefore lacked the ability to coordinate labor action that would allow them to leverage collective gains. Prior to deregulation, strong transportation and labor laws permitted the Teamsters to organize firm employees, which they did with great success.[253] However, their jurisdictional conflict with the ILWU over control of containers, and missed opportunity to merge, left the Teamsters more vulnerable to deregulation, which restructured the trucking market in ways that did not affect dock work. Throughout the trucking industry, deregulation introduced fierce competition and increased the number of firms, particularly in the drayage sector.[254] Drayage trucking companies systematically moved to a model of contracting out,[255] under which firms assigned work to nominally "independent" owner-operators,[256] who purchased or rented their own trucks and were paid by the load or trip, rather than by the hour.[257] This insulated companies from trucker liability and also significantly reduced labor costs by eliminating the need to pay employment taxes and benefits (such as health care and retirement). It also shifted the downside industry risks—particularly the cost of bottlenecks and delays associated with port clearance and cargo identification—to the drivers, who became responsible for truck maintenance, fuel, tolls, taxes, and other expenses. As such, trucking firms became "non-asset-based companies," shedding fixed expenses to increase their share value.[258] In addition, and most crucially, the move to independent contractors undermined unionization, since independent contractors were banned from union organizing under antitrust law.[259] Without the alliance of longshoremen—whose history of competition with the Teamsters and genuine concern for their own dwindling jobs made them disinclined to flex muscle to aid their trucking counterparts—port drivers' ability to break out of the independent contracting system and rebuild union density was severely restricted.

Figure 2.4
Trucks waiting on Figueroa Street, Wilmington. Photo by the author.

As a result, the conditions of port truckers deteriorated sharply.[260] In Los Angeles and Long Beach, the independent-contractor form came to predominate in the drayage trucking sector, with nearly 90 percent of truckers so designated.[261] For these drivers, the average annual salary, after expenses, was $28,000.[262] In part because of delays, they worked on average fifty-six hours per week, thereby earning an effective wage rate of less than $10 per hour.[263] Because they were designated independent contractors, they received no health insurance or pension contributions from their companies.

The drayage labor force also came to be defined by workers of color. By 2000, in Los Angeles and Long Beach, port truckers were almost entirely Latino and nearly half were immigrants.[264] Bonacich and Wilson describe the shift from white drivers at mid-century to predominantly nonwhite drivers beginning in the mid-1980s as a product of deregulation and immigration. Entrepreneurialism was long part of the trucker ethos and, in the immediate wake of deregulation, some white drivers became owner-operators. Yet the industry rapidly shifted. The increase in immigration during the 1980s, powered by Central American civil wars, brought more immigrant job seekers into the industry in part because "you didn't need a green card or an I-9 form."[265] Firms became smaller, more immigrants

Law in the Development of the Port

entered, and wages declined.[266] Bonacich and Wilson report that by 1985 the Teamsters had "lost the harbor."[267] They called a strike, but the "Central Americans did not want the union because of the green card issue," and the strike failed.[268] Tensions between truckers and longshoremen flared as truckers felt disrespected by the largely white longshoremen, whose hourly pay structure made them in no hurry to reduce the transport delays that plagued truckers.[269] Observers identified the drayage sector nationwide as the most problematic element of port logistics, characterized by delay, poor safety conditions, and pollution.[270] The "handoff" from ocean vessels to trucks was viewed as inefficient:[271] to pick up their cargo, truckers had to idle in long queues to enter the port, access the terminal, obtain their chassis, and load their containers.[272] Yet it was a system that benefited trucking companies (which externalized the cost of labor and pollution) and shippers (which were able to pay trucking firms less for their services).[273] Accordingly, those with economic power in the system had no incentive to change the arrangement.[274]

At the start of the new millennium, there were roughly sixteen thousand drayage trucks servicing the Los Angeles and Long Beach ports each day.[275] Because drivers could not generally afford to upgrade, this fleet was aging—the ports were the place where "old trucks went to die"[276]—and ran on diesel fuel, a known carcinogen.[277] Truck emissions, combined with those from ocean vessels and dock transport equipment, caused significant air pollution, which threatened trucker and broader community health. A 2007 NRDC report showed that the black carbon inside truck cabs increased "health risks by up to 2,600 excess cancers per million drivers."[278] Overall, the California Air Resources Board found that diesel particulate matter emissions from all port-related activities constituted roughly one-fifth of all such emissions in the Los Angeles basin.[279] Communities near the ports had cancer risk levels that "exceeded 500 in a million"; further from the port, the risk was less but still significant.[280] From the perspective of community and labor groups, law had contributed to these harmful effects—by disempowering local communities and truck drivers relative to the port. Their response would be to try to reshape the law to fix the problems it had produced.

3 The Port as a Unit of Legal Analysis

Changing law requires understanding the law at issue, what levers exist to change it, and what constraints are in place. Restructuring port trucking in Los Angeles and Long Beach has thus meant analyzing each port's distinct legal character as a local government entity bound to the national and global marketplace. The ports possess broad, locally derived regulatory powers over operations; yet those local powers shape nonlocal activities and thus overlap with state and federal regulatory regimes—particularly those that relate to transportation, labor, and the environment. Nonlocal regulation asserts minimum standards and demands uniformity. This could be a spur to local reform—requiring port action to comply with nonlocal regulatory mandates—but could also operate as a limit on any port legal change deemed inconsistent with federal authority. Reforming port policy to respond to environmental and labor problems has required evaluating and connecting three aspects of this legal regime: (1) a port's local power as a city entity to make law through its internal governance structure; (2) nonlocal governance schemes that could be leveraged to pressure the port to act; and (3) the potential scope of local authority under the federal preemption doctrine.

Local Governance

Port governance is a function of the spatial organization of port activity and the relation between its constituent parts. The essential unit is the port itself: defined as a facility at which ships dock and are loaded and unloaded (with cargo or passengers), providing a conduit between the sea and "hinterland."[1] As a geographic matter, a port is divided between maritime and land domains. Port waterways include an inner harbor, inside the breakwater, and in some cases, channels to access different areas of the port. In terms of landmass, a port may be constructed on the landside area of the harbor or adjacent islands. The port facility is broken down into smaller units. Ports are constructed with wharves—technically, structures that permit ships to dock—and these are divided into quays (parallel to the shoreline) and piers (perpendicular to the shoreline). Each type of structure is further divided into berths, which are the slips into which an individual ship fits for loading and unloading.

Overlaid on these structures is a basic unit of port organization: the terminal. A terminal is the place where freight and passengers either originate or terminate.[2] In practice, there are multiple terminals within a port that are distinguished by function. Contemporary ports are divided into terminals dedicated to different types of cargo: containers, break-bulk (goods packed in boxes or other noncontainerized forms), dry bulk (loose cargo like grain or coal), liquid bulk, automobiles, and passengers. Terminals may be run directly by the local port authority, but with increasing worldwide "port devolution," large global ports typically contract out to private "terminal operators" that coordinate the passage of cargo from marine to land transportation.[3] Within a given terminal, there may be multiple terminal operators leasing sections that encompass specific berths and the surrounding land, which contains loading equipment, access to road and rail, and storage. In most large ports around the world, terminals are run by transnational corporations within the global supply chain.[4]

There are different types of terminal operators. Some are vertically integrated with shippers or ocean carriers, while others are independent terminal operating firms that provide systematic logistical services at the ports—unloading to truck and rail (on dock and off), as well as warehouse transport. In the contemporary port industry, there has been increasing corporate consolidation such that there are a small number of logistics companies that offer comprehensive intermodal services.[5] In addition, some carriers have sought to integrate port services by setting up port terminal subsidiaries, as have a few shippers. For instance, the Port of Los Angeles leases terminals to shippers (e.g., ExxonMobil), carriers (e.g., China Shipping), and dedicated terminal operators (e.g., TraPac).[6] Workers for companies in the terminal areas are generally employees of the terminal operators or firms subcontracted by them. These include longshoremen and the clerks responsible for checking in goods from the steamships and ensuring they are conveyed to the correct railcar or truck.

How a port is legally structured depends on its relation to local government. Based on their peculiar history, the ports of Los Angeles and Long Beach are city departments with the power to control port property. This power ultimately derives from the public trust doctrine, codified in the

The Port as a Unit of Legal Analysis

Tidelands Trust Act, under which the state holds the tidelands in trust for the use and benefit of the people in promoting navigation and commerce.[7] The state has granted some trust lands to local governments, including Los Angeles and Long Beach, which hold the property as legislative trustees to advance defined trust purposes.[8] The cities, in turn, have created propriety harbor departments to manage trust property.[9]

Port governance is established by city charter. In Los Angeles and Long Beach, the current port structure—the product of amendments over the past twenty-five years that have reined in port independence—gives local officials significant control over port personnel and policy. The governance structure of Los Angeles is typical of both ports. The Los Angeles Board of Harbor Commissioners consists of five members appointed by the mayor, subject to City Council approval; board members may be removed by the mayor without council confirmation.[10] The board delegates day-to-day operations to a professional staff, particularly the executive director (also called the general manager), who is given supervisory authority.[11] The executive director has substantial enforcement and implementation power,[12] but is constrained by the board, which has the power to hire and fire the director.[13] This structure confers significant mayoral control since the mayor appoints the commissioners who then appoint a director subject to termination at will.[14] Creating and implementing new port rules may therefore be advanced through mayoral selection of harbor commission personnel. Those personnel are empowered to make port rules and enter into port contracts, subject to approval by the City Council, which thereby wields ultimate legislative authority. The board has the statutory power to "[m]ake and enforce all necessary rules and regulations governing the maintenance, operation and use of the Harbor District," and "[f]ix and collect rates and charges for the use of the Harbor Assets"[15]—in both cases subject to council approval.[16] It also has the power to enter into "any franchise, concession, permit, license, or lease" in furtherance of departmental purposes, subject to council approval for agreements of more than five years.[17] Certain decisions, including leasing large (more than three thousand feet) port space, must be approved by four-fifths of the board and two-thirds of the City Council.[18] As this suggests, significant port rule change can be effectuated through internal board approval validated by the City Council, thus requiring cooperation between the mayor and council members to change port policy.

The port operates, and generates revenue to cover costs and capital improvements, through the board's exercise of its charter powers. It generates revenue primarily from two sources: shipping income, which comes from fees imposed on cargo, and permit (or rental) income, which consists of charging port occupants for the right to use port property.[19] Permit income is negotiated via individual contracts with port users, which include terminal operators as well as ocean and ground carriers. Terminal operators enter into leases allowing them to build, operate, and maintain cargo-handling facilities and related infrastructure, while carriers enter concessions that give them the right to enter and use port property for specified purposes and under negotiated conditions.[20] In this way, the port's operations are defined through contract, with the board negotiating the terms of use and rates. Port rules set by the board establish the permissible scope and conditions of port contracts, thus giving the harbor commission—and ultimately the local officials to whom it is accountable—the power to define who can enter the port and under what terms.

Nonlocal Governance

Because they are linked to regional economies and globally networked transportation systems, when ports exercise local power, they invariably affect nonlocal interests. Ports thus act in a regulatory environment in which local authority intersects with—and ultimately is limited by—federal and state laws designed to promote minimum standards and uniformity. These laws can cut in two different directions. On one side, federal and state authority can force a port to take action and internalize costs that it may otherwise resist. Those seeking reform of port operations may turn to nonlocal law as leverage to do so. On the other side, federal law may preclude action a port may want to take—or at least limit action to specific circumstances in which it has a defined local impact. In this way, any legal change must be sensitive to the preemptive force of federal jurisdiction.

The effect of nonlocal law depends on whether it seeks to regulate or deregulate, and how its standards have been interpreted relative to the interests of specific constituencies. When nonlocal law regulates a field in a manner viewed by a constituency as harmful, that constituency is forced to seek alternative legal avenues of redress. Federal labor law fits into this category: a national scheme designed to promote worker interests, which has been interpreted over time in ways deemed hostile to those interests.[21] At its inception, New Deal-era legislation codified the collective bargaining system,[22] which was validated by courts,[23] ushering in a period of robust private sector unionization.[24] However, what began as a framework to enable worker collective action ultimately became a constraint.[25] Reactionary legislative amendments,[26] damaging judicial interpretations,[27]

The Port as a Unit of Legal Analysis

and industry capture of administrative processes reshaped the legal playing field.[28] As a result, the system of workplace elections established under the NLRA is now viewed by organized labor as disadvantaging unions, which lack strong tools to respond to employer retaliation. Organized labor has therefore generally eschewed the NLRA as the default entry point to win unionization. Instead, unions have sought to leverage other sources of legal and political pressure to secure agreements from employers not to interfere in union organizing and election campaigns before seeking to gain employer recognition and negotiate contracts within the federal system.[29]

The weakness of federal labor law is particularly apparent when one focuses on the plight of the growing numbers of nonstatutory employees: so-called independent contractors or owner-operators, like port truckers, who are excluded from the benefits and protections of federal and state labor and employment law. Despite the fact that many such workers toil in low-wage conditions under the de facto control of companies, like Uber, that offer take-it-or-leave-it contracts that specify worker conduct in exacting detail, their formal designation as contractors places them outside the umbrella of labor and employment law, which relies on a narrow technical definition of employee that turns on comprehensive legal control by employers. A crucial legal implication of workers' nonemployee status is that they are prohibited from collective action; because they are legally treated as independent business owners, any organizing for mutual benefit is deemed to be a combination "in restraint of trade" and thus in violation of federal antitrust law.[30] Federal law does confer immunity from antitrust liability for organized labor, but it only permits collective action to take wages out of competition (through collective bargaining or otherwise) by groups of employees or their union representatives—not independent contractors.[31]

In contrast, when nonlocal law provides strong regulatory standards and empowers constituency action, it can be an effective tool for reform. In the case of the ports, state and federal environmental policy has played this role. In 1970, Congress passed the National Environmental Policy Act (NEPA), covering projects funded or approved by federal agencies, while California passed a similar state law, known as the California Environmental Quality Act (CEQA), covering projects requiring state or local agency approval.[32] NEPA requires that new developments evaluate potential negative environmental impacts and mitigation measures; CEQA requires that environmental impacts be mitigated to the extent feasible, and may permit development despite negative impacts if it is determined there are overriding benefits.[33] These laws apply to port infrastructure expansion and add a layer of environmental review that can be asserted by stakeholders to try to mitigate harms.[34] Neither law can completely block expansion, but both can delay it (and increase costs) by permitting public comments on environmental review plans and potentially requiring that incomplete plans be redone.

The federal Clean Air Act—the key parts of which were also passed in 1970—requires compliance with air quality standards for pollutants from stationary and mobile sources.[35] The act requires compliance with national standards and sets up federal-state partnerships to establish state implementation plans, which may be enforced against regulated sources through citizen lawsuits.[36] California has its own state Clean Air Act, which also requires the creation of local air quality plans to regulate certain pollutants at more stringent levels than those mandated by federal law.[37] State standards are set by the California Air Resources Board (CARB), which oversees local air quality management districts (AQMDs).[38] AQMDs have authority to set and implement state plans in compliance with state and federal law, subject to approval by CARB, which is charged with submitting the state plans to the Environmental Protection Agency (EPA).[39] CARB is also responsible for regulating mobile sources of air pollution and sets specific motor vehicle emission standards.[40] AQMDs regulate fixed sources of air pollution, which require AQMD permits to operate.[41] Together, these federal and state environmental laws give government officials and local citizen groups tools to challenge port development and implement environmental standards that are higher than the regulatory floor.

Federal law designed to deregulate a marketplace—imposing a ceiling rather than setting a floor—has the opposite effect: disabling local regulation that proposes to raise standards above a minimum baseline. Federal transportation deregulation asserts federal law as one such ceiling. The Shipping Act of 1984 gives the Federal Maritime Commission jurisdiction over ports to promote competition, requiring "just and reasonable regulations and practices relating to or connected with receiving, handling, storing, or delivering property."[42] The commission polices interport coordination on rulemaking in order to ensure that it does not reduce competition by producing "an unreasonable reduction in transportation service or an unreasonable increase in transportation cost."[43] Any effort to set joint standards between the Los Angeles and Long Beach ports is subject to this check on anticompetitive measures, which can be enforced by the Federal Maritime Commission through a civil injunctive action.

For trucking, the deregulatory framework centers on the Federal Aviation Administration Authorization Act (FAAAA),[44] enacted in 1995 to prevent states and localities from passing trucking standards that would circumvent the deregulatory provisions of the 1980 Motor Carrier Act. The FAAAA explicitly preempts nonuniform state or local regulation of motor carriers.[45] Designed to be identical with the preemption provision contained in federal airline legislation,[46] the FAAAA provides that a state or locality "may not enact or enforce a law, regulation, or other provision having the force or effect of law related to a price, route, or service of any motor carrier ... with respect to the transportation of property."[47] Ports seeking to enact any policy change affecting trucking have to avoid the preemptive effect of the FAAAA.

Preemption

The possibility of FAAAA preemption highlights a fundamental challenge facing proponents of legal reform at the ports: to make local policy change affecting environmental and labor interests, the ports would have to position any rulemaking within the ambiguous space for local action afforded by federal preemption doctrine.[48] Federal preemption derives from the Supremacy Clause,[49] but courts have started from the premise that local laws are not automatically precluded by federal law without a clear showing of congressional intent to do so.[50] Such intent may be explicitly stated in the statutory text of federal law or may be implied from the purposes federal law serves.[51] Implied preemption occurs either when local law actually conflicts with federal law or the federal regulatory scheme is so pervasive that it is deemed to occupy the legislative field.[52] Federal labor law has been held to impliedly preempt local laws interfering with the NLRA's "integrated scheme of regulation,"[53] and precludes local regulation in areas left to be controlled by the free play of market forces.[54] Environmental laws such as the Clean Air Act are generally viewed as asserting "floor preemption"—prohibiting local laws that fall below minimum standards, but permitting local regulations exceeding federal minimums and giving states significant roles in regulatory development and enforcement.[55] The FAAAA explicitly preempts state or local laws contrary to the federal policy of trucking deregulation.[56]

Whether federal law is determined to preempt a specific local act depends not just on the scope of the federal law, but also on the nature of the local act itself.[57] A key doctrinal distinction is between local action that is regulatory and proprietary. Federal law only preempts local actions that are "tantamount to regulation,"[58] not market participation by a local entity in its proprietary capacity.[59] Thus, even when federal law has preemptive effect, a local government is not generally preempted if it directly participates in the marketplace as a proprietor, such as through the purchase of goods and services.[60] However, the line between regulation and market participation is vague and contested. Courts have recognized that even local procurement can be used in ways that constitutes regulation and thus may be preempted.[61] The seminal market participation case in the labor law context—which upheld a state-negotiated project labor agreement—asserted that the state was acting in a proprietary role when it had "no interest in setting policy" and when its action was "'specifically tailored to one particular job' and 'aimed to ensure an efficient project that would be completed as quickly and effectively as possible at the lowest cost.'"[62] The market participant exception cuts across substantive legal domains and gives space for local action despite other federal regulatory schemes, namely, environmental and transportation law. Thus, in theory, the market participant exception to the preemption doctrine provides a pathway for localities to pass rules affecting port operations designed to protect local investment and promote efficient operations.

Yet in the early 2000s, on the cusp of the Campaign for Clean Trucks, the precedent interpreting market participation in the labor, environmental, and transportation contexts was thin. Within the jurisdiction covered by the United States Court of Appeals for the Ninth Circuit (which includes California), no appellate case applying the market participant exception in the labor context had moved beyond permitting project labor agreements.[63] In 2004, the United States Supreme Court held that the Clean Air Act preempted an effort by the South Coast Air Quality Management District (SCAQMD) to set fleet emission standards;[64] however, it invited the parties to consider the market participant exception and, on remand, the district court held that the rules "as applied to state and local government actors, fall within the market participant doctrine and are therefore outside the scope of" the act.[65] In the only reported case on the issue, the Ninth Circuit applied the market participant exception to the FAAAA, upholding a Santa Ana law requiring vehicles be towed by city-approved trucks as an exercise of the city's proprietary power.[66] While market participation therefore offered a route for local action on port trucking, there was little doctrinal guidance on how to navigate that route in practice.[67] It was into this uncertain space that the Campaign for Clean Trucks cautiously stepped.

4 Resisting the Port: Activism in Separate Spheres

Although the legal road to clean trucks ultimately ran through the doctrine of preemption, the activism that generated the challenge to the port complex emerged from the specific grievances of local residents in Los Angeles and Long Beach affected by port expansion. Through the 1990s, community activists from those neighboring cities sought a greater voice in port governance to fight the local impacts produced by their ports' global role. This bottom-up mobilization came to focus on the dysfunctional drayage truck sector as a key source of community concern. It thus ran on a parallel—and independent—track relative to top-down planning processes within the labor movement, which were also directed toward trucking reform. Labor lawyers focused on legal strategies to transform the independent-contractor structure of trucking at the ports, but lacked the immediate legal and political hook to advance their plan. The entry of environmental advocates aligned with community interests—but also motivated by the regional effects of port pollution—altered the political balance. Wielding the power of environmental law, these advocates succeeded in blocking a crucial port expansion project. In so doing, they created the opening for a broader challenge to port trucking that would unite community, labor, and environmental groups in powerful coalition.

The Hundred Years' War: Community Mobilization Against Port Expansion

Activism against the growth of the Los Angeles and Long Beach port complex—and particularly against its environmental and community impacts—took root in the areas most affected by it: San Pedro, Wilmington, and adjacent neighborhoods through which the port transportation system cut. Tensions were long simmering and erupted around proposed terminal and transportation infrastructure expansion. Opposition emerged within different communities, responding to the encroachment of particular projects, and reflecting the differential impact of port development.

Relations between the port and surrounding communities were not always rancorous, and varied over time and by location. The early years of port development helped to build San Pedro and Wilmington, with each community receiving resources for schools and services, and experiencing housing and commercial development.[1] By the 1930s, the port communities enjoyed prosperity supported by tourism and local entrepreneurship.[2] The growth of postwar aerospace manufacturing further buoyed middle-class life in the South Bay.[3] But this began to change in the 1980s, and particularly with the recession of the early 1990s, when aerospace downsizing cost the area more than fifteen thousand jobs.[4] This economic decline coincided with continued port expansion, driven by growing Asian imports, and codified in the joint infrastructure projects encompassed in the 2020 Plan.[5] As port growth inexorably encroached on surrounding communities, tensions with residents grew.[6] However, the distinctive geographies and histories of Wilmington and San Pedro meant that the impacts of port development and the responses to them were different.

Wilmington, occupying nine square miles at the port complex's northern border, had long been the region's industrial workhorse. Perched on the massive Wilmington Oil Field, by the 1980s the community of 40,000 residents was home to dozens of oil refineries and over a hundred working oil wells, as well as numerous waste disposal facilities, auto-wrecking plants, and junk yards.[7] It also contained rail-switching yards and was well traveled by trucks coming to and from the ports. In a 1985 series, the *Los Angeles Times* described Wilmington in ravaged terms:

> It is planted atop one of the nation's most productive oil fields, and dozens of petroleum-related companies have interests here, but residents see few signs of the millions of dollars that those firms and other industries make. Instead, residents say, they see only industry's noxious fumes, noise and truck traffic. ... Situated near the geographical bottom of Los Angeles, Wilmington also appears to be at the bottom of government priorities ... [seen in its] debris-cluttered vacant lots and side streets, in its growing number of homeless people, in its withering business district, and in the hundreds of junked automobiles that line city streets.[8]

According to the *Los Angeles Times*, Wilmington served as a "regional dumping ground with 13 closed waste dumps—one of the largest concentrations in the city of Los Angeles—and six toxic-waste storage or treatment plants. It also [was] the proposed site of one of the largest hazardous-waste treatment facilities in the state."[9]

Many reasons were cited for Wilmington's decay. Land use planning had been haphazard and short sighted. Wilmington's community plan, a zoning blueprint governing local land use that had been promulgated in 1970, created a hodgepodge of development, with residential, commercial, and industrial uses mixing uneasily in the same areas.[10] Despite efforts to protect residents from industrial use, sixty residential dwellings were in areas zoned for manufacturing; new high-density apartment development was adding to a sense of overcrowding.[11] Other resident complaints were that toxic waste cleanup was poor despite a number of legislative efforts to secure financing; and the city-led redevelopment of a 232-acre industrial park near the waterfront, though designated in 1974, lagged due to lack of financing and government will.[12] Most of the community's ire, however, focused on the Port of Los Angeles, which owned substantial property (including 20 percent of decaying East Wilmington) and dominated land use decisions, often in disregard of community concerns.[13] Responding to that ire, the port's Executive Director Ezunial Burts declared: "The port does not have a responsibility to develop a community."[14]

Faced with these interlocking problems, community members began to take action in the 1980s. Wilmington residents—about two-thirds Latino, primarily blue-collar workers, and roughly one-half homeowners—drew upon existing institutions and a sense of cultural pride to begin challenging what many viewed as the community's subservience to the port's industrial needs.[15] As one activist put it, "We are subsidizing the existence of the harbor with our city streets and the air we breathe."[16]

Activism sought to advance an affirmative development agenda while simultaneously seeking to mitigate port externalities. On the affirmative side, residents attempted to assert greater control over community development and to promote city and port investment in community-sensitive ways. Overcrowding was an early target, with a homeowners' group pressing for a moratorium on high-density apartments,[17] and defeating a proposed 189-unit apartment development in East Wilmington.[18] But the community's major demand was commercial development and recreational access to the harbor, which was completely blocked off by port facilities south of Harry Bridges Boulevard, depriving Wilmington of a public beach. Despite a city-led effort in the mid-1980s to enlist activists and business interests to address land use problems,[19] tensions remained high. To defuse the situation, the city chose the former head of the Los Angeles City Planning Commission, Calvin Hamilton, to conduct a $35,000 study of waterfront development in Wilmington,[20] and identify ways that the port could be a "better neighbor."[21] His plan, released in October 1987, proposed a wish list of revitalization projects, totaling $1 billion, that among other things called for the creation of a Mexican-themed waterfront marketplace at Slip No. 5 (at the intersection of Harry Bridges and Avalon Boulevards), simultaneously promoting commercial development and achieving the goal of beach access.[22] The basic principles of the Hamilton proposal were adopted by the Wilmington Home Owners association in their twenty-eight-point proposal to the harbor commission.[23] The principles were also the foundation for a study plan offered by City Council Member Joan Milke Flores,[24] a former Los Angeles city hall secretary who lived in San Pedro and had represented the South Bay in the Fifteenth District since 1984.[25] However, the Hamilton proposal quickly ran into problems. Most notably, it clashed with the port's own Hazardous Facilities Relocation Plan, which proposed relocating several hazardous oil terminals (primarily in San Pedro) to a new landfill but pointedly did not propose moving the notoriously hazardous Wilmington Liquid Bulk Terminals, the existing tenant at Slip No. 5.[26]

Activists thus took a different tack. They lobbied Los Angeles planning officials to include zoning changes to facilitate waterfront development in a citywide rezoning project initiated in 1988.[27] They also pressured the city's Industry and Economic Development Committee (on which Flores sat) to recommend moving the Wilmington Liquid Bulk Terminals from Slip No. 5.[28] These efforts ultimately bore fruit, with the planning commission's new zoning scheme permitting commercial development on the waterfront alongside industrial use; as a result, the harbor commission agreed to create community access to the waterfront at Slip No. 5 and to relocate the Wilmington Liquid Bulk Terminals and its hazardous materials.[29] That sparked new activity: the city appointed a resident advisory group to plan the details of the waterfront development and jumpstarted its flagging industrial park redevelopment plan.[30] Some observers began to imagine a "revival" in Wilmington.[31] To achieve it, community energy poured into shaping a revision of the Wilmington community plan—a process initiated by Flores in 1983[32]—that residents argued should include new buffer zones to protect residential areas, preservation of the historic Banning Park neighborhood, traffic mitigation, downtown revitalization, and waterfront development.[33] ExxonMobil resisted a proposal to redesignate the Wilmington Oil Field as "urbanized," potentially requiring it to cap some active oil wells; however, the city ultimately backed the redesignation, paving the way for the plan's approval in 1990[34]—and raising hopes for a community renaissance.

Yet revitalization efforts stalled in the face of city retrenchment. The harbor commission rejected the resident advisory committee's call for incorporating the historic Heinz Pet Food Cannery, located near Slip No. 5, into the proposed waterfront development,[35] and the city ultimately agreed to allow the port to raze the cannery to make way for an equipment storage facility.[36] Although the port continued to agree in principle to a waterfront development, its resistance to the cannery underscored that its support was limited to a modest development that would not fundamentally interfere with port expansion.

Zoning changes in notorious East Wilmington—wedged between the Dominguez Channel and Terminal Island Freeway, adjacent to a sulfur processing plant at the port—could not unwind the damage created by decades of neglect and incompatible land uses, giving rise to its status as a "Third World" community marked by "unpaved dirt tracks," "garbage piles," "[f]eral dogs," prostitutes, and drug dealers.[37] Residents in the Far East Wilmington Improvement Association alleged that the port and city were conspiring to intentionally neglect the community to enable the port to purchase land for expansion at low prices.[38] As recession swept through the region in the early 1990s, the community's fortunes declined further, with the *Los Angeles Times* calling East Wilmington "arguably the most run-down section of Los Angeles."[39] The economic downturn also caused interest in redevelopment to falter. Plans for waterfront commercial

Figure 4.1
Road in East Wilmington. Photo taken by the author.

development waned, while the port offered to make good on its promise to provide waterfront access with a modest community center at the original port site, known as "Banning's Landing," which was already a public landing at the base of Avalon Avenue.[40] This concept was first proposed in 1988 and approved by the harbor commission in 1995; however, by 2000, Banning's Landing—beset by cost overruns and structural problems—still sat incomplete.[41] Although the port insisted the project would be completed the next year, residents were dubious. Longtime activist Gertrude Schwab remarked, "We don't put much faith in what the port tells us anymore. If it's ever finished, we will shout hallelujah."[42]

Alongside the push to promote a positive development agenda in Wilmington were resident efforts to mitigate the harms imposed by port activity and growth. The impact of transportation was a constant concern as truckers increasingly used Wilmington streets to access the port and dumped empty containers on vacant lots around the community.[43] In 1987, Schwab reported to the harbor commission that two trucks per minute passed through the intersection at Avalon Boulevard and Anaheim Street—Wilmington's main crossroads.[44] The commission promised to create a dedicated truck route that would bypass residential streets, but the timeline was over a decade long,[45] and the port's immediate decision to approve the Wilmington Liquid Bulk Terminals' concrete importing plant fueled further resentment about increased trucking.[46] Community efforts focused on keeping trucks off residential streets. Resident complaints that police neglected Wilmington resulted in sporadic traffic enforcement spikes,[47] followed by a city-ordered ban on heavy trucks from three of the community's main streets.[48] The city commissioned a half-million-dollar traffic study,[49] which resulted in a 1994 harbor commission traffic plan featuring a proposal to build a sound wall on Wilmington's southern border.[50] Yet as off-dock rail connections intensified drayage around the port, truck travel continued to increase.

Residents also contested other port externalities. As more containers came into the ports than went out,[51] residents complained that they were being haphazardly stored throughout the community, creating visual blight and physical risks.[52] City Council Member Flores introduced a motion requiring the port to track container movement in order to devise a plan to minimize community impact,[53] yet containers continued to pile up and increasingly became targets of theft.[54] Residents also voiced discontent about ongoing environmental degradation, particularly in light of the 2020 Plan, which the ports' own environmental impact report noted would worsen air quality and further restrict commercial fishing and public recreation.[55] As one leader of the Wilmington Home Owners put it, "The conclusion seems to be if it is economically beneficial for the ports, to hell with local communities."[56] Community groups also fought against the debris and noise created by scrap metal processors, beating back a port proposal to relocate one company from San Pedro,[57] but having their complaints about the renewal of a twenty-seven-year lease for a large processor on Terminal Island fall on deaf ears.[58]

San Pedro, to the west, also had similar complaints about scrap yards,[59] as well as others focused on the environmental impact of port tenants, like Kaiser International, the Los Angeles port's largest commodity exporter, which bulk loaded coal and petroleum coke. Residents and pleasure boaters lodged a complaint with the SCAQMD to prohibit coke storage on the ground that it spewed black dust, causing air pollution and sullying nearby boats and houses.[60] Although district officials initially denied Kaiser a key permit, an appeals panel decided to allow continued operations, citing the fact that the harbor was a "working port."[61] Residents carried on the fight, eventually prompting the City Council to move Kaiser's bulk loading facility away from San Pedro's recreational facilities.[62]

The fact that the Kaiser complaint emanated from recreational boaters highlighted a key difference between San Pedro and Wilmington. While Wilmington's zoning made it the "backland" for the port's industrial uses, San Pedro had emerged as the port's recreational and commercial hub—a fact that Wilmington residents often highlighted with internecine frustration to emphasize their differential treatment.[63] San Pedro's position resulted from a different geography and distinct history. Abutting the harbor's west and main channels, San Pedro's recreational development grew partly from the obsolescence of older port facilities, which became incompatible with the need for larger berths and calmer waters to accommodate container vessels; the federal government's handover of Fort MacArthur on the West Channel's Cabrillo Beach in the 1970s also provided the land necessary for recreational development.[64] With greater recreational use mandated by state law in 1976, the Los Angeles port adopted a strategy of bifurcation, with San Pedro the recreational choice given its geographic benefits and higher proportion of residentially and commercially zoned land, which precluded Wilmington-style industrial expansion.[65] Against this backdrop, the port and city of Los Angeles sought to exploit San Pedro's advantages. The Los Angeles port itself owned several properties on the tidal lands and in the late 1980s pursued aggressive development, investing over $3 million to upgrade Ports O' Call Village, the 1960s-era shopping center on the main channel, while also moving forward with plans to develop a $100 million marina (with over one thousand slips) and recreational complex on Cabrillo Beach,[66] as well as a $60 million facility for cruise ships that included commercial development and a hotel.[67]

The city also saw economic opportunity: in 1985, the Los Angeles Community Redevelopment Agency approved a hotel in the Beacon Street redevelopment area overlooking the main channel, envisioned as a place for corporate visitors to the port and tourists disembarking from the World Cruise Center.[68] The city also agreed to provide funding for a downtown revitalization plan.[69] Despite concerns that redevelopment would rob San Pedro of its ethnic distinctiveness,[70] for many residents, the overdue investment was producing a welcome boom, particularly as the cruise center—with day trips to Catalina Island—made San Pedro a tourist destination.[71] The recession, however, made sure the good times did not last and the promise of port-led redevelopment turned into another disappointment. A decade of financial and legal problems stalled the redevelopment of Ports O' Call Village,[72] which entered a steep decline,[73] while the opening of a Carnival cruise terminal in Long Beach undercut San Pedro's position as the cruise industry's regional hub.[74]

Beset with difficulties, San Pedro and Wilmington began serious efforts to secede from the city of Los Angeles in the late 1990s[75]—a threat made more credible by the simultaneous effort by Hollywood and the San Fernando Valley.[76] In a play to tamp down secessionist fever, and quell what activists called the "Hundred Years' War," Republican Mayor Richard Riordan and other city officials made gestures to promote greater community involvement in port planning.[77] As a harbor commissioner noted when rolling out the new community plan, "There has been unparalleled expansion and prosperity in the port. ... The one frontier not tackled has been port-community relations. We need to round out the mayor's tenure in order to be a complete success."[78] The secessionists were not placated. Despite conciliatory efforts by the newly elected mayor—South Los Angeles native and scion of a powerful Democratic family, James Hahn[79]—the secessionists proceeded to advance their bid in 2001 by making the economic case for independence to the Los Angeles County Local Agency Formation Commission (LAFCO), whose approval was required before secession could be put to city voters.[80] Despite an aggressive case by secession supporters, LAFCO ultimately concluded that an independent harbor city would not be economically viable,[81] particularly in light of a California State Lands Commission recommendation to keep the port with the city of Los Angeles in the event of secession.[82]

Resisting the Port

It was fitting that the secession movement's decline occurred against the backdrop of yet another dispute between the Los Angeles port and Wilmington residents that tested the genuineness of the port's new community partnership. In 2001, as part of its drive to add twenty-five acres to the TraPac container terminal in the West Basin by expanding it north across Harry Bridges Boulevard to C Street, the port proposed building a twenty-foot-high concrete wall to separate the community from the new terminal boundary.[83] The plan, which had been previously proposed in the early 1990s, brought a firestorm of controversy, as residents once again complained that the port's talk of community collaboration did not match its actions, which would further undermine the goal of harbor access.[84] One community activist put it bluntly: "We don't need the Berlin Wall."[85] At a community meeting in April, when the details of the port's plan to expand Harry Bridges Boulevard to accommodate six lanes of truck traffic was revealed, residents exploded.[86] At that meeting was Jesse Marquez, a former aerospace electrician born and raised in Wilmington. As a high school track athlete, whose lungs burned when he ran, Marquez was radicalized by a chemical plant fire that injured his family members.[87] At the moment the expansion plan was unveiled, Marquez recalled shouting, "Hell no, over my dead body. If anybody wants it, tell them that [at] my house this Saturday we'll form a committee. We're going to fight this project."[88] A working group of about fifteen residents met to establish the Wilmington Coalition to stop the wall.[89] In the weeks that followed, the meetings grew to fifty residents and the following year the group secured funding from the Liberty Hill Foundation to set up an independent organization that in 2003 changed its name to the Coalition for a Safe Environment—which ultimately succeeded in preventing the wall's construction and laid the foundation for growing anti-port activism.[90]

Other port-related projects caused similar disruption—and produced similar community responses. The Alameda Corridor rail line cut through East Wilmington, eliminating many of the gritty neighborhood's only businesses,[91] while intermodal truck and rail traffic disrupted community life in adjacent cities like Carson and around the downtown rail yards.[92] Commerce, home of the UP East LA and BNSF Hobart yards servicing the ports and bisected by the 710 freeway, experienced drayage truck increases as port traffic grew in the 1990s.[93] A series of town hall meetings brought out residents concerned with the safety and environmental impact of the trucks.[94] City neglect prompted a handful of families to begin meeting as an ad hoc group. An informal survey confirmed the extent of community concern with the impact of the rail yards and trucking on safety, health, and property.⁹⁵ With the leadership of Angelo Logan, an aerospace mechanic, the families formed East Yard Communities for Environmental Justice in 2000.⁹⁶ Galvanized by the Alameda Corridor project and plans for a massive expansion of the 710 freeway to accommodate more port trucks, East Yard Communities became focused on strategies to address the trucks' noxious byproduct: diesel exhaust.⁹⁷ Resident research into truck pollution revealed that while infrastructure design disproportionately impacted their community, the underlying cause of pollution stemmed from the nature of the port trucking industry itself.⁹⁸ This point was brought home at a community forum in 2005 to address diesel exhaust, at which resident truck drivers spoke. Logan recalled the event:

> They ... really laid out their situation in terms of the way in which they were being exploited and their hands being tied in terms of ... not being able to get into trucks that were safer, that were cleaner and that [allowed] them to be good environmental health stewards. And immediately thereafter ... [we] realized that there was a real issue in terms of the trucking industry and the way that the trucking industry was exploiting the drivers themselves.⁹⁹

Labor's Municipal Strategy: Contracting around the Independent-Contractor Problem

During this time, leaders within organized labor were also focused on ports, but from a distinct perspective. For the Teamsters, trucking deregulation decimated the ranks of what had been one of the strongest unions in the United States.¹⁰⁰ Whereas 46 percent of the country's approximately one million truckers were unionized in 1978, only 23 percent of roughly two million truck drivers were in unions by 1996.¹⁰¹ An even lower percentage of the nearly four million truckers nationwide were in unions by the early 2000s.¹⁰²

In the wake of deregulation, truckers tried to organize independent associations. Central American drivers, who comprised the vast majority of truckers at the Los Angeles and Long Beach ports, formed their own organizations in the 1980s as vehicles for community support and labor struggle. One of these associations, the Waterfront Rail Truckers Union (WRTU), formed in 1986, spearheaded a series of strikes to address delays and other disputes,¹⁰³ one of which involved WRTU members withholding containers until they received payment from a bankrupt trucking company.¹⁰⁴ Members were radical and militant. In the late 1980s, they began challenging their classification as independent contractors by roughly two dozen port trucking companies, including H&M Terminals Transport, Inc.¹⁰⁵ In tax filings with the IRS, the truckers argued that by not classifying them as employees, the companies were evading Social Security, state disability, and unemployment taxes.[106] The IRS agreed in some cases and in 1991 initiated an audit of H&M and other companies.[107] WRTU truckers also picketed H&M, highlighting the fact that although many drivers worked exclusively for one firm and even carried company identification cards, they were unable to organize unions or apply for workers' benefits.[108] As one organizer asserted, the drivers "don't want to be made fools of anymore."[109] Although the WRTU receded in importance, independent organizing continued into the 1990s,[110] with other groups such as the Latin American Truckers' Association protesting the impact of fuel costs.[111] Some of these independent groups reached out to unions, which were unwilling to invest the resources to support an organizing campaign.[112]

The situation changed in the mid-1990s when truckers initiated a large-scale union organizing drive—led not by the Teamsters, but the Communications Workers of America (CWA).[113] The truckers' connection to the CWA was partly driven by personal contacts with CWA organizers, but was also a function of the lack of interest shown by the Teamsters.[114] Indeed, some insiders speculated that the alliance between the CWA and the truckers may have been orchestrated by the ILWU to keep the Teamsters out of the harbor, which was the ILWU's base.[115] CWA Local 9400 began holding meetings for workers in 1995 and 1996, quickly attracting thousands.[116] To demonstrate this growing strength, in May 1996, CWA organized picketing in front of terminal gates, which was enjoined when the PMA filed suit in Long Beach Superior Court.[117] The truckers organized convoys from the ports to highlight their plight and labor leaders persuaded the Los Angeles City Council to pass a resolution in support of unionization.[118]

However, there remained the thorny problem of the drivers' predominantly independent-contractor status, which precluded them from organizing. To get around this problem, CWA launched a dual campaign. One part was a traditional unionization effort directed at the handful of companies that still used employees; the second involved an ambitious plan, to be financed by entrepreneur Donald Allen, to create a new trucking company, the Transport Maritime Association (TMA), which would hire truckers as employees and then contract them out to the existing companies at higher rates.[119] In May 1996, roughly four thousand truckers declined to accept contracts from their existing companies and instead signed up to be TMA employees with the promise of pay at $25 per hour.[120] Despite a diversion of some cargo, the trucking companies held fast and refused to contract with TMA.[121] When one of the lead organizers suffered a heart attack and it turned out that Allen lacked the resources to bring TMA to scale, the campaign died, with some faulting the CWA for not investigating Allen's finances and for lacking sophisticated knowledge of the port trucking industry.[122] Although the campaign failed to advance trucker unionization,[123] it did reveal a deep desire among the workers for change, their willingness to take action, and the economic vulnerability of the port to a trucking strike.

The CWA campaign also refocused efforts to address the independent-contractor problem. Labor leaders identified two approaches. One was to find a way to directly organize truckers as independent contractors without running afoul of antitrust law, which treated truckers as business owners prohibited from organizing in restraint of trade. Without amending federal antitrust law, which seemed politically impossible, the value of this approach was uncertain, since any state effort to permit independent-contractor organizing could be deemed preempted. The Teamsters did put some energy into this strategy, pursuing a legislative campaign to permit direct organizing of independent contractors; but the union's effort to pass a state law exempting independent contractors from antitrust failed when Governor Arnold Schwarzenegger vetoed it in 2005.[124] From there, the Teamsters abandoned that project.

Instead, the Teamsters pursued a second approach, in which truckers would be legally converted into employees and then organized under the NLRA. The CWA campaign attempted to do this by creating the labor-leasing firm, TMA, which was to hire drivers as employees, who would then be unionized—passing the increased costs onto trucking firms in the form of higher contract rates.[125] In the wake of that failed campaign, truckers adopted another strategy that echoed earlier WRTU efforts: litigation challenging the truckers' misclassification. In 1996, lawyer Fred Kumetz brought suit on behalf of thirty drivers who claimed that they had been misclassified as independent contractors by forty transportation companies.[126] The suit sought class action certification to represent a larger class of 6,500 harbor truckers claiming $250 million in damages—primarily to recover payments made for insurance coverage (which included workers' compensation).[127] Kumetz, a plaintiff's lawyer not associated with organized labor, was approached by truckers after TMA collapsed.[128] In filing the suit, Kumetz argued that "the drivers, nearly all Latino immigrants, frequently are coerced by 'fly-by-night' companies into signing exploitative contracts without understanding the contents and are duped into paying for workers' compensation and liability insurance without understanding the law."[129] Some plaintiffs alleged that having to pay for insurance (and other ownership costs) reduced their earnings to below the poverty level.[130] In response, Robert Millman, a lawyer from labor defense powerhouse Littler Mendelson, claimed that "[i]t would not be economically feasible to treat these people as drivers. The cost of goods would just go skyrocketing."[131]

The contracts at issue in the case gave truckers the choice of obtaining their own insurance or getting it through the trucking firms' less expensive group policies, the cost of which would be deducted from the truckers' compensation.[132] Under the group policy arrangement, companies charged the drivers more than the cost of premiums paid and also made the drivers responsible for a $1,000 deductible payment that was not specified in the contracts.[133] In 1999, the case—*Albillo v. Intermodal Container Services*—was certified as a class action and tried before a special panel of three retired judges appointed by the California Superior Court and the Workers' Compensation Appeals Board.[134] In 2000, after trial, the panel ruled that there had been no violation of the workers' compensation provision of the Labor Code, nor had plaintiffs proven a violation of the state unfair business practices law, or shown fraud or deceit.[135] However, the trial court did find that the firms failed to comply with insurance disclosure requirements and consequently awarded the truckers injunctive relief and restitution, although the court refused to award attorneys' fees. The Court of Appeal, in a 2003 published decision reversing the trial court in part, held that the firms did violate the Labor Code by electing to be covered by workers' compensation while "requiring [drivers] to bear the cost of obtaining workers' compensation insurance."[136] Yet even this success, while compensating workers for wrongful payments, did not achieve the large-scale goal of employee conversion; in fact, it produced a negative effect by making firms less likely to elect workers' compensation coverage in the first instance.

As the Teamsters watched the CWA campaign and *Albillo* lawsuit unfold, they began devising plans for their own initiative. In 2000, the Teamsters, through its Port Division, announced a nationwide port trucker campaign, run by Assistant Director Ron Carver.[137] Coming on the heels of the trial court setback in *Albillo*, lead Teamsters organizer Ed Berk was undaunted: "I don't think they're going to throw in the towel on this one court case."[138] However, the Teamsters had absorbed the lessons of that case, and the CWA campaign before it, concluding that the way to win was not through piecemeal organizing or lawsuits, but through broad change that could convert large numbers of independent contractors back into employees. Although many inside the union believed that independent contractors were misclassified as such, the barriers to individual enforcement were too high to justify a case-by-case strategy and, as *Albillo* highlighted, even success in court did not ensure change in industry practice. As Teamsters attorney Mike Manley reflected, the union's view was that "you really can't [address the industry] through a campaign of Board elections and slugging it out in representation cases. The industry is too vast ... you'd be doing it forever."[139] In Los Angeles, there were some unionized firms, like Horizon, and others that still hired truckers as employees, like Toll, but without addressing the independent-contractor problem, the industry would remain low wage. The key was to "transform the market."[140]

The Teamsters' national ports campaign initially pursued different tracks. In Miami, the Teamsters joined forces with an existing effort by independent truckers to gain recognition as employees. In 2000, the truckers canceled their leases and refused to sign new contracts, instead demanding that the companies hire them as employees through the Teamsters hiring hall.[141] The union had reached out to a handful of small companies that agreed to hire truckers as employees; the strategy was to send truckers to work for those companies, which, along with a few other companies that the ILA had already unionized, would gain market share, forcing other companies to follow suit.[142] However, the trucking companies held firm, and the campaign fizzled, resulting in a small increase in hauling fees after a campaign marred by lawsuits and allegations of harassment.[143] Organizers complained, "We didn't get even 25 percent of what we wanted"[144]—reinforcing the limited effect of striking without first securing employee status. In Los Angeles, the Teamsters focused on the policy arena at an early stage. Although the union suggested that it might reintroduce a new version of the failed TMA, only on more solid financial footing, its major efforts were directed toward disrupting port operations and pressuring local decision makers to act to address trucker conditions.[145] In February 2000, the Teamsters unveiled a port truckers' "bill of rights," and generated publicity for it by organizing truck convoys from the ports to Los Angeles City Hall.[146] However, these efforts did not illuminate a clear path around the independent-contractor problem.

Despite significant challenges, port trucker unionization remained one of the major prizes of Teamsters organizing in the new millennium. In raw numbers, the scale of port trucking was modest, with approximately forty thousand port truckers operating as independent contractors out of an overall trucking industry of almost four million.[147] However, port trucking was an area of historic strength, and there were practical and strategic reasons to pursue unionization in that sector. As a practical matter, there was a potential legal hook for organizing: the legal status of certain ports (like Los

Resisting the Port

Angeles and Long Beach) as proprietary departments under the umbrella of local government meant that they could potentially influence the nature of trucking through their contracting power. Discussions of how to make this happen were underway in 2004 when Mike Manley was hired in the Teamsters' D.C.-based office of general counsel, headed by Pat Szymanski.[148] Manley, originally from Kansas, worked as an organizer at the East Lawrence Improvement Association before deciding to become a lawyer.[149] In 1980, he enrolled in Kansas Law School, and then went to a Kansas City law firm, Blake & Uhlig, where he eventually became partner.[150] The firm was general counsel to the International Brotherhood of Boilermakers, and Manley remembered that when he was hired by the Teamsters, there was "great interest in the fact that I had done a lot of work for the boilermakers and shipyards—which I guess shows how deep the department was in the [ports] campaign at that point."[151] When Manley got the Teamsters job, and was assigned to the Port Division to help organize port drivers, he remembered incredulously asking Szymanski: "What are you doing? ... They are independent contractors."[152]

As Manley quickly learned, the plan was to change that status by focusing on the ports' role as market participants. In his conversations with Ron Carver, Manley said the question was: "Is there a way to kind of leverage the port to declare [truckers] to be employees?"[153] Figuring out an affirmative answer to that question was not only important on its own terms but also had broader strategic implications. The ports were key nodes in the larger supply chain that led from manufacturers to regional warehouses, and ultimately to large retail chains, such as Walmart. Some labor leaders believed that if unions could gain a stronger foothold at the ports, it would contribute to a longer-term campaign to organize retail giants.[154] This was something that the Teamsters had argued for, but "didn't get necessarily a lot of traction ... in terms of resources" from union leadership at the AFL-CIO.[155]

That changed with the formation of Change to Win—an alliance of progressive unions that broke away from the AFL-CIO in September 2005.[156] Change to Win—established with "little structure, [but a] big focus on organizing"[157]—set up the Strategic Organizing Center, which was built around different industry sectors, with the goal of identifying "how the pool of existing resources at Change to Win could expand the pace [of] organizing."[158] The center was "like a startup" that put "together experienced organizers and campaigners [to take] a fresh look at industries that were basically nonunion."[159] By bringing together organizers and researchers from different unions, the goal was to innovate according to "best practices."[160] The center divided up the economy into different industry sectors: transportation, retail, home construction, and food processing.[161]

John Canham-Clyne and Nick Weiner both volunteered to work on transportation.[162] Canham-Clyne was a former freelance writer who covered the Iran-contra affair for *In These Times* and wrote a book on single-payer health care.[163] He left journalism in 1996 to work as the research director for Congress Watch at Public Citizen, and from there was recruited to the Hotel Employees and Restaurant Employees Union (HERE) to direct campaign research for their hospital-organizing campaigns, first in Las Vegas and then in New Haven, Connecticut.[164] After Change to Win was formed, Canham-Clyne was recruited to its core staff.[165] There, he was joined by Nick Weiner, who also came from HERE, where he first worked with locals in Baltimore, Maryland, and Washington, D.C., and then joined the national hotel organizing effort at the UNITE HERE office in D.C.[166]

Together Canham-Clyne and Weiner set out to research the trucking industry, becoming the bridge between Change to Win and the existing Teamsters leadership. Through their research, Canham-Clyne and Weiner identified the ports as a potential target of opportunity for organizing. As one of the few publicly owned pieces of freight infrastructure, ports offered "potential hooks" for organizing: many were in friendly political jurisdictions and drayage was a relatively sticky industry because of the massive infrastructure investment at the ports.[167] On the basis of this research, Change to Win launched a national ports campaign, directed by Canham-Clyne, to build upon the Teamsters' existing organizing efforts and "move [that] work forward faster."[168] The larger goal, in line with that articulated by the Teamsters, was to "organize the supply chain."[169] The ports were the first link in this chain because they were a "chokepoint" that could be used to produce other wins.[170]

Although the Teamsters had invested in port trucking, the involvement of Change to Win was a crucial step forward.[171] Teamsters leaders believed that any port campaign had to be comprehensive: even though individual ports were immobile, the power of shippers to divert cargo to competitor ports meant that there had to be a unified national strategy, otherwise there was a "real risk" that any individual port campaign could be easily broken.[172] As Canham-Clyne reflected, "The Teamsters were trying to find a way to help poor drivers get out of the legal box that they were in ... since deregulation."[173] Change to Win attempted to build on "a foundation of real commitment from the Teamsters to try to figure out this knotty puzzle," recognizing that the workers had the will to strike, but because of the independent-contractor problem, "there had to be an additional lever

Resisting the Port

to discipline the industry."[174] Canham-Clyne understood that his job was to "essentially sew together a lot of really good work that had been done before and build the alliance in a much deeper way than had previously existed, so that we would make sure that as we went down the road, both politically and legally, we couldn't be divided."[175]

With Change to Win staff and resources in place, a working group was formed and serious planning commenced in 2006.[176] Within the Teamsters, Manley, Carver, and Chuck Mack (West Coast vice president of the Teamsters and head of the Port Division) met repeatedly with Canham-Clyne and Weiner to develop an organizing theory and strategy, which focused on "strengthening the local political control over the ports."[177] The plan was to target ports in "blue" states or localities with friendly political climates: Los Angeles-Long Beach, Seattle-Tacoma, Oakland, New York-New Jersey, and Miami. From Manley's perspective, if the campaign could get these ports to "adopt a model that made drivers employees," "more than 50 percent of what was coming into the country would be through facilities where port drivers are employees. And then you'd go to second tier targets ... [which] are harder nuts to crack."[178] During these meetings, organizers like Weiner drew on their past experiences dealing with local governments and airports to develop a strategy predicated upon a concession model: contractually linking the entry of trucking firms onto port property to the conversion of truck drivers from independent contractors to employees. As Weiner recalled, "once we started looking into port trucking, we kind of came up with [the concession] theory. And then we vetted the theory and that took a few months."[179]

An early question focused on the legality of different possible options for organizing port truck drivers specifically and low-wage workers in contingent work arrangements more broadly. In 2006, there was a meeting of the "legal tribes" in Washington, D.C., where prominent labor lawyers gathered to discuss new strategies.[180] This meeting included the Teamsters' Manley, as well as Szymanski, who had left the Teamsters to become general counsel to Change to Win; Brad Raymond, the new general counsel of the Teamsters; Judy Scott, general counsel to the Service Employees International Union; and long-time outside union counsel Stephen Berzon of Altshuler Berzon, and Richard McCracken of Davis, Cowell & Bowe.[181] One of the issues discussed was the legal feasibility of using the port's status as a government entity to convert truck drivers into employees.[182] It was a familiar idea to those present, discussed in various forms for some time. But at that point, with Change to Win backing, the moment had finally come to advance the strategy. The question was: "Is this doable?"[183]

The concession model held appeal for a number of reasons. For one, it had been tested in other forms and had proven an effective tool for organizing. In particular, lawyers associated with the unions had done work on airport organizing in which the airport authorities, as public agencies, had used concession agreements, or franchises, to require food and beverage vendors to remain neutral in union organizing campaigns.[184] Thus, from a mechanical point of view, union lawyers were familiar with the technical aspects of government contracting and how it potentially related to labor issues. Perhaps most crucially, the concession model was viewed as legally defensible against federal preemption. The union lawyers involved had experience with contracting models to create living wage laws and job training programs, and believed that the same concept could be adapted to the ports under a market participant theory.[185] For this analysis to work, there had to be a justification for the ports' market participation itself. That justification turned out to rest less on port trucking's effect on labor relations than its impact on the environment.

The Turning Point: *NRDC v. China Shipping* and the Clean Air Action Plan

That labor and environmental legal analyses would harmonize around the market participant exception to preemption was not clear on the cusp of the Campaign for Clean Trucks. Unlike the labor movement, mainstream environmental organizations could wield power through federal and state regulatory regimes, and had not yet clearly defined their relation to locally based initiatives. Judicial recognition of a market participant exception to the Clean Air Act did not occur until 2005.[186] It was at that point, at least in theory, that the legal interests of the labor and environmental movements in asserting the market participant exception were in unison. But there still needed to be a cause upon which to act jointly. That cause would be port pollution.

Throughout the Hundred Years' War, port communities had become all too familiar with the immediate reality of port pollution.[187] The storage and transport of hazardous materials created a number of risks.[188] Oil spills were a continuous problem.[189] Toxic chemicals stored at the ports occasionally sparked fires.[190] There were ongoing battles over debris and noise caused by scrap metal processors,[191] and struggles over the location of coal-exporting facilities.[192] Other port-related problems drew increasing environmental attention. South Bay traffic ranked among the city's worst,[193] and traffic congestion was a key cause of community discontent.

Slower to develop was the recognition of the ports as an environmental problem susceptible to challenge by the environmental movement. This recognition grew as port expansion increasingly bumped up against environmental regulation. As the early 1990s brought Plan 2020, large-scale port development came to hinge on environmental clearance.[194] The California Coastal Commission—a quasi-judicial body charged with approving all coastal development, including port expansion, under the land and water stewardship provisions of the California Coastal Act[195]—emerged as an important actor. The commission initially withheld approval of a dredging-and-landfill project deemed crucial to Plan 2020's expansion until the Port of Los Angeles agreed to offset the loss of 582 acres of waterways.[196] When the port agreed to replace the waterways on an acre-for-acre basis elsewhere in Southern California, the commission gave the project a green light, but retained authority to review each stage.[197]

Regulatory attention increasingly focused on the defining environmental problem of the Los Angeles basin: smog. Regulation was shaped by the different tools available to federal and state agencies to deal with key pollution sources. Ships, which used dirty "bunker" fuel high in sulfur content,[198] were known to be significant polluters, but the ports' power to regulate them was ambiguous. Shipping lines generally asserted that federal and local regulators did not have power to control vessel emissions while they were in international waters—a view that had some precedential support.[199] Within U.S. boundaries, the EPA had deferred the question of whether it had authority to regulate foreign-flagged ships.[200] There was a strong argument, however, that ports could impose rules on ships once docked under an exception to the Clean Air Act permitting local "in-use" regulations for nonroad sources of emissions.[201] This focused attention on what such on-dock rules should look like. A 1984 SCAQMD study revealed that 2 percent of the area's nitrogen oxide—a key component of smog—came from port ships burning diesel fuel while idling.[202] In response, the SCAQMD issued a proposed rule requiring ships to plug into dockside electric power[203]—an operation known in the industry as "cold ironing"—which became an important goal of environmental advocacy. Regulators also searched for ways to address diesel truck pollution. While CARB had been granted authority by the EPA to set vehicle emission standards, the "in-use" exception to the Clean Air Act also allowed the ports to regulate the "use, operation, or movement" of trucks,[204] which opened the possibility of imposing restrictions on idling.

The immediate battle revolved around setting regional air quality standards in the first instance. From the earliest period of federal clean air regulation, there were fights about regional compliance with the Clean Air Act. California's first State Implementation Plan for the South Coast Air Basin was rejected by the EPA in 1972.[205] After a series of delays and revisions,

California submitted a proposed plan for ozone and carbon monoxide in 1982, but conceded that even if implemented, it would not meet mandatory national air quality standards.[206] When the EPA nonetheless approved the state plan, a citizen lawsuit was filed, resulting in a Ninth Circuit order reversing the approval and ordering the EPA to "face up" to its obligation to implement national standards.[207] The Coalition for Clean Air and Sierra Club promptly filed another lawsuit to force the EPA to do so, which resulted in a settlement agreement committing the EPA to finalize its own plan.[208] After further backtracking,[209] in the early 1990s, the EPA was again ordered to finalize smog control regulations requiring the South Coast Air Basin to dramatically reduce the ozone ingredients hydrocarbon and nitrogen oxide by 2010.[210] When it was unveiled, the EPA's plan focused on reducing emissions from trucks, ships, and airplanes—mobile sources over which the SCAQMD did not have regulatory jurisdiction.[211]

Yet the federal plan's release provoked a local outcry over the scope of its proposed changes, which included taxing shipping and airline companies, requiring trucking companies to replace diesel engines, limiting out-of-state trucks to one Southern California stop, and forcing ocean liners to steam 100 miles offshore.[212] Government officials and industry representatives expressed concern over the plan's $5.4 billion total price tag and argued that, if fully implemented, it would preclude the construction of the Alameda Corridor project and effectively shut down the Los Angeles and Long Beach ports.[213] With that threat looming, officials negotiated a compromise that deferred federal compliance for another three years.[214] In 1997, the SCAQMD adopted a new state plan that relaxed approximately thirty control measures. The EPA disapproved this part of the state plan and environmental groups again sued the SCAQMD to enforce the previous standards—leading to a revised plan in 1999 to strengthen ozone control measures, accelerate the implementation timeline, and provide an explicit commitment to attaining the federal standard.[215]

While the ports thus received a temporary reprieve, it was clear that they could no longer expect to conduct business as usual. The environmental and health impacts of diesel fuel vehicles were receiving increasing attention.[216] In 1999, a front-page *Los Angeles Times* story reported on the danger of diesel-fuel-burning vehicles, whose higher fuel content and intense heat-burning engines produced greater concentrations of carbon, sulfur, and nitrogen oxide.[217] "At the Ports of Long Beach and Los Angeles—massive operations that are filled with trucks, ships, trains and cranes—workers breathe some of the most severe doses of diesel exhaust found anyplace in California."[218] A seminal 2000 SCAQMD study on the relation between air pollution and cancer, entitled the Multiple Air Toxics Exposure Study (MATES II), concluded that about 70 percent of the carcinogenic risk in the basin was "attributed to diesel particulate emissions."[219] The study made specific reference to the negative impact of diesel emissions coming from the ports and connected transportation networks.[220] The communities at greatest risk, unsurprisingly, were adjacent to the ports.[221] A map published with the study (figure 4.2) highlighted the increased cancer risk in harbor communities and galvanized residents who began to mobilize around environmental justice—distributing the map at official meetings and public actions.[222]

At a 2001 conference on air pollution at the University of Southern California (USC), Coalition for a Safe Environment Executive Director Jesse Marquez challenged assembled scientists to link their findings on air

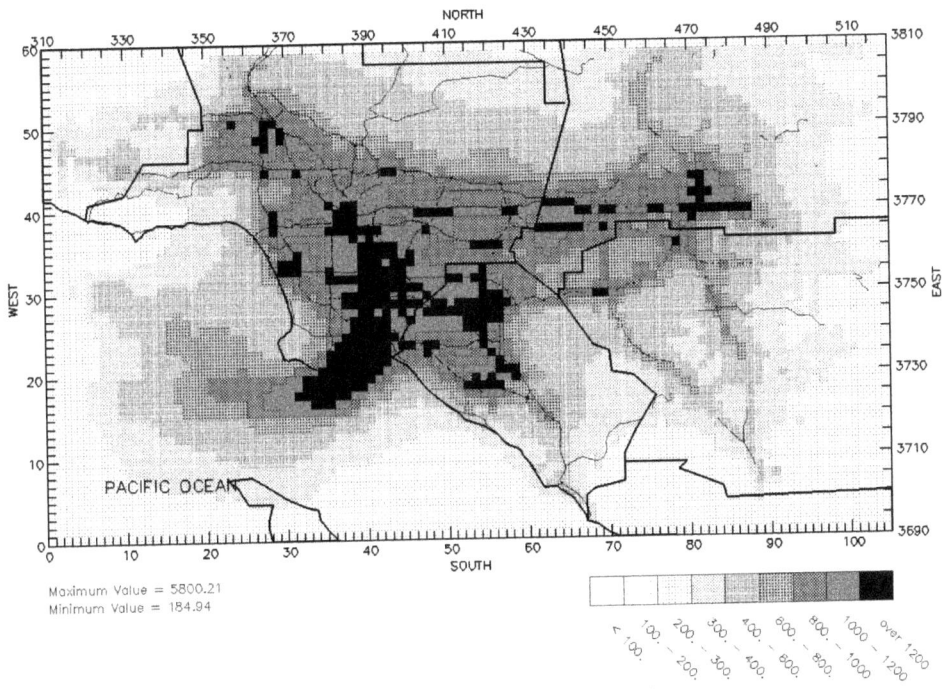

Model estimated risk for the Basin
(Number in a million, all sources)

Figure 4.2
Taken from South Coast Air Quality Management District, *Multiple Air Toxics Exposure Study Final Report (Mates II)* (2000), fig. 5–3a.

pollution to the unregulated growth of the ports and their impact on local low-income communities.[223] New partnerships between community activists and the scientific community began to develop. The stage for environmental action against the ports was set—though that action initially would play out once again in court.

It was fitting that in the pivotal environmental fight against the ports, the focal point would be containers arriving from China, which by the mid-1990s was the Los Angeles region's second-largest trading partner, accounting for $18 billion of cargo.[224] In 1996, the Los Angeles and Long Beach ports vied for the crucial business of China Ocean Shipping Company (COSCO), a state-run carrier handling a quarter of Chinese trade with the United States.[225] Though it had been operating at the Port of Long Beach since 1981, COSCO sought expanded terminal facilities.[226] Long Beach initially won an agreement by the company to lease space at the Port of Long Beach,[227] which officials predicted would create 300 to 600 jobs.[228] In September 1996, the Long Beach Harbor Commission approved the creation of a new $200 million terminal on the site of a naval station, with 140 acres reserved for COSCO.[229]

Yet opposition from diverse quarters derailed the plan. Conservatives argued that the lease created security risks, with Congressman Duncan Hunter—a Republican from military stronghold San Diego—proposing federal legislation to bar foreign entities from leasing the naval station property.[230] Objections also came from the Audubon Society, which wanted to preserve a habitat for black-crowned night herons, and the California Coastal Commission, which initially expressed concern about contamination from dredging (though it ultimately granted its approval).[231] The preservationist group Long Beach Heritage challenged the coastal commission's decision to permit the demolition of naval station buildings, which had been deemed eligible for listing on the National Register of Historic Places.[232]

Long Beach Heritage took its challenge to court. Represented by environmental lawyer Jan Chatten-Brown, the group filed suit under CEQA to oppose terminal development. The challenge rested on the timing of the Port of Long Beach's environmental approval, which was conducted after the port had already entered into a letter of intent with COSCO to lease a container storage facility on "Pier T in the former Naval Station."[233] The Board of Harbor Commissioners approved the project environmental impact report (EIR) on September 3, 1996, and two months later the city of Long Beach entered into a "Preferential Assignment Agreement" giving COSCO "a nonexclusive preferential assignment of the wharf and contiguous wharf premises" of more than one hundred acres on the Pier T site.[234]

Long Beach Heritage filed a petition for writ of mandate, which was consolidated with similar suits filed by the Audubon Society and the cities of Vernon and Compton.[235] In February 1997, after trial, Superior Court Judge Robert O'Brien rejected Long Beach's EIR as a "foregone conclusion" and ordered the city to "reconsider the project free and clear of any pre-commit[ment] ... and with a complete evaluation of the EIR *before* deciding on the project."[236] Long Beach held a public hearing and issued another approval, but Chatten-Brown argued it was still marred by the fact that it was made for property already encumbered by an existing lease.[237] When Judge O'Brien agreed, rejecting the EIR for a second time, the Port of Long Beach rescinded the lease and disavowed the letter of intent in order to reconsider the plan—which it promptly reapproved.[238] Stating that the entire environmental review process had become "simply something to get through," Judge O'Brien agreed once again to consider its adequacy.[239] On September 2, 1997, O'Brien—for the third time—rejected the EIR as a "post hoc rationalization for the Board's approval of the Project" and ordered a new review "without pre-commitment, pre-approval, or pre-disposition, as required by the California Environmental Quality Act."[240] The city appealed and the Navy began considering alternative uses for the site.[241] The CEQA process for the COSCO terminal plan, however, was rendered moot when Congress passed a defense bill that contained a prohibition on leasing the base to COSCO.[242] Representative Hunter, along with his colleagues James Inhofe (R-OK) and Randy "Duke" Cunningham (R-CA) had argued that China could use the base for "military purposes and intelligence-gathering."[243] This argument aligned anticommunists and veterans organizations, which joined with environmentalists and preservationists to permanently block the Long Beach port from permitting a company flying a Chinese flag from using the old naval base site.

With Long Beach thwarted, the Port of Los Angeles pursued its own China partner. Residents had made inquiries to the harbor commission about plans for the West Basin site of the former Todd Shipyard and Chevron area just north of the Vincent Thomas Bridge in San Pedro. Their answer came on March 28, 2001, when the Los Angeles harbor commission approved a lease with China Shipping Holding Company (China Shipping).[244] Under the terms of the $650 million lease,[245] China Shipping would occupy a 174-acre terminal built by the port, which would support entry of up to 300 vessels—carrying approximately 1.5 million containers—a year.[246] The terminal—to be located at berths 100 and 102—would be designed to accept 9,100 TEU container vessels, which were larger than any at the time.[247]

Community resistance was swift. San Pedro Peninsula Homeowners United, led by activist Noel Park, and the San Pedro and Peninsula Homeowners Coalition challenged the proposed terminal, which was to include two new wharves and ten massive cranes (up to sixteen stories high) 500 feet from resident homes, along with a backland area with new roads to accommodate traffic.[248] On May 8, 2001, at a tense meeting in which residents were given only five minutes to speak, the Los Angeles City Council rejected resident demands that it conduct an EIR prior to approving the project.[249] Instead, the council approved the project by a unanimous vote.[250] Residents thus turned to court, contacting NRDC to pursue legal action.

On June 14, 2001, NRDC—representing the resident groups San Pedro and Peninsula Homeowners Coalition and San Pedro Peninsula Homeowners United, as well as the Coalition for Clean Air and NRDC's own members—filed suit against the city, port, and harbor commissioners.[251] Petitioners—represented by Gail Ruderman Feuer and Julie Masters of NRDC along with Chatten-Brown and another private environmental lawyer, Roger Beers—argued that, in approving the project, the city had failed to comply with CEQA,[252] which required an EIR of significant developments that identified environmental issues and how they would be mitigated, and provided a period for public comment.[253] Petitioners thus sought a writ of mandate directing the city to conduct a new project-specific EIR.[254] "It's time for the port to consider the needs of local communities before it approves a massive expansion in their backyard," claimed NRDC's Masters.[255] The suit emphasized the environmental impact of the incoming ships themselves, as well as increased tugboat activity (over five hundred trips per year) and truck traffic (an estimated one million new trips) to support them.[256]

The technical legal issue focused on the port's effort to exempt the China Shipping project from CEQA review by arguing that its approval was encompassed within two EIR processes that predated the lease agreement. The first was a 1997 EIR conducted by the harbor commission that approved the development of a multifaceted West Basin Transportation Project to "optimize container transport capabilities," which included plans to deepen and widen the basin, create a new on-dock railway linked to the Alameda Corridor, and build a new wharf at berths 98 through 100 to accommodate the largest container vessels.[257] The second was a 2000 Environmental Impact Statement/Impact Review (EIS/EIR) conducted by the U.S. Army Corp of

Engineers to evaluate the impacts of a harbor dredging operation that proposed using dredged material to create a landfill from berths 97 through 109 as a potential site for container storage or docking in order to "accommodate the most modern vessels."[258] In preparing for the lease agreement, the harbor commission told China Shipping that the "elements contained in the lease have been adequately assessed in the [1997] West Basin Transportation Improvements Program EIR ... and have been adequately assessed in the [2000] Port of Los Angeles Channel Deepening EIS/EIR. ... As such, the Director of Environmental Management has determined that the proposed activity is exempt."[259] The city attorney's office approved a permit authorizing the construction of the China Shipping terminal from landfill taken from the harbor dredging project.[260] Under the final twenty-five-year lease agreement, China Shipping was granted the right to use berths 100 and 102 to construct terminal, wharf, and backland space, to be built in three phases: phase one included construction of the container terminal and first wharf at berth 100 by November 2002; phase two involved the extension of the first wharf and the completion of the second (at berth 102) by March 2005; and phase three involved the construction of backland space to support the terminal.[261] Apparently concerned that the scale of the new project might not be encompassed in the prior EIRs, the city also entered into a "side letter agreement," approved by City Council, which stated that the port and city "will use their best efforts to minimize negative environmental impacts" with respect to emissions from container ships, tugboats, trucks, and rail lines accessing the new terminal.[262]

Petitioners argued that the lease committed the port to all three phases of the development, and that the 1997 and 2000 EIRs did not even address the potential impacts of phase one—much less all three.[263] In particular, the 1997 EIR emphasized near-dock rail access, not container terminal construction, and did not contemplate the much larger scope of environmental impacts—including bigger wharves, larger operation space, more ships, and more trucks—while the 2000 review emphasized dredging.[264] The defendants—represented by lawyers from the city attorney's office and outside counsel Morrison & Foerster and McCutchen, Doyle, Brown & Enersen—denied these allegations.[265]

In 2002, after the suit was filed, activists held a protest in the Knoll Hill neighborhood of San Pedro, just next to the proposed China Shipping project.[266] In a convergence of political interests, harbor secessionists had seized on the project to rally support for their cause, inviting African-American leaders angry at newly elected Los Angeles Mayor James Hahn (a San Pedro resident) for breaking his promise to endorse African-American Bernard

Parks for a second term as police chief.[267] Hahn's sister, Janice, also a San Pedro resident, was a City Council member who represented the Fifteenth District (which linked Watts to the harbor communities of San Pedro and Wilmington) and supported residents pushing for a new environmental review of the China Shipping project.[268] In an effort to tamp down community controversy, Mayor Hahn had appointed a Port Community Advisory Committee, which a week prior to the protest had recommended that the U.S. Army Corp of Engineers conduct a new environmental review of the China Shipping site.[269] But that recommendation was rejected by the harbor commission and Hahn remained in political hot water.[270]

The state trial court in the NRDC suit provided the environmentalists no relief.[271] On May 30, 2002, the trial court rejected their challenge, holding that the first phase of the China Shipping project was within the scope of the 1997 EIR and therefore did not have to be redone; because the city and port had apparently conceded that an EIR would have to be done on subsequent project phases for berth expansion, the trial court held those challenges to be moot.[272]

On appeal, the petitioners asserted that the trial court lacked substantial evidence to conclude there were grounds for a CEQA exemption.[273] In doing so, their brief placed front and center the issue of air pollution caused by diesel-powered vehicles:

> [T]he transportation of ... containers to and from the site would generate a tremendous increase in the use of diesel trucks, diesel tugboats, and off-road diesel equipment, polluting the air and water and burdening the local streets and freeways. Of particular concern to Appellants who live nearby, and to members of the Appellant environmental groups, is the tremendous quantity of diesel exhaust—a known carcinogen—that would be pumped into the surrounding community.[274]

In response, the defendants focused on the scope of the earlier environmental reviews. They argued that the prior approvals clearly encompassed the development contemplated in phase one, and that the lease approval was conditioned on a subsequent environmental review for phases two and three—which were therefore not at issue.[275] In their brief, the defendants contended that berth 100, formerly the site of Chevron's wharf, was used for container storage and handling at the time of the lease and that the proposed change would actually "increase efficiency and reduce impacts because it puts the wharf closer to the Berth 100 backlands."[276] The defendants further contended that the selection of the West Basin site for China Shipping was the result of listening to residents, who had earlier proposed that any growth in container handling should be conducted there, and argued that resident failure to participate in the 1997 EIR process showed their acquiescence.[277] In addition, defendants asserted that the 1997 EIR explicitly contemplated the berth 100 wharf, and the 2000 EIS/EIR clearly proposed using dredged material to expand the China Shipping site "to provide more backlands and allow construction of an additional container wharf."[278] In the port's view, the impacts had already been accounted for and to the extent that they had not, the residents only had themselves to blame for not participating in the earlier processes.

The attorney general of California weighed in with an amicus brief on behalf of petitioners, arguing that by committing itself to construct all three phases of the development but only purporting to approve phase one, the defendants had improperly segmented the project in direct violation of CEQA, reducing it "to a process whose result will be largely to generate paper, to produce an EIR that describes a journey whose destination is already predetermined and contractually committed to before the public has any chance to see either the road map or the full price tag."[279] As part of its appeal, petitioners asked the appellate court to stay the terminal's construction,[280] which the court declined to do, although it did expedite hearing the case, setting argument for October 18, 2002.[281]

In the meantime, NRDC's Feuer went to federal court arguing that the Army Corps had failed in its 2000 environmental review to adequately evaluate the China Shipping project, again asking for an injunction against further development.[282] District Court Judge Margaret Morrow agreed, issuing a temporary restraining order (TRO) on July 24 "as work crews were pouring 100 feet of new concrete in a rush to complete the China Shipping Holding Co. terminal," which was over halfway done.[283] Fifty residents attended the court hearing, including community activist Marquez, who was astonished by the court decision: "This kind of thing has never happened before."[284] But the community's enthusiasm was short lived. Three days later, Judge Morrow refused to convert the TRO into a preliminary injunction, holding that the petitioners had not proved sufficient harm in letting the project proceed, while the port asserted that delay would cost it $1.2 million per day and undermine its reputation in the competitive shipping world.[285]

The state court took a different view. After hearing the case on October 18, the appellate panel decided to issue a stay, blocking construction of the key phase-one element: a 1200-foot wharf at berth 100, which had already been nearly completed.[286] Reversing the decision below, the appellate court curtly dismissed the port's contention that the China Shipping project was encompassed under the previous EIR as "supported neither factually nor legally."[287] Specifically, the court held that because the China

Shipping project did not arise until after the completion of either EIR, it could not "be considered part of the overall 'project' addressed in those documents."[288] "The fact that the port and China Shipping entered into a side letter agreement ... provides adequate support for appellants' argument that the port was required to prepare an initial study leading to either preparation of an EIR or a negative declaration for this Project. This was not done."[289] In a stunning blow to the port, the appellate court not only found the city to have violated CEQA and ordered a new EIR addressing all phases of the project, it also directed "the trial court to issue an injunction consistent with the stay we have issued precluding further construction or operation of the Project pending completion of the environmental review process."[290] The *Los Angeles Times* reported that the injunction "bars the pouring of 200 additional feet of concrete needed to complete the wharf," which was nearly 90 percent finished.[291] The Court of Appeal rejected the city's request for re-hearing by the city,[292] and the California Supreme Court denied a petition for review.[293]

Court victory did not end the terminal fight. Under the rules of CEQA, it simply required the city and port to go back and conduct an appropriate EIR.[294] Skirmishes continued, with the environmental groups failing to block delivery of four sixteen-story high cranes to the China Shipping site as the court held that unloading the already assembled cranes fell outside of the injunction.[295] Yet time was of the essence: "In the competitive world of global trade, the Port of Los Angeles did not want to lose an important customer such as China Shipping to another port."[296] The coalition therefore had important leverage, which it used to negotiate an unprecedented—and game-changing—settlement.

On March 5, 2003, in order to circumvent a lengthy battle over the project, the port and environmentalists entered into a $60 million settlement agreement, financed entirely through port revenue.[297] The agreement, approved by Superior Court Judge Dzintra Janavs, permitted the port to finish phase one of the project within weeks while it awaited completion of the EIR, which it was still required to do (the coalition reserved the right to challenge any inadequacy in the EIR).[298] In exchange, the port—in an unprecedented concession to environmental concerns—agreed to specific mitigation measures, which included requiring container handling equipment to use alternative fuels,[299] installing "low profile" cranes,[300] building facilities for "shoreside electrical power for ship hoteling," retrofitting China Shipping ships to use electrical power while docked,[301] creating a traffic mitigation plan,[302] and setting aside $50 million over five years for community-specific mitigation.[303] This community mitigation fund

Resisting the Port

included $10 million for the Gateway Cities Program to provide "incentives to replace, repower or retrofit existing diesel-powered on-road trucks," $20 million for air quality mitigation, and $20 million for aesthetic improvements to the community, including parks and landscaping.[304] In NRDC lawyer Gail Feuer's words, "Today is Day 1 in the greening of the Port of Los Angeles."[305]

Yet the greening project was nearly over as soon as it began. In a startling setback, it was quickly revealed that the port had not consulted China Shipping about the settlement terms, particularly the requirement that all docked ships turn off their diesel engines and plug into electrical outlets.[306] As it turned out, China Shipping leased its ships and would not commit to the retrofitting needed to convert them to electrical power, which it estimated would cost $300,000 per ship.[307] With a fleet of 100 ships, the cost would be well above the $5 million the port committed in the settlement for retrofitting. With the deal in jeopardy, city and port officials engaged in damage control. Port Executive Director Larry Kelly flew to Shanghai to meet with China Shipping representatives, while city officials indirectly blamed NRDC for the failure to notify China Shipping of the settlement terms—arguing that a confidentiality requirement imposed by the plaintiffs prevented city officials from revealing the terms of the settlement until after it was executed.[308] NRDC's Feuer responded in disbelief: "It never actually dawned on us that they weren't talking to China Shipping."[309] If the city had asked for permission to run the settlement by China Shipping—whose buy-in was obviously critical to effectuating the deal—Feuer was sure the plaintiffs would have provided permission to do so.[310] As the prospect of renegotiation grew, the Los Angeles city controller issued an audit stating that the true cost of the settlement would be twice as much as the city had advertised[311]—a figure that city officials vehemently rejected.[312]

A year after the landmark settlement, the completed terminal sat vacant as city officials worked to salvage the deal. As negotiations unfolded, NRDC agreed to a revised proposal under which China Shipping would commit to plug in 70 percent of docked ships rather than the 100 percent proposed in the original agreement, while only making two cranes low profile; however, NRDC held fast to its demand that the port include language in the EIR recognizing the project's "aesthetic impacts" on the surrounding communities.[313] At this the port balked, claiming it did not want to "prejudge" the outcome of the environmental review.[314] After a flurry of meetings, an amended settlement was hammered out, with the port agreeing to make clear that the original $20 million community fund was "being created in part to allow for the mitigation of the aesthetic impacts of the China

Figure 4.3
China Shipping Container Terminal. Photograph from Port of Los Angeles. Available at http://www.portoflosangeles.org/img/ChinaShipping_hi-res.jpg (accessed Jan. 24, 2017).

Shipping terminal off of port lands," while the environmentalists agreed to language that the port was "not prejudging whether these impacts are adverse or significant."[315]

This resolution cleared the way for China Shipping to take occupancy, which it did in May 2004[316]—marking the creation of what was hailed as the world's first green terminal, with cold ironing (i.e., dockside electrical plug-in) capability "expected to eliminate more than three tons of nitrogen oxides (NOx) and 350 pounds of diesel particulate matter for each ship that plugs in."[317] City Council Member Janice Hahn summed up what activists hoped would be the foundation for future change: "The China Shipping Settlement sets a precedent of how we do things at the Port today and into the future."[318]

Her claim turned out to be prescient, as the China Shipping victory reinforced other efforts to stem port-related diesel emissions. At the federal level, the EPA passed a series of rules setting more stringent diesel emission controls on heavy-duty highway vehicles (trucks) to take effect in the 2007 model year,[319] as well as similar standards on nonroad vehicles (trains and ships).[320] At the local level, resident and environmental groups, in newfound alliance, began pressing Mayor James Hahn for stronger city regulation. The ground was laid by Mayor Hahn, who as a candidate promised harbor residents that he would commit to a "no net increase" policy capping port emissions at 2001 levels.[321] In the face of an expected quadrupling of container traffic at Los Angeles and Long Beach by 2025,[322] particulate matter from port sources was predicted to increase from 1,000 to over 2,700 tons per year.[323] In a letter to San Pedro activist Noel Park, Mayor Hahn also committed to "review all past, present and future environmental documents in an open public process to ensure that all laws—particularly those related to environmental projects—have been obeyed, all city procedures followed, and all adverse impacts upon the communities mitigated."[324] As secession fever raged and the China Shipping fight was at its height, Mayor Hahn, in his first "state of the harbor address," promised to promote a greener port by moving industrial uses to Terminal Island and creating a recreational waterfront promenade from Vincent Thomas Bridge to the breakwater.[325] As part of his address, Hahn indicated that the port was developing green policies, such as a conversion of port machines to low-emission technology and the creation of a no-net-increase plan.[326] Some of the items in the speech, such as cold ironing, found their way into the China Shipping agreement. How Hahn planned to implement "no net increase" remained unclear.

The mayor's plan interacted with—and was pushed forward by—the continuous flow of evidence of port pollution and regulatory responses to it. In an environmental report card issued in early 2004, NRDC and the Coalition for Clean Air graded Los Angeles a C- and Long Beach a C for their environmental practices.[327] In its press release announcing the report card, NRDC stated that the two ports released as much diesel exhaust as 16,000 idling trucks per day.[328] In a subsequent report, NRDC systematically reviewed the negative health impacts of port emissions and made a number of proposals to mitigate them.[329] Among those recommendations were replacing extremely old trucks, retrofitting others, and mandating the use of cleaner-burning fuels.[330] In addition, following the China Shipping model, the report recommended moving ships to shoreside electrical power[331]—a proposal advanced by CARB in April. The NRDC report also identified the need for stricter rules on truck idling,[332] noting that a 2002 bill sponsored by Democratic State Assemblyman Alan Lowenthal from Long Beach, which banned idling for more than thirty minutes outside the port, had been largely circumvented by moving the trucking queue inside port property.[333] In July 2004, CARB approved a rule prohibiting diesel vehicles of 10,000 pounds or more from idling more than five minutes anywhere.[334]

During this period, Assemblyman Lowenthal upped the pressure on Mayor Hahn to make good on his no-net-increase promise. In February 2004, Lowenthal introduced Assembly Bill 2042, which required the ports of Los Angeles and Long Beach, in concert with the SCAQMD, to set the ports' air quality baseline at 2001 levels and to "ensure that all future growth ... will have a zero-net increase in air pollution."[335] The bill, backed by NRDC (fresh off its China Shipping win) and other environmental groups, was strongly opposed by shippers, local chambers of commerce, and the ports themselves. In objecting to the proposed bill, the Pacific Merchant Shipping Association (PMSA)—the industry trade group representing shipping lines and terminal operators—argued that the ports were already developing emission-reducing technology and that the bill "erects a vague and potentially prohibitive obstacle to future growth (that would) send a negative message to the international trade community."[336] The PMSA also tried to use the complexity of environmental agency jurisdiction to its advantage.[337] The legislative record stated that the PMSA "cautions against assigning mobile source emission regulation to a regional agency [i.e., the SCAQMD], a prospect that could create 'islands of divergent authority for sources that travel between air districts (and other state and federal jurisdictions).' For this reason, they believe authority should remain with the ARB and federal EPA."[338] Marching in lockstep, the Long Beach Chamber of Commerce similarly objected to the bill's "conflicting approach to mobile source emissions," while decrying the 2001 baseline as unrealistic.[339] Similarly, the Port of Long Beach rejected the baseline as "unachievable," particularly in light of its lack of authority over ocean-going vessels, and underscored the absence of regulatory clarity.[340] In response, Lowenthal amended the bill to reset the emissions baseline to 2002.[341] Nonetheless, the Long Beach Board of Harbor Commissioners voted to oppose AB 2042,[342] setting up a conflict with the Long Beach City Council, which the next day unanimously voted to support "AB 2042 in order to protect public health and safety by avoiding an increase in air pollution from the ports of San Pedro Bay."[343] The City Council thereby directed the city clerk "to transmit a copy of th[e] resolution to the Governor, to the members of the California Legislature representing the Long Beach and Los Angeles areas, and any other officials, agencies, entities, and individuals as may be deemed appropriate."[344]

It was against this backdrop that environmentalists and harbor residents, armed with the China Shipping victory, set their sights on Long Beach, which had begun to move forward with its own 115-acre expansion project at Pier J in a new attempt to accommodate COSCO—away from the forbidden naval station property.[345] Pier J, at the southern tip of the port

Resisting the Port

below the venerable Queen Mary, had nearly doubled in size in the early 1990s to accommodate Maersk's container vessels and a new on-dock rail yard.[346] A decade later, the site was targeted for further expansion via landfill designed to increase the pier complex to 385 acres in three phases, to be completed by 2015.[347] Toward that end—and careful to avoid the problems that beset the China Shipping project next door—the Long Beach Board of Harbor Commissioners circulated a draft EIR for Pier J in 2003.[348] In response, the SCAQMD staff issued two comment letters. The first, sent on February 7 by the CEQA section program supervisor, argued that increased port traffic could contribute to a carbon monoxide hotspot and proposed mitigation measures that included turning off idling trucks and installing electrical connections to plug in docked ships.[349] The second letter, sent by the planning and rules manager on October 8, homed in on what would become a key objection: that the port had not adequately accounted for increased diesel emission from the project.[350] Specifically, the letter argued that in modeling the health risk assessment, the port had assumed a 75 percent reduction of diesel emissions from heavy-duty vehicles based on the phase-in of the EPA's 2001 diesel rules and CARB's 2001 Risk Reduction Plan.[351] However, the SCAQMD contended that because those rules applied prospectively, with the EPA rule not fully phased in until 2010, the port had to factor in delays in emission reduction due to truck turnover—which it had not done, thus understating the impact of increased truck traffic caused by the expansion.[352]

The SCAQMD reiterated this central objection in its July 30, 2004 comments on the final EIR.[353] NRDC, pivoting from its negotiations on the amended China Shipping settlement, also provided comments critical of the expansion plan.[354] Although the threat of litigation was only thinly veiled, in August the port nonetheless approved the EIR, triggering an appeal to the Long Beach City Council by NRDC and other groups.[355] While awaiting the meeting, NRDC's position received further support. On September 9, USC researchers released a study in the *New England Journal of Medicine* finding that children who lived in smoggy areas, particularly those surrounding the ports, were more likely to have permanently underdeveloped lungs.[356] Another USC study found increased rates of cancer downwind of the ports.[357] Armed with this evidence at the City Council meeting on September 14, NRDC's Feuer offered a powerful critique of the Pier J EIR, emphasizing that it did not harmonize with the no-net-increase approach to which the council had already committed and that it incorrectly set 2015 as the air quality baseline despite the fact that phase one construction would be done in 2007.[358] She cited China Shipping as precedent, noting that the EIR did not address feasible programs to reduce emissions like cold ironing. In parrying City Council member questions about the port's legal authority to mandate cold ironing,[359] Feuer stressed the port's authority as the landlord: "[T]hat's the power the Port has. The Port can say as a condition of the lease that you need to have plug ins at this facility. ... I think there's no question there's legal authority to do it."[360] NRDC was supported by staff from the SCAQMD, but was opposed by some labor union representatives, who questioned the impact on jobs, as well as the Long Beach port's director of planning, who characterized the NRDC proposals as "pie in the sky."[361] NRDC, however, carried the day. In the words of Council Member Jackie Kell, Feuer had made the port's EIR "look like a complexion full of zits."[362]

The Long Beach City Council decided to delay a vote on the EIR and port staff recommended its rescission. This came on the heels of a harsh letter from the SCAQMD that reiterated its main technical objections, "strongly" recommending that the port "reconsider" the EIR in order to "ensure that requirements" under CEQA and NEPA were met.[363] Litigation was again threatened.[364] In light of this, port Managing Director Geraldine Knatz supported the EIR's rescission, stating that "[o]ur feeling is that we want to have the best document that we can have."[365] The board agreed, formally rescinding its approval at a September 29, 2004 meeting.[366] Keeping on the pressure, Feuer urged the board "not just to go back to address and analyze these issues but ... to please send the message [to staff] that what they have adopted is not enough, that this Board wants more and that more can be done."[367] In response, Commissioner James Hankla applauded NRDC for doing "this Board a great service," and said that "staff should consider the Board is directing it to evaluate this process de novo and evaluate every single aspect of the EIR from the standpoint of NRDC, Coalition for Clean Air, Earth Corps as well as AQMD."[368] In recalling this outcome, one NRDC lawyer emphasized its significance, noting that although China Shipping had received the most attention, the Pier J victory was a "pretty big deal because ... it's very rare to have an agency go back on its position and win at an administrative level."[369]

As a vindication of Long Beach's no-net-increase stance, Pier J also set the stage for the final battle over AB 2042. As the political debate neared its resolution, no-net-increase opponents succeeded in scaling the bill back—setting the air quality baseline at a more recent year (2004),[370] while weakening enforcement.[371] Even in this watered-down state, Governor Arnold Schwarzenegger was not willing to anger the ports and Chamber of Commerce on this issue. Although he asserted that "[i]mproving the quality of our air is a priority of my Administration," he stated that "this bill will not reduce pollution in any way," and instead directed the California EPA and CARB "to work with the ports, the railroads, other goods movement facilities, local air districts, the U.S. Environmental Protection Agency ... and local communities to develop such a program for our ports throughout the state."[372] In a nod to industry arguments of jurisdictional competence, he concluded that "[a]s most of the pollution is generated by federally regulated sources, I urge the federal government to provide the necessary incentives and regulations."[373]

The defeat of no net increase at the state level had the effect of galvanizing efforts around Mayor Hahn's local initiative. On July 7, 2004, residents were outraged by the release of a Plan to Achieve No Net Increase of Air Emissions at the Port of Los Angeles, authored by the port with the aid of Houston-based Starcrest Consulting Group.[374] The plan was presented to the Hahn-created Port Community Advisory Committee.[375] What upset residents most was the plan's claim that the port could achieve Hahn's promise of no net increase without any major new programs by assuming a sharp reduction in air pollution based on the China Shipping truck retrofitting program.[376] Residents objected that the program would only replace 400 of the more than 6,000 trucks that were over twenty years old—and noted even that would not be completed until 2008.[377]

Embarrassed by the blowback, Mayor James Hahn and City Council Member Janice Hahn instructed Los Angeles port Executive Director Larry Keller to establish a task force to develop a credible strategy.[378] In the wake of the conflict, Keller resigned.[379] Mayor Hahn appointed a twenty-eight-member No Net Increase Task Force, which began meeting in October 2004 and included Noel Park (also on the Port Community Advisory Committee) and Gail Feuer, along with representatives from industry, labor unions, and other community and environmental groups. Despite this, critics continued to blast the mayor for failing to keep his promise to remediate projects completed prior to 2001.[380]

The task force persevered, considering a range of initiatives to deal with port emissions, which included an ambitious (and, at $35 million, expensive) replacement program to convert 1,000 old diesel trucks to 2004 clean models.[381] Industry groups participated but were wary, with one terminal operator suggesting that a plan mandating cold ironing would not "survive a constitutional challenge."[382] Although the task force was supposed to present a plan to Mayor Hahn by December 31, 2004, election year politics appeared to intervene, with the group's draft proposal delayed until just before the hotly contested primary between Hahn and challenger Antonio

Villaraigosa in March 2005.[383] Yet the delay did not dampen the efforts of the task force, which received a boost from state and federal environmental regulators who began collaborating with members to produce a sustainable plan.[384] A draft plan was released on March 3, which contained proposals—without attending to cost or feasibility—for cleaner fuel, subsidized new truck conversion, and cold ironing.[385] The most controversial proposal—electrical rail lines—drew strenuous objection from BNSF and UP, whose attorney complained that "[i]t's a real stretch when you consider these things don't exist."[386] Despite progress, the plan was perhaps most notable for what it did not include: support from neighboring Long Beach, which rejected an invitation to participate and instead produced its own green port plan in January 2005 without input from air quality regulators.[387] Opining on the Los Angeles plan, Port of Long Beach Planning Director Robert Kanter reiterated the complaints he voiced in the Pier J fight, stating, "There are some radical ideas, pie-in-the-sky ideas, that I don't think are likely to take place in the near term."[388]

Nonetheless, the Los Angeles task force forged ahead, producing an emission-reduction plan projected to prevent 2,200 premature deaths over twenty years at a cost of $11 billion.[389] Yet industry resistance to aspects of the plan prevented consensus; as a result, the task force did not vote to endorse the plan,[390] but rather simply turned over its recommendations to Mayor Hahn one week before the end of his term.[391] The 600-page report was impressive in its detailed scientific analysis of emissions and in the scope of its policy proposals,[392] which included sixty-eight separate control measures for different emission source categories (ocean-going vessels, harbor craft, cargo-handling equipment, rail, and heavy-duty vehicles).[393] The plan's basic structure was to offer analysis and recommendations proposed by air regulators and environmentalists, while interlineating industry objections throughout.

In addition to fighting over specific regulations, industry and environmentalists clashed on the issue of the port's legal authority to implement the proposals—a harbinger of fights to come. Section 5 of the report provided a detailed legal analysis that focused primarily on the issue of federal preemption, particularly with respect to the Clean Air Act.[394] That analysis was drafted through rancorous negotiations between SCAQMD and NRDC (particularly Gail Feuer) on one side, and lawyers for the PMSA and rail lines on the other. The result was a carefully worded document that offered a sweeping review of preemption doctrine and a proposal-by-proposal legal analysis, which was impressive in its comprehensiveness, while exposing the deep differences between environmental and industry lawyers on the issue of local authority.

While the SCAQMD and NRDC asserted that the port's implementation of no-net-increase measures "could be characterized as proprietary conduct that is exempt from federal preemption under the market participant exemption,"[395] rail and PMSA lawyers were much more skeptical, arguing that the Port of Los Angeles "may not adopt a sweeping set of control measures through its contracts and leases in order to implement broad social policy regarding air quality under the guise of the market participant exception."[396] The legal gauntlet was thus thrown down. On July 30, 2005—his last day in office—Hahn endorsed the No Net Increase Task Force report and recommended "that the Villaraigosa administration adopt the report's finding to make sure that the Port of Los Angeles is the nation's leader in clean air standards."[397] Although many task force members had hailed the plan as a step in the right direction, some community representatives were disappointed with Hahn's failure to keep his no-net-increase promise, instead tossing the "hot potato" to the next mayor.[398]

For his part, the new mayor seemed determined not to drop it. To the contrary, Villaraigosa—a former union organizer and Democratic speaker of the California Assembly, who had campaigned on a platform of green growth and swept into the mayor's office with a progressive coalition of labor, environmental, and other liberal constituencies—appeared committed to aggressive action to meet the seemingly intractable problem of reconciling port expansion with environmental and community health. His first moves signaled the priority he was to give to greening the ports and building upon the ultimately inadequate Hahn no-net-increase effort. Attention focused on his choice of city commissioners, which constituted a critical exercise of influence that set policy direction for the powerful agencies that governed Los Angeles.[399] For Villaraigosa, filling vacancies on the harbor commission at the Port of Los Angeles was high on his priority list upon taking office. Against the backdrop of the China Shipping litigation and the sense that port expansion was threatened by ongoing environmental clashes, the mayor was committed at the outset to appointing board members with environmental experience and community credibility.[400] In addition, the recent resignation of port Executive Director Keller left a vacuum in leadership, which the mayor wanted to quickly fill.

The process Villaraigosa initiated to find qualified city commissioners was designed to not simply reward supporters or promote insiders. Upon his election, Villaraigosa convened an advisory group of seventy-five diverse stakeholders and asked them to create a pipeline of applicants for commission positions who were "not the usual suspects," but rather people "who think outside the box, who are creative, who come from all over the city."[401] One of those people was Jerilyn López Mendoza, a UCLA School of Law graduate and long-time environmental lawyer, who headed the Environmental Justice Project at the Environmental Defense Fund, where she had been for five years.[402] In addition to environmental expertise, Mendoza had a deep familiarity with labor issues and the complexity of Los Angeles's proprietary departments, having just been lead lawyer on the campaign that produced a multi-million-dollar community benefits agreement with the Los Angeles International Airport (LAX).[403] Mendoza was contacted by two members of the Villaraigosa transition team, Paula Daniels, former member of the California Coastal Commission, and Cecilia Estelano, a partner at Munger, Tolles & Olson.[404] With their encouragement, Mendoza filled out an application and was soon contacted by a screening firm that, she recalled, "asked me ... pointed questions, like what was my theory of social change and how did I define my work in terms of environmental justice?"[405] Mendoza made it to the final stage, where she met with the mayor, along with Bud Ovrom, deputy mayor for housing and economic development, and Sharon Delugach, who was coordinating commission appointments. At that meeting, Mendoza and the mayor engaged in a lengthy "exchange of monologues, where he would sort of explain things to me from his perspective and then I would sort of explain my perspective based on his perspective."[406] In this conversation, Mendoza recalled the mayor laying out his position, that the Port of Los Angeles is "always going to be a working port. ... It really is just one of our most important economic assets. It's never going to be Marina del Rey. It's never going to be a tourist location. ... My vision for the port is I want to see the cleanest, greenest port in the world. ... Do you think that's possible?"[407]

Mendoza's answer was yes—"if you have the political will."[408] Her selection as commissioner indicated that the mayor did indeed have the will—a view underscored by the appointment of David Freeman, who was former energy secretary to President Carter, general manager of the Tennessee Valley Power Authority, energy czar to Governor Gray Davis, and at the time head of the Los Angeles Department of Water and Power.[409] Freeman, a close adviser to Villaraigosa, was considered someone able to get things done.[410] Freeman and Mendoza were appointed in July 2005, and confirmed in September, along with Kaylynn Kim, a private attorney; Doug Krause, general counsel of East West Bank; and Joe Radisich, president of the Southern California District Council of the ILWU.[411] Radisich's appointment signaled the early support for greening the port by the ILWU, whose leaders understood that there would be no growth without cleaning up port operations.

The new board immediately took a different approach, holding its first scheduled meeting in an overflowing community center in Wilmington,[412] rather than its traditional spot in the San Pedro Harbor Administration Building.[413] There, Freeman, as board president, criticized the Hahn no-net-increase plan's 2001 emissions baseline, telling the crowd, "Surely, you can't settle on that."[414] He asked port staff to evaluate the Hahn task force plan, moving with a greater sense of urgency by requesting a review of which proposals could be accelerated and expanded.[415]

This urgency was heightened as multiple regulatory bodies vied to restrict port emissions. The SCAQMD's chairman calculated that the ports produced 100 tons of NOx a day, an amount greater than emissions from six million cars, while also producing 20 percent of the region's diesel particulate matter, responsible for 1,700 premature deaths a year.[416] In response, the SCAQMD sought guidance from its lawyers to find authority to regulate the port complex as a single stationary source.[417] CARB kept the pressure up, finding that the port increased cancer risk up to fifteen miles away,[418] while also linking cargo transportation, particularly near the port, to a host of health problems, which it estimated cost over $6 billion to treat.[419] CARB's study found that 2,400 people died annually as a result of port-related air pollution, many of them in surrounding neighborhoods.[420]

Public health care costs and ongoing community opposition pushed forward regulatory action.[421] In April 2006, as part of a Governor Schwarzenegger-sponsored initiative to meet federal clean air deadlines, CARB approved a plan to reduce goods-movement emissions to 2001 levels through a variety of proposals—including cleaner ship fuel, cold ironing, and replacing old diesel trucks.[422] Yet the lack of funding or mandatory requirements caused harbor commission President David Freeman to scoff: "Are they ordering people to do things? No? Then what the hell good are they?"[423] Community residents also complained.[424] Regulatory agencies and environmentalists pointed fingers, with state agencies contending that they had no authority to regulate the biggest polluters—ocean-going vessels and railroads—while NRDC disagreed.[425] The port, for its part, attempted to negotiate emission reductions into ocean vessel leases, while the shipping industry was developing its own market-based emission control plan.[426] The challenge for the Villaraigosa administration was promoting the Los Angeles port as a regional growth engine, while dealing with its "bad reputation" as a source of pollution and other community detriments.[427]

Commissioners Freeman and Mendoza explicitly viewed meeting this challenge as their primary goal. As Freeman remembered, tackling the green growth problem "was the reason the mayor named me and [Mendoza] and people like that to get the job done. ... I mean, obviously the exact details of how we were going to go about it were not preordained, but ... I was put on there because of my environmental credentials and the fact that the mayor knew me as a person [who] wasn't just a bullshit artist but kind of made things happen that we talked about."[428]

As Mendoza recalled, the board viewed its mission as executing the mayor's goal of making the Port of Los Angeles "the cleanest, greenest port in the world."[429] In discussing how to do that, she said, the commissioners quickly "realized two things: one was everything we did had to be in close coordination with Long Beach. [Without coordination,] the customers and other people who work and live at the port would just move over to Long Beach where they didn't have to deal with it. ... The second thing we realized was that we weren't going to get anything done unless we adjusted all sources of pollution ... even though we knew that trucks were of primary concern."[430] The commission moved assertively on both fronts.

To promote inter-port cooperation, the first order of business was hiring a Los Angeles port director who could reach across the bay to her Long Beach counterparts. That process was managed within the mayor's office by small group of advisors that included David Libatique, who became part of Villaraigosa's transition team and then was assigned to the Los Angeles Business Team as port liaison, under the supervision of Deputy Mayor Bud Ovram.[431] Libatique joined Villaraigosa's transition team in 2005 after working as a deputy to City Council Member Martin Ludlow.[432] With a master's degree in public policy from Harvard, Libatique was a policy generalist who was charged during the mayoral transition with preparing background memos on the ports. It was "a natural fit" and Libatique immediately found himself enmeshed in port-related air quality work.[433] Libatique helped vet the port director candidates, ultimately presenting three to the mayor.[434] The goal was to find a new director who would "focus on dealing with the environmental impacts but create a path forward for the port to continue to be an economic engine for ... the city."[435] In January 2006, the mayor selected Geraldine Knatz, formerly managing director of the Port of Long Beach, who held a doctorate in biological science and was viewed as a strong supporter of "greening and growing."[436] With twenty-three years of experience at Long Beach, Knatz was also seen as a bridge builder who could advance the coordination agenda.[437]

With Knatz in place, the commission reached out to Long Beach to advance a joint plan that would attempt to comprehensively address the port complex's multiple sources of pollution—recognizing that when it came to pollution, there was no "dividing line in the air."[438] Although the mayor made his harbor commission and port director selections with "green growth" in mind, Libatique recalled that "there wasn't that much advanced planning about how everything was going to roll out."[439] Instead, the mayor entrusted his new team to develop a plan, which it quickly set out to do. Shortly after Knatz was hired, she and Freeman met with their Long Beach counterparts to establish a framework for discussions that would lead to a comprehensive policy—to be called the Clean Air Action Plan (CAAP).[440] With the process and goals agreed upon, both ports' harbor commissions began holding joint monthly meetings to discuss the details. Mayor Villaraigosa reached out to union leaders to gain their support, arguing that enhanced environmental standards at the port were good for the health of union members.[441] Commissioners and port staff also met with industry leaders to get them on board. With labor and environmentalists aligned behind a new plan, industry was on the defensive. According to Freeman, the message to industry representatives was clear: the ports could promise expansion only if shippers and other industry players agreed to clean up the system. In Freeman's terms, the board said "you come to us with an expansion proposal and we'll approve it ... [if you] clean up what you're doing."[442]

Both commissions were in a position to facilitate growth plans provided that they complied with environmental goals. It was ultimately the ports' power to reject or delay expansion that provided the leverage needed to get industry buy-in. And although shippers and carriers had other ports they could use, those ports were generally not as attractive because of preexisting infrastructure investments in Los Angeles and Long Beach, as well as access to the lucrative regional market.[443] It was in this context that the Los Angeles and Long Beach harbor commissions developed the outlines of CAAP, a draft of which was circulated in July 2006.[444] The main approach was to regulate emission sources tied to the ports—by, for example, requiring docked ships to burn cleaner fuel or adopt cold ironing.[445] Other parts of the plan referenced ambitious goals for overall emission reductions, but the outlines were still tentative.[446]

By the time the final plan was released in November 2006, a focus on trucks had crystallized.[447] While the draft plan was vague on the trucks piece, the final plan emphasized replacing the diesel trucks that accessed the port and offered a clearer road map to effectuate that goal.[448] Although explicitly presented as a "living document," the CAAP Technical Report, through Control Measure HDV1, provided a clear emission control framework for "Heavy-Duty Vehicles"—which formed the foundation for what would ultimately become the Clean Truck Program. The report's central contribution was to recognize that port drayage trucks, on average over ten years old, were a significant source of air pollution and to call for the rapid greening of the entire drayage truck fleet serving the ports within a five-year period.[449] In order to cut diesel emissions by nearly half, the report focused on replacing and retrofitting what it estimated to be the 16,800 "frequent and semi-frequent trucks" that accounted for roughly 80 percent of all port calls.[450] The goal was to achieve "clean" standards, which meant replacing or upgrading all "frequent caller" trucks (those that made more than seven calls per week) to meet EPA 2007 emission standards; for "semi-frequent caller" trucks (3.5 to 7 calls per week), the goal was to replace or upgrade trucks that were model year 1992 or older, while retrofitting newer trucks with certified emission-reduction technologies.[451] To do this, the report proposed to provide "significant incentives to owner/operators to encourage accelerated turnover/retrofits, and on the terminal side to maximize the use of 'clean' trucks through lease requirements and/or other mechanisms."[452]

The financial impacts of various incentive programs were modeled, with the main proposal to replace roughly half the trucks and retrofit the other half estimated to cost approximately $1.8 billion.[453] The report acknowledged that even with ports contributing $300 million and the SCAQMD another $36 million,[454] "additional funding on a massive scale will be needed."[455] Only a tentative implementation framework was provided, with several options put on the table to move the ambitious plan forward, ranging from those imposing costs directly on drivers to those shifting all costs to the public.[456] Each plan was evaluated in light of emission goals but also taking into account "wages/quality of life" for truck drivers.[457] Proposals included requiring individual drivers to display an emblem indicating emission compliance; imposing an "impact fee at the gate" on dirty trucks; assigning exclusive franchises to clean trucking companies "that can document that their drivers are paid a 'prevailing wage'"; creating a joint powers authority that would buy trucks and hire drivers, thus competing with existing for-profit companies; and having the ports directly buy trucks and hire drivers, mandating that only city drivers would be allowed on port property.[458] The commissioners included a specific timeline for action because "we didn't want it to be just a clean air plan that implied it was going to be put on a shelf somewhere."[459] Thus, they asked port staff to develop "further program details" and an "implementation plan" for review and approval "by end of 1st quarter 2007."[460]

On November 20, 2006, after a raucous, four-hour joint session of the harbor commissions,[461] at which numerous residents and officials (including the Los Angeles mayor) testified, both ports approved CAAP by a unanimous vote.[462] As if to further underline the importance of the trucking piece, the presidents of both ports read a statement into the record, which directed their "respective staffs to work expeditiously to bring forward a plan" to tackle the "dirty truck problem."[463] The "skeletal outline" of this program included "a 5-year, focused effort to replace or retrofit the entire fleet of over 16,000 trucks that regularly serve our Ports with trucks that at least meet the 2007 control standards and that are driven by people who at least earn the prevailing wage."[464] The directive made clear that the ports were to restrict noncompliant trucks from entry and that the fees necessary to fund the program "would be imposed on 'shippers,' and not on the drivers."[465] Furthermore, the ports were instructed to "invite private enterprise trucking companies to hire the drivers on terms that offer the proper incentives and conditions to achieve the Clean Air Action Plan goals while resulting in adequately paid drivers."[466] The goal of CAAP was to reduce diesel truck emissions by 80 percent.[467]

Although the vote was hailed as a serious step toward addressing the "diesel death zone," large questions remained about CAAP's implementation and funding, despite pledges from the ports of $200 million and the SCAQMD of $48 million, as well as the passage of state Proposition 1B, which authorized $20 billion in bond funding for transportation projects, $1 billion of which was targeted to support air cleanup.[468] The focus on trucks previewed—and was pushed forward by—the emergence of a new environmental-labor alliance that saw clean trucks as a way to achieve emission reductions, while advancing the Teamsters' long-standing goal of unionizing port truck drivers. Evidence of this alliance was on display at the final joint port meeting on CAAP, where drivers testified and parked outside in solidarity, while NRDC lawyer Melissa Lin Perrella made the sustainability argument that would define the Campaign for Clean Trucks: "The problem is that if you give a poor truck driver a clean truck, he needs to be able to afford maintaining it."[469] Reducing pollution over the long term would require raising the standards of the truck drivers. The Campaign for Clean Trucks was thus born.

5   Reforming the Port: The Campaign for Clean Trucks

Compared to the decades of sustained conflict between residents of San Pedro harbor communities and the port complex around issues of expansion and community impact, the campaign to redress the port's intertwined environmental and economic harms would coalesce and culminate with relative speed. What had prevented community activism in the 1980s and 1990s from having a substantial effect on port governance and policy was the absence of genuine political opportunity to influence official decision makers and the dearth of resources to coordinate challenges and bring them to scale. Two decades of business-friendly political rule in Los Angeles and Long Beach left community protest against port encroachment to fall largely on deaf ears. And despite ad hoc local mobilization and ongoing legal skirmishes around environmental and labor issues, funding for a large-scale and head-on challenge to port power was nowhere to be found. The Clean Air Action Plan, or CAAP—itself a product of broader political trends ushering in more progressive local politicians, particularly in Los Angeles—thus presented a singular opportunity: providing a legal framework and a policy process in which community, environmental, and labor activists could insert themselves—and ultimately ally—in advancing the goal of reforming the port trucking industry.

In 2006, as the CAAP process began to rapidly unfold, precisely what that goal would look like and how it would be achieved was anything but clear. Although CAAP had spotlighted the problem of "dirty trucks" and thus drawn the attention of organized labor and environmentalists alike, it was by no means inevitable that these fractious and often antagonistic movements could identify and seize the opportunity—to say nothing of actually succeeding in mobilizing to achieve transformational policy reform. This chapter tells the story of how that policy reform—which came to be called the Clean Truck Program—was ultimately won, in the face of significant odds, while also revealing how this victory carried with it the seeds of subsequent defeat. The narrative arc of the chapter traces how campaign leaders used the legal opportunity presented by CAAP to build the Coalition for Clean and Safe Ports, which in turn mobilized its legal and political assets, inside and outside of local government, to draft and defend the Clean Truck Program as it successfully advanced through the policymaking process, culminating in its enactment (in different forms) in Los Angeles and Long Beach in 2008. In telling this story, the chapter provides an account of the critical role that campaign lawyers' legal interpretation of preemption doctrine played in creating the space for local government policymaking in the face of the constant threat of industry legal challenge. It then shows how that interpretation ultimately failed to prevail in court, leaving the campaign with a partial victory that split the coalition's interests, imposed new financial burdens on already beleaguered truckers, and channeled the struggle to transform port trucking in new directions.

The Alliance: Forming the Coalition

How CAAP came to focus on clean trucks was in part a story of regulatory efficacy. In the complex jurisdictional framework for air regulation, drayage trucks that serviced the ports were viewed as within port control in a way that ocean-going vessels and rail trains were not. Yet the move toward clean trucks was also a product of political opportunity and interest convergence. Opportunity was built upon the need to develop a growth plan for the future of both ports that accounted for environmental concerns. All stakeholders recognized the importance of a sustainable emission control framework. The question was: What would it look like? By highlighting the need to clean up 16,000 dirty diesel trucks, CAAP made a potential link between environmentalism and unionism—which the labor movement was eager to strengthen. Hence, cleaning up trucks was connected to the concept of transforming the structure of the drayage truck industry in a way that implicated drivers' employment status. For organized labor, environmentalists brought the regulatory leverage and community activists brought the grassroots credibility. For environmental and community leaders, labor brought political heft and the ability to move local power.

Personnel

In creating the coalition to challenge port trucking, there were both top-down and bottom-up processes at play. The top-down process was driven by Change to Win, which was in the midst of formulating its national ports strategy, focused on the concession model, at the very moment the CAAP

Reforming the Port

process was moving toward its approval. The intersection between Change to Win's national ports strategy and the Los Angeles-Long Beach CAAP process occurred by design, but the precise timing was somewhat fortuitous. The national Blue-Green Alliance, a formal collaboration between labor and environmental groups, was founded as way to overcome historic antagonisms to develop policies that created good jobs and a healthy environment. Carl Pope, director of the Sierra Club, announced an initial agreement between the Sierra Club and the United Steelworkers union in June 2006.[1] He then began meeting with other labor leaders to build out the alliance.

In July, Pope met with top officials at Change to Win to discuss potential collaborations. At that time, although Change to Win had begun to move forward with its five-port concession strategy, the ports team did not have a strong grasp of the local situation in Los Angeles. The Sierra Club, in contrast, had just completed a video about the China Shipping case—called "Terminal Impact"[2]—and, through the local chapter, was deeply engaged in ongoing efforts to stem port pollution. It was also around this time that news reports indicated that Dubai was trying to buy a terminal at the Port of Los Angeles, which raised security concerns.[3] In discussing Change to Win's ports campaign, Pope, who was closely connected to local Sierra Club activists, mentioned the CAAP process. Change to Win's Nick Weiner, who was at the July meeting, remembered that Pope's mention of Los Angeles, although "just happenstance," allowed the port team to key in on Los Angeles as an auspicious site and to "connect the dots" between the concession model and environmental concerns.[4] As Weiner recalled, "We discovered that, oh right, these are old polluting trucks and they contribute to the pollution in L.A. in particular. [The Pope meeting] kind of just happened ... around the same time so that we were able to then further develop [the concession] theory."[5]

From there, Weiner and his colleagues were asked to "figure out L.A."[6]—a task they undertook with speed and intensity. Weiner and John Canham-Clyne immediately reached out to Maria Elena Durazo, head of the powerful Los Angeles County Federation of Labor (County Fed), and Madeline Janis, director of the Los Angeles Alliance for a New Economy (LAANE), which was known for spearheading passage of Los Angeles's Living Wage Ordinance in 1997.[7] LAANE's mission, creating a "new economy that works for everyone," was advanced by "championing the role that local government can play in nudging either individual industries or the broader regional economy."[8] With LAANE's support, the local campaign "took off" toward the goal of passing a concession policy at the ports of Los Angeles and Long Beach.⁹ Although Change to Win launched its campaign nationwide, there was optimism about Los Angeles because "the politicians and politics lined up ... [and] our ability to build a coalition lined up" because the "infrastructure was already there."[10] For Canham-Clyne, the key factors leading to Los Angeles were the strength of the local labor movement, parts of which (particularly, the Teamsters, International Brotherhood of Electrical Workers, and the United Teachers Union of Los Angeles) had helped elect Mayor Villaraigosa; the "air quality crisis" and the work of environmental groups to address it; and the "very specific willingness of [the] drivers to fight."[11] It was on this basis that Change to Win focused its energy on Los Angeles.

The first order of business was to mobilize local infrastructure to support a campaign. "Change to Win always felt strongly that ... to be effective on the ground, you needed a lot of people who really knew the landscape."[12] Change to Win chose LAANE, known for its sophisticated campaign research and policy work, to house staff and be the focal point of the coalition-building process. As Canham-Clyne recalled, "We did want to make sure that LAANE was involved ... because they had demonstrated experience in bringing together community organizations and the labor movement in ways that actually functioned."[13] Change to Win thus made an initial funding grant to LAANE in order to support campaign hiring and administrative assistance.[14] Hiring was overseen by Change to Win's Weiner and Canham-Clyne, who sought to bring in personnel with skills necessary to move the port agenda. A key member of this team was Jon Zerolnick, who joined LAANE in 2006.[15] A Yale undergraduate who pursued graduate labor studies at the University of Massachusetts, Zerolnick was a researcher with deep experience in corporate campaigns.[16] During college, he worked in the dining halls as a member of HERE Local 35 (with which he went on strike). During graduate school, he interned with HERE Local 11 in Los Angeles. When HERE offered to hire him full time, Zerolnick dropped out of graduate school and went to Las Vegas to work on a culinary workers campaign with Local 226.[17] He then served as a researcher on a campaign to organize workers at the Venetian hotel. From there, Zerolnick went to Denver to join the AFL-CIO in a multi-union organizing campaign at the Denver International Airport and came to Los Angeles in 2002 to staff the research department of the United Farmworkers union.[18] When the Change to Win split occurred, he consulted with unions for a while until he received a call from Canham-Clyne in 2006, inviting him to become part of the ports team at LAANE, to which Zerolnick was already attracted because of "the overlap of policy and ... coalition building and organizing."[19]

Zerolnick was soon joined by Patricia Castellanos, who technically was hired first after an interview with Canham-Clyne but took some time off and thus started just after Zerolnick.[20] Castellanos brought a number of key experiences and skills as an organizer with the proven "ability to build coalitions."[21] She had roots in the South Bay after working there on a number of electoral campaigns in the early 1990s, including the fight against the anti-immigrant initiative, Proposition 187.[22] She then spent nearly a decade at AGENDA, a South Los Angeles-based community organizing group and progressive think tank, where she worked on policy and education campaigns with environmental justice groups around the country.[23] Castellanos also brought connections to the mayor's office. She had campaigned for Villaraigosa in 2005 and joined his staff once he was elected, working on goods movement policy under Larry Frank in the Neighborhood Services office, where she was "trying to build relationships for the mayor in that area."[24] Like Zerolnick, Castellanos was affirmatively recruited. She had "heard rumblings" about a ports campaign when Canham-Clyne called to ask if she was interested.[25] Roxana Tynan from LAANE also reached out to encourage Castellanos, who joined the staff in August 2006 and spent the first few months applying for foundation grants to staff the project at "a high level."[26] She succeeded in securing an initial grant from the Hewlett Packard Foundation and gradually increased funding to support two organizers and three researchers at the height of the campaign.[27] Although they were both housed at LAANE, which was the campaign's "glue," Castellanos and Zerolnick worked with Weiner and Canham-Clyne in an "integrated" relationship in which they considered themselves "all staff together."[28]

From the outset, their mission was to advance the concession concept designed by Weiner and Canham-Clyne. In its basic form, the concept was to use the port's legal authority as a market actor to require drayage trucking companies to effectuate a double conversion: of their fleet to clean trucks and of their drivers to employees. The market-based rationale, which formed the legal hook upon which the plan rested, was that the double conversion was necessary to provide sustainable emission reductions, which were, in turn, necessary to ensure stable port growth. Employee conversion was key to making the trucking companies internalize the long-term costs of clean fleet acquisition and maintenance. A short-term subsidy could incentivize the drivers to buy clean trucks. But to have those trucks maintained over time required that they be owned by the entities best able to bear that cost: the trucking companies themselves. Shifting ownership to the companies would thus prevent the cycle of truck deterioration, pollution, and litigation that had stymied port growth. When Weiner and

Canham-Clyne reached out to LAANE, they had already fully "hatched this idea" in D.C.[29] Thus, at the point of initial coalition building, Zerolnick and Castellanos understood that the plan, though still incomplete, would adopt the "essence" of what Change to Win had developed, in conversation with LAANE and key environmental groups, and that it involved the "port creating a direct contractual relationship with trucking companies."[30]

Weiner and Canham-Clyne advanced the concession model against the backdrop of careful legal analysis, which had been conducted by the Teamsters' Mike Manley and Andrew Kahn of the Teamsters' outside law firm Davis, Cowell & Bowe in San Francisco. The Teamsters retained Kahn because they needed California counsel and because Kahn and Richard McCracken, another partner at Davis, Cowell & Bowe, had been involved in the early conversations about port organizing—and were among the nation's leading labor lawyers on strategic campaign work. The legal question to Manley and Kahn was: "politically if we could pull this off, would it withstand challenge?"[31] Their analysis looked at the possibility of a lawsuit based on federal preemption and also researched potential actions by the Federal Maritime Commission under the Shipping Act. With respect to the commission, the lawyers concluded that the employee provision was not anticompetitive and met the Shipping Act's reasonableness test.[32] On preemption, their conclusion was that "we should be okay. A port would have authority, as a market participant and as a matter of its proprietary rights, to restrict who could come onto its property."[33] The lawyers were sure that the American Trucking Associations (ATA) would sue the ports if the Clean Truck Program passed, but they believed that the ports would ultimately prevail. With Manley's analysis of the program as a valid exercise of port authority, the campaign was given legal clearance. As Weiner recalled, "the attorneys thought we had a pretty good case in the Ninth Circuit" and the "likelihood was remote" that the Supreme Court would ultimately take the case.[34]

Partnerships

The Campaign for Clean Trucks' critical first steps involved bringing together a diverse range of partner groups with the expertise to shape policy and the power to move political decision makers. Key among these groups were labor, environmental and environmental justice organizations, public health advocates, and faith-based groups. For LAANE, the initial goal was to convince partner organizations that addressing environmental and community impacts meant transforming the port trucking industry in a way that achieved employee status for drivers.[35]

The campaign was built upon the political power of organized labor and thus solidifying local union alliances was a crucial starting point. Getting buy-in from the "blue" side of the blue-green coalition was important given historical tension between unions and environmentalists, particularly around the port where unions like the ILWU and Building and Trades Council viewed environmental roadblocks to port expansion as inconsistent with their members' economic interests. As the campaign got underway, LAANE met with union leaders from ILWU Local 13 and Teamsters Local 848, both of which had been active on port trucking issues.[36] Dave Arian from ILWU Local 13 and Miguel Lopez from Teamsters Local 848 were key leaders, who would come to play important roles in the CAAP implementation process. Lopez, as the Teamsters port division representative, was deeply involved in efforts to organize port truckers. In 2004, he led a campaign to petition the Port of Los Angeles to make shippers and terminal operators pay a fuel surcharge to compensate drivers for increased diesel costs.[37] The following year, he and ILWU Local 13 President Mark Mendoza organized a protest against the new Los Angeles and Long Beach PierPass system, which assessed a cargo fee during peak hours to permit ports to stay open four nights a week and Saturdays—forcing truckers to work extended shifts without more compensation.[38]

Yet despite this collaboration, there were tensions between the ILWU and Teamsters from the outset, reflecting long-standing interests. The Teamsters had nothing to lose in the campaign and everything to gain. With no port drivers under union contract, the Teamsters saw fixing the independent-contractor problem as a solution to one of the union's most intractable organizing dilemmas. For the ILWU, in contrast, the campaign posed serious risks to its already strong position at the port since any reduction of port activity meant a potential threat to its membership. In addition, ILWU leaders were concerned about the possibility of another union being able to shut down the port through strike activity and thus disrupt their members' employment; partly for this reason, the ILWU (though generally supportive of green initiatives) was not strongly supportive of employee status for port truckers. In line with these divergent positions, the Teamsters locals (848 in Long Beach and 63 in East Los Angeles) signed on to the campaign—with Miguel Lopez eventually joining the campaign's steering committee—while the ILWU declined.

To gain traction with the ports, the coalition had to send a "strong message ... that you can't expand unless you are going to clean up your pollution."[39] The environmentalists brought the "legal muscle" to make good on this threat and thus were crucial allies in the overall plan.[40] Castellanos was the point person for outreach and took the first steps toward building and deepening relations with environmental partners. Some of this groundwork had already been laid by LAANE's participation in an earlier campaign to negotiate a community benefits agreement with LAX, in which LAANE worked with environmental advocates—particularly Jerilyn López Mendoza of the Environmental Defense Fund—in crafting a half-billion-dollar community benefits package that supported noise mitigation, school upgrades, and job programs for communities adjacent to the airport. As a result of that campaign, Castellanos recalled that "there was some foundation for our relationship with our environmental partners already established ... [that we were able to] then use as a building block and go deeper."[41]

Doing so meant linking into preexisting port advocacy networks and capitalizing on areas of interest convergence. NRDC, which played a crucial role shaping port development since the China Shipping case, was an essential partner—already sharing some common political and legal ground with organized labor. Earlier blue-green collaborations built trust: NRDC was involved in the LAX community benefits campaign and had worked with the Teamsters on previous litigation to ban Mexican trucks from entering the United States.[42] There were also overlapping legal interests at stake. As the Campaign for Clean Trucks was taking shape, NRDC was simultaneously advancing a theory of market participation as an exception to federal preemption that supported labor's vision for the port concession model. In *Engine Manufacturers Association v. SCAQMD*, NRDC argued that the SCAQMD should be permitted to develop its own emission rules governing commercial fleet vehicles despite Clean Air Act preemption—"seriously pushing the courts" to recognize "local jurisdiction through the market participant exception."[43] In 2005, the district court in *Engine Manufacturers* recognized the exception under the Clean Air Act and that decision was affirmed by the Ninth Circuit two years later[44]—at the height of the Campaign for Clean Trucks.

It was against this backdrop that Castellanos initially reached out to Adrian Martinez, a staff attorney at NRDC, who had a deep background in environmental justice issues.[45] Martinez studied environmental science in college and received a full-tuition public interest law scholarship to attend the University of Colorado Law School, where he went to pursue environmental law.[46] A second-year internship at NRDC turned into a postgraduate fellowship; when Gail Feuer left to become a superior court judge, Martinez took over her position in NRDC's clean air unit. Soon thereafter, he switched over to environmental justice, which was his passion.[47] With experience in port trucking gained from his participation on Los Angeles

Mayor James Hahn's No Net Increase Taskforce, Martinez became the primary NRDC staff member on the coalition, charged with thinking about "how legally they could create a more accountable system."[48]

Martinez was joined by David Pettit, a former legal aid lawyer who came to NRDC in 2007 after a stint as a partner in a boutique litigation firm in Los Angeles. Pettit "came into [the job] thinking, in environmental justice terms, that an alliance of labor and environment, should it happen, would be extremely powerful."[49] Pettit's first meeting as an NRDC attorney was about CAAP. From there, he was "able to figure out fairly quickly that the interests all pointed in the same direction," which meant "shifting the costs and the economic burden of cleaning the trucks from the drivers to ... the trucking companies."[50]

Melissa Lin Perrella was another NRDC lawyer involved in the ports campaign. An ethnic studies and social welfare major in college interested in the intersection of "public health, civil rights, and low-income issues," Perrella had gone to Georgetown Law Center with a desire to pursue a public interest career, initially taking a job as an associate with a big law firm, Orrick, Herrington & Sutcliffe.[51] She was there for five years before applying to work on environmental justice issues at NRDC, where she started in 2004.[52]

For the NRDC team, joining the coalition was a chance to build "effective power" to protect the community from harmful pollution.[53] The alliance with organized labor helped NRDC better understand how "the economics of the port drayage system ... impact the environmental conditions."[54] Although NRDC lawyers felt "strongly that the economics of the system need to be changed" they "didn't take a position on whether or not drivers should be unionized."[55] Martinez became a member of the campaign steering committee, where his role was to put the legal issues "on the table" so that coalition members could understand the "legal constraints" before evaluating the policy issues.[56] In participating in the coalition, NRDC lawyers represented NRDC's own members, not the coalition, although Martinez would address legal issues that would "pop up."[57] In developing policy, NRDC lawyers would analyze issues from two perspectives: "[T]rying to do what's best for the environment [and] broader coalition, but [also] mindful of: if this ends up in the courtroom, how is this policy going to play out before a judge?"[58] Generally, other coalition groups did not have separate legal counsel and would rely on NRDC to help them understand the legal stakes.[59]

To expand the coalition, LAANE also built relations with other environmental and environmental justice groups that had begun moving toward similar strategies to reduce port emissions. The idea of using port concessions to reshape trucking was also percolating up from below. Convergence between labor and environmentalists occurred through the portal of CAAP, which provided the "perfect opening" for the concession plan.[60] Thus, in Weiner's terms, the creation of the Clean Truck Program occurred as strains of activism that had been running in parallel began to intersect.

> [O]n the ground in the environmental movement and ... separately with the Teamsters there was this clean truck concept. ... Everyone was kind of spinning around. We came up with a policy proposal that would unite the workers and the enviros. But there were folks on the ground who conceptually or intuitively were going there anyway ... people had been close to that idea, but hadn't quite nailed it. ... [T]here was a lot of work to do and a lot of meetings ... for people to sort of get it. What's the concept? How do we put meat on the bones? How do we get it implemented? That all had to be sort of worked through.[61]

To do that, Weiner and Canham-Clyne "had a bunch of meetings with people and got to know them, and build trust with them, and got them connected with the drivers and the organizers."[62] In connecting with environmental and community groups, Change to Win leaders sought to "deepen the community's understanding of the economics by bringing the drivers into the conversation."[63]

Connections to environmental partners were built through different networks and sought to be attentive to the tensions between mainstream environmentalism and the environmental justice movement. Environmental activism around the port itself had multiple sources. Tom Politeo, a computer programmer and software developer who was born and raised in San Pedro, was involved in early environmental activism in the harbor area.[64] Like Jesse Marquez, founder of the Coalition for a Safe Environment in Wilmington, Politeo ran high school track and became sensitive to the impact of air quality on his athletic activity; also like Marquez, he was moved to activism after two explosions in the 1970s revealed the dangers of chemical and oil storage around the ports.[65] In the face of projected port growth, Politeo and other San Pedro residents, including homeowner activist Noel Park, began regularly attending harbor commission meetings in the 1980s. After the seminal MATES II study was released in 2000, showing elevated cancer risk around the ports, residents discussed strategies to reduce air pollution.

Through their own analysis, the San Pedro activists also arrived at a concession model as a way to force trucking companies to have "consideration for the community where they are working."[66] In the early 2000s,

Reforming the Port

Park presented the concession model to the harbor commission based on what the city, led by City Council Member Cindy Miscikowski, had done at LAX to force concessionaires to meet codes of conduct.[67] Politeo, a Sierra Club member along with Park, argued that truckers "should be paid by the clock and not by the can."[68] Thus, the concept of a concession model to address port trucking pollution was born of "multiple inventors."[69] As Politeo recalled:

> [T]hese trucks were starting to queue up in fairly long lines. The trucks would sit there in idle. All the time they're idling, they're inching forward, and they're polluting. And they're noisy. The truckers can choose to come to the port anytime they want. But, if they want to move containers, they have to come when the containers are available to be moved. They end up lining up in these long lines, and sitting around for hours sometimes, three hours, four hours, before they get a can to move. They move the cans and they may end up moving the cans during rush hour. We're looking at this, thinking in terms of the way their sources are being managed. The trucking companies and the shippers who control the terminals don't see any of the costs associated with the truckers waiting in long lines. They don't pay for the extra fuel because the truckers pay for that. They don't pay them for sitting around for three hours because it's the truckers' time. We looked at this, and said, "This is an environment in which the people who have the decision-making power don't feel the effect of whether the decisions are smart or not."[70]

In 2001, Politeo, Park, and a handful of other members of the Los Angeles-Orange County chapter of the Sierra Club formed the Harbor Vision Task Force as a formal standing committee within the Sierra Club focused on the environmental impact of goods movement and how to grow the port "green."[71] The task force held its first meetings at the Long Beach Yacht Club (where one member happened to dock his yacht) and then moved to the San Pedro Public Library. The group was small but active, with a decidedly prolabor bent. There were "a couple of longshoremen" and two former Teamsters: Sharon Cotrell from Long Beach and Dr. John Miller, who had put himself through college in Tennessee by working on a truck loading dock.[72] In 2002, Cotrell arranged a meeting with Gary Smith, head of the Teamsters local who was working on Long Beach port issues; the groups collaborated to help gain passage of State Assemblyman Alan Lowenthal's anti-idling bill, which had little effect, but cemented a working partnership. The Sierra Club did not get involved in the China Shipping suit, because it "didn't have the resources to make that happen," and as a matter of triage decided "NRDC is doing that."[73] Organizationally, the Sierra Club did not support Mayor Hahn's no-net-increase initiative, which Politeo believed was insufficient, although Park was active on that task force.[74]

After Villaraigosa's election, his administration brought together stakeholders under the auspices of Green LA (funded by the Liberty Hill Foundation),[75] which formed a Port Working Group with Politeo, Andrea Hricko from USC's Keck School of Medicine, Candice Kim from the Coalition for Clean Air, and other environmental representatives.[76] Politeo suggested reaching out to labor, a move that resulted in a series of "brainstorming" meetings in Wilmington attended by Miguel Lopez from the Teamsters local, and representatives from the ILWU and International Brotherhood of Electrical Workers.[77]

Environmental justice and public health advocates became networked through these processes. Angelo Logan of East Yard Communities for Environmental Justice was a member of Green LA's Port Working Group, as was Jesse Marquez,[78] who was also part of the initial Harbor Vision Task Force convened by the Sierra Club in 2001.[79] In 2003, Marquez formed the Impact Project, along with Hricko, and produced a series of policy briefs on trade and transportation.[80] During this period, Marquez also began to meet with Latino truckers who, in his view, "because of the history of Teamster racism wanted nothing to do with the unions," but were worried about how port security projects implemented in the wake of 9/11 would affect their work.[81]

Colleen Callahan, manager of air quality policy for the American Lung Association of California, also became involved in the Port Working Group.[82] Callahan was an urban and environmental policy major at Occidental College, where she studied under prominent progressive faculty Peter Dreier and Robert Gottlieb.[83] After a stint at the Center for Food and Justice, in 2006 she joined the American Lung Association, where her charge was getting it "more involve[d] in the environmental health advocacy work locally."[84] As a member of Green LA's Port Working Group, Callahan linked up with other environmental activists and then with LAANE staff.[85] Elina Green, who was project manager at the Long Beach Alliance for Children with Asthma (LBACA), recalled meeting Teamsters leader Miguel Lopez and LAANE's Patricia Castellanos through advocacy on environmental mitigation in relation to a proposed intermodal rail yard for BNSF in West Long Beach called the Southern California International Gateway.[86] Community groups, including LBACA, contested the EIR in that project beginning in 2006, and through that process forged crucial alliances with organized labor. From Green's point of view, the rail yard fight "was actually how the Teamsters sort of started to see the community side

Reforming the Port

of things and they recognized that, well, if they supported us in our ask for that rail yard, then there would be potential for support in their campaign and we started to see the issues from each other's side."[87] For Green, the power of the coalition derived from this assemblage of "crazy-strange bedfellows."[88]

The connection between environmental and community groups, LAANE, and Change to Win occurred through these formal networks and outside of them. Politeo of the Sierra Club recalled being contacted by Weiner in 2006 asking for support in developing a concession plan. "I remember my thought was 'Holy shit! They want to do our work for us.' I'm delighted. I sent a slightly less effusive message back, saying that 'Yes, we're interested in these things and even more.'"[89] Politeo began meeting with Change to Win and LAANE staff. The opportunity, as he saw it, was to leverage the staff and political power that was lacking before. "So, here we've got Change to Win, the Teamsters, and LAANE, all interested in this. Okay, I'm not going to skip on this."[90] Politeo recalled that his meetings with LAANE, Change to Win, and the Teamsters flowed seamlessly out of the Port Working Group. "[I]t's almost as if Nick Weiner walked into the room and said at one of our other meetings, 'I'm taking over. It's my show now.' Over some short period of time, those who acceded to that remained, and the rest left."[91] In short order, ILWU "sort of disappeared."[92] And other groups began to join, including Clergy and Laity United for Economic Justice (CLUE), an interdenominational group closely aligned with LAANE, which organized clergy in Long Beach, making arguments for port reform based on principles of faith and morality.[93] In addition, the coalition added immigrant rights groups, the Coalition for Humane Immigrant Rights of Los Angeles and Hermandad Mexicana,[94] as well as the San Pedro-based Harbor-Watts Economic Development Corporation, a community-based group created in 1997 that focused on neighborhood capacity building and economic revitalization.

In assembling this coalition, LAANE staff did the bulk of the outreach work. Because of her prior environmental justice organizing and South Bay campaign work, Castellanos was particularly sensitive to being inclusive: "I … did not want to be caught in the scenario where we were just working with the NRDCs and [Coalition for Clean Airs] of the world and not giving equal footing to [for example] the East Yard Communities for Environmental Justice."[95] During July and August 2006, Castellanos and Zerolnick conducted a first round of meetings with a number of groups, including East Yard Communities, the Coalition for a Safe Environment, and LBACA, in which they asked the groups to "download" what they knew about trucks and provide input on the potential campaign.[96] "[W]e didn't come into this campaign thinking there is nothing happening out there. ... And so it was an opportunity for us to learn."[97] LAANE had already been in contact with some of the groups in connection with the CAAP process; others they met with for the first time.[98] It was during the second round of meetings that LAANE staff sought to enlist groups to formally join the campaign. During these meetings, LAANE focused on presenting the main conceptual analysis, emphasizing that "the employment status of the drivers had to be addressed" and the ports had to have a direct relationship with the trucking companies in order to create "accountability in the system."[99] According to Zerolnick, the frame was less "Are you with us?" and more "Here's our analysis. Does this make sense?"[100]

The general approach to coalition building was to emphasize the opportunity to create a "potential solution" that would be in the "mutual interest" of labor, environmental, and community groups—creating a platform for long-term benefits and progressive policy change.[101] At outreach meetings, some groups wanted to discuss policy details, while others focused on the working relationship with organized labor.[102] There was "some trepidation" among the environmental justice groups about working with a "humongous labor union."[103] Castellanos shared those concerns and promised to "figure it out together."[104] Although the meetings produced active engagement, Castellanos did not "remember much resistance."[105]

Organizations went through different processes to consider whether to join the coalition. LAANE's Castellanos and Zerolnick reached out to the American Lung Association's Callahan to ask if the association would join the emerging coalition.[106] Callahan recalled having to raise the issue up to "some pretty high channels" within the national organization to get approval to join since there "were some concerns about whether it was necessary to support the concessionary model or whether just pushing for the most current EPA [truck] standards ... was sufficient."[107] LBACA, itself a coalition of local residents and health organizations, had to get approval from the entire membership.[108] East Yard Communities' Logan was excited about the partnership but wanted details about how it was going to work. He recalled being a member of the CAAP stakeholder group when he was contacted by LAANE after "we had been trying to reach out to labor without success."[109] Although enthusiastic about the partnership, "our group's questions were: How's this all going to work out? What are the power dynamics? What is the decision-making structure? ... [W]e wanted ... a governance structure that was really democratic."[110]

The mission statement for what would become the Coalition for Clean and Safe Ports sought to meet this democratic demand, while emphasizing the main goals of the campaign:

> Our objective is to improve the condition of the trucking industry and of truck drivers operating at the San Pedro Bay Ports and along associated goods movement corridors. We are guided by the need to reduce associated health impacts on workers and local communities by resolving shortcomings associated with current port trucking practices. In doing so, we will address port trucking's many challenges that face industry, community, government, labor and the environment.
>
> To accomplish our objective, we will foster an appropriate role for trucking as part of goods movement planning and solutions. We will ensure trucks run cleanly, quietly, safely and efficiently with a stable, employee workforce that pays livable wages and offers drivers all the rights and benefits of an employee. We will make sure improvements adopted in the San Pedro Bay area help create systemic solutions that improve conditions overall and don't simply transfer problems to other areas, such as adjacent communities, our inland ports or other stops along the goods movement chain.
>
> We will act on a timely basis as part of a democratic, broad-based coalition to promote public awareness of trucking problems and solutions and we will seek to influence policy makers to put decisive solutions into effect as rapidly as possible.[111]

Policy

The intense period of initial organizing saw the first instance of coordination between members of the fledging coalition: the filing of written comments on the first public draft of CAAP. Released in July 2006, CAAP required its own EIR and thus both NRDC and LAANE filed comments during the period for public review.[112] Although the CAAP draft identified clean trucks as an issue, it did not make the connection to employment status, providing the coalition with an opening. Zerolnick recalled thinking that the CAAP provisions on trucking read as if the drafters were saying, "'We're not really sure how to do it. We'll come back to this.' So [the campaign] submitted public comment and said, 'Well, actually we have some ideas for how to do this. ... [A]nd the basic structural problems are independent-contractor status and the lack of a relationship between the port and this sector of the industry.'"[113]

Zerolnick drafted a comment letter and circulated it to all partners, who made editorial suggestions.[114] He also worked closely with Manley and lawyers at NRDC, particularly Adrian Martinez, as he fine-tuned the proposal.[115] The input was focused on sharpening the links between industry accountability, employee status, and emission reduction. In Martinez's terms, the focus was on remedying "the Wild, Wild West situation where there really weren't effective standards and there was no accountability. ... [Workers] weren't getting paid much, they were on the hook for all the insurance and the costs of the equipment, so it was this natural marriage that if you're going to fix the problem, you need to fix the systemic problem which is the lack of accountability from these trucking companies."[116]

The final letter seamlessly integrated these arguments, referencing the research that Change to Win had done as a basis to propose a Clean Truck Program built on the concession approach.[117] The letter, sent to the directors of both ports, was submitted on behalf of LAANE and its "coalition partners."[118] The comments were conceptual, focusing on the "real market forces operating on the Port truckers," as well as "the significant and persistent structural problems in the industry."[119] The bulk of the comments were devoted to detailing the economics of the drayage market and its dysfunctions, while explicating the concession model of transforming the industry. The letter emphasized the twofold problem of independent-contractor drivers and lack of port control over trucking companies.[120] It then proposed a "long-term solution" under which the ports would "jointly enter into a direct contractual relationship with responsible motor carriers to provide drayage services at both Ports, utilizing the same model employed by airports to provide food and other services to air travelers."[121] The comments contemplated a request for proposal process awarding port entry only to trucking companies that met "clear standards concerning capitalization requirements, revenues paid to the Ports, environmental standards for trucking equipment operating at the Ports, other environmental mitigation measures and benchmarks, employee status for drivers, employment preferences for the current workforce of owner-operators, and labor peace requirements to ensure that revenue streams to the Ports are uninterrupted."[122]

Under this plan, the letter emphasized that the benefits would be clean trucks maintained over the long term, achieving emission reductions while also promoting security and greater accountability.[123] The letter was short on specific policy proposals, but long on analysis and prescription, powerfully laying out the essence of what would become the Clean Truck Program. Although the details were still unclear, the key move was linking clean trucks to employee status through a direct contract between the ports and the trucking companies.[124] CAAP thus provided the critical opportunity to unite disparate labor, environmental, and community interests around a coherent policy program to attack diesel truck emissions.

The last step was to officially convene the Coalition for Clean and Safe Ports and formally launch the Campaign for Clean Trucks. The launch was timed to happen right before the joint ports' CAAP review meeting on November 20, 2006; in order to maximize publicity, the coalition staged a major press conference.[125] The coalition's first order of business was to mobilize for the November 20 meeting, which it did by organizing a "massive community driver turnout," which helped shape the electric environment leading to CAAP approval.[126]

Although the coalition grew over the two-year fight for clean trucks,[127] its initial composition reflected wide support that underscored the success of LAANE's outreach.[128] In the end, the Coalition for Clean and Safe Ports was broad and deep. As Martinez recalled: "We had community, we had faith-based groups, we had the environmental justice community, we had the environmental community, we had economic development groups. ... We had lawyers, we had scientists involved, we had economic experts, we had people on the ground."[129]

In keeping with its commitment to inclusivity and democracy, while also acknowledging the need for clear and efficient decision making, the coalition structured a tripartite governance system. Policy decisions were ultimately to be decided by a supermajority vote of the coalition members.[130] To facilitate operations, members agreed to create a steering committee composed of a smaller group of representatives from key organizational partners: three labor, two environmental, two community, two immigration, and one to two research/academic partners.[131] This committee—which "played to the coalition's strengths" by giving voice to the diverse groups involved[132]—was charged with agenda setting, providing strategic recommendations, and making day-to-day and urgent decisions.[133] The steering committee was created in recognition of the fact that the groups were part of a "live campaign" that required some quick decisions, but also was designed to vet policy and strategy ideas in order to make recommendations for full coalition approval.[134] As necessary, the coalition also agreed to set up working subcommittees to deal with various policy issues and give recommendations to the full coalition. These subcommittees were established to develop coalition policy with respect to specific community, environmental, and labor issues. LAANE staffed the subcommittees, but LAANE employees did not formally sit on them. Thus structured, the coalition was ready to take action.

The Affirmative Phase: Mobilizing Local Law

With the coalition in place, LAANE's effort shifted to rolling out the campaign to pass what would become the Clean Truck Program. The basic approach was twofold. First, the coalition would meet during an intense period to hammer out the details of the program—converting the model taken from Change to Win into a workable policy. During this time, the coalition would engage decision makers and stakeholders to build support for the program. Second, these elements—a clear policy draft and outside pressure—would be used to move the policy through internal city and port channels.

The Outside Game: Developing the Program, Exerting Pressure
At the outset of the campaign, both the Los Angeles and Long Beach ports were still aligned in the process, reflecting their ongoing commitment to implementing CAAP. In early 2007, the ports established a stakeholder group comprised of representatives from the ports, air agencies (CARB and SCAQMD), industry, environmental and labor groups, and academia.[135] Several coalition members participated, including Angelo Logan from East Yard Communities, Melissa Lin Perrella from NRDC, Jesse Marquez from the Coalition for a Safe Environment, Elina Green from LBACA, Candice Kim from the Coalition for Clean Air, Miguel Lopez from the Teamsters, and Patricia Castellanos from LAANE.[136] The stakeholder group was created to provide input into the ports' larger process of CAAP implementation, which included the development of a detailed Clean Truck Program.[137]

To inform that process—and ultimately shape what the final program would look like—the coalition moved quickly to build out the policy. Following on the heels of CAAP approval, which established the general framework for port truck regulation, "things really kicked into high gear."[138] In late 2006, the coalition set to work on filling in program details in order to shape the final rules. At the outset, the coalition had its basic "yardstick": any Clean Truck Program had to be "accountable, sustainable, and comprehensive," which meant that it would rest upon fleet *and* employee conversion—thus avoiding a short-term solution converting the fleet to clean trucks through a one-time public subsidy that left the trucking companies without responsibility for long-term maintenance.[139] The major question for the coalition members was: "What are the standards going to be?"[140]

To answer this question, the coalition engaged in external and internal discussions. Externally, LAANE and Change to Win organizers met with port staff and key elected officials to present the general framework provided by Change to Win. From there, Zerolnick—working closely with campaign lawyers, the Teamsters' Manley and NRDC's Martinez—began to draft the policy. This was an iterative process that connected to the coalition's internal discussions. Within the coalition, members broke into subcommittees charged with developing standards around labor, environmental, and community issues.¹⁴¹ To advance this process, the coalition initiated monthly standing meetings of the full membership, with the individual subcommittees engaged in intensive policy discussions that continued during the interim periods.¹⁴² Community partners responded to specific requests for evaluating provisions and came up with some of their own. For example, residents working with coalition member East Yard Communities proposed to make trucking companies park trucks off neighborhood streets and adhere to specified truck routes that would minimize community disruption.¹⁴³ Once vetted at the subcommittee level, provisions were passed onto the steering committee for incorporation into the working draft and then presented to the entire coalition for general approval. Although full coalition approval was technically by supermajority vote, Zerolnick recalled that decisions were all made by consensus.¹⁴⁴ As the draft details evolved, LAANE and Change to Win organizers would meet again with city and port officials, getting their feedback and buy-in.¹⁴⁵

What emerged from this process was a document that the coalition called a request for proposal (RFP) designed as a vehicle for implementing the concession model. The RFP was essentially a scoring system to rate potential concessionaires.¹⁴⁶ Scores were based on responses to application questions designed to ensure that trucking companies met criteria necessary to effectuate the Clean Truck Program.¹⁴⁷ The RFP model was chosen because the coalition assumed that for ease of administration the ports would limit entry to a handful of trucking concessionaires and the RFP provided a standard system to allow the ports to rank applicants.¹⁴⁸ The RFP document was primarily drafted by Zerolnick, shaped by extensive discussions among coalition members, and contained items the coalition viewed as a "bottom line"—phasing out old trucks and employee conversion—and others that were on a "wish list."¹⁴⁹

The RFP's main purpose was to ensure "that the most responsible entities operate at the Port."¹⁵⁰ Toward that end, the RFP designated responsible business, security, environmental, labor, community, and efficiency standards, though the overall plan hinged on converting old dirty trucks to new clean ones, while also converting the drivers to employees. The standards were to be implemented through the ports' contract power: "[s]uccessful applicants will enter into a contract with the Port mandating a turnover of the entire truck fleet over five years."¹⁵¹ Applicants were also required to "use only employee drivers (as opposed to independent contractors) to provide drayage services."¹⁵² The RFP was structured so as to assign a baseline qualification to applicants meeting minimum criteria, while then giving extra points to applicants that could demonstrate good business practices and community relations—which were the "wish list" items.[153] The minimum standards were framed to advance core elements of the Clean Truck Program. Applicants were asked, "Does the Applicant utilize only employee drivers to perform drayage services?" and were informed that they "must comply with the requirements of the Clear Bay Clean Air Action Plan (CAAP) regarding the reduction of pollution from diesel trucks."[154] Applicants also had to "provide an assurance of labor peace"[155]—an agreement that they would not disrupt unionization efforts in exchange for a commitment on the part of employees not to strike. The time frame for employee conversion was not specified, though applicants were told that they had five years to convert their entire fleets to EPA 2007 standards (by purchasing new trucks or through retrofit) with a minimum of one-fifth of the fleet converted each year.[156] Applicants were also asked "to make arrangements to provide off-street parking" for out-of-service trucks and to "work with the Port ... to develop a plan to minimize the impact of HDVs [heavy duty vehicles] on port-adjacent communities."[157] Concession fees were to be set at an initial level of $5,000 per truck in addition to a 10 percent monthly revenue fee.[158] In April 2007, the RFP was submitted to both harbor commissions, which responded by stating they would take it "under advisement."[159] Although it was not meant to be public, the RFP was leaked to the press.

On April 12, 2007, the ports jointly issued their own proposed Clean Truck Program, which gave the coalition most of what it wanted—adopting the concession model as its cornerstone—though in a very different format.[160] In what NRDC's Perrella called "a huge, huge step forward in our quest for clean air,"[161] the ports agreed to use their "tariff authority"—their power to pass port rules, called "tariffs"—to "only allow concessionaires operating 'clean' trucks to enter port terminals without having to pay a new Truck Impact Fee at the gate."[162] For the purposes of the program, a clean truck had to meet the so-called "CAAP standard," which meant EPA 2007-compliant new trucks, retrofitted trucks for those model year 1994 and newer, and trucks replaced through the Gateway Cities program created under the China Shipping settlement.[163] Older trucks would be progressively banned (with a 2012 target date), though could continue to enter if their companies—referred to as Licensed Motor Carriers, or LMCs—paid a Truck Impact Fee of $34 to $54 per container.[164] Proceeds from that fee and a $26 cargo fee, along with other sources of public funding, would be used to subsidize truck replacement and retrofit. Concessionaires would also have to commit to "require employee drivers (after a transition period),"[165] with the goal of achieving full employee conversion by January 1, 2012.[166] Following the coalition model, the ports proposed to confer concessions after an RFP process in which "applicants w[ould] be evaluated for financial strength and asset control."[167] The ports' proposed plan did not go as far as the coalition's in limiting entry to those companies that best met business practice standards. Nonetheless, from the coalition's perspective, "it really did contain most of what we wanted."[168] Industry representatives viewed it through the opposite lens and immediately asserted the threat of litigation. As Curtis Whalen, executive director of the Intermodal Carriers Conference of the ATA, put it: "We are looking at it now from our lawyers' point of view to see what we might do. I think we might challenge that. ... By definition, these containers represent interstate commerce. It would impact interstate commerce in a dramatic way. Can a port authority do that?"[169]

The coalition believed that the answer was "yes" and seized the opportunity to push forward. In May 2007, responding to the ports' proposal, Zerolnick (again with input from lawyers Manley and Martinez) drafted another comment letter—this time submitted under the formal auspices of the Coalition for Clean and Safe Ports.[170] Unlike the first letter, which was conceptual, this one was "more concrete," addressing specific policy details.[171] The letter, while commending the ports for their "leadership" and "hard work," sought to offer areas for the plan's improvement.[172] Although it addressed a variety of technical details, it emphasized the employee provision, which was not explicated in detail in the ports' draft. Specifically, it argued that—unlike the conversion to clean trucks—there should be no transition period for the conversion to employee drivers.[173] To do otherwise, the letter suggested, would create potential unfairness for companies that complied earlier and would impose insurmountable administrative problems.[174]

It was the spring of 2007 and negotiations over the terms of the Clean Truck Program had begun in earnest. As the negotiations developed, they would focus on three crucial elements of the program: (1) the nature and timing of the ban on dirty trucks and the related phase-in of clean trucks; (2) the amount and structure of fees imposed on truck cargo, and the related amount of financial incentives allocated to fund clean truck conversion; and (3) the structure and content of the concession agreement, with particular emphasis on the extent and timing of employee conversion.

To advance the coalition's positions on these issues, members sought to "debate it out in public," organizing around a series of harbor commission meetings to demonstrate that the coalition was "a force to be reckoned with."[175] The Los Angeles and Long Beach harbor commissions held regular public meetings to discuss policy development, at which coalition members, community residents, and truck drivers turned out to press the argument that "we need to fix the trucking system."[176] As NRDC's David Pettit described with wry humor, the coalition would turn these normally staid events into dramatic affairs by bringing hundreds of people "with torches and pitchforks."[177] There were also special meetings devoted specifically to the Clean Truck Program, which were in Zerolnick's memory "even longer and even more contentious."[178] In one, held in June 2007, 300 drivers turned out to support the program. Edgar Sanchez, a driver from Long Beach, pointed to coalition support as motivating him to speak out: "Before we didn't have the courage or the confidence to tell people how we feel out of fear we'd be fired or labeled as troublemakers. ... Not anymore. We see the smoke pouring out of our trucks and we breathe it all day, every day. ... But we also work long hours at minimum rates. We can be fired at any moment, like slaves without a voice.[179]

A few months later, on October 12, the ports held a Joint Public Workshop on the Clean Truck Program—a six-hour meeting at which the ports took "tons of testimony" from various stakeholders,[180] including LAANE's Castellanos and NRDC's Perrella, as well as numerous truckers and community residents.[181] As the Joint Public Workshop underscored, a primary function of the coalition was to turn out members at these meetings to testify in favor of the proposed program. These meetings were also often a focal point for circulating and responding to draft policies. Drafts would emanate from the ports and Zerolnick would work primarily with NRDC lawyers to craft a response; that draft would be circulated among coalition members for comments and then once finalized sent back to the ports for review. Meetings were opportunities for exchange and amplification. During this back-and-forth discussion, coalition members would shape program language and clarify objectives. For instance, a LBACA community resident working with the coalition developed the idea to put placards on trucks indicating a number to call to report any emission and safety issues[182]—an idea that was eventually incorporated into the working plan.

During this period, coalition pressure was applied in open spaces and behind closed doors. The coalition staged a number of public actions, including a caravan of 100 big rigs down the 110 freeway to the Port of Long Beach.[183] Coalition members also met privately with harbor commissioners, mayor's office staff, and city council staff in both cities—though the approach increasingly diverged between Los Angeles and Long Beach. In Los Angeles, the coalition had allies in key elected politicians and harbor commissioners and thus the outreach was designed to give them the materials and arguments necessary to hold the line against industry lobbying. The big push was convincing "people to understand that the employee provision was an environmental provision."[184] This was true at the commission level and in the mayor's office, where there were some divisions among the mayor's staff about whether the program should just focus on the green elements or should also include the blue focus on employee drivers. As a result, the coalition had to "fend off repeated attempts by ... forces within the mayor's office who wanted to jettison the labor components of the Clean Truck Program."[185] In Los Angeles, the coalition also had a powerful champion in City Council Member Janice Hahn, with whom members met regularly to work out strategy and policy details.[186] In Castellanos's view, Hahn "genuinely was supportive of workers and workers' issues. I think this was in her district and she cared about it."[187]

In Long Beach, the approach was different given the perceived skepticism of recently elected Mayor Bob Foster about the employee provision of the program. Foster, a Democrat who had headed Southern California Edison, won the Long Beach mayor's race in a run-off election in June 2006. He took office that next month, just as CAAP was moving toward approval and the battle for clean trucks was taking shape. Los Angeles Harbor Commissioner Jerilyn López Mendoza recalled having lunch with Foster early in his term to discuss the prospects for port coordination around CAAP. After the lunch, she called LAANE organizer William Smart to ask: "Have you guys talked to Bob Foster yet? ... I don't think he's on board with an employee mandate. ... I think you all have some work to do."[188] Coalition members were deployed to increase the pressure on Foster—since unilateral action by Los Angeles could undermine the entire project by diverting cargo to Long Beach. Colleen Callahan of the American Lung Association would "bring health professionals" to meetings with Long Beach harbor commissioners and Mayor Foster, to whom she laid out "why the policy proposal would address health."[189] Similarly, Elina Green of LBACA mobilized the group's community-based membership to share with Long Beach officials the challenges they experienced caring for children with asthma and how the Clean Truck Program would promote better public health.[190] As LBACA members worked to "pull any strings" they had with Long Beach officials, they also faced local reprisal: Green recalled one meeting with Mayor Foster and a small group of coalition members in which the mayor was "literally yelling at us the entire meeting."[191]

Coalition members played different roles in exerting outside pressure over the course of the two-year campaign. In private meetings and public hearings, LAANE and Change to Win made the case for industry restructuring, while NRDC emphasized the environmental benefits (and held out the implicit litigation threat). LAANE's Castellanos, Change to Win's Weiner and Canham-Clyne, and NRDC's Martinez, Perrella, and Pettit met regularly with port staff, both mayors' offices, and both city councils, though the emphasis was on the Long Beach City Council because of Janice Hahn's support in Los Angeles.[192] The goal of these meetings was to make the case for sustainability, while also demonstrating the power of the blue-green alliance. In this regard, Castellanos recalled the coalition's first meeting with the Los Angeles mayor's office and port staff: When NRDC showed up with the Teamsters, port director Geraldine Knatz was "a little confused" and there was a lot of "brow raising."[193] To complement these efforts, environmental justice organizers mobilized their base. Marquez and Logan turned out community members to attend commission meetings and meet with elected officials.[194] Other groups similarly engaged in turn out efforts, and everyone attended periodic public rallies.

Although all the groups played their roles, some also acknowledged that LAANE was in charge. While each coalition member spent considerable time and resources advancing the campaign, in the end, LAANE "had staff dedicated to this campaign" and was "really in the driver's seat."[195] Some members expressed concerns about being tokenized but generally praised LAANE's ability to "really listen" to coalition members and bring everyone on board.[196] With the coalition thus united, members worked to hold officials accountable as they attempted to move the program through internal political channels.

The Inside Game: Mobilizing Legal Expertise, Moving Policy

In Los Angeles, as the Campaign for Clean Trucks heated up in 2007, internal policy development proceeded along parallel, though deeply interconnected, paths. It started at the very top, with an effort by the Teamsters to obtain a commitment by the Los Angeles mayor and port officials to support some version of the Clean Truck Program. It then went through three phases of policy development. First, city lawyers—in conversation with campaign lawyers—conducted a legal analysis to evaluate and ultimately sign off on the policy, focusing primarily on the risk of preemption. Second, the mayor's office staff managed industry resistance by contracting for an outside economic analysis of the program's costs and benefits that set the framework for the final policy drive. Third, in that final drive, port staff took the lead in thinking through policy details and resolving conflicting industry and coalition views, producing the version of the Clean Truck Program that would ultimately be approved. During this final phase, the

Reforming the Port

Figure 5.1
Coalition actions at the Port of Los Angeles. Photograph (top left) from Patricia Castellanos and Doug Bloch, "A Smart California Port Policy for the Green-Growth Future, Spearheaded by Progressives—Part II," *California Progress Report*, May 28, 2008. Other photographs courtesy of Barbara Maynard, Teamsters communication director.

Long Beach harbor commission broke ranks with Los Angeles and pursued an independent policy.

Larry Frank, Mayor Antonio Villaraigosa's deputy mayor for neighborhood and community services, was the key internal point person. Frank was a long-time labor activist. He had come to California in 1976 to work with the United Farmworkers and from that point his resume read like a mini-history of efforts to bridge between labor and other progressive movements.[197] After stints with the Amalgamated Textile Workers Union and the Communication Workers of America, in 1982 he launched the Jobs with Peace campaign in Los Angeles, developing a plan to promote job growth and increased social services by reducing military spending—a campaign operated out of the Los Angeles County Fed. In 1984, the campaign succeeded in passing a local initiative to get more funding for jobs and services, and implemented an innovative strategy to mobilize occasional voters that laid the foundation for the rebirth of the Los Angeles labor movement in the early 1990s, when Miguel Contreras, as head of the County Fed, expanded the strategy to build electoral support for labor's agenda.[198] Around this time, Frank met an up-and-coming labor organizer, Antonio Villar (later Villaraigosa, after he married and merged names with his wife), who was co-chair of the Black-Brown Roundtable and had just been elected as the first Chicano leader of the predominantly African-American Local 3230 of the American Federation of Government Employees union. After deciding to get a law degree from UCLA, Frank ran a sentencing practice for a decade, and then was recruited to work for Kent Wong, as his number two at the UCLA Labor Center, where he worked from 2000 to 2005 and was the driving force behind opening the center's downtown office.[199]

When Villaraigosa was elected mayor of Los Angeles in 2005, Frank joined the transition team and was the only deputy mayor who served in the administration "wire to wire," ending his eight-year term as deputy chief of staff. Given Frank's labor background, it was natural that as deputy mayor he would immediately inherit the port portfolio and handle labor issues. As Frank recalled, outgoing Los Angeles Mayor James Hahn had the view that the city should first reduce port emissions and then seek to grow. Mayor Villaraigosa, in contrast, came in with a different "point of view": "At the same time we were attempting to grow the port, we were going to green the port ... by engaging with labor and environmentalists."[200] Frank's job was to advance this goal.

His first order of business was to convene a meeting of labor and environmental leaders at the port. Within the first couple of months of the mayor's administration, Frank organized a meeting at Banning's Landing in Wilmington, which included officials from the Los Angeles port, Jesse Marquez from the Coalition for a Safe Environment, a representative from the Environmental Defense Fund, and East Yard Communities' Angelo Logan, among others.[201] On the labor side were leaders from the Teamsters and the ILWU, including Dave Arian. Frank thought it was "a life changing meeting for a lot of people in the room."[202] The environmental justice advocates told "impassioned stories" about "kids with asthma" from the community.[203] When several of the labor leaders revealed that they or their children had asthma, "a shift took place at that meeting—and in the future conversations."[204] Growing out of that meeting, the ILWU became committed to the "green growth" platform advanced through CAAP, helping to develop a plan to reduce port emissions by 25 percent. The mayor viewed ILWU support as crucial because "they had not just the west coast ports of

Reforming the Port

the United States, but they also had Prince Rupert, Vancouver ... and had relationships with all the ports on the west coast of Mexico."[205] The ILWU could thus be a "helpful pressure point," serving as "ears and eyes on the ground."[206] Villaraigosa also recognized that the ILWU, if unhappy, could make the campaign more difficult: the union could "block headway [and] not just in Los Angeles," given that it controlled trucking inside the ports and its members were concerned that Teamsters' inroads could disrupt work flow.[207] Relying on ILWU support, the mayor's first big break was when the giant ocean carrier and terminal operator Maersk agreed to add a second tank for low sulfur fuel on its vessels—a change that the port then required of vessels at all other terminals[208] From that point, there were a "remarkable series of meetings that happened in the mayor's office."[209]

The first—and most important—occurred in November 2006,[210] when "all of a sudden we have the President of the Teamsters union coming in."[211] The mayor's meeting with James Hoffa was brokered by Maria Elena Durazo (head of the County Fed) and Jim Santangelo (the Teamsters International vice president), and attended by Larry Frank, John Canham-Clyne, and some other Teamsters officials. Frank's job was to produce the staff briefing for the mayor, which outlined what the Teamsters were seeking—support for the Clean Truck Program—and made recommendations. Frank's basic recommendation was: "This is a fascinating proposal. It is an extremely difficult piece of work. You need to listen."[212] Frank believed that the mayor's Chief of Staff Robin Kramer and Deputy Chief of Staff Jimmy Blackman would be uncertain about the proposal. Frank thus counseled Villaraigosa to proceed deliberately, advising him: "You should not make any commitments in this first meeting," given that the deal had the potential to "impact the mayor's relationships with the business community."[213] Yet after hearing from Hoffa and other labor leaders, the "mayor was in. First meeting."[214] For Frank, it was a reflection of the mayor's commitment: "He believed this was going to be good for L.A., for L.A. workers, and for the environment."[215] Having agreed to the concept, the goal of producing a solution to the port trucking problem was then tracked for policy development at the port, where the mayor's new appointments to the harbor commission and port directorship—made to advance CAAP—would play a key role in the approval of the Clean Truck Program.[216]

Inside the Los Angeles mayor's office, staff understood that a clean truck policy was a priority and worked to advance it. In Frank's terms: "The work was outrageously difficult ... the biggest lift I was involved in during my eight years." Frank was part of all the discussions going forward: "There was outside capacity that was going to get us to a cleaner port and there was inside capacity understanding how that would happen. ... I played the inside role."²¹⁷ On the inside, Frank was joined by additional staff, who knew about the campaign, and met directly with Weiner, who presented Change to Win's analysis of the drayage truck market and how the Clean Truck Program would affect it.²¹⁸ On the basis of this analysis and their own review, staff concluded that the drayage sector was "a perfectly competitive market with ... a strong negative externality."²¹⁹ As the outside pressure of the coalition—and industry opponents—scaled up, the mayor's staff faced multiple challenges. One was "to maintain the integrity of the internal policy-making process ... by keeping outside influence outside."²²⁰ Mayor staff member David Libatique, and later Sean Arian, provided a "buffer" against the coalition.²²¹ While the mayor's staff continued meeting with coalition and industry representatives throughout the process, they attempted to minimize the degree to which there was any perception of unfairness in the negotiation.

A key challenge was advancing the program in the face of increasingly intense industry opposition—and, partly as a consequence, some opposition within the mayor's office itself and at the port, where Knatz and Freeman were uncertain about the viability of the Teamsters' proposal and concerned about the political fallout.²²² The employee provision remained on the table—and remained controversial. To effectively engage the opposition—and to assess whether it was worth spending political capital to do so—Mayor Villaraigosa's staff and the Port of Los Angeles staff needed to be comfortable with the legal foundation for the program. In early 2007, the coalition's legal analysis was presented to the mayor's staff members, who wanted assurance that it had been done.²²³ Once it became apparent the program was really moving forward, port lawyers began "leading the charge" to make sure they had their "ducks in line" on the legal issues.²²⁴ As a result, there were several meetings attended by the port's general counsel, city attorney Thomas Russell; other city attorney lawyers assigned to the port, particularly Joy Crose; the Teamsters' Mike Manley; and lawyers from NRDC. These meetings focused on solidifying the legal argument for the concession approach. Manley circulated versions of the memos he had drafted for the Teamsters to the city attorneys, came out to meet with them, and responded to questions and concerns.²²⁵ In these discussions—also attended by LAANE and Change to Win organizers—Manley viewed his role as "trying to convince [port lawyers] not to recommend to the port commission to reject" the program.²²⁶ Manley's analysis when he talked to port counsel was "that the ATA is going to sue you," but "if you're sued you can win."²²⁷ He argued for the program as a "unified whole": "We had these allies and the thing about labor and environmental groups is we always accuse each other of ditching ... for our own interests. ... And so I was careful not to come across as if I were saying, 'Well, this 2007 truck stuff doesn't matter as long as they're employees.'"[228]

Labor and environmentalists converged around legal theory as well as political interests. NRDC was present at the meetings with port lawyers as the legal "hammer," but also to help make the case for local authority.[229] As Pettit remembered, NRDC and other environmental groups were independently "pushing using this market participant exception and at the same time, labor had been eyeing it as a potential approach to resolve several issues. And so it kind of came together where we were both saying" the same thing.[230] NRDC, like the labor lawyers, understood the legal risk of the concession plan and believed that there was "a unified view of how strong the arguments were."[231]

Ultimately, it was port counsel who had the last word on the legal analysis. Much of this work fell to city attorney Joy Crose, who was lead counsel to the Port of Los Angeles on the Clean Truck Program. Her role was to conduct a "legal review of the program" and prepare all "program implementing documents, including contracts, tariffs, ordinances and resolutions."[232] To do this, Crose worked with her counterpart in Long Beach, and also engaged outside counsel, Steven Rosenthal, chair of litigation in the Washington, D.C. office of Kaye Scholer. After interviewing a number of law firms toward the end of 2006, the city attorney's office hired Rosenthal and his team to advise the port. Rosenthal had deep expertise on "the commerce clause, federal preemption, and federal statutes relating to the regulation of commerce," gained in representing airports and ports over the course of his thirty-year career.[233] Together with the city attorneys, Rosenthal advised the port on the legal issues related to enacting the Clean Truck Program. Reflecting on his general approach to city policy, Rosenthal noted that when it came to reviewing "new, complex programs, you can identify risks" and suggest "this is why we think this approach is a better idea"— but always in a context in which the client understood that "there is no certainty."[234]

The policy makers and their staff were not seeking certainty, just credible assessment. Weiner felt that the campaign's legal groundwork helped to get the port attorneys to "buy into our analysis," which was basically: "yeah, there's a risk. But it's good policy. ... [W]e've got a good legal case."[235] NRDC's Adrian Martinez described the value of the legal analysis in similar terms. He believed that the legal analysis empowered the city and port to take a stronger position on the bottom-line policy details: If the ATA was going to sue on whatever policy passed, he argued, it freed the port to develop the most effective policy on its own terms and then to "go to court with the best program we have."[236]

Similarly, the initial legal analysis provided a ready response to the industry's legal pushback that occurred during policy formulation. Martinez recalled that industry groups had "a lot of legal power, so whenever the port or somebody would propose something, they'd give this very long, threatening legal letter that said you can't do this, you can't do this, you can't do this, you can't do this and here's the legal reasons why." But the coalition had "lawyers on the other side … firing back comment letters: oh, but look at this case, look at this case, and making these similar sophisticated arguments on why you can do it. And I think that was the big difference."[237] David Libatique believed that for the mayor's office staff tasked with advancing the program, the coalition's legal analysis was critical as a predicate to moving forward: "the legal analysis that was provided by the attorneys basically told us if … we're going to have an effect on port trade … we would have to act as a market participant and the way we would do that would be through a concession-based model."[238] The mayor's general counsel, Tom Saenz—who had joined Villaraigosa in 2005 after serving as director of litigation at the Mexican American Legal Defense and Education Fund—also reviewed the program and provided a legal opinion to his client.

The context of mayoral decision making was also shaped by politics. Mayor Villaraigosa's first major policy initiative—a controversial attempt to take over the Los Angeles Unified School District board through the enactment of a state law—was held unconstitutional by a superior court judge in late 2006, giving the mayor a stinging defeat.[239] The mayor needed a policy win and a strong pro-environment position at the port promised to deliver political dividends, while also solving a critical regional problem. While Villaraigosa supported employee conversion, he understood its legal and political vulnerabilities—and could not risk a signature policy going down in the courts twice in a row. In this context, Villaraigosa and his team discussed what would happen if a court struck down employee conversion while leaving clean truck conversion in place, although Frank made clear: "that was not what we were planning on."[240] Although the mayor's team knew the plan could fail and truckers could be left bearing the cost of clean truck conversion without the benefit of employee status, Frank remembered feeling a sense of cautious optimism: "We really were hopeful based on our analysis and what we were told."[241]

The urgency of solving the trucks problem was underscored as both ports faced community resistance to several massive expansion projects, delayed by CAAP, which were unveiled to the public in mid-2007. These included replacing the Gerald Desmond Bridge to permit entry of larger container vessels; expanding and upgrading facilities in several terminals, including TraPac, China Shipping, and APL; creating new rail and road access; and building a new terminal for crude oil.[242] The pressure once again was on the ports to accommodate growing container volume and local officials were eager to solidify the ports' position given their vital regional economic role—by one account, responsible for more than 250,000 jobs in Southern California and nearly $7 billion in state and local taxes.[243] In light of this, Los Angeles Harbor Commission President David Freeman vowed: "We're going to grow and we're going to clean up this place or my head will be served up on a silver platter in Los Angeles Mayor Antonio Villaraigosa's office."[244] Some coalition members seemed ready to sharpen their knives. The Sierra Club's Tom Politeo warned that the ports' growth rate would outpace mitigation efforts, while LBACA's Green put it more bluntly: "They say growing green means expanding terminals and putting more trucks on the road. What's cleaner about that? It's not logical."[245] Both ports, for their part, seemed to recognize the fight ahead, with the Port of Long Beach director of planning stating that he expected that "every one of the environmental impact documents for these projects will be challenged and end up in court."[246] The ports also sought to market to community members, attempting to "make the ports hip" through a "traveling educational exhibit" designed "to dazzle students with port facts"—at a price tag of $1 million.[247]

The ports simultaneously had to calibrate their response to increasing industry resistance to the Clean Truck Program, which focused on concerns about cost. After the ports released their joint program proposal in April 2007, "the real fight began. Once that was public ... industry came out strong and ... the ports, the mayors, the electeds reacted to that."[248] A report by the Los Angeles Economic Development Corporation in May warned that the cargo fee proposed to fund clean trucks might divert cargo to other ports.[249] In June, agricultural exporters complained that the program could make U.S. agriculture "uncompetitive."[250]

The ports' response was to conduct their own economic analysis of the proposed program, which was contracted to outside consultants at Economics & Politics, Inc. Completed in September 2007, the report (referred to as the Husing Report after its main author, economist John Husing), rested on extensive interviews with industry actors, as well as a statistical analysis of a variety of economic data.[251] The report compared the cost of converting the estimated 16,800 trucks regularly serving the ports to clean trucks through the existing structure of independent-contractor drivers to the cost of a plan based on employee conversion. It concluded that the proposal to convert to employee-operated clean trucks would cost LMCs nearly $150,000 per truck, which would include the cost of retrofitting or replacing the trucks and the cost of compensating the drivers—for a total cost of nearly $2.5 billion for converting the entire fleet.[252] This cost was calculated after factoring in port subsidies for fleet conversion, which were to be funded through truck fees and other public sources (including SCAQMD and state Proposition 1B transportation funds).[253] The report focused on two costs associated with driver compensation. First, the report analyzed the impact of the federal government's new security program, which required anyone accessing foreign entry points, including ports, to obtain a Transportation Workers Identification Credential—a biometric ID card also known as a TWIC card.[254] The federal regulations barred undocumented immigrants from obtaining a TWIC card and Husing estimated that this would reduce the supply of port trucking drivers by up to 22 percent, causing LMCs to raise their prices by up to 25 percent to cover the costs of luring new drivers.[255] The second type of driver-related costs were the payroll and benefit cost increases associated with the conversion of drivers to employees.[256] Combining these driver costs with the cost of clean truck conversion, the report estimated that LMCs would raise their prices by an average of 80 percent to offset the cost of implementing the Clean Truck Program.[257] Although emphasizing that this would be a "relatively insignificant" increase in overall shipping costs, it was notable that the price increase under a fleet conversion plan that continued to use independent contractors was significantly lower (at less than 50 percent).[258]

Worried that the ports would primarily focus on costs, Jon Zerolnick and others at LAANE set out to "quantify the benefits of passing the program."[259] Zerolnick thus took the lead in authoring *The Road to Shared Prosperity*—released a month before the Husing Report—which projected "direct and indirect financial benefits of over $4.2 billion" as a result of increased employee income and shifted taxes, as well as health care savings resulting from better community health and reduced taxpayer subsidies for driver health care.[260] When it was released, the Husing Report also made a nod toward the benefit side by acknowledging an SCAQMD estimate of a "cumulative economic benefit of $4.7 to $5.9 billion due to reductions in premature deaths, lost work time and medical problems."[261] However, its overall conclusions about employee conversion were negative. The Husing

Reforming the Port

Report suggested that shippers "will resist the LMC price increases due to their size" and "would delay such an increase as long as possible and explore other options."[262] For the LMCs themselves, the report warned that in the transition period, "there is the risk of the destruction of their firms and possibly bankruptcy. For those that survive, the question arises as to how they would recoup the accumulated loss created during the transition period."[263] Husing predicted that one-third of small LMCs would go out of business.[264] The report did not engage the issue of long-term sustainability emphasized by the coalition.

Predictably, industry reaction focused on the Husing Report's cost analysis, which strengthened opposition to employee conversion. The Pacific Merchant Shipping Association and National Industrial Transportation League—jointly representing Walmart, Exxon, General Motors, and other major importers—formally asked the Federal Maritime Commission to intervene to stop the Clean Truck Program.[265] Some trucking company owners threatened dire consequences. One family-run business owner said in response to the Husing Report: "Do the math. They want just a handful of companies to do business with. ... I am not interested in having 500 truck drivers as employees. If I have to remodel my business, I will probably walk away. I won't want to go through it."[266] Industry groups pressed their position and ratcheted up the litigation threat. In a letter sent to both harbor commissions and mayors, a coalition of business groups urged that the ban on dirty trucks be scrapped in favor of emission standards, and warned that the proposal was "anti-competitive," was outside the ports' "legal authority under state law," and thus "will result in litigation."[267] Against this backdrop, staff members within the Los Angeles mayor's office and port were legitimately concerned and a key question became why employee conversion was essential to a program that purported to advance environmental goals. Even Los Angeles Harbor Commission President David Freeman, a staunch program supporter, appeared to equivocate: "We all, of course, want to get the truck program up and running. ... But quite frankly, when we do the economic analysis it raises some questions."[268] In response, coalition members expressed frustration that the ports were mishandling program implementation and had lost valuable momentum. As NRDC's David Pettit put it: "The ports of Los Angeles and Long Beach get a failing grade for slipping behind in the implementation of their landmark Clean Air Action Plan."[269] It was fall 2007 and the program was at a crossroads.

The "turning point" was born out of tragedy,[270] when Los Angeles mayor's office staffer David Libatique was hit and seriously injured by a port drayage truck while coming out of a meeting at the TraPac Container Terminal.

Although he would recover to full strength and eventually return, his temporary absence left a personnel gap at the Los Angeles mayor's office. That gap was filled by Sean Arian, who was almost preternaturally well suited for the task ahead. Arian was the product of "multiple generations of longshoremen in San Pedro" (he was the son of ILWU leader Dave Arian), who as a boy suffered asthma and thus understood the Clean Truck Program in "very personal" terms.[271] He also had a unique combination of skills. A Columbia-trained lawyer, Arian had spurned the practice of law for the high-powered world of management consulting at McKinsey & Company, which he joined after a Fulbright fellowship in Latin America (where he focused on access to justice) and a federal court clerkship.[272] As a McKinsey analyst, Arian worked for Mayor Villaraigosa setting up a Project Management Unit to audit and track the mayor's accomplishments. In early 2007, Arian left McKinsey and became the city's director of economic development on the mayor's Business Team. Once Libatique was injured, the port portfolio was given to Arian. As Frank recalled, at this point, "I started playing linebacker and Sean Arian started playing quarterback." Arian directed the response to the Husing Report, while "any time someone went off in the wrong direction, … it was [Frank's] job to tackle them and get the mayor to say no."[273]

Arian entered a situation in which the foundation for a Clean Truck Program had been laid, but significant industry roadblocks remained. Arian took measure of the political context. He was told that the coalition had "done all the legal work," and understood that the coalition had also set forth the "big picture" in a way that made clear "what the community thinks" and what the "political upside and downside were."[274] This gave the mayor "political space" to advance a policy that would be "truly ground breaking."[275] But the case for the link between the environmental benefits of the program and employee conversion had not been persuasively made and industry arguments against it were gaining traction after the Husing Report. Arian sought to more forcefully make the case that port pollution was a "systemic problem" of "market failure," and thus not amenable to environmental regulation by itself.[276] Although the Clean Truck Program was fundamentally about environmental remediation, to get there, Arian argued it was crucial to attack "the root cause of the problem as opposed to just trying to attack one of the externalities."[277] The issue was how to convince the ports to move toward what Libatique called an "asset-based market," in which LMCs owned their trucks.[278]

Two decisions changed the program's course—at least in Los Angeles. First, Arian and others within the Los Angeles mayor's office recognized that for the program to succeed, staff at the port had to buy in. To achieve this, Arian—working closely with Castellanos and Weiner[279]—was able to convince the mayor to "put somebody high ranking" in charge of developing the program at the Los Angeles port.[280]

That person turned out to be John Holmes, who held what was arguably the second most important job in the port after the director: overseeing day-to-day operations as the port's deputy executive director. Holmes had spent nearly thirty years in the Coast Guard, the last three directing operations in Southern California—making him a "known quantity" at the port.[281] In Holmes's view, he was charged with designing the Clean Truck Program because figuring out how to get clean trucks in and out of the terminals was ultimately an "operational issue."[282] When he was assigned to the program by Knatz in late 2007, the "two things" he knew were that the program was "going to have a rolling ban ... to culminate in five years in having all the trucks ... be EPA 2007 or newer" and there was going to be employee conversion.[283] As Holmes recalled: "My role was to basically figure it all out."[284]

On the campaign side, it was Weiner's role to facilitate this process. Weiner's goal was to help Holmes credibly advance the employee conversion piece as an integral part of the environmental program and not just a union project.[285] In Weiner's analysis, because port staff were on the front line of dealing with industry opposition, they were under the most pressure to respond to industry claims that "the sky is going to fall."[286] That front line pressure was a constant challenge for the campaign, since port staff would report industry concerns up the ladder, ultimately landing back at the mayor's office, where the mayor's staff would get "nervous and weak-kneed" and the coalition would have to "prop up our supporters" and "beat back all these claims."[287] As a result, the coalition believed it was crucial to have "someone at the staff level at the port ... able to push back and move this agenda."[288] Weiner understood that the politics were fraught, since industry was "trying to tag the port as basically in bed" with the unions.[289] Weiner also knew that Holmes was politically astute and appreciated the stakes: that he technically worked for the mayor, while having to deal with industry as the port's primary constituency.[290]

Arian introduced Holmes to Weiner. As Weiner recalled, the pitch to Holmes was: "the mayor really wants this, so [Weiner] could help you understand why this makes sense and talk it through."[291] Weiner recalled that Holmes' initial posture was "skeptical"—asking "what does this have to do with the Clean Truck Program?"[292] To answer that question, Weiner initiated a series of "one-on-one discussions" with Holmes about the concession model and, specifically, the employee conversion piece. From Weiner's vantage point, these discussions were fruitful as Holmes eventually became comfortable with the idea that "the employee requirement is really so that the companies will be responsible ... to maintain the trucks and not these drivers."[293] Thus, Holmes accepted the main thrust of the coalition's argument: that if the maintenance costs were not shifted onto the trucking companies, the maintenance could not be sustained over the long term since the drivers could not shoulder the expense.[294] From there, Holmes began to work on the program details. Given concerns about cost, a key issue was determining the appropriate level of financial incentives, which trucking companies had advocated for as necessary to make the conversion "happen as soon as possible."[295] For the incentive piece, Holmes brought in the port's director of finance. But they were working off the basis of the Husing Report, which did not provide a strong framework for advancing the entire clean truck package.

This led to the second crucial decision. After Holmes was on board, Arian received the mayor's permission to bring in another consulting firm to reevaluate the economics of the Clean Truck Program. After reviewing the Husing Report, Arian believed that there was an insufficient "fact base" to convince stakeholders of the need for industry transformation and thus argued for what amounted to a more sophisticated cut at the economics done by "one of the top consulting firms."[296] Using his connections, Arian was able to bring in Boston Consulting Group (BCG) (his former firm's chief rival), which agreed to send in its "A Team" on a pro bono basis to analyze the economic impact of converting to clean trucks. To justify this, Arian dissected the Husing Report in a way that conveyed to port staff that Husing's "analysis didn't go far enough to give us the information we need to evaluate our options."[297] From Weiner's perspective, Arian was able to "basically rip apart" the analysis of John Husing—setting the stage for the entry of BCG.[298] Arian viewed BCG as "potentially a huge risk" because its consultants' reputational capital was based on telling clients "what they think the right answer is regardless of whether that's the answer you wanted to get from them," and their analysis would become part of the public record.[299] However, Arian firmly believed that "we needed to have a really strong fact base by a mutual third party respected organization."[300] Arian recalled that port counsel Tom Russell was the "very first person who hopped on board" with the BCG plan since for the market participation theory to work as a legal rationale for the program, there needed to be a strong evidentiary record of why the program made business sense.[301]

Holmes also worked closely with BCG analysts. Holmes recalled that a "key factor" was BCG "working a month with us, locked in a room basically trying to figure out how this could work."[302] Holmes's view at this stage was that to get where the Los Angeles port wanted to be "required a sea change in the drayage trucking industry."[303] To do that, he saw the basic choice as regulating or incentivizing the industry—which was not going to go green "just to do the right thing."[304] Industry's basic question was: "Are you going to pay me to do it or make me do it?"[305] The answer was: some combination of both.

Toward that end, Holmes's work with BCG was designed to promote trucking participation in the program and "get the numbers right."[306] BCG modeled industry responses to different program scenarios, in which the main elements were the amount of the cargo fee, the timing of the truck ban, the development of security structures, the nature of driver status (independent contractor or employee), and the amount of incentives (which varied by whether the new trucks would run on diesel or alternative fuel). The port wanted to encourage companies to buy liquefied natural gas (LNG) trucks, which were an average of $50,000 more expensive than diesel trucks. One issue Holmes grappled with was how much the incentive had to be to persuade companies to buy LNG trucks. In addition, Holmes was focused on working out the details of the concession arrangement and its impact on the community. In his view, the employee provision was closely associated with the off-street parking provision, since employees would "slip seat"—transfer their trucks to different drivers from one shift to the next—which required a place to park the trucks during the transfer period.[307] Holmes was sensitive to complaints about traffic and viewed the idea of truck placards with a phone number to report drivers improperly using local streets as an effort to respond to community concerns.[308]

Once the economic models were run, Holmes visited "the twenty-five biggest trucking companies in the country" to validate the results.[309] In these meetings—with major carriers like Schneider, Swift, and Knight, as well as shippers like Walmart—Holmes would say: "We're thinking of doing the program this way. What are we missing?" Through that process, Holmes and the consultants gained "a whole bunch of knowledge" about industry structure and equipment costs that were then factored back into the modeling analysis.[310] Holmes recalled that his meetings with industry were not all adversarial. To the contrary, many representatives of major trucking companies expressed support for minimum standards in an industry they viewed as built on "caveman economics," in which fly-by-night carriers forced a "race to the bottom."[311] Yet, although these firms supported many of the environmental elements of the program "and gave good feedback," they uniformly did not agree with employee conversion.[312] Holmes also understood that even those companies that supported the general approach were likely to join an industry lawsuit against the program if it passed, simply because industry rejected port regulation on principle.[313]

As Holmes and BCG carried out their analysis, the harbor commissioners were also working to iron out the policy details. Commissioner Mendoza recalled that her starting point was at odds with Freeman's, who thought that the Los Angeles port should simply "mandate the purchase of 500 LNG trucks" and thus become the direct owner of a portion of the port trucking fleet.[314] However, as the program developed around employee conversion within the mayor's office, Freeman embraced the concept and fought for it like a "momma bear."[315] Together, Mendoza and Freeman took the lead in moving the entire program forward. Mendoza described her approach to dealing with the city attorneys assigned to the program:

> [W]e didn't ask them, we would tell them, "This is what we're going to do. You guys have to figure out how to make it work." And ... we got a lot of push-back and a lot of "you know, we've never done that before," or "we have outside consultants who have done the analysis and they think that we don't have a really good chance." Our response was, "Okay, do we know for certain that this would not be successful in court? No. Well, if we don't try, we won't know."[316]

Freeman saw his role as motivating the staff, which "required, shall we say, inspiration. And my role was to inspire them ... by just telling them that if they didn't get this stuff done, there was going to be hell to pay."[317] In Freeman's view, "Holmes was the best staff person there in getting religion and helping to make it happen."[318] Policy issues brought up by port staff were hammered out, either in an ad hoc committee on environmental review staffed by Mendoza and Freeman, or in a closed session of the entire board. In the face of concerns that "truck companies would boycott" the port,[319]

> David Freeman and John Holmes would get on a plane and go fly to Walmart in Bentonville, Arkansas, or they'd fly to the different trucking companies that were in Texas and Virginia and places like that, and they would sit down and talk to the business owners and say, "Look, we know this is uncomfortable, we know this is different, we know this is the first time you've been asked to do something like that but at the same time if we clean up, it will allow us to expand in a way that the community will not rise up and riot the way they have in the past. If you want to grow, if you want the port to grow, if you want your goods to get in and out faster, we also have to be as green as possible so that we can grow."[320]

The information gained in these trips allowed Freeman to respond to staff concerns about costs. And Mendoza used her lawyering skills to respond

Reforming the Port 141

to issues raised about the legality of employee conversion, asking: "Well, why aren't we a market participant? ... Why can't we make that argument? What do we have to do to make that argument compelling?"[321] Freeman knew that "our furthest reach under the law was to require the truckers to have employees."[322]

The End Game: Passing the Clean Truck Program

As BCG worked on its analysis, the Los Angeles and Long Beach ports initiated a sequential approach to program implementation. Based on a political calculus that it was best to lock in elements of the program in stages—ranging from least to most controversial—the ports began moving forward specific elements of the Clean Truck Program: from clean truck conversion, to industry incentives, to employee conversion. This order tracked the key elements of the program debated from the outset and set an agenda for phased implementation: first, the progressive ban on dirty trucks; second, the Clean Truck Fee; and third (and most controversially), the concession plan with employee conversion. In the first two steps, the ports of Los Angeles and Long Beach moved in sync, but in the third, they diverged. Throughout the process, industry pressure mounted to split off the environmental standards from the employee conversion piece—and to sever the environmentalists from organized labor in the coalition.

The first step was, in relative terms, the easiest. On November 1, 2007—as the BCG team was just getting under way—the Port of Los Angeles Board of Harbor Commissioners unanimously approved a progressive dirty truck ban.[323] Following a strong staff recommendation,[324] the board approved Order 6935 to phase in the ban over five years.[325] The legal structure of the ban, drafted by city attorney Crose, highlighted the legal authority of the port, while the operational structure bore the imprint of Holmes's expertise. Legally, the order amended port Tariff No. 4, which was originally adopted in 1989 to govern the rates and terms of terminal operations.[326] As amended, Tariff No. 4 imposed a prohibition on the terminal operators—not the trucking companies or the truck drivers. Specifically, the ban stated that "no Terminal Operator shall permit access to any Terminal in the Port of Los Angeles" to nonconforming trucks.[327] In the order's findings section, the justification for the ban was made in market participant terms. There, the case was made for the port as a proprietary entity with business interests in pollution reduction:

> Independently, the failure of the Port to adequately address air pollution impacts, including diesel truck emissions, would threaten future Port growth both because of legal constraints under the California Environmental Quality Act (CEQA) and

the National Environmental Policy Act and the opposition of surrounding residents and communities to further expansion without an actual improvement in environmental conditions surrounding the ports.[328]

The findings focused on trucks as "a critical element in the efficient operations of the Port," and concluded: "Reasonable environmental measures are simply good business practices."[329]

On the operations side, the order established a new system for efficiently identifying clean trucks—requiring all trucks to install a Radio Frequency Identification Device by August 1, 2008. The device would contain a unique identification number that could be electronically read by terminal operators, which could cross-reference the number against records showing the vehicle model year and compliance with clean truck standards.[330] The Long Beach Board of Harbor Commissioners unanimously approved an identical ban five days later,[331] prompting Long Beach Mayor Foster at a joint news conference with Los Angeles Mayor Villaraigosa to tout the two cities' effort "to lead the world in pushing for cleaner air and healthier environment with our shared goal of having the cleanest ports in the world."[332]

With the dirty truck ban legally and logistically in place, the battle immediately turned to the issues of new truck financing and employee conversion. Upon passage of the Los Angeles dirty truck ban, port director Knatz struck a stern negotiating posture, acknowledging the need for short-term program funding, but making clear "we can't subsidize it forever."[333] With respect to employee conversion, trucking companies again emphasized the litigation threat. Cecilia Ibarra, assistant operations manager for trucking company Total Distribution Service of Wilmington, was explicit—and articulated industry's particular hostility toward the employee provision. "We want clean air as much as anyone, but the board's actions may drive us into litigation. A concession program is a step toward unionization. I can already hear the ka-chink, ka-chink, ka-chink in union coffers."[334] Despite this effort to isolate employee conversion from the program's environmental elements, coalition members continued to assert a unified front. LBACA's Green pressed the case: "I don't understand why the board decided to vote on just the clean-truck portion of the clean-air plan. ... It's hard not to think that they were pandering to the environmental community by throwing us a bone, as though we would be happy with just a progressive ban."[335] In an official statement released after the Long Beach ban was adopted, the coalition kept up the pressure to move forward: "Without reform, the Los Angeles and Long Beach ports remain unprepared to meet ever-increasing trade demands, and they will be unequipped to compete in today's rapidly changing economy."[336]

On the financing side, the issue from the outset had been at what price to set the Clean Truck Fee (a charge on containers carried by drayage trucks to be imposed on shippers) in order to create a fund for clean truck conversion. The ports' initial proposal contained a range—from $34 to $54 per container—which set the bargaining zone. Industry sought to push the ports toward the lowest end of the range, while other stakeholders sought to keep up the pressure for the ports to act aggressively. In 2006, state Senator Lowenthal from Long Beach advanced a bill to impose a $60 fee on cargo loaded in forty-foot containers to fund emission reduction, but Governor Schwarzenegger vetoed it on the ground that it would hurt U.S. exports.[337] In 2007, Lowenthal reintroduced the bill, but agreed to withdraw it in September as the Clean Truck Program appeared to advance.[338] Other public funds were potentially available for truck conversion, but not at the levels needed.[339] The financial viability of truck conversion thus hinged on the cargo fee.

Long Beach made the first move, approving a cargo fee of $35 per loaded twenty-foot container on December 17, 2007.[340] The Clean Truck Fee would be "assessed on containerized merchandise entering or leaving the Ports by Drayage Truck," to be paid by the "Beneficial Cargo Owner," and collected by the terminal operator.[341] The fee was expected to raise $1.6 billion for a Clean Truck Fund,[342] to be used by the port "exclusively for replacement and retrofit of Drayage Trucks serving the Ports of Los Angeles and Long Beach."[343] Still marching in lockstep, the Los Angeles harbor commissioners approved an identical fee four days later.[344]

Finally, the stage was set for the showdown over employee conversion. Both sides in the debate pushed hard. In Long Beach, Mayor Foster worked to isolate organized labor. As Weiner recalled, Foster "kept ... wanting to meet with the environmental folks in our coalition without the labor folks ... [asking:] "Why can't we just do this? Why do you need this employee thing?"[345] Commissioner Freeman perceived Foster as set against employee conversion: "He just did not believe that we had the right to force these [trucking companies] to have employees."[346] And Foster was worried that "the Teamsters will take over" and undermine future environmental programs.[347]

The coalition kept up the pressure with a series of public demonstrations—against the backdrop of ongoing private negotiations. On the day that Long Beach approved its Clean Truck Fee, the coalition organized a rally of 150 truckers at the port entrance to stress the need for employee conversion. "We all support cleaner air, but none of us wants a loan or a grant to buy a new truck," said driver Miguel Pineda. "If these plans become law, I won't be able to put food on the family table."³⁴⁸ He added that "a lot of truckers have stopped spending money on repairs because they aren't sure they will still have jobs next year. ... It's a terrible situation; we live like slaves in the 19th Century."³⁴⁹ A month later, the *Los Angeles Times* ran a front-page story on "unsafe trucks" coming out of the ports, focusing on the plight of low-income independent contractors who could not afford to replace tires on their big rigs and frequented "llanteros" (Spanish for tire repairer) who would use hot blades to carve new grooves into seriously worn tires; the story highlighted other drivers engaged in desperate measures like "lashing bumpers to chassis with bungee cords and smearing mud over cracked parts to hide the problems from CHP officers."³⁵⁰

In the face of pending expansion plans—the *Los Angeles Times* noted that fifteen port projects had been held up since the China Shipping case in 2001—the coalition also pressed to underscore the legal stakes.³⁵¹ NRDC's Adrian Martinez and Teamsters Port Division Assistant Director Ron Carver sent a letter to the ports stating: "Unless we are assured that your plans include reasonable proposals for mitigating the environmental harm of your existing facility, let alone your proposed expansion, we cannot see how we could let the process continue without a challenge."³⁵² The message was clear—and industry reacted strenuously, with the vice-president of the TraPac Terminal calling it a "shakedown."³⁵³ NRDC's Martinez observed that "things are getting nasty out there," while Mayor Villaraigosa tried to give a positive spin: "In the interests of green growth, historic adversaries have become part of a very delicate coalition. It's as though everyone is coming to this party holding hands but reluctant to get on the dance floor. But they will, eventually. They have to."³⁵⁴

Before that could happen, the coalition sought to make clear that the legal consequences of inaction would be further logjam. Markers were laid down at both ports. In Los Angeles, the proposed TraPac Terminal expansion was the legal flashpoint. On December 6, 2007, the Los Angeles harbor commission, in a statement of overriding considerations balancing the environmental harms and economic benefits,³⁵⁵ unanimously approved the EIR for a $1.5 billion upgrade projected to create 6,000 jobs and generate $200 million annually in taxes.³⁵⁶ Four individuals and sixteen groups—including the Coalition for Clean and Safe Ports, NRDC, the Coalition for Clean Air, the American Lung Association, Sierra Club, and LAANE—appealed the approval to the Los Angeles City Council on the ground that it did not adequately address the air pollution impact.³⁵⁷ One of the individual appellants was Kathleen Woodfield, president of the San Pedro and Peninsula Homeowners Coalition, which had been a lead plaintiff in the

Figure 5.2
Mayor Antonio Villaraigosa speaking to a coalition member at a rally. Photo courtesy of Barbara Maynard, Teamsters communication director.

China Shipping litigation.[358] As she recalled, the appellants were not represented by legal counsel: "NRDC was clear that they were not representing us."[359] Yet NRDC did flex its own legal muscle, indicating its intent to file a CEQA lawsuit if the council appeal was unsuccessful.[360] In response, City Council Member Hahn blocked the EIR from getting out of a key council committee and began negotiating with environmental and neighborhood groups.[361]

Two months later, in Long Beach, NRDC and the Coalition for a Safe Environment filed an intent-to-sue letter with the port, asserting an innovative legal theory: that the port was an entity subject to federal oversight as a hazardous waste site under the federal Resource Conservation and Recovery Act.[362] NRDC's Pettit claimed: "We want the court to take over the whole thing at once in order to enforce a new priority of public health over profit . ... We think that will require court appointment of a port czar to force the port to use currently available technology to fix the problem."[363] The letter requested that the port stop expansion projects until it could prove they would not "at any time increase the level of hazardous diesel particulates emanating from the port."[364]

If the NRDC letter was meant to put pressure on the Port of Long Beach as the clock ticked down toward a resolution on employee conversion, it did not have the desired effect. To the contrary, on February 16, 2008, the Port of Long Beach officially broke ranks with its Los Angeles counterpart, announcing a meeting to approve the final element of its Clean Truck Program—a concession plan—without employee conversion.[365] "Their announcement caught us all by surprise," said LAANE's Castellanos.[366] It was, as Frank recalled, "one of the earliest signs there was trouble."[367] In Long Beach's concession plan, trucks would be granted a "right of access to port property" in exchange for LMCs entering into contracts that ensured compliance with existing laws (including the preexisting elements of the Clean Truck Program), as well as "local truck route and parking restrictions."[368] The ordinance also modified the Clean Truck Fee for cargo moved on trucks not purchased with port subsidies, waiving the fee for containers transported on alternative fuel trucks and halving it for clean diesel trucks.[369]

Released on the cusp of a three-day weekend, the plan was slated for vote the following Tuesday, February 19, by the Long Beach harbor commission. Despite the short turnaround, the coalition mobilized to attend the six-hour hearing, with public comments given by several residents, drivers, and coalition members—including Zerolnick, Martinez, Politeo, Kim, Logan, Green, and Callahan.[370] Nonetheless, board approval was unanimous,[371] and port officials touted their program as a "victory for clean air"[372]—one that cleaved apart the environmental and labor elements.[373] Mayor Foster's chief of staff, responding to the mayor's break with Los Angeles, said: "It doesn't scare us that there is a difference of opinion. ... What scares us is not acting to clean the air as quickly as possible. If their board is not ready to go yet, fine. ... Ours is."[374] NRDC, unsurprisingly, disagreed in its letter to the board: "Perhaps the most glaring flaw in the port's program is the lack of its key partner and neighbor, the Port of Los Angeles. If Los Angeles decides to go in a different direction in its clean-trucks program, the result would be chaos at the ports."[375]

All eyes therefore turned back to Los Angeles. Industry opponents of employee conversion sought to cast it as a union ploy. NRDC's Martinez perceived that opponents were "freaked out" by the unified front maintained by NRDC and the Teamsters and recalled a lot of "fearmongering."[376] The *Los Angeles Times*, in reporting on the ongoing battle, emphasized the unionization angle: "Critics of the employee provision of the clean truck program ... are concerned that it could be used by the Teamsters as a springboard to launch unionization efforts at ports nationwide."[377] It also noted that Change to Win had donated $500,000 to Mayor Villaraigosa's local telephone tax initiative, Proposition S, insinuating that the unions expected a quid pro quo.[378] A month later, an editorial began: "Pollution, death and economic stagnation. These catastrophes are being brought to you by the Natural Resources Defense Council and the International Brotherhood of Teamsters."[379] It argued that the clean trucks deal was being jeopardized over a "dispute that has nothing to do with pollution and everything to do with an unholy alliance between environmentalists and organized labor."[380] Noting that Long Beach had already passed its plan without employee conversion and that Los Angeles seemed on the verge of doing the opposite, the editorial board suggested that the Clean Truck Program of both ports would be tied up for years in litigation—in Long Beach by NRDC and in Los Angeles by the ATA.[381] The editorial argued that it did not have to be so, since "[u]nder a lease-to-own program, a nonprofit or other organization could buy new trucks and lease them to the truckers, charging low fees that would be subsidized by the ports."[382]

Coalition members fought back. Coalition for Clean Air director Martin Schlageter responded to the *Los Angeles Times* that he was "baffled at your editorial placing the blame for delay on advocates of clean air," noting that it was an NRDC lawsuit that produced CAAP in the first instance and that the Long Beach plan suffered from the "glaring weakness" of failing to influence driver working conditions, which was a concern "for environmentalists and labor advocates alike."[383] Coalition members also sought to make fun of the argument that there must be "something wrong" in a collaboration between labor and environmentalists, with Pettit and his colleagues donning T-shirts with "The Unholy Alliance" printed on them.[384] In retrospect, Pettit thought "we, NRDC—did a poor job of messaging what [the program] had to do with clean air. And I think we came off poorly ... in all the media attention."[385] Yet in a sign that political support was holding firm, the Los Angeles City Council formally adopted the Board of Harbor Commissioners' progressive ban and Clean Truck Fee at the end of February.[386]

By March 2008, on the brink of the Los Angeles port's final decision on employee conversion, the political back-and-forth reached a fever pitch. The coalition had taken out five full-page color advertisements in the *Long Beach Press-Telegram* denouncing Mayor Foster, while Los Angeles City Council Member Janice Hahn was privately urging him to reconsider.[387] For their part, the Teamsters were seeking to use their political clout to block state funding for Long Beach to complete renovation of the Gerald Desmond Bridge.[388] LAANE sent Foster a public records request seeking all his communications with industry representatives.[389] Yet Foster remained defiant: "So the end result is, if this happens to be the only office I ever hold and the only term I ever serve, I'm comfortable with that."[390] Industry groups also struck a strident tone in advance of the Los Angeles decision. Curtis Whalen, executive director of the ATA, stated that Villaraigosa's "biggest problem is he has good intentions, but they are not legal."[391]

It was at the height of this debate, on March 7, 2008, that BCG released its long-awaited report—which had been delayed in the wake of the Long Beach decision. At its heart was an analysis of the economics of employee conversion—the sole remaining piece of the Los Angeles Clean Truck Program. In clinical terms, far removed from the supercharged rhetoric of campaign adversaries, the report evaluated three program options: the first would permit the continued operation of independent contractors and give them a share of the incentive financing to acquire clean trucks; the second would also permit continued operation of independent contractors, but limit incentive financing to LMCs; and the third would require LMCs to make what it called an "employee commitment."[392] The report's key move—and what made it different from the Husing Report—was that it compared short-term (one to five year) and long-term (more than five year) outcomes in relation to stated environmental, port operations, and safety and security goals.[393] The report's bottom line was that the employee model provided "the best path to long term sustainability" although it posed some "near term risks."[394] Specifically, the report concluded that the employee model would "maximize the likelihood of creating a market in which the reciprocal obligations between the Port (granting a commission) and LMCs (providing drayage services) create a sustainable reliable supply of truckers attracted to stable and relatively well paying jobs in an operationally efficient and orderly drayage market."[395] It went on to recommend that a "100% employee driver requirement, phased in over five years" was the best option: "transparent, aligning incentives and easiest to administer."[396] Financing support for clean truck conversion was set at 80 percent for new diesel trucks and up to 80 percent for LNG trucks, with $5,000 given to scrap pre-1989 trucks.[397]

The report was not all rosy. It predicted that under the "employee commitment" scenario, there would be more cargo diversion—approximately 3 percent—than under the other options, but that this cost would likely be outweighed by overall benefits.[398] The "key risk" was that shippers would divert cargo over and above this 3 percent threshold based on factors other than increased price,[399] such as fear of "future disruption or instability."[400] This risk would be exacerbated, the report stated, if the Port of Los Angeles and the Port of Long Beach adopted different programs[401]—which, of

Reforming the Port

course, had already happened, though the report seemed to hold out hope that there was still a possibility that Long Beach might reverse course. In reporting on the plan, the *Los Angeles Times* highlighted the cargo risk, beginning its article by emphasizing the BCG report's conclusion "that 'substantial diversions' of the Los Angeles port's business probably would shift to the neighboring port of Long Beach or to other harbors."[402]

Despite these market concerns, the BCG report ultimately did the work it was designed to do. In Frank's recollection, the Husing Report had fostered concern that the Clean Truck Program would reduce the supply of trucks and "stop the U.S. economy."[403] The BCG report showed that as the market was rationalized, it would actually "move the trucking industry toward the port and there would not be a shortage of trucks."[404] According to Frank, the report's analysis was "indisputable."[405] At the staff level, the cost analysis permitted Holmes and other port managers to solidify the employee provision and calculate the precise level of financing. On March 12, 2008, a staff report authored by Holmes and Deputy Director of Finance Molly Campbell recommended approval of the concession and incentive plans analyzed in the BCG report.[406] At the board level, Arian recalled that the report was the "key study" that persuaded skeptical commissioners to move toward approval.[407]

On the day that the Los Angeles board was set to meet on the concession plan, the *Los Angeles Times* took one last opportunity to hammer home the litigation threat. In an editorial titled "Harbor No Illusions: L.A.'s Plan to Clean Up Port Pollution Is Sure to Wind Up in Court," the editorial board argued that "[i]n the real world of lawsuits and endless court proceedings, [the truck plan] would stall progress on cleaning the air indefinitely."[408] Yet at this stage in the game, the early legal work on the port's authority to pass the program seemed to fortify city staff. As Libatique recalled, staff did not shy away from the concession model: Because of "the amount of work and time we put into the legal underpinnings of the program ... I think there was a level of confidence here that we had a strong legal case."[409] If litigation came—which it would—the city was prepared to defend its program then, rather than back off it now. The program that the board did adopt on March 20, 2008, contained the signature element of what the coalition had spent almost two years working toward: the concession plan.[410] Resisting industry pressure, the board in Order 6956 (again drafted by Crose) further amended Tariff No. 4 to "require parties who access Port land and terminals for purposes of providing drayage services to the Port of Los Angeles to have a Concession Agreement" with the port.[411] Following the same format used in the earlier order banning dirty trucks, the amendment placed enforcement responsibility on the terminal operators, stating that "no Terminal Operator shall permit access into any Terminal in the Port of Los Angeles to any Drayage Truck unless such Drayage Truck is registered under a Concession."[412] Terminal operators that violated the order were subject to criminal sanction under the tariff's general penalties provision, which made any violation of the tariff a misdemeanor punishable by a fine of up to $500 and imprisonment of up to six months.[413]

The order set forth the legal justification for the concession plan. Expressing concern about the "environmental, operational, and safety and security objectives of the Port" (language adapted from the BCG report), the findings emphasized the goal of encouraging "evolution of the Port drayage market towards an asset-based market in which Licensed Motor Carriers that hold the motor carrier concessions also own the truck assets used to perform under the concession."[414] After noting that the port "currently has no business relationship with the thousands of trucks, drivers or licensed motor carriers" hauling cargo, the order mandated a concession program "that specifies conditions that must be met in order to provide drayage services" at the port.[415] It stated that a drayage company "must enter into" an agreement, which would last for a term of five years, "in order to access" the port.[416] Although the order itself did not explicitly discuss employee conversion, a separate transmittal by staff (circulated to the full board) proposed a "non-exhaustive list of the main Concession requirements," which included that LMCs "[t]ransition to 100% employee drivers for Port of Los Angeles drayage in five years, according to a schedule specified by the port."[417] The list also included the concessionaire's commitment to create an "off street parking plan"; to affix "placards on all Concession controlled trucks referring to a 1-800 phone number to report concerns regarding truck emissions, safety and operations"; and to pay a concession fee of "$2500 plus an annual fee of $100 per truck."[418] The order also modified the Clean Truck Fee—exempting cargo transported by concessionaires that used 2007-compliant alternative fuel trucks (even if purchased with port subsidies) and any that purchased clean diesel trucks without port subsidies[419]—while deferring collection of the Clean Truck Fee from June 1 to October 1, 2008.[420]

In boilerplate terms that acknowledged the legal vulnerability of at least some elements of the concession plan, a severability provision was also added. It stated: "If any provision of Port of Los Angeles Tariff No. 4 shall be determined by court or agency of competent jurisdiction to be unenforceable, unlawful or subject to an order of temporary or permanent injunction from enforcement, such determination shall only apply to the specific provision and the remainder of the provisions ... shall continue in full force and effect."[421]

The purpose of the severability provision was to reduce legal and political risk to the overall Clean Truck Program. Legally, the severability provision would permit a court to excise some provisions, while keeping intact the broader plan. Commissioner Freeman recalled "the meeting in the mayor's office where we decided, well, we want a severability clause ... so that if we lose that one, it doesn't contaminate the rest of the case. And we all went ahead with that."[422] For the mayor, there were also political implications. Severability maximized the possibility of sustaining some aspects of the Clean Truck Program—and thus being able to declare political victory while also addressing an important policy concern. Perhaps most crucially, if the employee provision was ultimately struck down, progress could still be made toward greening the port and organized labor would be no worse off than if the effort had not been made at all.

In connection with the board order establishing the concession plan, a separate resolution drafted by Crose and approved by the board laid out the market participation rationale, while also adding in the core financial incentives. The resolution noted that the objective of the plan was to "create and sustain an efficient, reliable supply of drayage services to the port," which "as land owner of the Harbor District land and assets has the right and the obligation to manage and control the access to its land by tenants and invitees to ensure that operations thereon maintain safety and security of Port operations on a sustainable basis."[423] The resolution went on to stress that the "air quality, port security, and safety goals" of the plan were "more likely to be achieved and sustained over the long term" through concessions. It also pointed to the "[s]erious and long-standing problems" produced by "inadequate maintenance" and "unsafe, negligent or reckless driving of trucks" at the port, and asserted that the concession model was the "most efficient," "greenest," "safest," "most community friendly," "most responsive and flexible," and "easiest model to administer."[424] In addition, the resolution—following the staff report and BCG analysis—authorized a Truck Funding Program, to "provide funding of up to 80% of the value" of clean trucks (in the form of either a lease-to-own agreement or up-front grant to purchase or retrofit, available to LMCs and independent owners) "in order to make the transition ... more affordable."[425] Other incentives included a Truck Procurement Assistance Program to "provide volume discounted pricing" to concessionaires and a Scrap Truck Buyback Program providing $5,000 to owners of pre-1989 trucks who turned them in.[426] With that, the centerpiece of the Clean Truck Program was approved.

The following day, the ATA's Whalen called it a "scheme to unionize port drivers" and vowed that: "We're going to go after Los Angeles with everything we've got so their plan goes to hell in a handbasket. We will win and we will win handily."[427]

Yet a few steps still remained. The Los Angeles port wanted resolution of the dispute over the TraPac expansion—which had served as the coalition's final "bargaining chip."[428] As homeowner activist Kathleen Woodfield recalled, the port "wanted to move forward with the TraPac project. For as long as it was appealed it was in limbo."[429] To move this along, Woodfield, Martinez, and other coalition members met with City Council Member Hahn, Mayor Villaraigosa, Port Executive Director Knatz, and Commissioner Freeman to hammer out a deal.[430] Working closely with Hahn—whose support coalition members thought was "strong and pretty much unequivocal"[431]—the coalition negotiated a settlement in early April.[432] Under its terms—negotiated primarily by Martinez and port general counsel Tom Russell[433]—the port agreed to create a Mitigation Trust Fund with an immediate $12 million contribution toward community improvements and an air filtration system for local schools.[434] The agreement also provided—as a means of avoiding future litigation—that more funds would be contributed for mitigation in conjunction with future expansion projects at the port:[435] $1.50 for each additional cruise ship passenger and $2 for each additional container.[436] The funds were to go to a nonprofit group, the Harbor Community Benefit Foundation, created for the purposes of disbursing the money.[437] The agreement also committed the port to continue working to support the Clean Truck Program.[438] It was the first EIR approved at the port since China Shipping, and Commissioner Freeman touted the agreement as a model for pending expansion plans. "The entire environmental community is giving its blessing to Mayor Villaraigosa's green growth program. ... We will work together on all future [projects] and not resort to litigation."[439] The NRDC's Pettit, though pleased with the outcome, did not go quite that far, emphasizing that: "The agreement does not give up our right to sue on any project other than TraPac."[440]

The accord did, however, clear the way for the concession agreement itself to be approved. The draft agreement, circulated in early May, spelled out the mechanics of "driver hiring" in detail, setting forth a phased implementation in which a concessionaire "shall be granted a transition period ... by which to transition its Concession drivers to 100% Employee Concession drivers by no later than December 31, 2013."[441] Under the transition plan, 20 percent of drivers had to be converted by the end of 2009, 66 percent by the end of 2010, 85 percent by the end of 2011, and 100 percent by the end of 2013.[442] The agreement also required concessionaires to "submit for approval ... an off-street parking plan," "post placards" on their trucks, agree to regular maintenance, attest to financial capability, and pay the concession fees.[443] The agreement further specified enforcement procedures, identifying (though not defining) minor and major defaults, and imposing sanctions, which included revoking the concession agreement itself.[444] The final loose ends were quickly tied. The board approved the concession agreement on May 15.[445] Then, at a meeting on June 17, the City Council approved the final concession plan.[446] With the mayor's signature, the Clean Truck Program was city law.[447]

In the end, after nearly two years of struggle, the coalition had won a program in Los Angeles that included the twin goals of fleet and employee conversion. Both were achieved through a legal structure (depicted in figure 5.3) that linked the port's contractual and police powers: trucking companies were required to sign concession agreements to enter the port and terminal operators were required to bar entry to any company that

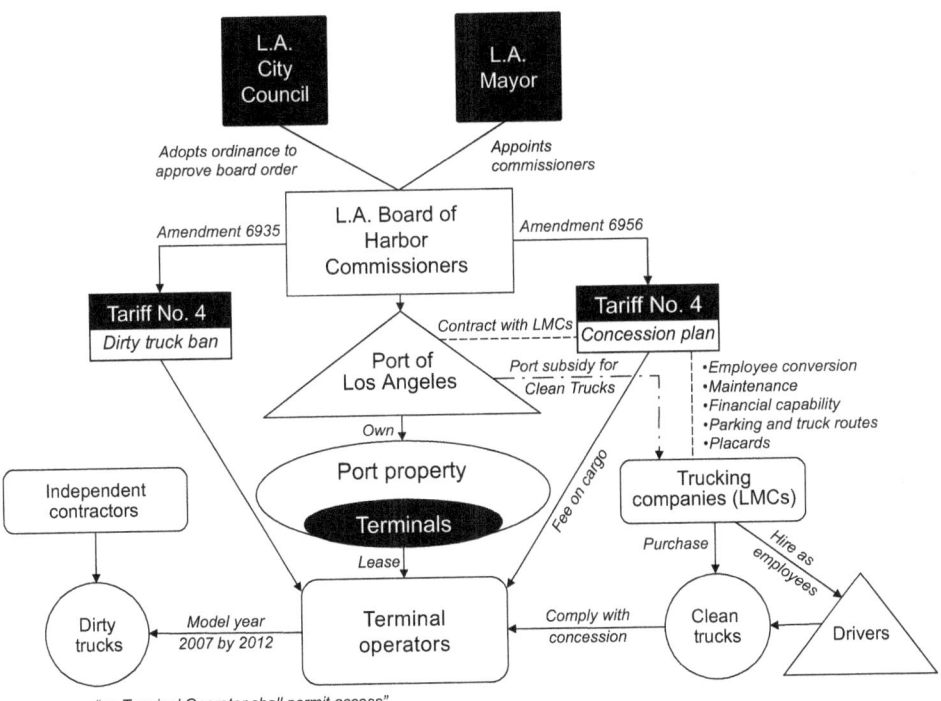

Figure 5.3
Structure of Los Angeles Clean Truck Program.

did not comply with port rules—or incur criminal sanctions. The program was to be financed primarily through the imposition of cargo fees, which would shift the costs of clean truck acquisition to shippers, while providing incentives for trucking companies to upgrade their fleets. Looking at the final program as a whole, there were some details with which the coalition could quibble. At the outset, the coalition had set its sights on full conversion to alternative fuel trucks, a higher fee, and immediate employee conversion.[448] What it got was partial alternative fuel conversion, a lower fee with significant exceptions, and a five-year phase in for employee conversion. However, these were minor sacrifices to achieve the ultimate program, which closely tracked the outline proposed in the coalition's RFP a year earlier.

The Defensive Phase: Responding to Federal Law

The ink was barely dry on the Clean Truck Program when legal challenges to it commenced. From the outset, this had been anticipated by the coalition as part of the overall process of enacting and defending the law. As LAANE's Jon Zerolnick described, from the beginning, it "was always a part of the timeline" that the "ATA sues here."[449] Teamsters counsel Mike Manley's early legal opinions had predicted lawsuits on preemption and maritime law grounds—precisely what ended up occurring. The campaign, premised on enacting local law, now turned to defending that law against efforts to negate it on federal grounds.

Private Litigation I: The Injunctive Phase

As it had long threatened, the ATA was first to the federal courthouse.[450] On July 28, 2008, the ATA filed a complaint in federal district court in Los Angeles for declaratory judgment and injunctive relief.[451] The lawsuit took aim at the concession plans of both the Los Angeles and Long Beach ports, arguing that they were preempted by the FAAAA's prohibition against municipalities enacting "a law, regulation, or other provision having the force and effect of law related to a price, route, or service of any motor carrier."[452] The complaint further alleged that both plans violated the Commerce Clause by imposing "invasive regulatory requirements on virtually all aspects of the business of a federal motor carrier, including truck maintenance, on-street and off-street parking, employee wages, employee benefits, hiring practices, truck signage, recordkeeping, auditing, frequency of service to the Ports, and even upon sale or transfer of the motor carrier's business."[453] Although challenging both ports, the complaint used the fact that Long

Beach had broken ranks on employee conversion to buttress its Commerce Clause argument: "The Port of Los Angeles *prohibits* motor carriers' use of more than 10,000 independent owner-operators of trucks on *their side* of the city line that bisects the San Pedro Bay port complex, while the Port of Long Beach *permits* such subcontracting on *its side* of the line—a textbook case ... for federal preemption to prevent [the] patchwork of service-determining laws, rules, and regulations from disrupting the motor carriage of property in interstate commerce."[454]

The ATA complaint also sought to distinguish the environmental provisions of the Clean Truck Programs from what it called "extraneous, burdensome regulations regarding wages, benefits, truck ownership, preferences for certain types of trucks, and frequency of service to ports, which have no material environmental impact."[455] Making clear that it did *"not* challenge the Ports' truck engine-retirement programs"[456]—that is, the progressive ban on dirty trucks—the ATA set its sights specifically on the operational provisions of the concession plans. As the ATA's CEO put it, "the litigation is not aimed and should not interfere with the ports' clean air efforts. We are challenging only the intrusive and unnecessary regulatory structure being created under the concession plans."[457]

That the ATA was especially concerned about employee conversion was revealed in its prayer for relief, in which it first sought to enjoin both plans in their entirety (Count I) and then *separately* sought to enjoin just that portion of the Los Angeles plan that precluded "independent owner-operators" from port entry (Count II).[458] The ATA was thus giving the court a choice: Even if it did not think the concession plans as a whole were preempted, the court could decide to simply enjoin the employee provision. Overall, the ATA's complaint deftly carved lines: distinguishing Long Beach from Los Angeles, and within Los Angeles isolating the employee conversion piece from the other concession provisions. Two days later, the ATA moved for a preliminary injunction to bar implementation of the concession plans.[459] Speaking later about the injunction, the ATA's Whalen sounded what would be the industry group's talking points: "Let's be clear: We are not against clean trucks. ... We are objecting to concession plans that are going to squeeze out a lot of existing motor carriers and thousands of independent owner-operators."[460] From the coalition's point of view, this argument rang hollow. As NRDC's Martinez recalled, the ATA was "trying to just throw monkey wrenches after monkey wrenches" to stop the plans, while all the time insisting it was "supportive of clean air."[461]

Confronted with a new effort to split the environmental and labor elements of the program, and facing the prospect of the cities' legal interests diverging from their own, the coalition—not formally party to the suit—had to decide how to respond. Playing to its legal strength, and emphasizing the connection between the environmental and labor pieces of the Los Angeles program, the coalition turned again to its environmental partners. On July 31, the NRDC, Sierra Club, and Coalition for Clean Air moved to intervene in the case, arguing that its members had "significantly protectable interests" that would not be adequately represented by the defendant ports, which "as public proprietary entities, ... must balance resource constraints and the interests of various constituencies—some of which (such as those of Plaintiff) are at odds with the proposed intervenors' interest—and have often taken positions on port-related policy and regulatory matters contrary to" the intervenors.[462] The court agreed and the intervenors proceeded to make their case in support of the concession plans.

In their brief opposing the ATA's preliminary injunction motion, NRDC's Pettit, Perrella, and Martinez sought to lay out the environmental case for market participation, devoting the first part of their brief to explicating why the plans were "necessary to protect public health."[463] In addition to reviewing the evidence of air pollution and public health impacts, the intervenors' brief stressed how all elements of the concession plans, including employee conversion (not mentioned by name) were "intertwined and ... all necessary to achieve the Ports' clean air goals."[464] NRDC's Perrella recalled that although it was "very clear that we were in favor of L.A.'s program," NRDC intervened to protect both the Los Angeles and Long Beach plans since "the heart of [the] ATA's argument really attacked the fundamentals of both programs. And so we felt like it was important to protect" them.[465] As Martinez put it, "Part of [our] goal is to defend the whole damn thing, so ultimately we think the ports have rights to do this."[466] In the litigation, NRDC represented its own interests, which meant that the organization's lawyers "work[ed] together to figure out the best approach."[467] In also representing the Sierra Club and Coalition for Clean Air, NRDC's retainer identified one person within each organization that served as the representative for client relations purposes, keeping NRDC out of the other client organizations' internal deliberations.[468]

Throughout the litigation, NRDC worked with counsel for the Port of Los Angeles, with whom it had a much closer relation than with counsel for Long Beach—based on the Los Angeles city attorneys' supportive posture during the policy phase.[469] NRDC lawyers had a "joint defense agreement" with the Port of Los Angeles lawyers,[470] with whom they would discuss "who would take what approach to what issue,"[471] and exchange drafts of briefs.[472] Although the city defendants were definitely in the lead, NRDC did not "shy away from ... being parties in the case."[473] Toward that end, NRDC lawyers sought to make arguments that built upon their comparative advantage as environmental experts. In this vein, NRDC pressed the argument that the ports needed to implement the concession plans to avoid potential environmental liability. As the "troublemakers" that initiated port litigation in the first instance, NRDC could make that case "more forcefully" than the city defendants, particularly since the ports were not going to concede that they had ongoing environmental legal risk.[474] In addition, NRDC was well-positioned to make the strong case for showing how the employee provision "benefits the environment."[475] Particularly to the extent that the ATA was seeking to paint the Los Angeles plan as a sop to unions, having NRDC make the labor argument was a way to keep the entire plan within an environmental frame. During the lawsuit, Teamsters counsel Manley recalled that the plan was for him to "recede as much as possible to the background," although he believed that the ATA representatives were "going to paint it as a Teamster case anyway—which they did."[476] Manley participated "indirectly" by providing his analyses of the market participant doctrine to NRDC lawyers, to whom he told: "here's the argument, here's why it supports us."[477]

Asserting a unified front in the preliminary injunction phase, both ports filed a joint opposition that strongly asserted local authority for their respective concession plans—though with a different emphasis than the environmental intervenors' brief. On the Los Angeles side, outside counsel Steven Rosenthal was the defendants' lead attorney, supported by his colleagues at Kaye Scholer, who worked closely with city attorneys Tom Russell, Joy Crose, and Simon Kahn. Long Beach was similarly represented by its city attorneys and outside counsel from two law firms.[478]

In their opposition, the ports made three arguments against preemption. First, they asserted that the ports had sovereign control over the tideland areas under the state Tidelands Trust Act.[479] Second, they argued that the programs fell squarely within the market participant exception to federal preemption as the product of "a proprietary action of the Ports in their capacity as commercial enterprises and landlords."[480] Here, the ports emphasized the market dimensions of the concession plans:

> The [Clean Truck Program] is a product of the Ports' recognition that in order to grow and to continue to compete successfully in that market, they need to address major environmental and security issues. The concession programs reflect the Ports' efforts to secure trucking services—services critical to their commercial operation—in a way that will further those objectives. Hence those programs fall squarely within the market participant doctrine.[481]

In making this argument, the ports relied heavily on the *Engine Manufacturers* case, litigated by NRDC, applying the market participant exception to the Clean Air Act in upholding SCAQMD rules setting emission standards for vehicles acquired by the state.[482] Based on that case, the ports emphasized that they were acting as a market participant in the "efficient procurement" of trucking services and that the plans' "narrow scope" defeated an inference that their "primary goal was to encourage a general policy rather than address a specific proprietary problem."[483] As a third ground of opposition, the ports argued that the concessions fell within a statutory exception to FAAAA preemption, which stated that the law "shall not restrict the safety regulatory authority of a State with respect to motor vehicles."[484] Specifically, the ports claimed that the plans were designed to enhance port security and eliminate unsafe trucks[485]—drawing on arguments that were laid out in the findings sections of the Clean Truck Program legislation. To buttress this argument, the ports submitted an affidavit from John Holmes averring that the Los Angeles program was, in addition to addressing environmental harms, "designed to address other problems that result from the truck activities at the Ports—safety and security."[486]

Weighing these arguments was District Judge Christina Snyder, who had been appointed in 1997 by President Bill Clinton after serving as a partner in a Los Angeles law firm.[487] In the early part of her career, Judge Snyder was a founder of Public Counsel Law Center, one of Los Angeles's largest legal services organizations, serving on its board and as its president.[488] Her first opinion was an early victory for the ports. In an order dated September 9, 2008, she denied the ATA's motion for a preliminary injunction.[489] Yet the scope of the order highlighted the challenges that the ports would face ahead. First, the court held that the concession plans likely regulated the "price, route, or service" of trucking companies and thus fell within the scope of FAAAA preemption.[490] The issue was therefore whether an exception to preemption would save the plans. Of the ports' three arguments against preemption, Judge Snyder only agreed with one: that "the defendants have shown that there is a significant probability that the concession agreements fall under the safety exception to the FAAAA, and that they may therefore be saved from preemption."[491]

Although acknowledging that there "does not appear to be any case law addressing the question of whether security concerns analogous to the concerns identified by the Ports fall within the safety exception," Judge Snyder ruled that it was likely they did even though the concessions were not passed for the "exclusive purpose of promoting safety."[492] However, the court rejected the state tidelands argument, noting that it was "not convinced that the fact that the Ports rest on sovereign tidelands renders them immune from preemption under the FAAA[A]."[493] Perhaps more alarmingly for the ports, the court ruled against application of the market participant exception on the grounds that the programs were not about "efficient procurement" but rather were "akin to a licensing scheme," and they were not sufficiently narrow in scope to qualify as meeting a "specific proprietary problem."[494] Because it lost the motion, the ATA appealed.[495] While both ports separately filed answers to the ATA's complaint, the case would now move—for the first time—to the Ninth Circuit.

As it did, the coalition continued its organizing and media work outside of court to make the case for employee conversion. Part of this strategy involved criticizing Long Beach's program. Speaking for a coalition of civil rights and consumer groups, NAACP President Julian Bond compared Long Beach drivers, who were being forced to take on debt to finance low-emission trucks, to sharecroppers in the Deep South, warning of a wave of "foreclosures on wheels."[496] Under the Long Beach plan, the Mercedes-Benz/Daimler Truck Finance Company was administering a lease-to-own program in which drivers would be given loans to purchase new clean trucks worth over $100,000.[497] However, as the coalition stated in a report delivered to Daimler headquarters, even with port incentives many drivers would not be able to meet the payments and would thus be subject to Daimler's aggressive collections department.[498] While the coalition urged Long Beach to adopt the Los Angeles employee conversion approach, some workers were confused, with one quoted as stating: "A lot of truckers have no idea what's going on with all these different plans and protests, so they are just going with the flow."[499] Yet the protests continued, with sixty truckers—some chanting "Clean trucks, yes! Bankruptcy, no!"—gathering to condemn the opening of Long Beach's Clean Trucks Center, charged with administering the financial incentive component of the Long Beach program.[500]

Concerns that trucking companies would boycott the Los Angeles program when it officially began on October 1, 2008 were allayed when two large national firms—Swift and Knight—agreed to participate in exchange for financial incentives.[501] This was followed by an announcement that 120 carriers—many members of the ATA—had applied to service the port under program criteria.[502] With the Great Recession taking its economic toll, some viewed both ports' Clean Truck Programs as coming at the most inauspicious economic time as container traffic had dropped sharply from the previous year.[503] However, the recession also made clean truck conversion more attractive for trucking companies dependent on diesel fuel.[504] As diesel prices soared in 2008 and hundreds of companies failed nationwide, some firms welcomed subsidies to convert trucking fleets to cleaner technology.[505] When the Los Angeles and Long Beach programs were formally initiated on October 1, both mayors hailed them as a success: with 95 percent of trucks entering the ports meeting program standards, only 2,000 had to be turned away.[506] Coalition member Martin Schlageter praised the programs' arrival: "Powerful institutional forces representing billions of dollars had for years urged that the ports not do anything. But the mayors and the ports stood firm."[507]

It was against this backdrop that the ATA's appeal of Judge Snyder's September 2008 denial of injunctive relief took place. On appeal, the industry group—represented by the same lawyers as in the lower court below—contested the lower court's interpretation of the preliminary injunction standard and its ruling that the FAAAA's "safety exception" likely protected the concession plans from federal preemption.[508] The ports countered these arguments and also presented their own claims for local authority under the market participant exception.[509] As they had in the lower court, the environmental intervenors reiterated the public health benefits of the plans, defended the lower court's application of the safety exception, and then reasserted the case for market participation.[510] The appeal attracted wider attention: the U.S. government (Department of Transportation, Federal Motor Carrier Safety Administration, and Department of Justice) filed an amicus brief in support of the ATA,[511] as did the National Industrial Transportation League[512] and National Association of Waterfront Employers,[513] all arguing against the lower court's application of the safety exception. The California attorney general came in on the side of the ports.[514]

The stage was thus set for a ruling by a three-judge panel of the Ninth Circuit. Although they differed in ideological orientation—Robert Beezer and Ferdinand Fernandez were conservatives appointed by Ronald Reagan, while Richard Paez was a liberal former legal aid lawyer appointed by Clinton—they spoke with a unanimous voice. In an order issued on March 20, 2009, the panel reversed the district court's denial of injunctive relief.[515] The court's key move was to distinguish among the various provisions of the concession plans, holding that "the district court legally erred in not examining the specific provisions of the Concession agreements, and it is likely that many of those provisions are preempted."[516] The panel's analysis rejected the ports' claims with devastating thoroughness.

Although it agreed with the district court on the preliminary injunction standard and its application to the tidelands and market participant arguments,[517] the panel disagreed on the safety exception. Noting that "the mere fact that one part of a regulation or group of regulations might come within an exception to preemption does not mean that all other parts of that regulation or group are also excepted,"[518] the court proceeded to highlight the nonsafety rationales for the plans, including "an extensive attempt to reshape and control the economics of the drayage industry" and to "ameliorate ... adverse economic effects."[519]

The court then turned to analyzing the plans' individual provisions. Homing in on the employee conversion piece in the Los Angeles plan, the court used the very findings the port had included in the ordinance to support market participant status to undercut the safety exception argument, asserting that "the record demonstrates that the Ports' primary concern was increasing efficiency and regulating the drayage market."[520] On that basis, the court held that "as we see it, the independent contractor phase-out provision is one highly likely to be shown to be preempted."[521] The court expressed similar skepticism about provisions in both plans that required job posting, hiring preferences for experienced drivers, and financial disclosure, as well as Los Angeles's off-street parking ban and Long Beach's driver health insurance requirement.[522] Moving through the preliminary injunction test, the court went on to suggest that a company faced with a concession plan would suffer irreparable harm by being put to a "Hobson's choice": if it refused to sign, its port drayage business likely would "evaporate" with the result for a small carrier probably "fatal," while if it did sign, a company would "incur large costs," which would "disrupt and change the whole nature of its business."[523] Noting that the public interest in deregulation further supported issuing a preliminary injunction, the court stopped just short of complete reversal, stating that in light of the severability provision: "we are not prepared to hold that every provision must be preempted."[524] Accordingly, the appellate court remanded the case back to the district court for "further consideration of the specific terms of each agreement and for the issuance of an appropriate preliminary injunction."[525]

Smelling blood, the ATA renewed its motion for preliminary injunction against both ports' plans on the ground of FAAAA preemption.[526] It was at this stage that the Los Angeles and Long Beach ports once again parted ways, each filing separate oppositions. In its brief, the Los Angeles port decided to play defense, conceding the provisions the Ninth Circuit had explicitly addressed (employee conversion, off-street parking, job posting and driver preferences, and financial capability), while arguing against preemption on safety grounds for those that remained.[527] There was a tension in this position given that it conceded the severability of a plan that the port had long argued constituted an indivisible scheme to redress port pollution. Yet the terms of the Ninth Circuit opinion seemed to require this tactical position to salvage any part of the plan. Long Beach was more aggressive, arguing that none of its plan's provisions, including those singled out by the Ninth Circuit, were preempted either because they simply duplicated already existing law or were related to port safety.[528] The ATA responded that duplicative provisions should be removed from the plans and because the preempted provisions were *not* severable, the plans "should be enjoined in their entirety."[529] The *Los Angeles Times* editors, anticipating the district court decision, weighed in against the Los Angeles program, asking Mayor Villaraigosa "and his union backers" to just "[l]et it go."[530]

On April 28, 2009, the district court ruled for the second time on the ATA's motion for preliminary injunction, issuing a split decision.[531] Taking a provision-by-provision approach, the court preliminarily enjoined key elements of the ports' plans—including Los Angeles's employee conversion provision, as well as both ports' provisions on hiring preferences, financial capability, and parking and route restrictions[532]—while letting stand other provisions the court held to be related to port safety.[533] Rejecting the ATA's claim that the plans should rise or fall as unified agreements, the court held that the preempted provisions were severable and that the safety provisions could be effectively implemented on their own.[534] NRDC released a statement criticizing the ruling: "Without the employee program, port cleanup goals could be severely delayed because most independent owner-operators cannot afford to maintain and repair their trucks."[535]

The ATA once again took appeal, disputing the district court's decision not to preempt the provisions deemed safety related.[536] By the time the Ninth Circuit decided the case, however, Long Beach was no longer part of it. With the appeal pending, the Port of Long Beach decided to strike a deal with the ATA, in which the port would permit truck access under a new "Registration and Agreement" to supersede the concession plan.[537] The new agreement stripped away provisions deemed unrelated to clean truck conversion—including those dealing with parking, truck routes, and financial capability.[538] Under the final agreement, concessionaires would certify truck compliance with basic registration, identification, safety, and security standards, while also certifying compliance with the environmental provisions of the Clean Truck Program.[539] On the basis of that settlement, the court dismissed the Long Beach defendants on October 20, 2009.[540] While Long Beach port director Richard Steinke affirmed his port's commitment to going green, Los Angeles director Geraldine Knatz asked, "Who will pay for the next fleet of clean trucks when today's new trucks will need to be replaced?"[541] The NRDC and other environmental groups challenged this settlement on CEQA grounds, arguing that it "substantially weakened the environmental benefits of the Port's Clean Truck Program" and that the port violated CEQA by failing to conduct an appropriate environmental review.[542] The NRDC's concern was that the agreement would water down enforcement of the program's maintenance provisions, thus presenting a novel legal issue: whether a government entity could change a program via settlement agreement without conducting another EIR. The district court held they could not, ordering an initial environmental study in July 2011.[543] The city conducted a study and issued a negative declaration, stating that the settlement agreement would not have a significant effect on the environment.[544]

Left to fend for itself, the Port of Los Angeles confronted an invigorated adversary. In court, the ATA sought to compel the disclosure of internal port documents reflecting staff deliberations over the concession agreement, to which the ATA claimed to be entitled in order to rebut port assertions that the agreement was motivated by safety concerns.[545] Although the ATA would ultimately lose this argument on privilege grounds,[546] it suggested the extent to which the ATA was willing to go to win. Out of court, ATA allies sought to take a page from the coalition's book, organizing drivers in opposition to the Los Angeles program. In November 2009, the National Port Drivers Association, claiming to represent independent-contractor drivers, staged a protest in which 400 truckers drove up the 710 freeway to Los Angeles City Hall, criticizing the fact that clean truck funding had not been given to independent-contractor drivers.[547] Around the same time, the Long Beach port began a marketing campaign to promote its program to community residents, featuring the president of ILWU Local 11, George Lujan, stating that his "union supports the Port of Long Beach Clean Trucks Program."[548]

On appeal, the Port of Los Angeles lost a little more ground in front of a different three-judge panel of the Ninth Circuit.[549] On February 24, 2010, the panel agreed with the district court's decision to enjoin the provisions it did but also decided to enjoin one more—the placard provision, which the court found to be specifically outlawed by a separate section of the FAAAA to which the safety exception did not apply.[550] The Ninth Circuit opinion contained a slight bit of good news for the Los Angeles port: holding that the port could exclude motor carriers that did not comply with the concession agreement and permitting implementation of the nonpreempted provisions.[551] However, the overall picture for Los Angeles looked grim. With Long Beach out of the case and the core features of its concession plan subject to a preliminary injunction, the Port of Los Angeles appeared to face long odds heading into trial,[552] which was set for April 2010.

The dramatic turnaround—from the district court's initial support of the concession plan prior to its official launch date in the fall of 2008 to its preliminary injunction in the spring of 2009—affected the rollout of the Clean Truck Program. Although scheduled to begin on October 1, 2008, it was widely assumed that the Los Angeles port would not strictly enforce the concession plan until April 2009 as it worked out technical issues and permitted trucking companies to purchase new trucks and begin the phase-in of employee conversion.[553] Despite the ATA lawsuit filed in July 2008, program implementation proceeded apace. As Jon Zerolnick recalled, at first, "nothing changed."[554] Trucking companies drew upon port and state funding to begin converting their fleets, while some, like Southern Counties Express and Swift, took initial steps to convert their drivers to employees.[555] By early 2009, the coalition was supporting Teamsters organizing at companies that had moved to employee drivers.[556] The district court's preliminary injunction ruling in April 2009 changed everything. With the threat of port sanction lifted, companies promptly moved back to independent-contractor drivers. Zerolnick described his impression of the industry's position this way: "Now once it's clear that I don't have to do this stuff and in fact by doing this stuff I'm putting myself at a pretty serious competitive disadvantage, then why the hell am I going to keep doing this?"[557]

As trucking companies kept or moved back to independent-contractor drivers, they also shifted to them the new costs associated with purchasing clean trucks. The ban on old, dirty trucks—the second phase of which was set to go into effect on January 1, 2010—forced drivers to purchase or lease new and expensive low-emission vehicles. Although the port offered subsidies, some drivers could not take advantage of them to buy their own trucks because they could not qualify for loans—or simply could not afford to pay off the loans, even with port incentives.[558] As one driver ominously predicted: "The first of the year will probably be the end of my family."[559]

Even for those who could acquire their own trucks, the total amount of the loan payments, higher maintenance and insurance costs, and higher registration fees placed new burdens on drivers who struggled before the program was implemented—and who faced even more intense challenges as the recession reduced work opportunities. In this context, trucking companies were cutting contract rates, putting the drivers in even more economic peril. The *Los Angeles Times* profiled one driver who sold his old truck and joined a company that still hired some employee drivers:

"We're like slaves. We've lost our freedom."[560] Those drivers who stayed independent owners appeared to have no more autonomy. Trucking companies that had directly purchased clean trucks made drivers lease them back, deducting insurance and maintenance costs from their pay.[561] The *Los Angeles Times* reported that these new trucks cost 50 percent more to operate on top of lease payments of $1,000 per month.[562] "Things were bad enough when we owned our trucks, but I would say the situation is desperate now," one driver concluded.[563] Another, with twenty years of trucking experience, reported that his take-home pay was $7 an hour—less than the minimum wage. His assessment was harsh: "This program has been a great deception to us. ... We no longer have hope to be in the middle class. We are all poor now."[564]

Public Litigation: Federal Agency Intervention
In the midst of the program rollout—and before the ATA trial—another legal altercation over the Clean Truck Program would be resolved. On the heels of Judge Snyder's first preliminary injunction denial in 2008, a second front opened in the litigation battle—this one initiated by the Federal Maritime Commission (FMC) exercising its jurisdiction to prevent anticompetitive agreements between U.S. ports.[565] On the verge of passing the CAAP program back in 2006, the ports asked for FMC approval of their plan to "promote cooperation, openness and joint action through means of discussion, development of consensus and agreement" in order to "decrease port-related air pollution emissions."[566] In August 2008, after passage of the Clean Truck Programs, the ports filed an amendment further detailing their agreement to "discuss, exchange information, cooperate and, to the extent each Port in its sole discretion deems appropriate, coordinate the adoption and implementation of programs to reduce truck emissions and improve Port safety and security."[567]

Responding to this amendment, on September 12, the five-member FMC (staffed with a majority of members appointed by President George W. Bush) issued a nine-page Request for Additional Information to determine the competitive impact of the coordinated programs—sparking a strong dissent by Commissioner Joseph Brennan, who argued that the commission was "making a monumental mistake in delaying, yet again, the overall environmental plan of the ports."[568] From the coalition's point of view, the FMC action was the result of industry representatives going to Washington, D.C. to pressure "a more favorable agency" to create problems for the ports in the wake of the industry's initial district court loss.[569] After an exchange of documentation with the ports, the FMC formally voted to seek the "surgical removal of substantially anticompetitive elements of the agreement, such as the employee mandate."[570]

On October 31, 2008, three months after the initial ATA lawsuit was filed—and in the waning days of the Bush administration—the FMC filed a complaint in the D.C. district court to enjoin both ports' Clean Truck Programs pursuant to section 6(h) of the Shipping Act of 1984,[571] which authorized the FMC to bring a civil action if an "agreement is likely, by a reduction in competition, to produce an unreasonable reduction in transportation service or an unreasonable increase in transportation cost."[572] The complaint, though framed broadly, homed in on the Los Angeles employee driver provision, challenging as "substantially anticompetitive" any program "to discuss, agree or implement a concession plan or plans ... that requires, directly or indirectly, the use of only employee drivers to perform truck drayage service" or that "prohibits, directly or indirectly, the use of independent owner-operator drivers."[573] In elaborating its legal claim, the complaint stated that "the Commission determined that the [Clean Truck Program]-induced changes to the drayage market and corresponding reduction in competition caused by the requirements to use employee-drivers exclusively ... will give rise to substantial transportation cost increases, beyond what is necessary to generate the public health and environmental benefits asserted by the Ports."[574]

The case moved quickly—and came to an abrupt resolution after President Barack Obama, who wrote a letter in support of the Clean Truck Program as a candidate,[575] took office in early 2009.[576] On November 17, 2008, the FMC moved for a preliminary injunction against that portion of the Clean Truck Program that "(1) requires the use of employee-drivers by LMC concessionaires; or (2) establishes truck purchasing incentives, subsidies and clean truck fee exemptions that disadvantage Independent Owner Operators" at the ports.[577] Because this motion was the first time since the enactment of section 6(h) that the FMC had sought a preliminary injunction, there was no precedent for the standard to be applied. In its pleadings, the FMC argued for a less onerous rule than that typically applied in preliminary injunction cases, urging the court to adopt a "more flexible" standard based on section 6(h)'s test for permanent injunctions in which the court would only consider whether the FMC had a substantial likelihood of success on the merits.[578] The FMC argued that it was likely to succeed based on an "extensive economic impact study" performed by its chief economist showing that "the net cost impact of the explicit POLA employee mandate, as presently confined to that port alone, likely will range between $3.0 billion and $4.6 billion through 2025—without offsetting benefits."[579]

Both ports—Los Angeles still represented by Kaye Scholer's Rosenthal and Long Beach by outside counsel from Troutman Sanders—vigorously disputed both the characterization of the doctrinal test for preliminary injunction and the assessment of the FMC's likelihood of success.[580] In a ruling on April 15, 2009, the district court for the District of Columbia, in an opinion by President George W. Bush appointee Richard Leon, squarely sided with the ports.[581] Calling the FMC's interpretation of the preliminary injunction standard a "stretch," the court ruled that the FMC could not meet it anyway.[582] Although the commission had shown a potential increase in transportation costs, it had not demonstrated that the increase was the product of reduced competition—both because any trucking cost increase seemed to result from compliance with the programs' terms (rather than market concentration) and because Long Beach's rejection of employee conversion showed that the ports were "actually in competition."[583]

The FMC swiftly retreated—perhaps motivated more by political change than legal defeat. In early June 2009, President Obama appointed Commissioner Brennan—the lone dissenter from the initial lawsuit—as acting FMC chair.[584] One week later, the commission moved to dismiss the proceeding, arguing that a number of "events resolve the issues underpinning the Plaintiff Commission's decision to bring this action and render unnecessary an injunction by this Court."[585] One basis that the FMC raised in moving to dismiss its lawsuit was the fact that the district court in the ATA suit had by then already enjoined the employee conversion provision of Los Angeles's concession plan—thus mooting the FMC challenge.[586] Another factor motivating the FMC's dismissal was the Great Recession, which had significantly reduced cargo shipments and imposed financial hardships on both ports as they sought to implement their truck conversion programs.[587] When it was initially filed, the FMC action had blocked the ports' power to collect the fees they had planned to use to fund clean truck purchases, contributing to a shortage of promised incentive financing.[588] After the district court rejected the FMC's preliminary injunction motion, the Port of Long Beach sought to bring its incentive program in line with the more generous program instituted in Los Angeles. Toward that end, on April 20, 2009, the Long Beach harbor commission harmonized its clean truck incentives with those of Los Angeles.[589] In its motion to dismiss, the FMC argued that this harmonization mooted the case to the extent that it had "sought to enjoin the disparities between the Ports."[590] The court agreed and the action was formally dismissed the following month.[591]

Private Litigation II: The Merits Phase

The dismissal of the FMC suit in the summer of 2009, at the moment the ATA was taking its appeal from Judge Snyder's second preliminary injunction ruling, raised the stakes of the ATA lawsuit, which was now the only legal barrier to the Los Angeles port's Clean Truck Program. As the case moved toward trial, following the 2010 Ninth Circuit ruling that preliminarily enjoined the employee driver provision (as well as the provisions on hiring preferences, financial capability, off-street parking, truck routes, and placards), each side focused on strengthening their arguments. While this effort concentrated on crafting legal briefs and assembling evidence, the coalition also sought to reinforce its arguments for clean trucks in the public domain. Seeking to shore up the economic case in favor of the vulnerable employee provision, LAANE issued a report on the cusp of trial demonstrating that "the combined costs for clean truck leases and vehicle maintenance are out of reach for individual port drivers" thus undermining the "heart of the environmental policy."[592] Timed to correspond with the parties' final briefings, the stage was set for trial.

The trial briefs laid out the now-familiar pattern of disagreement. The ATA argued FAAAA preemption of the Los Angeles concession plan; contended that although it did not have to show each provision was unrelated to safety, it could; and then argued against market participation and in favor of finding that the plan unduly burdened interstate commerce.[593] Narrowing its focus for trial, the ATA only challenged five key provisions of the Los Angeles concession plan related to: (1) employee conversion, (2) off-street parking, (3) maintenance, (4) placards, and (5) financial capability.[594] Reprising a back-up argument first made at the preliminary injunction hearing, which had gained increasing court attention throughout the case, the ATA also sought to apply the Supreme Court's 1954 decision in *Castle v. Hayes Freight Lines*, which held that states were prohibited under the Motor Carrier Act from interfering with a carrier's right to operate in interstate commerce.[595] Although the district court had earlier rejected this argument on the ground that the Motor Carrier Act was passed forty years before the FAAAA—whose safety exception permitted states to suspend carrier service—the Ninth Circuit had cryptically stated that "this issue is not finally resolved and may be reconsidered in further proceedings for a permanent injunction."[596] The ATA used this opening to argue that *Castle* stood for the proposition that only the federal government could determine a carrier's safety fitness and thus the port could not enforce the safety provisions of its concession agreement by barring truck access.[597]

Reforming the Port

For its part, the Port of Los Angeles took a slightly different approach than in the preliminary injunction phase. First, the port contended that the concession plan's individual provisions did not have the "force and effect of law related to a price, route, or service" and therefore were not preempted by FAAAA section 14501(c) at all. The port then alternatively argued that—assuming preemption did apply—the provisions fell within the market participant exception and also were permitted under the FAAAA's safety exception.[598] This order reflected the port's sense that the safety exception was a relatively weaker argument, but one that it "was kind of stuck with" after the district court upheld several provisions on that basis.[599] Sending a similar message, the port devoted only a page to the ATA's commerce clause argument and relegated the *Castle* claim to a footnote.[600] Not feeling as constrained by the lower court ruling, the NRDC's position at trial emphasized the validity of the concession plan under the market participant exception.[601]

The trial lasted seven days. The ATA's chief counsel, Robert Digges, appeared on behalf of the plaintiffs, alongside outside counsel Christopher McNatt, Jr. from Scopelitis, Garvin, Light, Hanson & Feary. On the Los Angeles side, city attorneys Tom Russell and Simon Kahn appeared with outside counsel from Kaye Scholer (Steven Rosenthal and his team). NRDC lawyers Melissa Lin Perrella and David Pettit appeared on behalf of the environmental intervenors. At trial, port counsel and NRDC each made opening statements, emphasizing distinct themes that foreshadowed counsels' division of labor throughout the remaining litigation. Port counsel Rosenthal sought to lay out the case that the Clean Truck Program was adopted to "address specific proprietary concerns at the port."[602] He suggested that the evidence would show the need for the port as an "enormous commercial enterprise" to address environmental and community impacts that had "brought significant expansion and improvement at the port to a screeching halt."[603] He also stressed the port's need to respond to security risks in laying out the case for application of the safety exception.[604] For the environmental intervenors, NRDC's Perrella focused on "[w]hether the remediation of port-generated air pollution by the Clean Truck Program and specifically by the concession agreement is protected under the market participant doctrine."[605] As she made clear, the intervenors' case would stress the public health impacts of port pollution and tie them directly to the independent-contractor status of the drivers.[606]

During trial, Rosenthal and his team presented evidence to show how the Clean Truck Program responded to proprietary concerns. Rosenthal questioned key port decision makers: Executive Director Geraldine Knatz, who discussed program details;[607] Commissioner David Freeman, who emphasized that the program was designed to facilitate port expansion;[608] and Deputy Executive Director of Operations John Holmes, who discussed how the program was intended "to provide a level of accountability, but also to insure the program was sustainable and that it met the environmental and security goals of the port."[609] Rosenthal's colleagues elicited testimony from witnesses highlighting how the program responded to local port traffic problems and transportation security concerns.[610] Perrella took the lead in questioning Dr. Elaine Chang, who outlined the case for port-induced air pollution and the need for the Clean Truck Program to address it,[611] and Long Beach resident Bernice Banares, who testified about her own asthma and safety concerns raised by port trucks.[612] NRDC's role was to "get into the record what the public health and environmental problem is and then to draw out facts related to how the port sought to address those problems and how addressing those problems was really intertwined with its pursuing its commercial interests."[613]

With the evidence thus tendered, the parties waited for Judge Snyder to rule, which she did on August 26, 2010, in a decision that gave a surprisingly sweeping victory to the port. In the court's detailed Findings of Fact and Conclusions of Law, Judge Snyder ruled that none of the provisions challenged by the ATA were preempted—contradicting the Ninth Circuit's assessment during the preliminary injunction phase. The trial court's decision—with an eye on the ATA's inevitable appeal—set forth alternative grounds for different provisions. With respect to the maintenance, placard, and financial capability provisions, the court ruled that there was insufficient evidence that they would affect truck prices, routes, or services and were thus not preempted under section 14501(c);[614] alternatively, the court concluded that even if these provisions were preempted, the maintenance and placard provisions fell within the safety exception (though the financial capacity provision, enacted to ensure "the Port will not lose its investment in truck grants," did not).[615] With respect to the employee driver and off-street parking provisions, the court held both preempted and neither within the safety exception.[616]

However, here the court moved to a different rationale: market participation. Rejecting the ATA's narrow definition of proprietary action, the court stated that "where restrictions are placed on services essential to the functioning of a government-run commercial enterprise, the market participant exception applies to non-procurement decisions."[617] The court then ruled that the entire concession agreement was "essentially proprietary"[618]—and thus not preempted—because it was passed "in response to litigation and the threat to ... [the Port of Los Angeles's] continued economic viability by community groups ... as a 'business necessity,' in order to eliminate obstacles to its growth."[619] Wanting to cover all its bases, the court also found each individual provision to fall within the market participant exception. Focusing on employee conversion, the court agreed with the port that it was "designed to transfer the financial burden of administration and record-keeping onto the trucking companies," which was "clearly an economically motivated action, and one that a private company with substantial market power—such as the oligopoly power of the Port—would take when possible in pursuit of maximizing profit."[620] The court further found that the off-street parking and placard provisions were "designed specifically to generate goodwill among local residents and to minimize exposure to litigation from them," while the financial capability and maintenance provisions were "aimed to ensure that the trucking companies had the resources to sustain the Port's investment in cleaner trucks."[621] Finally rejecting the ATA's *Castle* and dormant commerce clause claims,[622] the court resoundingly validated the port's Clean Truck Program and the legal strategy that produced it—at least for the moment.

That moment passed quickly. In September 2010, the ATA appealed—and back up the ladder the case went. In doing so, the ATA requested that the court stay the implementation of the Clean Truck Program pending the appeal.[623] Judge Snyder agreed to temporarily enjoin the employee provision, which she held was likely to produce irreparable harm to plaintiffs that was not outweighed by other equities, while permitting implementation of the rest of the concession plan.[624] By this point, the issues dividing the parties were fully crystalized and their briefs reflected well-worn arguments for and against preemption.[625] However, to underscore the stakes, a number of new amici weighed in on the side of the ATA's appeal: the Intermodal Association of North America, asserting the negative impact of the concession plan on the intermodal industry;[626] the National Right to Work Legal Defense Foundation, which argued that the concessions forced independent-contractor drivers to sacrifice their right to work as such;[627] the Owner Operator Independent Drivers Association, which argued that the concession plan was not responsive to the issues facing non-short-haul drivers;[628] and the Center for Constitutional Jurisprudence, which argued that market participation could only be exercised through procurement.[629]

The judges assigned to the appellate panel hearing the case were Betty Fletcher, an iconic liberal appointed by President Jimmy Carter as only the second woman judge in the Ninth Circuit; Randy Smith, a strong conservative from Idaho appointed by President George W. Bush; and Rudi Brewster, a senior district court judge from San Diego who had been appointed by President Reagan. In a two-to-one decision, with Smith in dissent, the panel upheld the bulk of the concession plan—but dealt the decisive blow to employee conversion.[630] Taking an expansive view of market participation, the court held that "when an independent State entity manages access to its facilities, and imposes conditions similar to those that would be imposed by a private landlord in the State's position, the State may claim the market participant doctrine."[631] Here, because the "Port has a financial interest in ensuring that drayage services are provided in a manner that is safe, reliable, and consistent with the Port's overall goals for facilities management," the court concluded that "the Port acted in its proprietary capacity as a market participant when it decided to enter into concession agreements."[632] However, the court stopped short "of holding that every provision in the concession agreement" was therefore valid, instead opting to "examine whether the provisions at issue further the State's interests as a facilities manager, or whether the provisions seek to affect conduct unrelated to those interests."[633] The court also made clear that the safety exception was available for appropriate provisions, despite the ATA's reading of *Castle* to the contrary.[634]

Turning to the specific provisions at issue, the court upheld four on diverse grounds: it concluded that the financial capability provision did not affect rates, routes, or services and thus was not preempted; it found the maintenance provision to fall within the safety exception; and it upheld the off-street parking and placard provisions as proprietary acts of the port as a market participant.[635] Yet the court could find nothing to save employee conversion, which it concluded sought "to impact third party behavior unrelated to the performance of the concessionaire's obligations to the Port."[636] Recognizing the port's interest in providing higher wages to attract drivers lost to the TWIC program, the court nonetheless concluded that the port could not achieve market stability "by unilaterally inserting itself into the contractual relationship between motor carriers and drivers."[637] Further recognizing the port's interest in protecting its investment in clean trucks, the court concluded that the concession agreements swept too broadly by binding all LMCs, "not merely those who drive Port-subsidized trucks."[638] Finally, acknowledging the port's interest in "streamlined administration" over a smaller number of LMCs, the court found it "insufficient to outweigh the Port's avowed desire to impact wages not subsidized by the State."[639] With this, employee conversion—the linchpin of a monumental campaign and innovative local policy—was held to be "tantamount to regulation" and thus preempted.[640]

It was September 2011, five years after the Campaign for Clean Trucks had begun, and the piece that had held the labor-environmental alliance together was gone. LAANE's Patricia Castellanos, responding to the decision, stated that it would "have devastating consequences for working families and port communities plagued by dirty air and dead-end jobs."[641] The port's (and coalition's) early calculation that it could win at the Ninth Circuit level—made with full knowledge of the uncertainty—turned out to be wrong as to employee conversion. The editorial board of the *Los Angeles Times*, on the eve of the Ninth Circuit decision, had expressed hope that it would "end the city's misguided attempt to team up with the Teamsters."[642] And, indeed, the port decided not to appeal the ruling—though not for the reasons suggested by the *Times*. On the negative side of the ledger, the Ninth Circuit ruling sent an undeniable signal: unable to persuade one of the circuit's most liberal judges, Betty Fletcher, it seemed fruitless—even reckless—to press the case for employee conversion in front of the conservative majority on the Supreme Court. On the positive side, it also was possible for the port and coalition to count the Ninth Circuit ruling as a win and walk away. As NRDC's Pettit saw it, the Ninth Circuit ruling endorsed the idea that a port "can have a concession plan and can put conditions on trucks ... even [those] in interstate commerce."[643] With the basic foundation of the concession concept thus left "intact,"[644] NRDC lawyers "viewed what we got from the Ninth Circuit as a victory."[645] As it turned out, so did the ATA.

Private Litigation III: The Supreme Court Phase

Refusing to settle for the victory over employee drivers, the ATA again set its sights on gutting the concession plan, this time by appealing to the Supreme Court.[646] NRDC saw the appeal as a statement by the ATA that "state and local government should not be able to place ... really any requirements on motor carriers, so it was: 'If we can show that you can't even do this placard provision, then that means that you can't do anything.'"[647] In pressing this case, the ATA retained new counsel for the appeal: Supreme Court specialist Roy Englert, an assistant solicitor general under President Reagan who had started his own appellate firm in 2001 and boasted a nearly perfect record in front of the Court. NRDC's Perrella noticed the difference, recalling that the ATA's Supreme Court counsel was "phenomenal," framing the briefs in a way "that was just really compelling."[648]

In its petition for certiorari, the ATA asked the Court to resolve what it characterized as three significant circuit splits: one over the application of the market participant exception to preemption, a second over the scope of FAAAA preemption, and a third over the vitality of *Castle*. With respect to market participation, the ATA argued that the Ninth Circuit created a conflict by saving the port's concession plans from preemption based on its status as a property owner when it did not "actually participate in the market" for drayage trucking and when the plan's restrictions were "unrelated to the efficient procurement of services."[649] The ATA further asserted that the Ninth Circuit read the FAAAA "rates, routes, or services" preemption too narrowly and that *Castle* still barred a state from "enforcing its laws through even a partial suspension of the motor carrier's ability to operate in interstate commerce."[650]

In response, the port sought to minimize the legal stakes, arguing that the Ninth Circuit's decision was in fact congruent with those of other circuits on market participation and thus no circuit split existed. Furthermore, the port suggested that the other issues presented were not substantial enough to warrant Court review: specifically, the Ninth Circuit's single ruling that the FAAAA did not preempt the financial capability provision and its narrow application of the safety exception to the maintenance provision were too minor to justify granting cert.[651] For their part, the environmental intervenors emphasized that the Ninth Circuit's decision on market participation was based on the trial court's extensive factual findings on the port's health and community impacts, which the ATA did not challenge on appeal.[652] They also continued to emphasize the business benefits of the port going green, which NRDC as an environmental organization believed it was best positioned to do.[653] In March 2012, the Supreme Court invited the U.S. Solicitor General to express its views on the matter, which it did, coming in on the side of the ATA.[654] In the U.S. brief, Solicitor General Donald Verrilli argued for an expansive concept of preemption, rejecting the idea that a port should be able to impose special rules, claiming instead that the port was "akin to a publicly managed transportation infrastructure" and thus should not be able to impose restrictions that were inconsistent with other ports.[655] However, the solicitor general also threw a line to the port by recommending that the Court not grant certiorari based on the limited significance of the Los Angeles case.[656]

If supporters of the Clean Truck Program were surprised when the Ninth Circuit struck down the employee driver provision, they were shocked when the Court agreed to consider the ATA's market participant and *Castle* claims: setting up review of the concession plan's placard and off-street parking provisions.[657] NRDC's Perrella described her response: "Surprised? ... It completely devastated us. ... I don't think I've had that really horrible feeling in my stomach ... the way I did when I found out the Supreme Court had decided to take the case. ... [H]ere we go again with a wacky environmental case in the Ninth Circuit before a bunch of conservative judges ... probably not the best forum for us."[658]

The Court decision to take the case signaled an interest in perhaps curtailing the market participant doctrine at the core of the Ninth Circuit's decision. Accordingly, the port adjusted its market participant argument in its merits brief. Parrying the ATA's claim that the port did not participate directly in the drayage market, the port emphasized its right as a property owner to enter into agreements affecting access to its land. In summarizing its core position, the port asserted that "absent a statement of clear congressional intent to the contrary, the courts should presume that proprietary state conduct dealing with the management of state-owned property is within the market participant doctrine."[659] Because the port as property owner had a clear "commercial motivation" in the contested provisions—promoting truck safety and improving community relations—they fell within the scope of the market participation doctrine.[660] NRDC shaped its merits brief in response to the city's draft, emphasizing the commercial benefits of the port's "green growth" strategy.[661] In making the market participant argument, NRDC drew upon its now-deep well of expertise and also exchanged views with port counsel and the Teamsters' Mike Manley.[662] In preparing for oral argument, Perrella helped organize two moot courts, one at the University of California, Irvine School of Law, and another at Public Citizen in Washington, D.C., where she persuaded former Solicitor General Seth Waxman to be on the panel.[663]

At oral argument, it was immediately clear that the port's position would be greeted with skepticism. Almost as soon as port counsel Steven Rosenthal began his opening statement, Justice Antonin Scalia pounced: "What exception do you appeal to? There are a number of exceptions there."[664] Barely letting him finish a sentence, Scalia insisted that Rosenthal was asking for "an exception for private contract operations as opposed to public matters," adding, "There are exceptions to the preemption [sic] and that is not one of them."[665] For Perrella, this was "difficult" because the port's argument hinged on the Court recognizing, as a first step, that there was in fact a market participant exception to the FAAAA and she "felt like at least a few of the justices couldn't even get past step one."[666] Contending with Justice Scalia and Chief Justice John Roberts over whether the concessions were enforced through the port's criminal sanctions, Rosenthal shifted toward

Figure 5.4

NRDC lawyers Melissa Lin Perrella, David Pettit, and Morgan Wyenn (from left to right) in front of the U.S. Supreme Court.

the question of whether the provisions carried the "force and effect of law" under the FAAAA.[667] This took the conversation away from market participation, meandering through an analysis of *Castle* and then back to questions of concession enforcement[668]—in this way, previewing the grounds for the Court's ultimate resolution.

If supporters of the Clean Truck Program were bracing for a sweeping curtailment of the market participant exception, what they got on June 13, 2013—in a unanimous decision by Justice Elena Kagan—was a narrow, technical reading of FAAAA section 14501(c)(1)'s operative language, preempting a local "law, regulation, or other provision having the force and effect of law related to a price, route, or service of any motor carrier."[669] Stating that the parties agreed that the provisions at issue related to a motor carrier's price, route, or service, Kagan's decision focused on the "force and effect of law" language. While the Court acknowledged that the FAAAA's terms exempted "contract-based participation in a market,"[670] it concluded that the placard and off-street parking provisions, though contained in a contract, were "part and parcel of a governmental program wielding coercive power over private parties, backed by the threat of criminal punishment."[671] Specifically, because the objectives of the agreement were accomplished "by amending the Port's tariff" to impose legal liability on terminal operators—a violation of which was subject to criminal sanction— it did not stand alone as a contract, but rather was part of a comprehensive regulatory scheme backed by "the hammer of criminal law."[672] Although the Court acknowledged that the "line between regulatory and proprietary conduct has soft edges," this case was "nowhere near" it since "the threat of criminal sanctions" showed the government acting "*qua* government, performing its prototypical regulatory role."[673] In the Court's view, the fact that the port may have passed the Clean Truck Program to "turn a profit" was irrelevant to the question of whether it had acted with the "force and effect of law."[674] What mattered was not intent, but rather the means used, which here involved enforcement of the placard and off-street parking provisions through "a coercive mechanism, available to no private party"—thus bringing them within the FAAAA's preemptive scope.[675] With this characterization of the concession provisions and reading of the statute, the Court sidestepped deep analysis of the scope of market participation.

The court then punted on the ATA's *Castle* claim. The ATA argued that *Castle* prevented the port from enforcing the remaining concession provisions on financial capacity and truck maintenance, which operated to deny noncompliant trucks port access; this denial, the ATA claimed, infringed the power to enforce carrier safety reserved by statute to the federal government.[676] However, the Court read *Castle* to only prevent a state actor from punishing "an interstate motor carrier for prior violations of trucking regulations," rather than "taking off the road a vehicle that is contemporaneously out of compliance."[677] Because the port had not yet begun to enforce the provisions, the Court concluded that it was not clear whether enforcement would be for past violations of the agreement—which would possibly be barred by *Castle*—or for ongoing violations—which the Court noted that even the ATA agreed would be permissible.[678] Because "the kind of enforcement ATA fears, and believes inconsistent with *Castle*, might never come to pass at all," the Court—threading a very fine needle—decided simply not to decide.[679]

In the end, although the port lost, it did manage to limit the doctrinal damage. The Court, in a footnote, stated that the port had occasionally framed the question as whether "a freestanding 'market-participant exception' limits §14501(c)(1)'s express terms."[680] However, mirroring the shift to the "force and effect of law" discussion in oral argument, the opinion's text did not mention the market participation doctrine by name, though it asserted that "contract-based participation in a market" was not preempted by the FAAAA.[681] In that sense, the port—along with the environmental and labor groups that had urged it on—dodged a doctrinal bullet.

Yet that was cold comfort to the city staff and coalition personnel who had struggled so mightily to win passage of the Los Angeles Clean Truck Program. Reflecting on the litigation loss, the port's John Holmes—one of the key architects of the concession plan—saw a contradiction at the heart of the legal outcome. Although the port remained legally liable for environmental compliance, he read the Court's analysis as depriving the port of full legal power to achieve it. The result, in his view, was that the port had "accountability without authority."[682]

For the coalition, the Court's opinion officially laid to rest the boldest aspirations of the Clean Truck Program. Built upon interconnected labor, community, and environmental claims for redress, what remained of the program—clean truck conversion and port financing to achieve it—spoke most directly to the environmental concerns. Although community residents would benefit from clean trucks, their other proposals—off-street parking and placards to enable reporting of bad drivers—had been legally excised from the concession plan. And one of organized labor's pre-eminent goals—changing truck drivers from independent contractors to employees—once tantalizingly close, was again a distant dream. Litigation, which had been such a powerful tool in bringing together the coalition, had been used by its adversary to tear its accomplishment apart.

The Aftermath: Maneuvering around Preemption

The ATA litigation reinforced the outlines of the difficult legal box that the labor movement was in. Although federal labor law did not generally serve the movement's strategic interests—at least when it came to port trucking—federal transportation law was held to preempt local efforts to change the balance of power. The environmental movement, on the other hand, was able to wield strong federal and state law to carve out space for local action. In this regard, the ATA's own strategic behavior helped the environmental cause by foregoing a challenge to the truck ban in favor of focusing its arguments on the concession plan and, specifically, employee conversion. This was both a tactical and economic choice. Supporting the ports' green initiative demonstrated industry social responsibility and sharpened the union critique. And facing the reality of increasing fuel costs and the promise of a local subsidy, it made economic sense to convert the port fleet.[683] From the industry's point of view, employee conversion was a dead-weight loss to be fought tooth and nail.

As a result of the litigation, environmentalists could claim short-term victory in reducing emissions, but at the cost of long-term uncertainty about maintenance and the effect on truckers, who continued as independent contractors—only now with the added burden of having to acquire and maintain costly new clean trucks.[684] In the face of federal preemption, the coalition pursued a strategy to maneuver around it that centered on amending the FAAAA to explicitly permit the Clean Truck Program.

A Legislative Window—Closed
By the time the Supreme Court finally resolved the ultimate fate of the concession plan, the coalition had long since turned to "Plan B."[685] After the Ninth Circuit's first preliminary injunction ruling in March 2009—in which it opined that the employee conversion provision was "one highly likely to be shown to be preempted"[686]—the coalition made a strategic decision not to wait idly by for the court process to wend its way toward resolution. As the district court issued its preliminary injunction against key elements of the concession plan a month later, the coalition had already set in motion a legislative campaign to moot the litigation. As Change to Win's Nick Weiner put it after the district court ruling: "We need to talk to our friends in Congress and see what our options are. ... We've come this far, and we are not going to give up because there are crummy laws."[687]

The campaign's first move was to the federal government, where it mobilized to amend the FAAAA preemption rule to allow the Clean Truck Program—employee conversion and all. LAANE's Jon Zerolnick remembered the coalition's calculation in mid-2009 this way: "So we thought, okay, we're pretty sure that we're going to win the court case. But maybe we're not. Maybe we're wrong. We don't think so. But, you know, if the F-quad-A—the Federal Aviation Administration Authorization Act of 1994—does preempt what we're doing, well, then let's just clarify the F-quad-A."[688]

The legislative campaign started with promise. In mid-2009, President Barack Obama had recently taken office and the Democrats controlled both houses of Congress—with a filibuster-proof majority in the Senate. The political stars were thus aligned and, while the focus was on health care reform, the coalition sought to capitalize on the opportunity.

A key early step was drafting language to modify the FAAAA to carve out an exception for the Los Angeles program. To advance that piece, the Los Angeles city attorneys, with input from NRDC and Teamsters counsel, generated an early draft.[689] The draft modified section 14501(c) of the FAAAA, which prohibited state or local laws "related to a price, route, or service of any motor carrier," to make clear that preemption "does not apply to the authority of a State, or a political subdivision of a State or other municipal authority of a State, to condition entry to Port Facilities for the purpose of ... improving the environmental, safety, security or congestion conditions of Port facilities or in nearby areas."[690] With that tentative language in hand, the port and coalition sought to build political support to move the policy forward. They would need to secure sponsors in both houses and to persuade the relevant committees—the Committee on Transportation and Infrastructure in the House, and the Committee on Commerce, Science, and Transportation in the Senate—to take up the bill.

In a display of its seriousness, in May 2009, the Port of Los Angeles paid $150,000 to hire former House Majority Leader Richard Gephardt's high-powered political consulting firm, the Gephardt Group, to lobby Congress on behalf of the amendment.[691] To support this effort, the coalition, spearheaded by LAANE, reached out to Southern California Congress members to educate them about the lawsuit and the proposed legislative fix, and to gain their support and potential cosponsorship.[692] To make the coalition's case, LAANE put together a comprehensive Briefing Book entitled, "Clearing the Roadblocks: A Map to Green and Grow a Key American Industry to Create 85,000 Middle-Class Jobs at Our Nation's Ports," which included an analysis of the "positive impacts" of the Los Angeles port's Clean Truck Program, an overview of the ATA litigation, key reports from the campaign (including the Boston Consulting Group report), press clippings, statements of support from prominent elected officials, and a list of organizational partners.[693] The coalition also assembled a two-page background paper, proclaiming that the "trucking industry, under the leadership of the American Trucking Association ... is attacking" the Clean Truck Program.[694] This effort bore fruit. On November 4, 2009, twenty-four members of the California congressional delegation wrote to James Oberstar (D-Minnesota), chair of the House Committee on Transportation and Infrastructure, to urge him to "consider making changes to the FAAAA so that California ports can successfully implement and enforce needed truck management programs."[695]

Supporters of the amendment also made their case in the media. On the first anniversary of the Los Angeles Clean Truck Program, Mayor Villaraigosa touted its accomplishments and urged "lawmakers in Washington to update federal law and allow a first-of-its-kind emissions reduction initiative like the Clean Truck Program to flourish."[696] Port director Geraldine Knatz sounded a similar note,[697] while other port city politicians lent their support.[698] New York City Mayor Michael Bloomberg threw his weight behind the campaign: "Today, I'm calling on Congress to support legislation that will empower ports to implement the L.A. Clean Truck Program, an innovative initiative that will create good, green jobs and improve the quality of the air that New Yorkers breathe."[699] In what came as no surprise, the nation's biggest trade associations were not persuaded: "We strongly oppose the efforts of the port to support changing long-standing federal law ... to include a provision within the Clean Truck Plan that has nothing to do with reducing truck emissions."[700]

For such a small change to an esoteric law, supporters assembled a powerful coalition to make their case on Capitol Hill. That coalition included familiar players from the Los Angeles campaign—including Change to Win, CLUE, East Yard Communities, LBACA, LAANE, and NRDC—as well as other powerful supporters (the national Blue-Green Alliance, the Steel Workers union, UNITE HERE, and Sierra Club) and community partners from around the country. Thus united, the coalition coordinated congressional visits in late 2009 and early 2010. Change to Win's Weiner helped bring in truck drivers from Los Angeles and coordinated a lobbying day on which Mayor Villaraigosa, consultant Gephardt, and Teamsters president Hoffa met with key lawmakers.[701] Jonathan Klein, director of CLUE, recalled making the faith case for the amendment, arguing "how unfair" the current law was "in the struggle for working people."[702] After delays in negotiating support due to ILWU resistance, the coalition eventually won crucial backers, including House Speaker Nancy Pelosi and Secretary of Labor Hilda Solis, and—in a key advance—persuaded Democratic Congressman Jerrold Nadler from New York to sponsor the House version of the bill. Nadler was, in Weiner's terms, a "good progressive guy," who had worked on port issues in New York and New Jersey and had strong ties to the Teamsters.[703]

As this push was underway, the politics began to unravel. The first thread came loose in January 2010, when Republican Scott Brown, in a surprise victory, won the Massachusetts Senate seat vacated by the passing of liberal stalwart Ted Kennedy. Deprived of its filibuster-proof majority in the

Senate, the amendment's supporters nonetheless fought on. Their strategy was to attach the amendment to a must-pass transportation reauthorization bill that could be carried in both houses on a Democratic majority.[704] The enactment of Obama's health care reform law in March 2010 fueled Tea Party resentment and drew predictions of a Republican House majority after the November midterm elections.

With Nadler's support in place, there was a vigorous campaign to pass the amendment before the midterms. In a letter dated April 22, 2010, the coalition—now with over one hundred organizations representing labor, environmental, and community groups from port cities around the country—again urged Oberstar's transportation committee to take up the amendment. After detailing the state of the litigation and challenging the ATA's "erroneous claims, not the least of which is that they really support the environmental goals of Los Angeles's Clean Truck Program," the letter promoted a concession model to "ensure trucks are adequately maintained," "eliminate bad actors," and "prevent fraud."[705] It concluded: "We can have both high trade volume and clean, safe communities, but only if ports are able to implement programs that give them the tools to address and solve the pollution problem in the ports, including enforcing compliance by bad actors."[706] The House Committee on Transportation and Infrastructure agreed to move forward, and on May 5, 2010, held a hearing on port trucking conditions,[707] at which NRDC's Perrella spoke about the status of the litigation.[708] The draft amendment then was circulated to lawyers in the Department of Transportation, who were charged with FAAAA enforcement (and who opposed the Port of Los Angeles in the ATA litigation). With their "wordsmithing," the draft went back to Nadler and the coalition, which signed off.[709] With a final "big push," the coalition gained commitments from key cosponsors—Democratic members of the California delegation, plus progressive allies from Maryland, Virginia, Wisconsin, Florida, New Jersey, New York, and Massachusetts—and the bill was ready.

On July 29, 2010, Nadler introduced the Clean Ports Act in the House as H.R. 5967.[710] The bill—in language that echoed the coalition's early draft—proposed to revise FAAAA section 14501(c)(2)(A) to declare that federal preemption of local laws related to "a price, route, or service of any motor carrier" would *not* apply to "the authority of a State, political subdivision of a State, or political authority of 2 or more States to adopt requirements for motor carriers and commercial motor vehicles providing services at port facilities that are reasonably related to the reduction of environmental pollution, traffic congestion, the improvement of highway safety, or the efficient utilization of port facilities."[711]

The campaign effort now turned to getting sufficient votes to pass—which meant more congressional lobbying—in a harrowingly narrow time frame. To do this, the coalition circulated the results of a survey of driver conditions one year after the ATA litigation, which emphasized that "[m]any port drivers, in order to compensate for new clean truck expenses, are working significantly longer hours, earning less, and feel considerably less optimistic about the future."[712] Noting that "trucking companies have seized greater control over drivers' work and the trucks they operate through drastic changes in methods of compensation,"[713] the coalition feverishly lobbied to gain support for an omnibus transportation bill. But with Republican electoral chances looking good as the midterm elections drew nearer, House Democrats were reluctant to make a strong push for a labor-backed bill, which was therefore delayed until after November. "And that's when it all fell apart."[714]

Although launched with great hope, the bill died an untimely death—undone by the catastrophic midterm election loss in November 2010 that negated the Democrats' majority in the House. Deprived of the ability to pass legislation along party lines in either chamber of Congress, Democrats could not persuade any member of the newly energized and more conservative Republican caucus to cross the aisle in support of a union legislative priority. Although the coalition went through the motions, the legislative point was effectively moot. On February 9, 2011, with fifty-nine cosponsors, Nadler's House bill was reintroduced as H.R. 572, with the operative language virtually unchanged.[715] Putting on a brave face, Nadler proclaimed that the Los Angeles model for clean trucks should be promoted, emphasizing evidence showing that most port truckers earned too little to afford new rigs.[716] The coalition once again revived its legislative outreach. In its electronic briefing packet, the coalition presented "media coverage, reports, and other materials" to "show why local governments need action from Washington to reduce emissions, create green jobs, improve public health, and help responsible businesses grow and compete as part of our national economic recovery strategy."[717] In a one-page overview, the coalition reiterated its argument that the ATA appeal was "preventing key portions of LA's Clean Truck Program from being enforced, threatening job-creating expansion and infrastructure projects from moving forward."[718] The coalition thus urged passage of H.R. 572, which "will empower, but not mandate, local ports to adopt requirements for motor carriers and vehicles that are reasonably related to the reduction of environmental pollution, traffic congestion, improving highway safety, or for the efficient utilization of port facilities."[719]

Yet it would not be. By late 2011, the coalition had secured a Senate sponsor, newly elected New York Senator Kirsten Gillibrand, who introduced an identical bill, S. 2011, on December 16.[720] The Senate bill was referred to the Committee on Commerce, Science, and Transportation, where it died stillborn.

In a last-ditch legislative effort, the coalition turned to the California state legislature in the wake of the federal midterm congressional election defeat. On February 18, 2011, with the cosponsorship of staunch labor ally and Assembly Speaker John Perez (representing parts of east and south Los Angeles), and Labor and Employment Committee Chair Sandre Swanson (from Oakland), the coalition helped to introduce the Truck Driver Employment and Public Safety Act, labeled Assembly Bill 950. The bill sought to amend the California Labor Code "for purposes of all of the provisions of state law that govern employment," to declare that "a drayage truck operator is an employee of the entity or person who arranges for or engages the services of the operator."[721] As drafted, the bill would have effectuated by state legislative mandate what the coalition had failed to achieve through the port concession plan. However, industry push-back was swift and decisive. Although the bill was read into the Labor and Employment Committee record, and subsequently received a full hearing in May 2011,[722] it was ordered to the inactive file in June after Perez met with representatives of the California Trucking Association.[723] In arguing against the bill, the industry group claimed it would harm the state's transportation industry and raised the problem of conflicting state and federal standards for employee status.[724] In addition, the California Trucking Association pointed to the 2008 investigation of drayage trucking companies by state Attorney General Jerry Brown—who found five small trucking companies misclassifying employee drivers as independent contractors at the Los Angeles and Long Beach ports—as evidence that misclassification was not a significant problem.[725] The irony of this claim could not have been lost on coalition advocates, who knew that the five prosecuted companies represented a lower bound, not an upper limit on violators as the industry suggested. Indeed, as the campaign moved into its final phase, advocates had already begun to challenge trucker misclassification as a systemic problem—intent to prove the industry wrong and salvage the effort to reform port trucking.

6 Reshaping the Industry: The Challenge to Misclassification

In any large-scale social movement challenge to entrenched power, there is a significant risk of failure and a moment—if failure does in fact come—when leaders must make a choice of whether to walk away or forge ahead, though perhaps on different grounds. In the aftermath of the unsuccessful local policy campaign to achieve employee status for port truck drivers, leaders of the Campaign for Clean Trucks responded to this existential choice and out of the ashes of one failed legal strategy fashioned an unexpectedly potent new one. In exploring the final encounter between advocates of port trucking reform and industry, this chapter spotlights the essential and underappreciated role of movement resilience and adaptation. Following the catastrophic loss of the Clean Truck Program's employee provision, the blue-green coalition's labor partners—the Teamsters, Change to Win, and LAANE—seriously considered cutting losses and abandoning the Los Angeles and Long Beach port complex. Why they chose to persist and how labor leaders devised a new advocacy approach to hold the campaign together is focus of this chapter. As the labor wing of the coalition made a resurgence—and reinvented itself as the Justice for Port Drivers campaign—its leaders developed a novel strategy built upon a model of integrated advocacy, in which multiple sources of leverage (litigation, worker organizing and striking, and policy reform) would be pursued in accordance with an overarching plan and defined endgame. A key point is that the plan and endgame evolved through the process—it was not clear precisely where the campaign would go in the face of the Clean Truck Program loss. But as the new strategy crystallized around the problem of *driver misclassification*, the edges of the path began to emerge.

In this new phase of the nearly two-decade long effort by the environmental and labor movements to hold the ports accountable to community stakeholders, labor would largely go it alone. This was both a practical result of the bifurcated Clean Truck Program (which constituted an environmental win and a labor loss) and a strategic choice. A crucial lesson from the Clean Truck Program litigation was that, if a municipal government entity was going to act as a market participant—and thus avoid federal preemption—in passing local law to change the employment status of drivers, it would have to make sure that the market problem it was addressing was in fact a *labor* problem. The Ninth Circuit, in blocking the employee provision of the Clean Truck Program, had emphasized that converting truck drivers into employees did not address the fundamental market problem claimed as the basis for the program: stopping environmental litigation impeding port growth. Accordingly, this time around, labor leaders understood that they could not advance a policy solution to the independent-contractor problem under a market participation theory while crying environmental harm. That blunt legal fact pushed the coalition's labor partners in their own direction, while the environmental partners continued to support employee status for drivers from a distance, seeking to build upon and enforce the clean truck mandate won at both ports.

It was in this context that, as the Justice for Port Drivers campaign developed, driver misclassification became the centerpiece of its mobilization strategy and legal theory. The campaign's central thesis was that the port trucking industry, by systematically misclassifying truckers as independent contractors when they should be employees, was premised upon structural illegality; that this structural illegality caused pervasive instability and disruption, as workers sued and struck offending trucking firms, producing ripple effects along the supply chain; and that, to correct the structural illegality and rationalize trucking operations, industry stakeholders and government at all levels, including the ports, had to be involved in crafting a solution. Formulating this thesis and executing the strategy to operationalize it once again would rely on a substantial campaign investment in top-down planning, which involved devising a legal theory to simultaneously authorize truck driver strikes (despite drivers' nominally independent-contractor status) and build the foundation for a potential labor peace policy at the ports. In coordination with Teamsters and Change to Win leaders, labor movement lawyers would therefore be critical actors in laying out the campaign design and avoiding the myriad pitfalls of federal labor law and preemption. As this chapter describes, legal planning (in coordination with a revised grassroots strategy to recruit truckers to strike) created the possibility to test an innovative misclassification theory in the federal labor law system that proved pivotal in opening up the legal space for robust union organizing in the port trucking industry. Its success depended critically on the accumulated legal expertise of movement lawyers in the intricacies of labor law, but also on the cultivation and support of newly empowered political officials in government agencies—particularly the federal Department of Labor and the California Department of Labor Standards Enforcement—authorized to enforce employment law. A critical theme in this chapter is therefore the novel use of litigation to advance a law reform strategy through the administrative state—one aimed at mobilizing individual and systemic legal claims to create precedent in favor of misclassification, while also putting economic pressure on offending firms to change their practices.

In this last phase of the campaign, there have been three distinct stages. The first—which started soon after the Ninth Circuit blocked the Clean Truck Program employee provision in 2009 and the federal legislative effort to amend the FAAAA unraveled in the wake of the 2010 congressional elections—was one of experimentation and transition, as the labor movement rallied from the edge of defeat to initiate an exploratory project of supporting misclassification litigation, engaging potential governmental allies, and laying the groundwork for a systemic approach to union organizing in an industry formally defined as outside the scope of labor law. In the second stage, beginning in 2013, once it became clear that misclassification litigation and organizing could produce concrete wins for drivers, labor leaders formulated the Justice for Port Drivers campaign, which hinged on an integrated advocacy model that sought to incrementally build the legal case for unionizing misclassified truck drivers and the political case for the ports of Los Angeles and Long Beach to pursue a new policy solution to the independent-contractor problem. In the third—and ongoing—stage, the campaign has mobilized a tripartite strategy of litigation, striking, and political advocacy to advance a layered plan to gain employee status for drivers at company targets of escalating significance—winning a surprising string of union agreements in a port trucking industry that for decades had none.

Transition: Mobilizing from the Ashes

As chapter 5 detailed, in the spring of 2009, after the employee provision of Los Angeles's Clean Truck Program was preliminarily enjoined on the ground that it was preempted, the campaign's immediate response was to go to Washington, D.C., in an effort to amend the FAAAA to explicitly permit what the court had denied. That effort—"Plan B"—which promised to take advantage of the heady post-Obama moment of Democratic Party control of Congress and the presidency, ended abruptly: undone by the unexpected loss of the Democrats' filibuster-proof majority in the Senate in January 2010 and the Republican takeover of the House of Representatives in the November midterm elections. With both the local and federal policy routes to reform foreclosed by a combination of judicial resistance and political misfortune, the campaign was forced to pursue a different, and quite uncertain, strategy—now "Plan C"—turning back to litigation to challenge trucking firms for misclassifying drivers as independent contractors. At the campaign's nadir in early 2010, whipsawed by the double indignity of the Senate loss (dooming the planned FAAAA amendment) and the Ninth Circuit decision affirming the preliminary injunction of employee conversion, Plan C was in reality less of a strategy and more of a stopgap—a way to buy time and consider next steps, or whether to proceed at all, in light of low morale and a diminished appetite among national labor leaders for continuing to fund what many perceived to be a failed campaign.

However, unlike earlier ad hoc efforts to sue misclassifying trucking companies, this time there was the potential for a more systematic approach given labor movement investment in the campaign and the possibility of leveraging public enforcement. It was in this context that the post-injunction program pursued by the coalition after the failure to amend the FAAAA—though still inchoate in its early transitional stage—sought to advance affirmative litigation against employers for misclassification-related legal violations to pressure contractor-based companies to accept employees (and win financial benefits for drivers),[1] while simultaneously attempting to organize unions at the handful of companies that, for their own idiosyncratic reasons, already hired employee drivers. The first steps were tentative: on one side, seeking to enlist government agency resources in the labor enforcement fight, encourage private law firms to take on class actions, and support individual drivers filing wage claims; on the other side, identifying and mobilizing against the major employee-based trucking company at the ports—Toll Group—in an effort to reclaim lost momentum. What remained to be seen was whether these incipient efforts would bear fruit and, even if they did, whether they could somehow be knitted together to reconstruct a strategy for transforming port trucking on a new legal ground.

Misclassification Litigation

Advocates had long believed that drayage truck drivers were illegally labeled independent contractors by companies that nonetheless exercised employer-like control—directing where and how work was done, precluding work for multiple companies, dictating schedules, and determining payment in ways that effectively prevented drivers from realizing a genuine profit.[2] As detailed in chapter 4, challenges to misclassification had its roots in the early 1990s campaign by the Waterfront Rail Truckers Union and the *Albillo v. Intermodal Container Services* class action at the end of that decade. Yet by the late 2000s, the political environment had changed in ways that refocused attention on the potential to make the misclassification case. The difference was twofold. First, after nearly five years of advocacy around the Clean Truck Program, there was widespread awareness of misclassification as a port trucking industry problem—among drivers, company owners, and policy makers—which helped initial organizing and litigation efforts gain traction and persuade sympathetic officials in relevant government agencies to consider contributing additional enforcement resources. Second, although the Teamsters and Change to Win teetered on the edge of withdrawal after the failed attempt to amend the FAAAA, the fact that they formally remained committed to the Los Angeles struggle meant ongoing support for planning and coordination to promote a systematic enforcement campaign[3]—something that had been lacking in previous litigation efforts.

The path was not easy. In an industry of hundreds of small companies, misclassification litigation was necessarily a piecemeal approach. Moreover, the legal argument for misclassification was not straightforward. The test for whether a worker was a statutory employee hinged on the degree of employer control. As a matter of legal doctrine, employer control was determined under a murky standard that varied by statutory scheme (e.g., tax versus employment law), but generally looked at the "economic realities" of the working relationship: such as whether the worker was engaged in a distinct business, supplied the materials, provided a special skill, worked without supervision, set the work schedule, and was paid by the job.[4] Failing to properly classify an employee as such was not an independent legal violation, but rather a predicate to showing an employer violation of other laws—for example, illegally deducting business expenses (like lease payments) from worker paychecks, or failing to pay minimum wage, keep appropriate records, or provide workers' compensation and unemployment insurance. For private lawyers, bringing misclassification suits often depended on the extent to which the legal violation would generate sufficient legal fees. Cases for back pay involving small numbers of low-paid workers did not always provide fees large enough to entice private lawyers to make the investment. Most importantly for the campaign, even successful legal cases could not force companies to hire drivers as employees—and often had the effect of simply making companies more stringent about following the independent-contractor rules.

It was against this backdrop that the coalition simultaneously pursued public and private enforcement options. The public option avoided the private attorney's fees problem by shifting the cost of litigation to government agencies responsible for enforcing employment law: the Department of Labor (DOL) at the federal level and the Department of Labor Standards Enforcement (DLSE) at the state level. Both agencies were empowered to investigate and bring enforcement actions against violators of employment laws—and had the resources and staff to conduct large-scale operations. For its part, the U.S. DOL housed the Wage and Hour Division, a legal department authorized to conduct investigations and bring cases under the Fair Labor Standards Act guaranteeing workers federal minimum wage and overtime payments. At the California DLSE, individual workers could file claims for violations of state employment law, such as the nonpayment of California's more generous minimum wage or the improper deduction of business expenses from employee pay; DLSE staff were empowered to investigate individual wage claims and could refer them to DLSE hearing officers authorized to take evidence and issue binding enforcement orders (appealable to civil court). Larger, more complex claims against an employer for labor violations involving multiple workers fell to the DLSE's Bureau of Field Enforcement, empowered to investigate and bring group enforcement actions. The DOL and DLSE's legal authority only extended to enforcement on behalf of statutory employees—they had no authority to enforce rules against companies for their treatment of independent contractors. Accordingly, at both the federal and state levels, there were, in theory, powerful enforcement mechanisms available to hold trucking companies to account for employment law violations perpetrated against employee drivers. The key was persuading decision makers at both agencies that port truckers were, in fact, misclassified employees—and to exercise their vast power to enforce truckers' rights.

There were reasons to be hopeful. In February 2008, during the height of the Campaign for Clean Trucks, California Attorney General Jerry Brown appointed a task force to investigate port trucking misclassification, which "uncovered numerous state labor law violations committed by several trucking companies operating at the ports."[5] As a result, Brown filed lawsuits in state court,[6] which alleged that port trucking firms had illegally avoided paying employment taxes and workers' compensation benefits, and also gained an unfair business advantage over companies that followed the law.[7] Brown won judgments against five small companies,[8] though all eventually went out of business.[9] Another suit against Pac Anchor Transportation in 2008 elicited a strong response,[10] with the company criticizing the attorney general for seeking "political gain" by currying favor with the Teamsters to win their support in his planned 2010 run for governor.[11] Pac Anchor fought the suit and won a 2009 superior court decision, which held that the state's unfair business practices claim was preempted by the FAAAA. The case was reversed on appeal, but that opinion was superseded as the case went up to the California Supreme Court—where LAANE, represented by Davis, Cowell & Bowe's Andy Kahn, filed an amicus brief in support of the state's position.[12] The California Supreme Court affirmed the appellate court's decision that the state's unfair competition claim was not preempted by the FAAAA, prompting Pac Anchor to file a petition for review with the United States Supreme Court.[13]

The Pac Anchor case underscored that the problem of port driver misclassification was starting to garner official attention—and the FAAAA remained a potential impediment to addressing it. The Clean Truck Program itself had been premised on the belief that drivers were systematically misclassified: employee conversion was to bring their formal legal title in line with their lived reality. However, after the program was enjoined in 2009 and the campaign focus shifted toward FAAAA amendment, labor leaders concluded that it was time to test whether their belief in systematic driver misclassification—which had been taken as an article of faith—was in fact empirically accurate.[14] As Change to Win national campaigns organizer Nick Weiner remembered thinking: Were "we drinking our own Kool-Aid? Would a government agency agree with us [that port truckers were misclassified] or not?"[15] To answer this question, the campaign commissioned a study to determine whether driver relationships with trucking firms would meet the "right-to-control" test used by the Internal Revenue Service to establish employee status.[16] Conceived in part to provide empirical support to buttress the campaign's argument for FAAAA amendment, the study was ultimately released in December 2010, just after the amendment process unraveled. Although it therefore could not be used in relation to FAAAA amendment (a.k.a. "Plan B"), the study—entitled *The Big Rig*—would provide compelling data relevant to the emerging "Plan C" focus on mobilizing public resources to attack misclassification. Written by Rebecca Smith at the National Employment Law Project (NELP), along with Professor David Bensman of the Rutgers School of Management and Labor Relations, and Paul Marvy of Change to Win, *The Big Rig*'s finding were based on a reanalysis of 10 surveys of 2,183 drivers at seven U.S. ports, along with original interviews of 54 drivers from the Ports of Seattle, Oakland, Los Angeles, Long Beach, and New York–New Jersey.[17] Based on this data, the report concluded that "the typical port truck driver is misclassified as an independent contractor" since drivers were subject to "strict behavioral controls," while being "financially dependent," and "tightly tied" to particular trucking companies.[18] On the day that the report was released, the Spanish-language daily, *La Opinión* ran a story highlighting its findings on misclassification that quoted a driver at the Port of Los Angeles lamenting: "We are at the mercy of God."[19]

The Big Rig's evidence of systematic misclassification landed in a political context that increasingly seemed to point toward the genuine possibility of public agency enforcement. One of the first actions taken by the Obama DOL in 2009 was to launch a misclassification initiative to investigate the problem in various industries, including trucking. Making the link between wage theft and the government budget, the Obama administration estimated in 2010 that stopping misclassification could generate $7 billion in tax revenue.[20] In California, after Jerry Brown's 2010 election as governor, he appointed Julie Su to head the DLSE. Su was a prominent workers' rights lawyer, who directed the litigation department at the Asian Pacific American Legal Center (APALC) in Los Angeles, where she had worked since joining the group as a Skadden Fellow in 1994.[21] Su had gained national recognition for her groundbreaking advocacy on behalf of Thai workers enslaved by garment contractors in El Monte, California—a case in which she had pressed for garment manufacturer and retailer liability for contract worker abuse under the "economic realities" test.[22] She thus came to the DLSE in early 2011 with directly relevant experience litigating employment cases in industries defined by contracting that gave her a deep understanding of the port truckers plight.

It was at this moment that labor veterans of the Campaign for Clean Trucks—from the Teamsters, Change to Win, and LAANE—initiated a plan to persuade federal and state labor officials to devote more resources to target misclassification in port trucking. With *The Big Rig* report as empirical support for the misclassification theory and fresh prolabor leadership at the premier federal and state labor enforcement agencies, the port labor team, still smarting from the clean truck campaign loss, allowed themselves a small degree of optimism. As Weiner recalled thinking at the time: "maybe we could get these labor laws enforced."[23] Toward that end, Weiner and colleagues launched outreach efforts in 2011 to "educate government agencies to make [misclassification] a priority and systematically figure out how to help workers file claims. ... It took a huge amount of resources, dedication, and focus to get it off the ground."[24]

To help achieve lift off, the campaign made the decision to complement its effort to generate top-down agency enforcement with a bottom-up strategy of providing legal education and support to misclassified port truck drivers. In order for agency enforcement to work, there had to be drivers with well-documented misclassification claims that could be addressed in the federal and state administrative systems. Thus, in early 2011, LAANE hired Sanjukta Paul as legal coordinator to advance the bottom-up strategy.[25] Paul had worked at civil rights litigation boutique Hadsell & Stormer and then opened her own solo civil rights and employment firm. On the verge of taking a hiatus from practice, she heard from a colleague that LAANE was looking for a lawyer on a short-term contract to help address misclassification in port trucking.[26] Attracted to being "part of a larger movement," Paul took the job, which involved supporting the coalition's effort to promote agency enforcement while also developing a bottom-up strategy to link individual enforcement to driver organizing.[27] She quickly set about "getting up to speed on the legal issues," which involved drafting memos to Change to Win's Weiner to "inform the top-down enforcement approach against the industry."[28] Toward that end, Paul developed legal theories to strengthen the case for agency enforcement and looked into types of available damages.[29]

Paul's legal analysis helped to frame meetings between campaign members and agency officials that occurred beginning in mid-2011. Because truck drivers' alleged independent-contractor status formally excluded them from the protection of federal and state employment laws, campaign representatives had to explain to labor officials the legal foundation for enforcement and develop new protocols to promote it. LAANE's Patricia Castellanos led a delegation to talk with California Labor Commissioner Su; Teamsters' attorney Mike Manley and Port Division Director Chuck Mack also joined in some of the meetings. As Manley recalled, the thrust of these discussions was: "Here's the evidence. Here's what we found. This is a misclassification. ... You're losing a whole lot of money by not going after these people."[30] This last argument was echoed by Jonathan Klein of CLUE, who recalled attending some meetings and arguing that misclassifying companies "cheat[ed] the government" by depriving it of tax revenue collected on properly paid wages.[31] At the DLSE meetings, in addition to exploring the possibility of increasing enforcement resources, discussions also touched on practical issues of implementation. The problem of trucker misclassification was so distinctive that the campaign "had to help create a whole form for [drivers] to file a claim,"³² because the DLSE did not have an appropriate document for enumerating illegal business deductions under California state law, which constituted the most significant financial liability facing trucking firms.³³

A similar approach was taken at the federal DOL, where Weiner coordinated meetings in 2011 to urge Secretary Hilda Solis and top enforcement officials to undertake parallel federal action. Manley attended some of these sessions, as did NELP's Smith, who discussed *The Big Rig* findings.³⁴ Manley described the overall goal of the agency meetings as "trying to move them to really do something other than just sit with us ... and say, 'Oh, my gosh, it's awful.'"³⁵ As a result of campaign outreach, both the DOL and DLSE made commitments to ramp up investigations. The wheels of government bureaucracy, however, moved slowly—and industry resistance was strong.

To reinforce the significance of the misclassification drive—and turn up the political pressure—Teamsters President James Hoffa visited the Los Angeles port in December 2011, just after the Ninth Circuit had issued its final ruling in the American Trucking Associations (ATA) case invalidating employee conversion. There to meet with striking workers at Toll Group, Inc., Hoffa's general message to port drivers was that the Teamsters were still in the fight despite the court setback: "We didn't think we were going to lose. ... We have to go a different way now."³⁶ Hoffa stressed the systematic nature of misclassification, stating that drivers did not have the power to set their own rates or choose where to haul cargo, while also emphasizing the tax loss to the government that resulted.³⁷ The industry response to potential misclassification liability was sharp, as companies complained that it would penalize trucking firms for establishing relationships with drivers that complied with Clean Truck Program requirements. "It doesn't seem fair. We are following a government mandate and now we have that mandate being used against us," said Vic La Rosa, president of Total Transportation Services, Inc.³⁸ Robert Millman, a lawyer from Littler Mendleson representing trucking companies, was similarly dismissive of the misclassification campaign: "The short story is nothing (like this) has worked. ... This is nothing new. The question is: Are they going to be able to come up with some new game plan?"³⁹ The answer, it turned out, was a definitive: yes.

Part of the new game plan focused on leases between companies that had purchased low-emission trucks under the Clean Truck Program and drivers with whom the companies contracted.⁴⁰ Driver advocates argued that lease arrangements between a company and driver in some cases made so many demands that the driver was effectively precluded from working for other firms, thus suggesting a degree of control tantamount to an employer-employee relationship.[41] Companies deducted lease payments and other business expenses from driver paychecks—a practice that was illegal under state law if the drivers were, in fact, employees.[42] In a legal research memo to the campaign, Paul suggested that the "documentary evidence" of deductions in paychecks could provide the "monetary hammer" for private lawsuits seeking damages, since such deductions were illegal under state labor law section 2802 and could add up to substantial amounts, thus enticing private lawyers to take on cases.[43] Paul also suggested that a new state law sponsored by the Teamsters, the California Willful Misclassification Law (S.B. 459)[44]—passed in October 2011—provided additional legal leverage.[45] An outgrowth of the legislative effort that had stalled around the more robust A.B. 950, which would have simply declared all port drivers employees, S.B. 459 made willful misclassification itself an independent state law violation, subjecting employers to substantial financial penalties.[46] It also made it illegal to make any "deductions" from the pay of an individual "willfully misclassified" "for any purpose, including for goods, materials, space rental, services, government licenses, repairs, equipment maintenance, or fines. ..."[47] In a memo to the coalition, Paul concluded that "the Willful Misclassification Law represents a bold and important advance in the fight against employers' misuse of the 'independent contractor' form to deny employees their basic legal rights."[48]

To push the legal fight forward, the coalition's misclassification strategy also sought to complement public enforcement with private litigation—and to coordinate the private litigation with driver organizing efforts. This strategy had its roots in two high-profile state court class actions filed by plaintiff-side attorneys at the Law Offices of Ellyn Moscowitz, which targeted labor violations by port trucking firms. The first, in November 2009, alleged that after the rollout of the Clean Truck Program, Total Transportation Services, Inc. committed numerous employment violations by failing to pay minimum wage and overtime, provide meal and rest breaks, and reimburse expenses for drivers that the firm purported to hire as employees through staff agencies.[49] The second suit was brought in June 2010 against Sun Pacific Trucking and Pacific Green Trucking, alleging similar violations.[50] As truckers' attorney Adam Luetto put it: "Port drivers consistently claim that they are forced to drive long hours without breaks and required to perform work they never get paid for. ... These drivers, unsurprisingly, are simply tired of working for free and we are working hard to hold their employers responsible for such unlawful employment practices."[51] The

Teamsters coordinated with the lawyers to provide evidence of violations.[52] Both of these cases settled.[53]

On the heels of these suits, individual drivers began to file their own wage claims with the DLSE, challenging their misclassification as independent contractors under state employment law. Individual wage claims allowed drivers to take advantage of the DLSE adjudication process, which included informal conferences and formal administrative hearings—but could be navigated by drivers on their own or supported by nonlawyers familiar with the process (though drivers had a right to have lawyers present during hearings and many chose to do so). In 2011, four Long Beach drivers acting on their own filed claims against Seacon Logix, which resulted in a January 2012 DLSE ruling ordering the company to pay over $100,000 in back wages and penalties.[54] Seacon Logix appealed the DLSE order and retaliated by suing the workers for breach of contract under their lease terms.[55] David Gurley, the DLSE attorney assigned to the ports, knew of the coalition's misclassification effort and reached out to Paul to help the workers, which she did (along with private employment lawyer Stephen Glick) by assisting them in filing retaliation actions—and thus forcing Seacon Logix to drop the breach of contract claims.[56] The Seacon Logix case was prosecuted on appeal by Labor Commissioner Su and upheld by the state superior court.[57] In commenting on the victory, Su echoed her garment advocacy past in stating:

> In this case, drivers had signed agreements labeling them independent contractors but the Court saw the truth behind the label.... This case highlights the critical need for labor law enforcement, particularly where misclassification cheats hardworking men and women like these port truck drivers out of the full pay to which they were entitled.... This is wage theft and we will do everything in our power to stop it.[58]

The DLSE's involvement refocused attention on the push for greater public enforcement. In February 2012, the DOL and DLSE signed a memorandum of understanding outlining their partnership to reduce misclassification.[59] In spring, the DLSE sent out subpoenas to several trucking companies and initiated investigations; the DOL launched a similar enforcement effort, resulting in approximately fifty investigations in total.[60] Some companies noted that the subpoenas did not list specific violations, but only mentioned potential problems.[61] In August 2012, the DOL filed a federal Fair Labor Standards Act lawsuit against Shippers Transport Express Inc. (called Shippers, for short), a large port drayage company owned by a larger parent company, Carrix, Inc. which also owned SSA Marine, a terminal operator on Pier A of the Long Beach port.[62] Industry decried the investigations, with a California Trucking Association representative complaining: "We have a problem when companies are harassed or targeted unjustifiably simply because they use independent contractors."[63]

In response, industry lawyers conducted trainings—styled the "Teamster and worker misclassification update"—instructing trucking companies on how to avoid running afoul of misclassification rules. One such "update" recommended that companies "DO NOT Use a Driver Handbook that looks like an employee manual," or require a driver to "Wear company logo," "Paint the truck a particular color," or "Display a company ID card."[64] Industry representatives tried to characterize the misclassification effort as another Teamsters ploy, with the executive director of the Harbor Trucking Association pointing to an alleged "smoking gun" letter from Hoffa to Governor Brown in April 2012, in which Hoffa stated he was "glad to know that California, in collaboration with the U.S. Department of Labor, is seeking to end this practice."[65] In October, ten trucking companies (calling themselves the Clean Truck Coalition) escalated the fight, filing suit directly against Commissioner Su. In the complaint, the companies sought a declaration that their "pooling agreement" to share clean trucks and lease them to independent contractors, because it was authorized by federal law, precluded Su from pursuing state enforcement actions against them; the companies also requested an injunction against further misclassification investigations.[66] Responding to industry attack, Su insisted that trucking firms were not being singled out and placed the blame for misclassification squarely on industry's shoulders:

> I think too often that entities have kept everything the same about their operation, but they once had employees and converted them to contractors to cut cost. It's bad for employees, it's bad for the competitors, and it cheats the public out of millions of dollars a year because they're not paying taxes. ... I reject the notion that we should blame hardworking people for the abuse they might suffer from the people who break the law. ... That's not the way our legal system is structured; that's not the way our labor laws work.[67]

Yet the blame game continued, with industry groups insisting that they were being unfairly targeted by government agencies and ratcheting up the political pressure to tamp down the investigations. In response to lawsuits seeking to recover expenses deducted from driver pay, trucking companies made sensational claims that drivers could end up overcompensated, since their contract pay was as high as $60,000 (failing to mention that deductions could bring take-home pay down below the minimum wage). In this way, industry tried to undermine political support for the DLSE by making it seem as if the agency was investing enforcement resources to help drivers who were not sympathetic low-wage workers. In an effort to navigate around this problem, LAANE's Paul drafted legal memos arguing for targeted enforcement against market-leading trucking firms, like Harbor Express, Inc. (owned along with Gold Point Transportation by Peter Kim), which would focus on violations of minimum wage and overtime laws (and thus avoid the charge of potential overcompensation).

As the trucking industry pushed back against agency enforcement actions, the campaign focused resources on generating more bottom-up energy among workers to file wage claims. Campaign leaders hoped this effort would put greater pressure on the DLSE to make findings of driver misclassification in adjudicating wage claims coming from workers on the front lines, while operating "as an organizing strategy to help educate drivers" about what misclassification was and how they were affected by it.[68] To bolster driver wage claims, the campaign turned to Paul to help "figure out" what to do.[69] In response, Paul planned an eight-week legal rights clinic, coordinated with state and federal enforcement agencies, beginning in September 2012. To prepare for these clinics, Paul reached out to partners in the labor movement, as well as government agency officials, for whom she provided an analysis of legal violations in the port trucking industry.[70] In August, Change to Win organizers passed out leaflets to stopped trucks inviting them to attend an initial meeting to be held at the Teamsters Local 848 office in Long Beach.[71] At the meeting were representatives from DLSE, DOL, and the California Division of Occupational Safety and Health, as well as labor lawyers and Teamsters organizers—all of whom provided information and encouraged port truckers to pursue their rights.[72] At this meeting, Paul facilitated a know-your-rights training for workers and organizers[73]— explaining what facts to look for in support of misclassification.[74] A few private attorneys who specialized in employment litigation were also in attendance to make contact with drivers and determine if there were possibilities for larger-scale lawsuits. After this initial meeting, Paul instituted a regular legal clinic, open two nights per week, which helped drivers identify employment violations, provided counseling on legal options, assisted in the preparation of administrative claims, calculated wages owed, and connected drivers to private attorneys.

Roughly fifteen cases were filed through the clinic.[75] Although Paul did not represent drivers directly, once they filed claims, she helped calculate damages, coordinated with DLSE attorneys, and used her private bar connections to help find plaintiff-side lawyers to represent drivers in the ensuing proceedings.⁷⁶ Paul also provided private attorneys with supporting legal analysis and developed creative theories for company liability. Through this process, a number of lawyers (many of whom members of the campaign knew directly or by reputation as allies and experts on employment issues) took on DLSE wage claims for drivers. These included solo practitioner Stephen Glick, who had been working on port driver employment issues prior to the campaign and accepted a number of cases, including against the major firm Container Connection; lawyers from Los Angeles-area labor law firms Gilbert & Sackman and Rothner Segall & Greenstone; civil rights lawyers from Traber & Voorhees (which worked closely with Paul's former firm, Hadsell & Stormer); and lawyers from the boutique public interest-oriented law firm Strumwasser Woocher. The number of these "first-generation" cases was deliberately limited to those with strong legal claims of misclassification in order to create good precedent for high-volume filings later.⁷⁷ The campaign's goal in supporting these cases was to seek redress for drivers who had been wronged, while demonstrating to the industry, port officials, and the wider political community that misclassification was an industry-wide problem that demanded attention.

In bringing attention to the misclassification problem and supporting drivers' individual wage claims in front of the DLSE, the campaign also sought to inspire the plaintiff's bar to start bringing major litigation targeting misclassifying companies, against whom the lawyers could win significant damages for large groups of drivers and collect attorney's fees—putting economic pressure on the companies and shining more light on industry practice. Although most of these cases would be brought by lawyers from private employment law and class action firms, on one occasion early in the misclassification effort, campaign leaders helped directly launch a suit that came out of the legal clinic. In February 2013, as Sanjukta Paul was preparing to leave LAANE, she joined the Wage Justice Center in a class action lawsuit against a family-owned network of port trucking firms controlled by defendant Erick Yoo (QTS Inc., LACA Express, and WinWin Logistics), seeking over $5 million in damages for violations including unpaid minimum wages, willful misclassification, unlawful pay deductions, and unfair competition.⁷⁸ Unlike the other large private lawsuits that were being filed around the time, this one (called *Talavera v. QTS*) was developed with the cooperation of campaign staff, namely Paul, and was advanced by a nonprofit legal organization, the Wage Justice Center, known for its cutting-edge employment litigation in the low-wage sector and its ability to collect judgments from recalcitrant employers (the nonprofit legal group, Asian American Center for Advancing Justice–Los Angeles, formerly APALC, also signed on as co-counsel in part to help represent QTS's Korean American drivers). Because of the legal groups' nonprofit status and philanthropic support, collecting attorney's fees, though salient, was not a primary driver of the QTS suit.

This put it in contrast with the growing number of misclassification cases being brought by lawyers from the private bar. The earliest ones predated the campaign. In January 2012, a group of plaintiffs' lawyers—led by Matthew Hayes and Joe Sayas, each of whom headed small labor law firms—filed a class action against Shippers Transport Express (*Taylor v. Shippers Transport Express, Inc.*) claiming failure to pay minimum wage and overtime, illegal business deductions, unlawful coercion and business practices, and waiting time penalties.[79] The Irvine-based law office of James Hawkins filed a state law class action suit in Orange County against Gold Point in March 2012.[80] Just after the *Talavera v. QTS* case was filed, in April 2013, lawyers from the Los Angeles employment law firm of Spiro Moore filed a class action against Container Connection (which had been determined by DOL to have misclassified workers in January 2013), alleging violations of deductions law—both under regular employee rules (California Labor Code section 2802) and those applying to "willfully misclassified" employees under the new A.B. 459—as well as various wage and hour violations.[81] Then, in May 2013, two drivers sued Wilmington-based Harbor Express (Gold Point's affiliate) on behalf of a broader class of drivers for misclassification. That suit was brought by lawyers at the Los Angeles boutique litigation firm, Kabateck Brown Kellner LLP, specializing in consumer class actions. Describing the suit, lead counsel Brian Kabateck stated: "It looks like a traditional employment, but they slap the title of independent contractors on them."[82] That idea, of course, resonated with Jon Zerolnick, still leading the LAANE ports campaign team, who reached out to Kabateck after learning about the lawsuit to explain how it related to the campaign's goals. Zerolnick had similar conversations with Hayes and Sayas, and shared with the lawyers some of the campaign's strategic research on Shippers and Harbor Express to support litigation efforts.

By early 2013, as private misclassification lawsuits and individual wage claims began to accumulate while hopes for a massive increase in public enforcement waned in the face of stiff industry opposition, a notable shift occurred within the campaign. The explosion of private legal actions, and the taste of early success in some individual DLSE claims, revealed that rights enforcement through litigation had the potential to be more than a tactic to discipline misbehaving trucking firms: It could—perhaps—be an essential part of an integrated strategy to yet again pursue systemic transformation of the port trucking industry. Specifically, if enough workers could win enough damages against misclassifying trucking firms—through class and individual actions—they could exert real economic pressure that, in tandem with sustained driver organizing, might begin to persuade firms that hiring employee drivers and accepting a union was more stable and economical than litigation with no end in sight. This seemed to be the emerging view as campaign organizers began supporting en masse driver filings with the DLSE against targeted trucking firms starting in the spring of 2013.[83] Stepping back from the precipice of defeat, the message from the campaign was: "We're going to continue to be here and be a problem."[84]

Union Organizing

Misclassification litigation was always understood as a means to an end—a way to pressure companies and thus create a pathway to union recognition.[85] As such, it was meant to complement the other key component of "Plan C": organizing drivers toward the goal of winning union contracts. Unionization of port truckers, of course, was the prize that drove Teamsters involvement in the Campaign for Clean Trucks. And in the immediate wake of the Clean Truck Program's passage in Los Angeles, the Teamsters initiated union organizing efforts at the companies that had converted to employee drivers. In early 2009, the union protested driver terminations at Swift and Southern Counties Express, lodging complaints with the regional office of the National Labor Relations Board (NLRB, or board) that employees had been fired in retaliation for union organizing.[86] However, when the district court in the ATA litigation preliminarily enjoined the employee provision of the Clean Truck Program in April 2009, trucking companies that had converted to the employee driver model shifted back to an independent-contractor format and the union campaigns fizzled. As a result, the Teamsters refocused organizing on the handful of employee-based companies that remained.

Of the hundreds of trucking companies that serviced the ports of Los Angeles and Long Beach, only a few had employee drivers. Experts pegged the port trucking industry around this time at 12,000 drivers, only 10 percent of whom were employees.[87] Pursuing unionization at companies with employee drivers posed obvious risks. Employee drivers could lose their jobs, the Teamsters could lose the campaigns, and the companies could decide to do what all the other companies already did—contract out their driving. Yet there were also significant potential benefits. In the wake of the stymied Campaign for Clean Trucks, a victory was badly needed to show drivers that the Teamsters—and the broader labor movement—could deliver tangible benefits. In addition, a unionized company could be held out as a successful model for others to follow—proving that employee-based drayage trucking could be economically viable. As CLUE's Klein put it: "We needed to have a win. And we knew that it was important for all of the port truck drivers to see us win ... to make people aware that this effort is brought to you by the Teamsters."[88] The union, with the support of allies like LAANE and CLUE, thus sought to "build some density,"[89] however modest, in the port drayage sector, with the hope of creating a foundation for further growth.

To advance the union strategy, Teamsters Local 848 took the lead, with a "big investment from Change to Win" and the Teamsters' national office.[90] The Teamsters Organizing Department assigned organizer Jason Gateley, who had organized Coca-Cola workers in Las Vegas, to run the campaign. The union's crucial first decision was selecting an initial target, which had to be a firm against which the union could exert maximum pressure without risking its withdrawal from the port market or contractor conversion. The campaign thus chose Toll Group, Inc., an $8.8 billion Australia-based logistics company whose main U.S. activity was importing retail goods and shipping them to warehouses and retail outlets throughout the country. In Los Angeles, Toll employed seventy-five local port drivers. The key leverage against Toll was that the company was heavily unionized in Australia, where the Transport Workers Union represented 12,000 Toll employees, thus forming a powerful block that could pressure management to support the U.S. workers.[91]

To organize Toll drivers, the Teamsters ran a campaign that sought to gain union certification through an NLRB-sponsored election.[92] After several months of Teamsters organizing, punctuated by James Hoffa's visit in December 2011, drivers at Toll filed an NLRB petition for a union election in January 2012.[93] When a female driver was fired in February for stopping to eat at McDonald's (following several other terminations), the organizing campaign kicked into high gear. Local Teamsters organizers, alongside their Australian union counterparts, protested in front of Toll's Los Angeles office in March,[94] while Hoffa and Los Angeles Mayor Antonio Villaraigosa made strong statements in support of the workers at the Los Angeles Good Jobs, Green Jobs Regional Conference a few weeks later.[95] Clergy, community members, and environmental activists lent organizing support, standing by workers during protests.[96] CLUE's Jonathan Klein joined a delegation to Toll's Los Angeles office to protest driver terminations, asking management to "rehire these people" and emphasizing the "injustice of firing them."[97] Another rally featured speeches by Congresswoman Janice Hahn and

Figure 6.1
Toll driver at the Port of Los Angeles. Photo courtesy of Barbara Maynard, Teamsters communication director.

Teamsters Local 848 Secretary-Treasurer Eric Tate. Teamsters lawyer Mike Manley filed claims at the NLRB in response to the employee firings and, when Toll refused to agree to a bargaining unit limited to drivers, successfully litigated that issue at the board.[98]

On April 11, 2012, in a historic vote, Toll drivers voted 46–15 in favor of representation by Teamsters Local 848.[99] As Weiner recalled, it was "the first Teamster contract in L.A. in 30 years.... Big deal!"[100] Sounding a call to action, one driver cast the vote in these terms: "Our victory means we are finally getting closer to the American Dream. If we can win, I know other port truck drivers across the U.S. can unite just like we did."[101] The contract, unanimously ratified on December 30, 2012,[102] gave drivers a nearly $6 per hour pay raise with additional increases over the life of the contract (to take effect on January 1, 2013); made all Toll drivers part of the company's retirement plan and guaranteed company pension contributions of $1 per hour in the contract's first two years; reduced driver payments for health insurance; provided paid vacation; and prohibited Toll from subcontracting out driving services.[103] Highlighting the hope that the Toll contract would be only the first, the campaign announced that to "encourage a more level playing field and wide-scale unionization, the contract provides drivers the ability to re-negotiate for higher wages when a simple majority of the Southern California market is organized."[104] In exchange for accepting the union, Toll received some benefits: the Teamsters agreed not to make Toll follow the more stringent national Master Freight Agreement and also sought to promote the trucking company as a model worthy of additional business.[105]

While practically and symbolically important, the Toll victory highlighted both how far the campaign had come—and the distance it had yet to travel. The success at Toll, though worthy of celebration, was a reminder of just how close the campaign had been to industry transformation and underscored its existential crisis. While Toll was the first union victory at the port in a generation, it did not contain a solution to the fundamental driver-contracting problem that sparked the monumental Campaign for Clean Trucks in the first instance. Despite the labor movement's enormous investment on their behalf, port truckers continued to bear the cost of port growth—which now included the added cost of clean truck conversion. Nonetheless, the Teamsters' successful effort to unionize Toll, though limited in scope, combined with the groundswell of misclassification litigation to signal the campaign's dramatic revival.[106]

Formulation: Labor Organizing in the Shadow of the Administrative State

The resurgent campaign would strike at the heart of the port trucking problem: attacking misclassification head on. What started as an alternative response, born of the failure of employee conversion as a legislative campaign, coalesced into a proactive strategy with a clear end in sight. Out of the welter of DLSE claims and class action lawsuits, and on the heels of the Toll contract, the labor movement confronted a crossroads at the ports. Without a policy mechanism for achieving employment for port truck drivers, the campaign was forced back to square one, formulating an organizing strategy that could surmount the industry's fundamental barrier: the independent-contractor status of drivers. That required reanalysis of opportunities and threats, and soul searching about whether it made sense to continue committing resources to a campaign that had already consumed significant time and money. It was at this transitional moment that the administrative system of labor law governing union-management relations and collective bargaining—housed under the NLRB, which was often viewed as ineffectual and unreliable—came in for reconsideration as a potential asset.

Although it began in fits and starts, this new stage of the campaign would develop as a boundary-pushing strategy to organize key targets—high-volume trucking firms whose essential business model depended on independent-contractor drivers—selected by the campaign to build union density in the heart of the industry. Litigation would play a supportive role in this effort, but unlike earlier efforts, this phase would tightly pair misclassification litigation with an innovative organizing strategy designed to yield new union contracts through the NLRB system. Specifically, the campaign was to proceed under the auspices of an overarching theory of unionization that ultimately hinged on the novel idea that misclassified workers could affirmatively assert labor rights under federal law: *they did not need their company to reclassify them first*. This position circumvented the legal problem that bedeviled earlier efforts. Thus, in the words of one organizer, the "genius" of this new strategy was "not saying reclassify us and make us into employees, but saying we are employees and we are going to act like it."[107]

With this approach, the campaign's goal was to negotiate union contracts with key trucking firm targets, using the pressure of ongoing litigation and significant monetary damages, combined with strategic picketing and strike activity that disrupted trucking firms, shippers, terminal operators, and the ports. Once successful contracts were in place, unionized firms could be held out as models that showed others that it was not only more stable and economical to operate as union shops, but also that such shops could actually thrive. With sufficient density and success, and with the support of port and city leadership eager to avoid ongoing disruption, the campaign could once again begin to imagine a master solution to the independent-contractor problem at the port complex: going back to lobby Los Angeles and Long Beach to exercise their legal authority as proprietors of port business to legally require trucking firms to hire employee drivers as a way to quell labor unrest and thus promote operational efficiency. Now, however, the legal ground for municipal action would rest squarely upon labor, not environmental, peace.

But before any of this could happen, the campaign needed to retool and reformulate. As with other phases of the port trucking struggle, this one moved forward with creative thinking and careful planning, unanticipated windfalls and setbacks, and enormous courage on the part of workers who put themselves on the line to imagine a better future. The strategy itself would be shaped by many of the same players on the labor side, led by the Teamsters, which had driven the employee conversion model at the port, and LAANE, whose port campaign staff was transitioning after the grueling fight for clean trucks. Although the environmental partners remained supportive, they did so at a distance, not involved in the planning and execution of the new Justice for Port Drivers campaign.[108]

Unfreezing Labor Law: How the NLRB Matters

The popular image of organizing is scrappy and bottom-up. One thinks of charismatic grassroots organizers, urging collective action, and ultimately inspiring workers fed up with mistreatment to find courage in solidarity: standing up to authority with sweat and tears to demand dignity and claim respect. And that image is in many respects entirely accurate. But when it comes to union organizing, there is another facet that is less familiar, removed from the action in the street and preoccupied with legal arcana and technical planning.

Since the New Deal, collective action by workers has been governed by federal law, which structures labor organizing down to the minutest detail. The National Labor Relations Act (NLRA) says who can undertake collective action and when; it says who can ask for union representation and how; and it says what happens when employers or unions violate its terms. Navigating these rules matters profoundly to union leaders and workers on the ground. Running afoul of them can jeopardize hard-fought campaigns for unionization. And workers risk losing their jobs if employers choose to fire them for violating the NLRA or engaging in conduct not protected by its terms. Figuring out how to successfully run a union campaign requires careful research and predictive judgments about how employers will respond and how legal challenges will be resolved. Winning therefore takes courage and stamina on the part of workers and union leaders—but it also takes smart strategy and legal theory.

Union campaigns run through a complex bureaucratic process that is housed in the federal administrative state. The NLRB, a product of the 1935 Wagner Act's effort to encourage collective bargaining and protect workers' right to association,[109] is charged with administering labor law: overseeing the process for union elections and handling disputes over "unfair labor practices" (or ULPs) that occur during union campaigns.[110] For employees, the key provision is Section 7 of the NLRA, which states that employees "shall have the right to self-organization; to form, join, or assist labor organizations, to bargain collectively through representatives of their own choosing, and to engage in other concerted activities for the purpose of collective bargaining. ..."[111] Although facially broad, this language has been interpreted over the years by the NLRB and courts to apply to only certain types of protected concerted action by employees. For example, it protects strikes over the "terms and conditions" of employment, but not strikes conducted for other reasons,[112] for instance, to force an employer to provide information about a proposed corporate merger.[113] This is important for workers because, if their activity does not fall within the scope of Section 7, they may be disciplined or discharged by the employer.[114]

Section 8 defines what constitutes a ULP (by both employers and unions), crucially asserting that it is unlawful for an employer to "interfere with, restrain, or coerce" employees exercising rights protected by the act—by, for example, firing a worker for engaging in protected union organizing.[115] A union, employee, or employer may bring ULP charges to the NLRB, typically at one of the regional offices, where staff may investigate, bring formal complaints in cases it deems to have merit, and seek injunctive relief to stop ULPs. If cases are not dismissed or resolved informally, they are assigned to an administrative law judge, who conducts a full hearing, issues a ruling, and can order remedies for violations.[116] Such rulings may be appealed to the full NLRB, the decisions of which, in turn, may be appealed to the U.S. Court of Appeals. Despite its authority to prosecute and adjudicate ULPs, the NLRB lacks enforcement power: It must petition the court of appeals to enforce remedies against employers and unions, and must ask the district court to issue injunctions.

Over the past half-century, this federal system has been converted—through judicial interpretation and administrative action—from one designed to support workers' rights to one generally seen as, if not hostile, at least not helpful to union organizing. At the turn of the millennium, experts viewed labor law as "ossified"—unduly cabined by the cumulative effect of legal precedent such that the NLRB, even if it wanted to, was constrained in how much it could support unionization.[117] It was against that backdrop that the Justice for Port Drivers campaign emerged as part of a broader effort to "unfreeze" the labor law system.[118] To fully appreciate the final phase of the port trucking struggle, which would focus on unionization and therefore ultimately run through the NLRB, it is important to understand three aspects of this system as context.

First, the main power workers have in union campaigns—the power to stop work and picket employers—is circumscribed by labor law. Most significantly, in the context of a campaign for union representation, workers are *prohibited* from picketing employers (i.e., carrying signs at the entrance of their workplace) to force union recognition prior to an election,[119] yet workers are *permitted* to picket in protest of an employer's ULP.[120] This is critical because it means that workers who picket in response to an employer ULP are technically protected from employer retaliation or discharge. Similarly, workers who *strike* (i.e., stop working for) an employer *in protest of a ULP* have more legal protections than economic strikers (those demanding recognition or economic concessions) insofar as they cannot be locked out or permanently replaced (as they could in an economic strike). ULP strikers may only be temporary replaced, which means that they have to be allowed back to work when they present themselves to the employer. If an employer were to fire a worker engaging in a protected ULP picket or strike, that employer action in itself would constitute a ULP, which a union or employee could petition the NLRB to remedy.

Second, a critical point for campaign planning is that *NLRB staffing profoundly affects the possibility for and strategy around union organizing*. Because ULP strikes may be conducted alongside unionization campaigns, the line between ULP-based strikes and their economic counterparts is often blurry. Employers thus regularly challenge the legality of any collective action and discharge workers who engage in it. This employer conduct then comes back to the NLRB, which must adjudicate whether the worker action in question is protected and thus the discharge is unlawful. *Who is authorized to make these decisions therefore has a substantial impact on the outcome of campaigns*. The key NLRB personnel are the members of the board itself—five members appointed by the president to five-year terms with Senate consent—and the NLRB general counsel—also a presidential appointee who serves a four-year term. The role of the NLRB is to conduct union elections and adjudicate ULPs; the role of the general counsel is independent from the board and consists primarily of investigative and prosecutorial functions, such as issuing and pursuing ULP complaints, which are within the general counsel's unreviewable discretion.

As a practical matter, the business of the NLRB (which is headquartered in Washington, D.C.) is delegated to twenty-six regional field offices that undertake the vast majority of its work (the Port of Los Angeles and the Port of Long Beach are both in Region 21). These offices are run by regional directors—appointed (and removable) by the NLRB general counsel (subject to board approval)—who are empowered to conduct union elections and to prosecute and enforce ULPs.[121] Regional staff investigate ULP charges and the regional director decides whether to issue a complaint and prosecute it before an administrative law judge; decisions not to pursue a ULP are reviewable by the general counsel, whose decision is final.

Third, and crucially in the context of the campaign to unionize port truckers, the entire NLRA scheme and its protections *only* apply to employees as defined by federal law—independent contractors are excluded. This means not only that independent contractors can be terminated for collective action without violating labor law, but also (as discussed in Chapter 3) that they are *affirmatively prohibited* from collective action under federal antitrust law (which does not apply to employees through the so-called labor exemption). That is, authentic independent contractors are deemed legal businesses precluded from organizing to act collectively "in restraint of trade" under the Sherman Act.[122]

The issue of whether port truckers were, in fact, authentic independent contractors was the heart of the port trucking misclassification campaign. This, of course, was the key problem that the Campaign for Clean Trucks had sought to address: moving port truckers from independent contractors into the protected status of employees from which they could pursue rights protected under the NLRA. Thus, the critical legal question after the failure of employee conversion in the Clean Truck Program was: *Could there be another legal pathway through the labor law system itself to achieve employee status for drivers and thus enable unionization?* The strategy that developed sought to test an innovative legal theory of misclassification in front of a newly sympathetic NLRB.

An Integrated Strategy: Suing and Striking

The NLRB strategy formulated by the Justice for Port Drivers campaign hinged on recalculating the landscape of opportunities and resources after the Clean Truck Program, which meant, first and foremost, convincing Teamsters leadership to stay invested at the Los Angeles and Long Beach ports and sustain funding essential to organizing success.[123] Making that case required presenting a new analysis of how ongoing Teamsters support could yield tangible union benefits. Developing and executing that analysis would fall to a core labor leadership team drawn from the clean trucks fight and supplemented with new organizing, research, and legal personnel. It included from Change to Win, Nick Weiner; from the Teamsters International office, attorney Mike Manley, Port Division Director Fred Potter, and Western Regional Director Manny Valenzuela; from Teamsters Local 848 in Long Beach, Secretary-Treasurer Eric Tate, and lead port organizer Carlos Santamaria; and from LAANE, Sheheryar Kaoosji, who would assume the port campaign director position at LAANE. In addition, LAANE would hire an attorney, Jean Choi, for "legal firepower,"[124] and an organizer, Ernesto Rocha, who among other responsibilities would be charged with ensuring that Los Angeles's new minimum wage law, passed in 2015, was applied to port truck drivers.[125]

By early 2013, this leadership team surveyed the distance traveled since the loss of the Clean Truck Program's employee provision and the failure of FAAAA amendment a few years earlier. The campaign had not only staunched the bleeding but also could point to real progress: the Toll contract at the close of 2012, combined with the growing number of individual driver victories at the DLSE and class action filings against major port trucking firms. What remained was for the team members to mesh their diverse skills and weave together these fragmentary pieces of progress into an integrated misclassification campaign built on a plausible theory for mobilizing legal and grassroots action toward the audacious end of unionizing independent-contractor firms through the NLRB.

The outlines of this integrated campaign began to sharpen around a two-pronged strategy of "suing and striking," with the ultimate goal of gaining union contracts at "top 50" trucking firms, which did roughly half the work at the port complex and which could succeed economically with employee drivers and thus serve as "consolidators" in the port trucking industry.[126] Suing was an important first step in this process because it would not only put financial pressure on target firms, but would also disrupt the relationship between drivers and employers, thereby helping drivers believe that they could achieve broader change. As Nick Weiner described the strategy, in order to support driver claims, Teamsters would organize a "critical mass" of drivers at a particular company to "file claims together" in order to "beat back retaliation."[127] Targets were selected by asking, "Where do we have contacts, what do they look like?" in a process that was "some science, some art."[128] With the litigation activity starting to gain steam in 2013, the question for LAANE's campaign point person, Sheheryar Kaoosji, was: "How do we actually turn it into a victory? Is there a way to use this effort to get the companies to change their business model and actually hire these folks as employees as they are treating them?"[129] The leadership understood "that litigation alone is not going to do it and there needs to be something more."[130]

As the "suing" part of the campaign's strategy gained increasing momentum, the focus therefore shifted to "striking"—particularly figuring out how to engage in collective action that would force "the industry, the port, and government officials to respond to the labor unrest" with a solution that systematically addressed the misclassification problem.[131] The theory was that suing and striking "combined [would] persuade the industry that they are better off dealing with us than fighting us."[132] The key legal question was how to pursue striking at independent-contractor port trucking firms that maximized pressure on firms while minimizing risk to drivers—a question that shifted the spotlight to the NLRB.

The pivot toward the NLRB occurred as the board and the general counsel—and through them the NLRB regional offices—had finally begun to change in prolabor directions after a decade of gridlock. Given the enormous power of NLRB officials, their appointments have historically been among the most controversial in the entire federal system. Republican presidents have regularly made appointments of industry representatives with antilabor views and have sought to delay or block Democratic appointments of prolabor members.[133] The partisan refusal to confirm appointees to the NLRB reached back to the George W. Bush administration, when the Democratic Congress blocked the president's nominees in 2008.[134] That fight escalated with the election of Barack Obama, who inherited a nonfunctioning NLRB when he entered office (comprised of one Democratic appointee and one Republican appointee with three vacancies). President Obama's attempt to fill the vacancies to create a majority of Democratic appointees was thwarted by Senate Republicans, who (though in the minority) were able to filibuster the nomination of Craig Becker, who was associate general counsel of the Service Employees International Union (SEIU). Obama's response was to use his so-called "recess appointment" power under the Constitution—allowing a president to "fill up all Vacancies that may happen during the Recess of the Senate"[135]—to appoint Becker and other subsequent members. In 2012, an employer dissatisfied with an NLRB ruling challenged the validity of these recess appointments (and thus the legitimacy of the ruling), igniting a case in which the U.S. Court of Appeals for the D.C. Circuit ruled in January 2013 that the appointments were invalid because they technically occurred during a formal session of the U.S. Congress and not between separate sessions.[136] Although the NLRB appealed, Obama withdrew the appointments and nominated three new members.[137] When Senator Harry Reid threatened in July 2013 to exercise the "nuclear option" and eliminate the filibuster altogether, President Obama and Senate Republicans struck a deal that paved the way for a full slate of NLRB confirmations. As a result, in August 2013, there was finally—after five years of political wrangling—a fully functioning five-member NLRB with a majority of its members appointed by a Democratic president.[138] The timing proved auspicious.

In the context of this political change, legal planning for NLRB-protected strikes by misclassified drivers—which had grudgingly started as a last-ditch effort after the failure of employee conversion—progressed with greater alacrity.[139] As Teamsters counsel Manley recounted, the misclassification phase of the campaign "centered on two ideas": first, "drivers had to act like employees" and, second, that independent contractors had "the right to strike over their misclassification."[140] Taken together, these ideas pointed toward a strategy in which drivers would assert their claim to be misclassified employees as a basis to gain labor law protection of their collective action against trucking firms. There were a number of open legal questions and strategic considerations. In mulling those over, Manley went back to reconsider misclassification research, including papers presented at earlier conferences that had just been "sitting around," not directly relevant to the initial legislative approach advanced in the Clean Truck Program.[141] This research focused on old labor cases in which workers "who were assumed to be independent contractors struck because they wanted to be employees. ... We began talking about that idea: what if we did that?"[142] The legal question became: *Could the drivers show they were misclassified and use that showing to both justify ULP strikes and provide a foundation for convincing the NLRB to permit an election for union representation?* No one had ever tested this theory, which articulated a novel legal ground authorizing workers to engage in strikes. In this uncharted territory, the campaign decided to move carefully forward on two fronts.

One was to accumulate evidence of misclassification that might prove persuasive to the NLRB in later cases. Manley described the legal reasoning behind this move: "[We thought,] 'What are some easy, relatively cheap ways to get determinations of employee status?,' which leads you naturally to look at administrative agencies because a court case is never cheap. And so that led us to DLSE and the board. The DLSE is particularly attractive because ... if you established employee status, there were tremendous economic damages attached to it because of the California statute that says you can't charge an employee for ... business expenses."[143] In addition, the DLSE process was easy and cheap; and because it did not require attorneys, there were no professional concerns about conflicts of interest in asking drivers to trade off damage awards in exchange for union contract benefits. With the potential for large damages for unreimbursed business expenses as the hammer, the campaign began executing the plan: "target companies and seek determinations of employee status."[144]

As Teamsters organizers continued to help drivers file DLSE claims, the campaign also moved forward on a second front, which involved building legal precedent at the NLRB Region 21 office in Los Angeles (or "the region" as the lawyers often called it) to advance the misclassification strategy. As Manley recalled, "The first thing at the board level was just to get the region to find one of [the port drivers] to be an employee."[145] That meant, first, educating the regional director and staff about the nature of port trucking and familiarizing them with the Teamsters organizing efforts in the industry. Manley believed "[t]hat was important because we were going to go to the region a lot and we needed an introduction."[146] The Toll campaign provided that introduction. The Teamsters leadership deliberately targeted Toll because its employee model and connection to a unionized international company provided a strong basis for organizing success. But, in addition, Toll allowed the Teamsters to pursue a traditional union election through Region 21 that presented the port trucking campaign in the most conventional light—as part of a familiar union representation case on behalf of employee drivers who wanted to join the Teamsters. This familiarity was important for impressing on regional staff—who liked "to see something they've seen before"—that the Teamsters port campaign was not "weird and crazy," but rather "just another campaign like any other."[147]

Building on the Toll campaign, the Teamsters believed that the time was ripe to push the misclassification theory to its legal conclusion: claiming it as an independent basis for a ULP under NLRA section 8, justifying legal picketing and strike activity, and paving the way for an NLRB-sponsored election. This had never been done before. The idea grew out of discussions within the Teamsters leadership. Manley remembered "sitting around with some folks and the director of organizing, Jeff Farmer, says something like, 'I don't know why misclassification isn't just a violation of the law?' And I thought at the time [sarcastically]: 'Yeah, OK, fine great, that's helpful. ... Crazy idea.' Then it was probably two months later when it did just come to me sitting at my desk, but I don't claim it was my brilliance. ... All of a sudden I thought to myself: 'He's right.'"[148] The campaign would thus advance a novel legal argument: misclassification on its own violated NLRA section 8(a)(1) by restraining and coercing drivers in the exercise of their protected rights as employees under the act. As Manley put it, "nothing is more coercive than telling someone who is an employee they're not an employee. Because if they are not an employee, they have no rights under the act. What is more coercive than that? So ... misclassification in and of itself violates the NLRA."[149]

The challenge was getting the board to agree.

Resurgence: Building a Test Case

With the legal argument in place, the Justice for Port Drivers campaign leaders deliberately orchestrated the next steps to gradually introduce NLRB Region 21 staff to the misclassification problem toward the ultimate end of persuading the full board to issue a formal ruling that misclassification violated section 8 of the NLRA. It was a classic test case approach—building legal precedent to lay the ground for major legal reform—though reconfigured to effectively operate within the particular administrative space of the NLRB. The short-term goal was to "come up with a strategy to get [port trucking] companies to start reclassifying,"[150] by familiarizing the

NLRB with the concept of misclassification. As Weiner recalled, "we knew we couldn't start" with the argument that misclassification was a ULP on its own terms. But leadership believed that if the campaign could get the regional director to issue a complaint on behalf of drivers that rested on a traditional ULP ground, like retaliation, that would validate the campaign's underlying claim that drivers were misclassified in the first instance. As Weiner laid out the logic, "If [the drivers] are already employees, then it is easier for the board to come to the conclusion that misclassification is a ULP."[151]

To help guide the board toward that conclusion, the campaign settled on an approach of incrementally pushing boundaries both in terms of the firms it targeted for organizing and the tactics it deployed. With each success, the campaign would grow more ambitious: from conventional organizing against employee firm Toll; to integrated suing and striking, first against a "hybrid" firm with a mix of employee and independent-contractor drivers; then on to a full assault on contractor-only firms, starting with easier targets before pursuing the major players. At each stage, as the campaign ratcheted up the pressure, it would correspondingly push the boundaries of its striking (both in terms of duration and location), while also advancing the misclassification as ULP argument: moving from asserting misclassification as a legal predicate for a traditional ULP in the early cases to ultimately claiming it as an independent legal violation against major independent-contractor firms.

Targets: From Toll to Total Misclassification

To roll out the test case approach after the Toll victory, the campaign began to pursue companies with increasing proportions of misclassified (i.e., formally independent-contractor) drivers, while expanding the scope and intensity of strike activities. Although the targeting of trucking firms to organize was "organic,"[152] in each instance, the campaign made a deliberate effort to move more forcefully toward the "issue of misclassification."[153] In Weiner's terms: we "pushed the envelope each time ... we started small and worked our way up to avoid making mistakes," learning from each strike how to go one step further on the next.[154] Before taking any steps, the campaign leaders paid an obligatory visit to the Teamsters Joint Council (the board overseeing regional Teamsters affiliates) and the Los Angeles County Federation of Labor to request that the misclassification strike campaign be "sanctioned," which was a "way of formally getting the endorsement of labor" so that other unions would honor the campaign's bona fide strikes.[155]

With the labor movement's official blessing, the campaign moved forward under the steady guidance of labor veterans buoyed by the fresh energy of new team members. A fixture of the veteran group was Teamsters Port Division Director Fred Potter, who had stood by the campaign in the face of internal dissent after the failure of employee conversion. Born and raised in New Jersey, Potter started working as a Teamsters driver in the construction industry in June 1970 and rose through the ranks to become president of Local 469 in the mid-1980s.[156] In 2006, he was elected as international vice president at large, serving on the Teamsters executive board as a voting member. Two years later, after the retirement of Chuck Mack, Potter was appointed by President Hoffa to be port division director to "organize port drivers and workers in the ports nationwide."[157] From that position, Potter's job was to "assist local unions that work in and out of the ports ... [and] to go out and try to win justice for [truck] drivers."[158] Although Potter's home was in New Jersey, the "epicenter" of the port trucking work was in California by virtue of its "very good laws," the fact that the state was "very aggressive in enforcing those laws," and given that there was political will and union support.[159] As the Justice for Port Drivers campaign got underway in Los Angeles and Long Beach, zeroing in on the goal of winning more Teamsters contracts, Potter was an essential figure on the ground since he had "the final say" on any union contract with a port trucking firm—subject to approval only by Hoffa.[160]

Potter was joined by a new addition to the Teamsters, Carlos Santamaria, who was hired by Local 848 as an organizer for the campaign against Toll—a company that he had just left as a driver.[161] An immigrant from El Salvador, whose father was a port driver, Santamaria grew up in Wilmington and joined the family business soon after graduating from high school, dropping out of community college to enter port drayage in 2000.[162] As an independent contractor, Santamaria worked primarily for Calko Transport—leaving to take an employee driver job in the mid-2000s, but disappointed with the pay, returning to Calko in 2006. Though Santamaria opted for the independent contractor's life, taking out a loan to purchase a ten-year old truck for $12,000, he chafed at his treatment, organizing several strikes against Calko, one for the company's failure to share fuel surcharges paid by customers: "They were stealing our money."[163] Other drivers looked to Santamaria as a leader, especially because he was one of the only fluent English speakers. His leadership skills also attracted the attention of Teamsters organizers, who had reached out to Calko workers in 2004 and came back again "full force" in 2007 seeking to "open up a campaign" around the Clean Truck Program.[164]

However, the Teamsters' recruitment of Santamaria would take time. In 2008, after giving his contact information to Teamsters organizers at Calko, Santamaria left for Swift Transportation, the national trucking company that had agreed to participate in the Clean Truck Program in exchange for financial incentives from the Port of Los Angeles. Santamaria appreciated the benefits of employee status, having previously held an employee job, and reasoned, in joining Swift, that: "I want to be the first one in line. ... I should just jump on this wagon before everybody else does."[165] When the Teamsters started organizing at Swift, they visited Santamaria at home to gauge his support. Though initially undecided, he started attending meetings and "little by little" became involved in the campaign.[166] Because he "had a good story" as a "second-generation truck driver" who had been both an employee and independent contractor, and lived blocks from the port and thus experienced congestion and pollution, the Teamsters asked Santamaria to talk to other drivers. He agreed, believing that he "was doing good for others."[167]

After the employee provision of the Clean Truck Program was enjoined, the Teamsters campaign fizzled as Swift converted its drivers back to independent contractors. Santamaria recalled the meeting when management made the announcement, telling drivers that they could "reach the American Dream" by becoming independent owners.[168] Santamaria stood up and said:

> What the hell are you talking about?. ... Hold on. ... You are telling me *you* are not making any money because we are sitting at the port [waiting for containers] and you can't pay me 18 bucks an hour because I'm not making enough loads. And you want to put me on a truck that costs 140,000 bucks. And you expect *me* to make money? What kind of a lie is this? I ain't stupid! If you aren't making any money how am I going to make money? The system doesn't work.[169]

Santamaria quit. But the experience drew him closer to Teamsters organizers, who invited him to a Coalition for Clean and Safe Ports meeting with Mayor Villaraigosa about the need to keep fighting for employee status and then to Congress to advocate for amending the FAAAA in 2010. On the heels of this activity, Santamaria went to Toll as an employee driver, working the night shift. When his third child was born shortly thereafter, he left Toll, and in late 2010 was recruited to join the Teamsters—where he was tapped to mount a union campaign against his former employer.[170]

Working with lead ports organizer Jason Gateley, Santamaria reached back out to his former colleagues at Toll and started to meet with groups of drivers: "it became my campaign, and basically that's when I learned how to organize. ... And it was my first victory."¹⁷¹ This victory was crucial, in Santamaria's view, because other drivers "started to believe" in the Teamsters and "there was a lot more interest" in their message.¹⁷² Such belief was essential because drivers were being asked to strike despite the fact that companies were telling them that their independent-contractor status prevented them from doing so. As campaign leaders began planning to support strikes by misclassified drivers, Santamaria worked closely with Gateley to figure out an organizing approach on the ground that could disrupt trucking operations at the ports, while assuring drivers protection against retribution.¹⁷³ When Gateley left as lead ports organizer after the Toll contract, Santamaria took over his position at the moment the campaign began to widen its net.

In directing the campaign's ground game against trucking firm targets, Santamaria would have a new partner from LAANE, which in the summer of 2013 hired Sheheryar Kaoosji as its ports campaign director, replacing Jon Zerolnick, who transitioned back to overseeing LAANE's research agenda. Kaoosji's interest in the labor movement was sparked after college, while working at an economic development organization in the rapidly gentrifying Mission District of San Francisco, where he collaborated with the SEIU and HERE on antidisplacement efforts and became "really intrigued by the labor movement as an entity that has its own autonomy."¹⁷⁴ After four years in the Bay Area, he decided to venture south to pursue a master's degree at UCLA's Public Policy School, which was known for its connection to the labor movement and focus on "strategic research," in which Kaoosji immersed himself in connection with coursework on economics and quantitative analysis.¹⁷⁵ Upon graduation in 2006, he joined newly formed Change to Win, where after a stint in farmworker organizing, he transitioned into a position conducting strategic research on the warehouse industry—"the next step in the supply chain from the port."¹⁷⁶ His warehouse work was centered on the Inland Empire, which had a massive and understudied warehouse industry, but few progressive institutions with capacity to organize. The warehouses were hiring workers through staffing agencies (thereby avoiding direct employment relationships) and moving goods "for the biggest companies in the world."¹⁷⁷ Armed with his research, Kaoosji supported the United Food and Commercial Workers Union in launching a campaign to pressure Walmart to permit union organizing at its warehouses. When that effort stalled in 2013, Kaoosji took the LAANE ports director position, which was a logical fit with his research skill set and focused on an industry that both resembled and was intrinsically related to warehousing. As ports director at LAANE, Kaoosji took over strategic research for the Justice for Port Drivers campaign: investigating targets (especially top companies) and industry trends, tracking litigation, and supporting driver organizing.[178]

The first target on the campaign's list was Carson-based Green Fleet, a "hybrid" company with a minority of misclassified drivers and the rest employees.[179] Green Fleet was selected by the campaign as the critical test of its suing-and-striking strategy, designed to build evidence of misclassification through the DLSE process and gradually extend the reach of NLRB union organizing protections to drivers at independent-contractor-based trucking firms. The DLSE yielded the first success, ordering Green Fleet to pay $280,822 in back wages and penalties to four misclassified drivers in early 2013.[180] With that order in hand, the campaign took the next step to develop its misclassification theory at the NLRB.

With Green Fleet, the campaign sought to ever-so-slightly raise the stakes at the NLRB by introducing regional staff to a different port trucking structure: a firm that generally resembled Toll's employee driver model—except for the minority of drivers that Green Fleet treated as independent owners. Apart from that important structural difference in the target, the NLRB campaign would essentially follow the Toll template: ULPs would be filed against Green Fleet on traditional grounds, like retaliation and intimidation, justifying a strike that would stay close to the Green Fleet yard—not yet seeking to disrupt operations at the ports. This was to be a gradual first step, answering the critical question campaign leaders were asking themselves: "Is the board going to come to the same conclusion that the U.S. Department of Labor and the California Labor Commissioner have regarding the employment status of these workers?"[181] The novelty of seeking to unionize independent contractors raised concerns among front-line organizers. When presented with the idea of misclassified drivers engaging in a "minority strike," Weiner recalled the incredulous reaction by Local 848 leader, Eric Tate: "Won't they all get fired?"[182] Tate "just thought we were crazy," but after talking it through with Weiner, Tate was game: "OK, let's give it a try."[183]

Advancing the NLRB strategy around misclassified port truck drivers meant engaging Region 21 staff in Los Angeles—which required finding local counsel to partner with campaign general counsel, Mike Manley. That job fell to Julie Gutman Dickinson of the boutique labor law firm, Bush Gottlieb, who was hired to represent the Teamsters during the Green Fleet campaign—and would play a key role in strategic planning, litigation, and settlement activities in all of the ensuing port trucking cases. Gutman Dickinson brought deep experience in Region 21 with an extensive network of labor and political contacts. The product of a "Jewish family very big on social change," Gutman Dickinson was called to social activism an early age.[184] Her grandmother was involved in the labor movement in Chicago, and her cousin was a Holocaust survivor who impressed upon her that her life mission had to be not simply about "just us but justice."[185] As a student at Stanford Law School, Gutman Dickinson was an acolyte of Professor Gerald López—famous for articulating a "rebellious" vision of lawyering that placed subordinated communities at the center of social change—and threw herself into social justice practice in East Palo Alto, then a low-income, predominantly African-American community adjacent to Stanford's bucolic campus. From these experiences, Gutman Dickinson came to think of her work in terms of "lawyering for social change": "working in bigger coalitions," "combining dynamic research with grassroots organizing," "looking at the bigger problem and analyzing how to solve it," and mobilizing law as "part of strategic campaign planning with the team."[186] She saw herself as "an organizer in my heart," committed to using "law as a tool" to empower subordinated communities to "take the lead."[187]

After graduating from law school in 1989, Gutman Dickinson clerked one year for a judge, studied Spanish in Guatemala, and was awarded a prestigious Echoing Green Fellowship to start a community economic development project in East Palo Alto, where she focused on fighting the displacement of local residents because of gentrification. In 1994, burned out from the intensity of her East Palo Alto work (for which she reported receiving death threats), she moved to Leonard Carder, a labor law firm in San Francisco, where she represented ILWU warehouse workers and the HERE, while conducting trainings on immigrant rights.[188] A "work-a-holic," she started searching for "bounds in her life" and in 1996 made the jump to Los Angeles, taking a position as a bilingual trial attorney at NLRB Region 21. There for a decade, Gutman Dickinson "really learned the ins and outs of the NLRB," and had "a lot of power and influence" to make labor law work for workers.[189] Her specialty was filing petitions for preliminary injunctions—so-called 10(j) injunctions in reference to their NLRA code section—in order to get workers who had been illegally fired back on the job and back organizing. She became known as the "10(j) queen."[190]

In Antonio Villaraigosa's second year as L.A. mayor, Gutman Dickinson left Region 21 to serve as his senior labor advisor and a member of the public works commission, where she spent three years brokering labor agreements between companies and unions representing janitors, security officers, airport service workers, and trash and recycling workers. In that capacity, she served with deputy mayor Larry Frank and played an "insider role" advising the mayor on the Clean Truck Program—urging him "to support the employee mandate."[191] Upon leaving the mayor's office, and after a stint as director of a human rights group, Gutman Dickinson became partner at Bush Gottlieb, whose lawyers had been recruiting her since her early days in Los Angeles. Through her work on the Clean Truck Program, Gutman Dickinson knew Nick Weiner and others from the Teamsters leadership; she also had political relationships that extended to the new mayor, Eric Garcetti. On top of it all, she "knew how to bring a case" at the NLRB.[192]

With Gutman Dickinson on board, the campaign began to move aggressively against Green Fleet. On March 8, 2013, the Teamsters filed ULP charges against the firm with Region 21. Those charges were subsequently amended several times to include allegations of Green Fleet supervisors engaging in unlawful conduct by pressuring employees to sign an anti-union petition, interrogating employees about union sympathies, threatening job loss and plant closure, and intimidating drivers with other threatened reprisals.[193] With these ULP allegations serving as legal justification, on August 26, 2013, approximately thirty of Green Fleet's ninety drivers went on a one-day strike.[194] Although limited in size and scope—it was designed only to last twenty-four hours and stay at the Green Fleet headquarters—the effect was explosive. Confronted with drivers and prominent leaders, including Congresswoman Janice Hahn, the company locked the drivers out. As Weiner remembered: "Hell was breaking loose."[195] Green Fleet accepted the union's offer to return to work after the strike ended, although additional ULP charges were filed against the company for illegal surveillance during the strike. From the campaign's standpoint, the Green Fleet action was a resounding success—suggesting it was time to turn the screw slightly tighter by targeting a trucking firm with a 100-percent independent-contractor fleet.

That target was another Carson-based firm, Pacific 9 Transportation (Pac 9), a top ten company with over 150 drivers, all of whom were labeled independent contractors. This represented an escalation: a direct assault on a company relying solely on the independent-contractor business model. It would also be the case in which the new misclassification-as-ULP theory would finally be tested in Region 21. Pac 9, whose drivers were already standing up and fighting back against their status, was a firm that appeared ripe for challenge. In June and July of 2013, forty-seven drivers filed DLSE claims against Pac 9 alleging more than $6 million in damages.[196] In Kaoosji's telling, this was an example of the drivers saying: "We are employees … so we are going to act like it by filing DLSE claims en masse … and taking action on the ground as a group and using the NLRB as protection."[197]

Reshaping the Industry

Following its now battle-tested approach, on November 4, 2013, the Teamsters union filed a ULP charge against Pac 9 alleging that the firm had threatened to close in retaliation for driver organizing and unlawfully interrogated a driver for union activity.[198] These charges were still traditional ULPs—they did not allege misclassification itself constituted a ULP, but they did require the NLRB to determine that the drivers were in reality misclassified employees under the NLRA in order to rule in their favor. In this way, during round one of the Pac 9 case in front of the NLRB, the campaign closely adhered to the playbook used in the Green Fleet strike.

Doing so set the stage for the campaign's first coordinated action, on November 17, 2013, which combined ULP strikes of Pac 9, Green Fleet, and American Logistics—the latter, a 100-percent employee firm targeted so that the strike would include drivers protesting the full range of port trucking companies, along with Pac 9's 100-percent independent-contractor model and Green Fleet's hybrid firm. In addition to expanding the list of targets, the campaign stretched the length of the strike, which began at 5 a.m., from twenty-four to thirty-six hours; the picketing started in front of Green Fleet's Carson yard, and at American Logistics in Carson and Pac 9 in Long Beach.[199]

This time, however, picketing would not be limited to the struck companies' truck yards as the campaign decided to test another tool in its tactical arsenal: the practice of so-called "ambulatory picketing." In the trucking industry, courts had interpreted the right to picket under the NLRA as permitting a union to essentially follow the trucks, which were deemed to be a mobile workplace.[200] When trucks entered other work sites to conduct pick-ups and drop-offs—such as multi-firm container loading yards, rail yards, warehouses, and even the ports—the union was permitted to picket on the premises of those sites.[201] This rule gave the Teamsters the legal authority to picket at crucial supply chain entry points, potentially creating disruption and delay for other trucking companies and terminal operators. In the first Green Fleet strike, campaign leaders opted to refrain from ambulatory picketing, not wanting to move too far too fast. However, because one of the campaign's ultimate aims was to demonstrate to powerful port clients and city leaders that driver strikes could impose systemic costs, it was necessary to unveil the potency of ambulatory picketing at some point. The November 17 combined strike, campaign leaders concluded, would be a good opportunity for a preview.

Not yet comfortable taking pickets all the way to the ports, the Teamsters opted for a more limited demonstration: putting up picket lines at the Shippers Transport Express storage yard, where SSA Marine—a major terminal operator at the Port of Long Beach owned by Shippers' parent company—would send containers for pickup and transport. Because struck company trucks frequented the Shippers yard, it was an open target for the campaign's first wave of ambulatory pickets. The pickets, in Weiner's recollection, "created complete mayhem."[202] As the pickets started to delay trucks, the yard operators called police to shut it down, which the police refused to do because campaign lawyers had already provided them with a legal memorandum establishing the legality of ambulatory picketing. However, when Teamsters attorney Mike Manley inadvertently stepped across the yard's property line, he was promptly arrested for trespassing. Weiner recounted humorously that of all the people there, the only one who was arrested was the lawyer.[203] Joking aside, the action at the Shippers yard was also the moment that the campaign realized the power it held with ambulatory picketing: "That's when we figured out: We could do this at the port."[204] As the strike concluded, Weiner declared: "It's tangible. We could win a good pension, win overtime pay for nonunion truck drivers. ... Getting paid piecemeal by-the-load, that's just what's fueling our claims."[205] And as Weiner articulated the drivers' grievances, long-time political stalwart Janice Hahn urged them to "keep on trucking."[206]

In the wake of the November 17 action, additional ULPs were filed, including some alleging the unlawful termination of two Green Fleet drivers in January 2014 in retaliation for engaging in union activity and filing DLSE claims. One of the drivers, Mateo Mares, issued the following statement on the Teamsters website:

> I was a driver for Green Fleet Systems hauling Skechers Shoes and other cargo from 2008 until I was fired in January this year. ... When I began work at Green Fleet, the company misclassified me as an independent contractor. A few years ago, I heard about the Teamsters organizing campaign. I started talking to the Teamsters and to other drivers about working conditions, and I learned that I was in fact an employee and that, by misclassifying me as an independent contractor, Green Fleet was making unlawful deductions from my paycheck and denying me many benefits, including the right to organize. I began assisting the Teamsters organizing efforts, talking to other drivers about the benefits of unionization, and taking part in various campaign activities, such as wearing a union vest and participating in two unfair labor practice strikes. Green Fleet's owner, Gary Mooney, and others interrogated me about my union activities and threatened to sue me. I organized a group of Green Fleet drivers to go to the California Department of Labor's Division of Labor Standards Enforcement to file wage claims. After Green Fleet learned that we filed wage claims, owner Gary Mooney told my co-worker Amilcar Cardona and me that we couldn't continue to work at Green Fleet unless

we dropped our wage claims. When we refused to drop our wage claims, Mooney fired Amilcar and me.[207]

In April 2014, in response to Cardona and Mares' termination, Gutman Dickinson filed a petition with Region 21, requesting that it seek interim injunctive relief against Green Fleet—to reinstate the drivers—under NLRA section 10(j), which authorized such relief pending the board's final resolution of a ULP.[208] Region 21 staff agreed with the "10(j) queen" and brought a petition in U.S. district court against Green Fleet soon thereafter.[209] As the 10(j) petition was pending, Gutman Dickinson joined campaign leaders in meetings with mayor's staff, port officials, and board commissioners to discuss possible policy responses to the misclassification problem. Gutman Dickinson's role thus shifted between direct representation of the Teamsters in board actions and broader advocacy work.

As the NLRB and policy advocacy progressed, driver litigation against trucking firms continued to expand, owing in significant part to campaign efforts to support drivers filing claims. Teamsters organizer Carlos Santamaria was key in this regard, leading outreach efforts to misclassified drivers at port trucking firms. Santamaria's starting point in talking to drivers was to tell them: "Look guys, you are employees, that's what we believe, no matter if you own a truck or don't own a truck ... doesn't really matter. You are under the control of the company."[210] Santamaria and his team would advise drivers they were entitled to reimbursement for expenses and would help them fill out DLSE forms and photocopy documents, walking "them through it, step by step."[211] Next, the organizers would take drivers in groups to the DLSE, where Santamaria would emphasize that driver wage claims could be used "to negotiate things that are worth more than money."[212] When claims were filed by other lawyers, Santamaria and his team tried "to make ourselves available to the lawyers," sharing documents and advice in order to maximize the potential for successful claims, since any legal finding that drivers were not misclassified would hurt the campaign.[213]

By early 2014, that work was starting to pay off. In February 2014, roughly a year into the Justice for Port Drivers campaign, NELP, Change to Win, and LAANE published an updated version of *The Big Rig* report,[214] which surveyed the scope of misclassification litigation at the ports up to that point. Looking at cases since January 2011, the authors reported that "[s]ome 400 port drivers have filed labor law complaints" with the DLSE, resulting in "19 decisions finding that drivers are employees" and assessing "more than a million dollars in wages, unlawful deductions, and penalties on behalf of at least 19 drivers against at least five companies: Green Fleet Systems, Seacon Logix, Western Freight Carrier, Total Transportation Services, and Mayor Logistics."[215] In addition, based largely on coalition efforts to place driver cases with private lawyers, the report noted that there were nine pending private lawsuits against trucking companies for violations related to driver misclassification: eight of which (including those against Gold Point, QTS, Harbor Express, Pacer Cartage, and Southern Counties Express Transportation) were class actions.[216] The suit against Pacer Cartage was bought in August 2013 as a state court class action by lawyers at Kabateck Brown Kellner (which had already filed the class action against Harbor Express earlier in the year and with whom the campaign had established some communications). The Pacer Cartage suit alleged a litany of employment violations related to driver misclassification, including willful misclassification under recently enacted California labor law, along with claims for unlawful failure to provide meal and rest breaks, pay minimum wage and overtime, and furnish timely and accurate wage statements; the complaint also alleged that Pacer was engaged in unfair competition in violation of California Business and Profession Code section 17200 for gaining an unlawful cost advantage through misclassification.[217] Pacer Cartage promptly petitioned to remove the case to federal court, where—after significant legal jousting over appropriate venue—it ultimately remained.

The same month the updated *Big Rig* report was released, another class action was added to the list against prime target Pac 9. The case, *Castro v. Pac 9 Transportation Inc.*, was filed on February 24, 2014 in Los Angeles superior court on behalf of "[a]ll current and former California employees of Defendant Pacific 9 Transportation Inc., employed in California, at any time beginning four (4) years prior to the filing of the Complaint ... who drove a truck as an 'independent contractor' for Defendant."[218] It was, like the Pacer Cartage class action, brought by the Kabateck Brown Kellner firm, which followed the same essential legal template—with the added wrinkle of filing the misclassification claims simultaneously under the state's class action law and the Labor Code "Private Attorney General Act" (PAGA), which permitted a single plaintiff to enforce state labor law and recover civil penalties "on behalf of himself or herself and other current or former employees" as an alternative to enforcement by the DLSE.[219] This was a way to preserve the misclassification claims in the event the court declined to certify class action status for the case, since PAGA claims did not require class certification, while adding additional legal leverage since PAGA provided for attorney's fees in successful cases and PAGA claims could not be waived in advance via mandatory arbitration agreements. In December 2014, a second class action suit was filed in Los Angeles Superior Court against Pac 9, adding a claim for failure to reimburse business expenses under Labor Code section 2802, the "monetary hammer" for damages not included in the *Castro* suit.[220]

The campaign's assiduous legal work at the NLRB was also beginning to yield results. The day after the *Castro* class action was filed, on February 25, 2014, in a major breakthrough, Region 21 notified Pac 9 of its intent to issue a complaint against it: not only accepting that there was sufficient evidence of ULPs, but—most significantly—agreeing for the first time in a case against an independent-contractor trucking firm that the drivers were, in fact, "statutory employees."[221] Realizing the enormous implications of this position, Pac 9 quickly entered into an informal settlement agreement,[222] which required Pac 9 to post notice of employees' right to organize, although Pac 9 did not admit fault or agree to pay penalties or fines.[223] As the *Los Angeles Times* reported, "The agreement comes after repeated victories at the state Labor Commissioner's office, where 30 drivers have won decisions against 11 port trucking firms, awarding them $3.6 million in wages and penalties."[224] Eric Tate of Teamsters Local 848, reacted with enthusiasm: "The 30-year debate is over. The misclassification lie has been busted. The port drivers are, in fact, employees. ... The NLRB has said so. ... Pac 9 has said so. Now every port truck driver who wants to end their sweatshop conditions can bargain collectively to climb the economic ladder into the middle class."[225] In a press release, the Teamsters' Fred Potter echoed the same sentiment: "Now every port driver who wants to end their sweatshop conditions—get fairly compensated for every hour they work so they can drive safely, have sanitary bathroom facilities, clean drinking water, and medical insurance for their families—can bargain collectively to climb the economic ladder into the middle class."[226] Alex Cherin, executive director of the Harbor Trucking Association, tried to minimize the significance of the Pac 9 settlement: "This is not new. ... It's the latest iteration of the Teamsters trying to unionize the industry over the last six years. This is just an administrative settlement, not a judicial settlement."[227] However, with the agreement in hand—and over five hundred more port driver DLSE claims in the pipeline, including newly filed claims against major player Shippers Transport Express—the Teamsters' Tate could proclaim, "We are on a big roll here."[228]

Yet that roll slowed appreciably as Pac 9 promptly proceeded to violate its own agreement by issuing a memorandum to its workers on March 28, 2014 stating "there have been news about Pac 9's relations with its drivers have been changed [sic]. We would like you to know that this is not true."[229] In response, NLRB staff sought to persuade Pac 9 to withdraw the memo and recommit to the settlement terms,[230] while the Teamsters filed more ULP charges. However, despite this backsliding, the NLRB's decision to pursue Pac 9 was from the campaign's perspective a "huge step," constituting the board's first signal to the drivers that, as Kaoosji put it: "Yes, you are misclassified and, yes, you have the right to organize—despite what everyone has told you."[231]

Campaign leaders sought to build on this argument to exert pressure against another all-independent-contractor firm, Total Transportation Services, Inc. (TTSI), based in Rancho Dominguez. TTSI drivers had started filing DLSE claims against the firm on their own as early as 2011. In February 2013, the Labor Commissioner ruled in favor of two drivers, ordering repayment of over $170,000,[232] and ordered an additional award of over $100,000 to a driver in October 2013.[233] In all three cases, the largest portion of the recovery was for illegal business deductions. Buoyed by these early successes, more drivers filed DLSE claims: eight in May and June of 2013, followed by twenty in July, nine in September, four in January 2014, and one in March. Then, on April 24, 2014, with evidence of misclassification firmly established at the DLSE and economic pressure mounting, the Teamsters—represented by Manley—filed ULP charges against TTSI for the first time.[234] With TTSI thereby placed in the strike queue, by mid-2014, the stage was set for the campaign to make a concerted push against a trio of misclassifying port trucking companies: the hybrid firm, Green Fleet, and the all-independent-contractor firms of Pac 9 and TTSI.

Tactics: Picketing between the Headlights and Misclassification as a ULP
With the initial trio of major port trucking firm targets selected, and 100-percent independent-contractor firms now in play, the next stage of the Justice for Port Drivers campaign focused on testing the boundaries of the drivers' key source of leverage: the power to strike. The November 17, 2013 combined strike against Green Fleet, Pac 9, and American Logistics had revealed the potential of ambulatory picketing to create disruption. Whereas the campaign had decided to contain that strike to the Shippers Transport Express storage yard, it was now ready to extend ambulatory picketing all the way to the ports. Because a union was authorized under the ambulatory picketing rules to follow the struck company's vehicles wherever they went, the campaign could set up pickets at port terminal entries and exits. As a result, in Weiner's colorful metaphor, trucks could "get into the roach motel but can't get out. ... No one is happy."[235]

The idea of strikers following the trucks, or "picketing between the headlights" as Manley called it, was an old strategy for the Teamsters—about to be applied in a new context.[236] To carry this out, the campaign converted the Teamsters Local 848 office in Long Beach into "strike headquarters."[237] As Weiner described it, Local 848 became the campaign's "war room," staffed by Local 848's Eric Tate, Carlos Santamaria and his organizing team, and communications guru Barbara Maynard. They were joined by LAANE's Sheheryar Kaoosji and his ports' project colleagues, local counsel Julie Gutman Dickinson, and national labor leaders who would come in at various points, including the Teamsters Mike Manley, Fred Potter, and Jeff Farmer, who was the union's national organizing director, as well as Change to Win's Nick Weiner. In that space, Weiner recalled that everyone came together in "a union-wide effort."[238]

That effort came to a critical juncture in April 2014. TTSI was in the spotlight. In addition to filing ULP charges against the firm with NLRB Region 21, the Teamsters helped TTSI drivers file complaints with the California Division of Occupational Safety and Health for unsafe, unsanitary, and environmentally dangerous truck yards on April 16,[239] while drivers spoke to the Los Angeles Board of Harbor Commissioners about their concerns the next day.[240] Recounting problems related to "the lack of dust control and personal protective equipment and unsanitary and inadequate restrooms and water supply at the company's truck yard," drivers asserted: "We want [the board] to take action, to see what's going on with this company. ... As drivers, we are tired of this."[241]

The same month brought new developments on the litigation front. Lawyers filed a class action in Los Angeles County Superior Court against Coast Bridge Logistics Inc. of Compton, raising misclassification claims over a period of four years covering roughly 200 drivers.[242] Around the same time at the DLSE, the labor commissioner ruled against the all-independent-contractor trucking firm Pacer (which had just been acquired in March 2014 by XPO Logistics, an integrated global supply chain conglomerate that owned trucking firms, containers, warehouses, intermodal transfer facilities, and other logistics assets); in its order, the DLSE found that the control Pacer asserted over its independent contractors demonstrated an employer-employee relationship and awarded the drivers over $2 million dollars in damages.[243] The decision provoked dueling assessments. David Arambula, the drivers' attorney, stated, "The industry has to take notice that these folks are getting together. ... They're organizing to exercise their rights."[244] The Harbor Trucking Association's Alex Cherin responded that a "vast majority of truckers in this harbor still want to remain independent contractors," claiming that there were "hundreds and hundreds of job openings for company drivers" if they wanted to "be paid hourly."[245] However, Rich Dines, vice president of the Port of Long Beach harbor commission, acknowledged slow "turn times" (how long it took for drivers to pick up cargo at the port) were hampering independent-contractor efficiency since drivers were "not being paid while they wait."[246]

Efficiency, of course, was precisely what the campaign sought to disrupt by striking—which was about to enter a new phase. On April 28, 2014, the campaign raised the stakes once again, launching a ULP strike against its trio of central targets: Green Fleet, Pac 9, and TTSI.[247] The strike of roughly one hundred drivers represented a pointed escalation in every dimension: Dropping American Logistics from the mix, it was the first that exclusively targeted misclassifying firms (though Green Fleet still had some employee drivers); the duration of the strike was extended from thirty-six to forty-eight hours; and the campaign pushed the ambulatory pickets to port terminals and giant retail warehouses.[248] In a showing of solidarity, approximately one hundred longshoremen represented by the ILWU honored the picket line, causing the strike to briefly shut down the Long Beach Container Terminal,[249] and prompting Potter to announce: "This is ground zero."[250] An arbitrator ordered the longshoremen back to work at fourteen terminals around 11 a.m. on the first day of the strike, after the Pacific Maritime Association (PMA) invoked a provision in their collective bargaining agreement preventing the ILWU from honoring nonunion ULP picket lines (although the arbitrator did recognize the validity of the truckers' strike).[251] After the longshoremen returned to work, Teamsters counsel Manley, surely mindful of the historic tension with the ILWU, was quick to point out: "That doesn't mean they didn't support our drivers."[252]

As the strike continued, picketing for the first time reached terminals in the Port of Los Angeles, while fanning out to Costco, FedEx, and Skechers warehouses in Mira Loma and the Moreno Valley.[253] As this suggested, the campaign was seeking to put pressure on retailers with the power to set shipping rates, thus folding in a corporate public relations dimension to the strike. As Kaoosji recalled, it was around this time that the campaign began "looking at bigger players in the industry ... making sure they [were] held accountable."[254] Toward that end, fired Green Fleet driver Mares and an active coworker, along with a delegation of community and faith-based leaders organized by LAANE, presented a petition with 25,000 signatures to Green Fleet's largest customer, casual footwear designer Skechers, asking it to hold Green Fleet "accountable to labor laws," while also appealing to Skechers' shareholders at the company's annual meeting.[255] In the midst of the strike, the Harbor Trucking Association's Cherin reiterated that there were "literally hundreds of unfilled vacancies for company [i.e., employee] drivers throughout Southern California," while adding: "Outside interest groups like LAANE and the Teamsters are continuing to spend their members' hard-earned money to battle an issue that a vast majority of harbor truck drivers have soundly rejected time and time again."[256]

As the strike came to an official close after its second day, picketing was reported at the Evergreen, APL, and Yusen terminals at the Los Angeles port, and at the Long Beach Container Terminal at Pier F and Total Terminals International at Long Beach's Pier T.[257] At these sites, drivers, organizers, and volunteers from LAANE, the County Fed, and the community picketed terminal entrances and exits for every struck company truck that came and went. Because terminal operators took the position that they could not turn away any truck, "everything [would] slow down to one truck" entering or exiting every few minutes.[258] The takeaways were twofold. First, the campaign's use of ambulatory picketing at the ports had passed its first critical test, succeeding as a controlled experiment in serious disruption. As LAANE's Kaoosji reflected, the strike's success sent a message to trucking companies: "We could put you potentially in a situation where you are not able to do business."[259] Second, exercising the power to disrupt port operations seemed to further empower the drivers, who had taken a courageous stand that had a real impact on their companies and the ports. Writing a few weeks later, well-known *Los Angeles Times* columnist Steve Lopez profiled Dennis Martinez, a driver from an April picket line, who stated he was unable to pay for his truck after five years, as he had originally anticipated, because of the extra cost of fuel and maintenance. When asked why he did not leave the industry, he responded: "Because I want to finish this fight."[260]

With head-spinning speed, that fight shifted once again from the streets to the legal system. On June 18, 2014, California's labor commissioner issued a series of orders awarding nearly $1 million in damages to fourteen TTSI drivers. That same day, NLRB Region 21 filed a complaint against Green Fleet[261]—thus accepting the litany of nearly fifty Teamsters' charges against the company and, equally important, accepting the argument that Green Fleet's independent-contractor drivers were misclassified.[262] For Manley, the Green Fleet complaint validated the campaign's test case strategy, which had deliberately combined ULP charges on behalf of employee and independent-contractor drivers and, for the latter, tried "very hard to have quality control in terms of the evidence of employee status."[263] Although the Teamsters did not argue in the Green Fleet case that driver misclassification was itself a ULP, instead relying on claims of retaliation and harassment, the region's decision to issue a complaint that included charges on behalf of Green Fleet's independent-contractor drivers rested on the legal conclusion that they were, in fact, employees. In Manley's view, the union's success in this regard "set the pattern for the region" and gave the campaign the green light to target all-independent-contractor trucking companies at the NLRB. The complaint, however, did not represent a legal ruling by the board and only underscored the importance of making sure that the record was airtight as the campaign moved toward trial in Green Fleet—its first critical case in Region 21 on the legal status of misclassified port truck drivers.

In addition to issuing the Green Fleet complaint, Region 21 was also busy ruling on skirmishes coming out of the two-day April strike against Green Fleet, Pac 9, and TTSI over the appropriate scope of picketing, holding that some of the Teamsters-led pickets were "excessive" under the NLRA.[264] Green Fleet counsel Thomas Lenz of the law firm Atkinson, Andelson, Loya, Ruud & Romo in Cerritos, applauded the ruling: "This action finally calls the recent tactics by the Teamsters for what they are—intimidation pure and simple."[265] Teamsters attorney Julie Gutman Dickinson had a different view: "We do not agree with the conclusions reached by the NLRB on this issue; however, our focus is on providing support to Green Fleet drivers who have been victims of more than 50 egregious unfair labor practices that the NLRB Region is prosecuting."[266]

By July 2014, campaign striking showed signs of having an impact—and prompted a new and more intense round of collective action. In a concrete example of striking's effect on industry, TTSI lost the lease to its main truck yard in Wilmington, owing to the disruption caused by "multiple activities that have taken place both inside and outside the facilities yard/gate."[267] When TTSI told its drivers to "find a new place to park your trucks/tractors," the Teamsters responded in a letter to TTSI Chief Executive Officer Vic La Rosa: "It is clear to us that you are taking these steps in retaliation for drivers asserting their rights as employees under the law."[268] Yet the TTSI lease drove home an important lesson from the April strike: The campaign could impose real economic costs on trucking firms and the port itself through its actions.

The question was how to widen the impact of trucker strikes: concentrating the costs of striking on the trucking industry without alienating longshoremen, whose solidarity with port truckers was strained by concerns over how work stoppage would affect their pocketbooks. Although the ILWU remained generally supportive of the port trucker campaign, as demonstrated in its members' refusal to cross the truckers' April picket line, longshoremen were contractually hamstrung in how far they could go to help. Because the longshoremen's collective bargaining agreement with the PMA prohibited them from supporting the truckers' nonunion strikes, the campaign had to figure out an alternative approach to disrupting the port that did not "depend on the longshoremen" to pull it off "and empowered drivers to have an impact."[269] However, the possibility of joint trucker-longshoremen action—and a mass shutdown of the ports—reemerged as the ILWU's contract with the PMA expired on July 1, 2014, potentially freeing longshoremen to honor nonunion picket lines as they attempted to negotiate a new agreement. Campaign leaders thus plotted their next move to strike a delicate balance: careful not to make it hinge on ILWU support, but hopeful that such support would come.

On Monday, July 7, 2014, the campaign mobilized a new "self-reliant" strike[270]—though one that still held out the specter of longshoremen work stoppage. Once again focusing firepower on the campaign's three major targets, over one hundred drivers from Green Fleet, Pac 9, and TTSI (representing 4 percent of port drayage trucks, about four hundred in total) went on a strike without a definite end in sight.[271] In order to avoid a showdown at the ports on the strike's first day, the campaign decided to limit picketing to company yards and held a rally at Wilmington Waterfront Park (across the street from the Los Angeles port), attended by public figures including Congresswoman Hahn. Nonetheless, according to the *Long Beach Press-Telegram*, "The strike left some 400 truckers unable to drive to the ports because they were reportedly told not to come to the terminals for fear that the International Longshore and Warehouse Union would have to honor the picket line."[272] This caused the California Trucking Association's CEO to indict the Teamsters action: "At a time when the International Longshore and Warehouse Union and the Pacific Maritime Association have demonstrated great restraint during their contract negotiation, the Teamsters have instead chosen to create disruption. If these disruptions spill over into marine terminals, the national economy could lose hundreds of millions of dollars."[273]

Late Monday, the hope of longshoremen support for the truckers' strike dimmed as the ILWU and PMA agreed to a short-term contract extension to start at 8 a.m. the next morning—although the question of whether the longshoremen were bound by the no-sympathy strike provision of the prior contract remained legally murky.[274] Taking advantage of this ambiguity, on Tuesday, July 8, the campaign moved pickets to the ports and the longshoremen honored the lines—walking off at the Evergreen, APL, and

Figure 6.2

Congresswoman Janice Hahn speaking with picketing drivers. Photo courtesy of Barbara Maynard, Teamsters communication director.

Yusen terminals in Los Angeles and at the Long Beach Container Terminal, thereby causing a shutdown that ended after the PMA invoked the temporary contract's arbitration provision to order the longshoremen return to work at 11 a.m.[275]

With the longshoremen back inside the ports, the campaign moved more assertively to disrupt the port terminals from the outside and, using that leverage, to pressure political officials to support their cause. In addition, as the strike expanded, the campaign sought to strengthen driver resolve. On Tuesday morning, the Teamsters announced a Justice for Port Drivers Hardship Fund, which raised $50,000 within "the first few hours."[276] On Wednesday, July 9, Rage Against the Machine guitarist Tom Morello performed at the Los Angeles port's Evergreen Container Terminal picket.[277] However, on July 10, Day Four of the strike, fifty drivers called a press conference to protest against the Teamsters, with one Green Fleet driver claiming, "The majority of us don't want to be represented by them."[278] As striking nonetheless continued, drivers and organizers showed up at a Los Angeles Board of Harbor Commissioners meeting, asking the commissioners to help resolve a range of driver health and safety issues.[279] Mayor Eric Garcetti, to whom campaign leaders had been talking throughout this period, took the opportunity to get involved: ordering the board to investigate and report on the drivers' health and safety claims, while urging "both parties to work with Port executives and the Harbor commission to ensure that this vital economic engine continues to serve this nation."[280] The following day, after a series of meetings brokered by the mayor, the union and trucking companies assented to a "cooling off period" (that the drivers unanimously voted to accept), during which the companies agreed to take back drivers without reprisal while the harbor commission agreed to continue its investigation of health and safety issues.[281]

This truce did not hold for long. By late July and August, drivers were making new allegations of retaliatory terminations against Green Fleet, Pac 9, and TTSI—as attention moved back to the legal arena. On July 29, 2014, a class action was filed by San Francisco-based labor law powerhouse Davis, Cowell & Bowe seeking to enjoin TTSI from following through on its announced plan to terminate drivers on August 1 "if they did not supply trucks themselves and waive DLSE claims."[282] On August 29, after nearly six months of trying to persuade Pac 9 to follow the terms of the settlement agreement it had entered with the Teamsters—which required Pac 9 to recognize its drivers' right to organize—Region 21 gave up negotiations, rescinding the agreement and issuing a formal complaint against Pac 9.[283] Although it received little fanfare in the press, the decision struck like a lightning bolt at campaign headquarters, electrifying leaders who read the Pac 9 complaint as the most clarion signal so far that the region was buying into its misclassification theory. Indeed, the Pac 9 complaint was the first ever issued by the NLRB that sought redress for ULPs committed by a trucking firm that purported to hire *only independent contractors*. It therefore constituted a stinging repudiation of the firm's effort to justify its treatment of drivers in the language of arm's length bargaining.

That very same day, the DLSE "ordered Total Transportation to give back pay to 14 truck drivers,"[284] totaling nearly $1 million.[285] The campaign held a press conference across the street from TTSI decrying the post-truce actions of the firm, as well as those of Green Fleet and Pac 9, which they accused of breaking their promise to Mayor Garcetti by forcing drivers to sign new leases in exchange for dropping DLSE claims.[286] Threatening to go back on strike, TTSI drivers delivered a petition to the company requesting that it recognize the Teamsters Local 848 as their bargaining representative.[287] TTSI did not take the request well. Within a few weeks, the company fired thirty-three drivers, all of whom had filed misclassification cases with the DLSE (and four of whom had recently received significant damages awards).[288] In the heat of this exchange, the campaign opened up a new front in the battle, reaching out to Saybrook Capital, the Los Angeles-based private equity firm that owned TTSI and promoted itself as a paragon of green growth.[289] Campaign leaders believed that appealing to the enlightened self-interest of socially conscious capital investors—who might value the reputational benefit of embracing an employee driver model and better appreciate the financial advantage of normalizing labor relations—could serve as effective leverage in their skirmishes with TTSI.

The campaign also used other modes of communicating its message. On September 29, 2014, port truck drivers took part in a day of prayer and fasting to raise awareness of wrongful misclassification by the three primary trucking firm targets.[290] The event was held in front of Pac 9 and lasted from dawn until 3 p.m.; it was organized by the Teamsters, along with LAANE and CLUE, which sent representatives from United Church of Christ, St. Joseph Catholic Church, St. Cyprian Catholic Church, and Pacific Unitarian Church.[291] The event marked nearly a year and a half of protest against misclassification at Green Fleet, Pac 9, and TTSI.[292]

Although the campaign remained laser focused on this trio of key targets, Shippers Transport Express—part of the Carrix-SSA Marine global logistics enterprise—remained on the legal hot seat. The *Taylor* class action filed in state court in 2012 was removed to federal court in 2013 after the plaintiffs had added terminal operator SSA Marine as a defendant and sought damages against it as a joint employer. In federal court, Shippers and SSA Marine moved for summary judgment on the theory that the FAAAA preempted the drivers' state labor law claims. On September 30, 2014, in a decisive victory for the plaintiffs, district court judge Beverly Reid O'Connell (also presiding over the DOL suit against Shippers) denied defendants' motion for summary judgment on the basis of FAAAA preemption and refused to dismiss plaintiffs' claims against SSA Marine as a joint employer; in addition, the court granted the plaintiffs' motion arguing that the drivers were, in fact, employees on the grounds that Shippers exercised sufficient control through its power to terminate drivers' contracts, dictate which loads to accept, and set rates, among other factors.[293] This ruling was relevant to the ongoing DOL lawsuit seeking prospective reclassification, in which *Taylor* class counsel, Joe Sayas, was permitted to formally appear on behalf of a *Taylor* plaintiff designated by the court as an interested party.

On October 10, 2014, pressure mounted against Green Fleet. Responding to the firm's January firing of drivers Amilcar Cardona and Mateo Mares, U.S. District Judge Philip Gutierrez, ruling on Region 21's 10(j) petition, issued a temporary injunction against Green Fleet ordering the drivers back to work.[294] It was another legal milestone: the first time a court in an NLRB case had effectively sided with the argument that port drivers were misclassified—an essential legal predicate for meeting the temporary injunction standard that there was a likelihood of success on the merits of the drivers' ULP charges. Reviewing the parties' clashing claims of legal status in careful detail, and using the lower standard of evidence required in ruling on the injunction, Judge Gutierrez concluded: "Because Respondent [Green Fleet] has not provided the Court with a compelling reason to disbelieve [the NLRB] the Court finds that [the NLRB] has provided some evidence that Cardona and Mares will be seen as employees under the Act."[295] Although it was not a ringing endorsement, it was another critical step forward in the legal campaign to build favorable precedent of misclassification at the NLRB. As Gutman Dickinson later recalled,[296] when Judge Gutierrez, a conservative with a law-and-order reputation, was assigned to the case, she drafted her amicus brief in support of the region's 10(j) motion to emphasize that issuing the injunction would advance the "public interest in law enforcement": arguing that allowing Green Fleet to "flagrantly violate" its legal obligations would "not go unnoticed" by other employers that "may be emboldened to flout the Government's authority with similar unlawful conduct."[297] That shift in legal framing seemed to work to the Teamsters' advantage, as Judge Gutierrez, in weighing the equities, agreed that it was in the public interest to issue the injunction to "deter continuing violations" and "preserve the Board's remedial power."[298] When Green Fleet's appeal from Gutierrez's ruling was denied, the company was forced to rehire Cardona and Mares in early November—as employees.[299] Mares proclaimed: "This ruling is a great victory not just for me ... but for all drivers at Green Fleet Systems who have been too fearful to openly support our effort to become Teamsters."[300]

With the campaign's "suing" strategy in Green Fleet proving to drivers they had genuine legal protections against termination if they challenged their misclassification, momentum started to build for more "striking." As Weiner recalled, the success of the July 2014 strike against Green Fleet, Pac 9, and TTSI felt like a "breakthrough": drivers began to have confidence that something could be achieved. The campaign had "a lot more credibility than we did three years [before] and politically, it was sort of like, well, this ain't going away."[301] The port trucking industry seemed to take notice, ramping up efforts to eliminate bottlenecks and backlogs—which more port strikes would only exacerbate.

In this regard, on November 6, 2014, the Harbor Trucking Association announced it would petition the Federal Motor Carrier Safety

Figure 6.3
Julie Gutman Dickinson with fired Green Fleet drivers Amilcar Cardona and Mateo Mares returning to work upon court order. Photo courtesy of Julie Gutman Dickinson.

Administration to temporarily waive its rule requiring a thirty-four-hour rest period for drivers in order to allow the Long Beach Container Terminal (LBCT) "to open its gates on Sundays to move backed-up cargo."[302] As reported in the *Long Beach Press-Telegram*: "The association's petition and LBCT's proposal to open on Sundays are the latest in a multipronged effort by the supply chain to move cargo clogged up within the nation's two busiest seaports."[303] This effort also included talks to address the increasingly acute shortage of truck chassis—the trailers used to load containers onto trucks. Since drivers typically owned or leased only the truck cab, they needed to hitch chassis in order to pick up cargo. And because shipping lines had mostly stopped providing chassis as part of their intermodal services, port truck drivers had to rely on a complex network of independent leasing companies for access. This had created a sense of chaos at the Los Angeles and Long Beach port complex, where there was no central location to pick up or drop off chassis, while some cargo owners would hoard chassis to ensure their availability during peak shipping season. A key component of the ports' effort to streamline trucking operations and reduce turn times hinged on discussions with chassis providers to create a pooling system to address chassis shortages.[304]

As these discussions between trucking firms and the ports got underway, truck drivers held their own meeting with the Los Angeles harbor commissioners, asking them to take action against the drayage industry's wrongful misclassification of drivers as independent contractors—and warning if they did not that drivers were on the verge of returning to the picket line after four months of ongoing retaliation by trucking companies.[305] Gene Seroka, Los Angeles port executive director, told drivers that "we will always find ways to help," and said port and city officials were working on "plausible solutions."[306] Drivers and campaign leaders were not reassured. Fed up with the lack of port action, drivers went back on strike the following week, although this time the number of targets shrank to two. With Green Fleet, at the mayor's urging, in "productive conversations" with the Teamsters, the campaign limited its picketing to TTSI and Pac 9's company yards in Compton and Carson,[307] while placing additional picket stations at several terminals at both ports.[308] By the end of the first day of picketing on Thursday, November 13, as some terminals turned away TTSI and Pac 9 trucks, TTSI announced it would also meet to discuss a solution with the Teamsters.[309] That weekend, expertly timed to push the discussion forward, *Los Angeles Times* columnist Steve Lopez ran a piece spotlighting Martin Davis, the owner of Pace Freight Systems, a successful port drayage company operating on an all-employee model, with driver pay starting at $16 dollars per hour. The article presented a vision of shared sacrifice that spoke directly to the campaign's struggle. As Davis reflected: "I guess I could've made a lot more money. ... But my wife and I have led a pretty good life. We're not wanting for anything. We're not living in Bel-Air, but our life is good."[310] The next Monday, November 17, Pac 9 also agreed to talks with the Teamsters, marking a pivotal moment in the misclassification campaign with the initial three target firms all in serious discussions with the union.

That same day, instead of riding the wave of progress, the campaign opted to expand the scope of the strike by adding five more trucking firms—all of which had features making them prime targets for the well-honed suing-and-striking machine. Pickets first went up against QTS, LACA Express, and WinWin Logistics, Inc.; the campaign also promised to add Pacer Cartage and Harbor Rail Transport the following day.[311] Every firm met the campaign's targeting criteria. All had 100-percent independent-contractor drivers and were also all subject to ongoing legal action challenging misclassification.

QTS, LACA Express, and WinWin Logistics were defendants in the pending class action suit initiated by Wage Justice Center, while Pacer Cartage faced a federal court class action and Harbor Rail Transport was fighting a state court mass action (on behalf of a large group of named plaintiffs not certified under state class action rules) brought by the Gomez Law Group, a plaintiff-side law firm based in San Diego.³¹² In addition, all companies faced significant liability from individual driver DLSE claims. Pacer Cartage, in particular, had appealed orders for more than $2 million on behalf of seven workers, who were also being represented by lawyers at the Gomez Law Group. Furthermore, the trucking firms were connected to larger corporate structures that made them appealing targets. QTS, LACA Express, and WinWin Logistics were part of the network of companies owned by the family of Erick Yoo, while Pacer Cartage and Harbor Rail Transport were subsidiaries of XPO Logistics, a public company that was a "big player" in the goods movement sector, with subsidiaries around the country in rail drayage and the lucrative "less-than-truckload" market (moving goods from multiple customers on one truck), which the Teamsters were attempting to organize.³¹³

In the wake of this expanded mobilization, the campaign amplified its message and pressed its advantages. A few days into the enlarged strike, Alex Paz, a former TTSI driver, declared: "We're showing the companies we're winning the fight."³¹⁴ Campaign supporters at a Wilmington press conference held up a sign that read: "You are in The State of Misclassification. ... Where highway robbery happens every day."³¹⁵ Seeking to further extend its ambulatory picketing, the campaign moved from the port terminals to the rail yards served by Pacer Cartage and Harbor Rail Transport.³¹⁶

Also extending its political reach, the campaign vowed to take its case to the Long Beach City Council,³¹⁷ one of whose members published a sympathetic op-ed in the *Long Beach Press-Telegram* denouncing trucking company "wage theft."³¹⁸ After drivers appeared at a Long Beach Board of Harbor Commissioners meeting on November 19,³¹⁹ conversations with struck company representatives accelerated, with QTS, LACA Express, and WinWin Logistics agreeing to discussions with the Teamsters the next day.³²⁰ On November 21, after eight days on the picket lines, drivers decided to come back to work after Pacer Cartage and Harbor Rail Transport agreed not to retaliate and to continue meetings with the Teamsters.³²¹ Pacer Cartage driver Humberto Canales buoyantly asserted, "The days of driver misclassification are numbered. We are employees and the law is on our side."³²² A perceptible shift had occurred. Observers noted that "the companies' willingness to now meet with drivers differs from their sentiments a year ago, when companies dug in their heels and refused to meet with drivers over the labor issues."[323] The involvement of Los Angeles Mayor Garcetti and Port of Los Angeles Executive Director Seroka were credited as important reasons for the companies' change of heart.[324]

The campaign's key leverage with the city officials—the power to disrupt the ports and delay operations—was augmented by the tumult surrounding port growth. Congestion at the ports remained a fundamental problem, driven by its own internal logic, which campaign strikes only served to exacerbate. The key causes of port congestion continued to be "the lack of available chassis, the arrival of larger ships and unresolved contract talks between longshore workers and their employers."[325] The chassis problem, in particular, was becoming more severe, with the absence of a central repository forcing drivers to wait prolonged periods to locate and attach a chassis under contract with the specific shipper from which the drivers were to receive cargo—which was the only way the cargo would be released.[326] As turn times suffered accordingly, the Long Beach Board of Harbor Commissioners kick started a plan to help finance a shared pool of 3,000 chassis on port property; the Long Beach port also proposed closer collaboration with its Los Angeles counterpart "without fear of violating antitrust" by asking the Federal Maritime Commission for permission to coordinate responses to congestion, focusing on truck turn times, gate hours, and financial incentives for companies to enter the ports during off-peak hours.[327] After the chassis pool plan was approved by unanimous vote of the Long Beach harbor commission,[328] the Long Beach port temporarily opened a thirty-acre facility at Pier S where drivers could drop off empty containers, avoiding the backlog at other container yards and thereby freeing up their chassis to accept more cargo.[329]

The flurry of political activity at both ports augured a new direction in the campaign. As campaign leaders sought to expand the ground game to pressure major port trucking firms, it also began to channel energy back toward the local political arena as a site of negotiation and potential policy resolution. That the campaign could even consider a new local political strategy to address misclassification was itself a monumental achievement. Five years out from the devastation of the Clean Truck Program defeat, the campaign had steadily rebuilt its leverage and regained its political clout. In hindsight, there appeared to be a clear progression: the campaign started small, with narrowly defined trucking firm targets made vulnerable by a battery of legal claims; through these claims, the campaign built precedent for misclassification strikes, which exerted further pressure on the targets and disrupted port operations; using this leverage, campaign leaders put public officials on notice that they were back in the game, enlisting sympathetic officials to help win preliminary agreements against first-wave targets; with this success in hand, the campaign's political credibility grew even stronger, permitting its leaders to once again imagine the possibility of systemic policy reform and unionization.

Transformation: Unionizing Misclassified Drivers

By the beginning of 2015, the Justice for Port Drivers campaign had much to crow about. As Teamsters Port Division Director Fred Potter reflected looking back at drivers' success at the DLSE: "when you win every complaint that you file, ... it says there is something wrong with the industry."[330] The numbers seemed to bear him out. A statement issued by the campaign indicated that there had been 705 DLSE claims filed since 2011, with 155 orders in favor of drivers.[331] Of the more than one hundred orders that the campaign had reviewed, *drivers had been deemed misclassified in all of them*, winning an average award of $110,000.[332] Extrapolating from these figures, the campaign estimated that the industry continued to face more than $60 million in pending DLSE liability.[333]

The campaign statement summarized additional legal success and spotlighted new trends suggesting its impact. On top of positive developments at the NLRB in the Green Fleet and Pac 9 cases, the statement pointed out other agencies clamping down on port trucker misclassification. For its part, the California Employment Development Department, responsible for collecting payroll taxes, had found that drivers at Green Fleet, QTS, LACA Express, and Pac 9 had been misclassified, and had ordered Pac 9 to pay three years of back taxes plus penalties to make up for what it should have paid for its employees.[334] The campaign statement also counted twenty-one pending class action lawsuits covering approximately three thousand drivers.[335]

All of this activity seemed to be taking its toll. In a sign that legal action could produce more than just monetary damages, Hub Group Trucking, after settling a class action suit against it in 2014, reclassified all its drivers as employees. In a more startling—and uncertain—trend, the port trucking firm QTS filed for bankruptcy protection in August 2014 and other companies appeared poised to follow suit. Although bankruptcy provided potential leverage, since judgment creditors (like drivers with outstanding wage claims) could push for substantive corporate changes (like reclassification), the question that remained was whether the campaign could translate the industry's legal liability and financial vulnerability into union contracts—which had eluded it since the Toll victory in 2012.

On the Brink: Bankruptcy and Employee Reclassification

The answer to the unionization question came quickly as Teamsters Local 848 rang in the New Year on an auspicious note: celebrating its first organizing success with an independent-contractor port trucking firm. Shippers Transport Express had been battered by campaign pickets and multiple lawsuits. On the legal front, in addition to the September 2014 ruling in favor of the *Taylor* plaintiffs, there were other cases pending. In advance of the *Taylor* federal court ruling, the *Taylor* plaintiffs' attorneys had filed another state court class action suit against Shippers (dropping SSA Marine), which also included a PAGA claim[336] (again seeking to keep the suit in state court and avoid dismissal if the class was not certified). In November 2014, civil rights law firm Hadsell & Stormer filed another lawsuit against Shippers on behalf of a driver killed in a fatal accident, which included claims of willful misclassification and wrongful death.[337] On November 17, 2014, Shippers entered into a consent judgment with the federal government in the DOL lawsuit, in which they were ordered to "properly reclassify ... all Drivers at Defendant's ... California facilities (including its facility located in Carson ...)."[338] This was a transformative victory, which reflected the organizational power of the DOL to litigate for injunctive relief (without having to worry about recovering fees), as well as the campaign's productive relationship with *Taylor* class counsel, Sayas, who supported the drivers' desire for reclassification in the DOL proceedings (without having reclassification become central to the *Taylor* damages lawsuit). Bowing to the pressure, Shippers notified drivers a week later that it intended to reclassify them as employees.[339]

In early January 2015, Shippers made good on that promise, while entering a labor peace agreement with the Teamsters under which the company agreed not to interfere with driver organizing and to recognize the union as the drivers' bargaining agent when a majority of drivers signed authorization cards—paving the way for them to vote for Teamsters representation.[340] As the drivers mulled over this vote, the *Long Beach Press-Telegram* editorial board struck a notably neutral tone: "While this editorial board may not always agree with the tactics of organized labor, it does advocate for workplace rights. Some workers may be better off with unions while other independent drivers may prefer to be their own bosses."[341] On January 9, the Shippers drivers resoundingly expressed their belief that they were better off with unions. At a press conference at Banning's Landing Community Center, it was announced that roughly 80 percent of Shippers' 111 drivers had voted to be represented by Local 848.[342] The campaign's reaction was euphoric. The Teamsters' Potter heaped praise upon the company: "It's about Shippers, who have done something that other trucking companies are unwilling to do in Los Angeles, and that's to face the reality that misclassification is the wrong direction for employees and it's the wrong direction for the future of this port."[343] Gutman Dickinson, wearing her lawyer hat, predicted that "other companies will follow to avoid huge legal liability from their unlawful 'independent contractor model' and to avoid strikes and debilitating labor disruption."[344] Driver Leonardo Mejia spoke about the impact of voting for the union: "No matter where I have worked in this industry, the problems are the same everywhere in every company—no respect, no dignity, bad pay. ... These problems are getting worse every day for the drivers. For me, now those days are over."[345] For his part, Shippers' General Manager Kevin Baddeley put on a brave face: "This is the future of the industry, I think. ... I think we're just the first. I'm OK with it. It's what the drivers want."[346] Driver Mejia confirmed this desire: "To be part of the union, it means respect and the most important thing to me is to have a voice in the workplace."[347]

Listening to that voice, Shippers and the Teamsters inked a one-year collective bargaining agreement in which drivers would earn from $18 to $21 an hour with medical insurance, paid leave, and overtime.[348] The Shippers victory was important but distinctive: Since Shippers was part of the network of businesses (including terminal operator SSA Marine) owned by Carrix, it was not competing head-to-head with other small port drayage companies and could draw upon corporate resources to buy labor peace. In addition, the company confronted both a committed DOL seeking injunctive relief to prevent prospective wage-and-hour violations and a tenacious private law firm pursuing a class action for past damages—which worked in tandem to put enormous legal pressure on the company to settle on terms that involved both reclassification and monetary compensation. Nonetheless, the Shippers agreement constituted the second union contract won by the Teamsters at the ports—and the first with a formally all-independent-contractor firm. In the wake of the agreement, Shippers settled the *Taylor* class action for $11 million in June 2015; the related state law class action was dismissed the following month.[349]

Seeking to seize the momentum on the heels of the Shippers vote, the campaign ratcheted up its political strategy. While officials in both Los Angeles and Long Beach had gotten deeply involved in brokering discussions between industry and campaign representatives, helping to resolve strikes and advancing systemic responses to port congestion, they remained reluctant to put their full weight behind another large-scale policy reform project to address misclassification. Still feeling the lingering burns from the Clean Truck Program's flame-out, officials in both cities seemed to be waiting for more evidence that a new policy proposal to promote employee reclassification could survive inevitable political and legal challenge. In this regard, the campaign's labor movement leaders seemed to deliberately chart their own independent course, consciously keeping their distance from the environmental partners at the center of the Campaign for Clean Trucks—and instead developing the labor-specific facts on the ground to build a strong legal foundation for any potential city action to avoid preemption.

In the meantime, campaign leaders pressed port officials to exercise the legal authority they already had: urging them to bar trucking firms engaged in misclassification from entering port property under existing concession agreements. On January 13, 2015, the Teamsters' Potter wrote a letter to the executive director of the Port of Long Beach, Jon Slangerup, urging him to take action against three firms facing "over 100 final rulings in misclassification-related wage claims": Pac 9, Win-Win Transportation, and Fargo Trucking. In strong terms drawing on the language of the very concessions adopted as part of the Clean Truck Program, Potter urged the port to enforce its contractual agreements with the companies to bar them from entry: arguing that they were "currently in violation of the spirit if not the letter of the *Port of Long Beach Motor Carrier Registration and Agreement*."[350] Specifically, Potter highlighted language that stated the port had "the right to deny access to the Port to any drayage truck that has not been registered, lacks proper identification devices, or does not meet applicable environmental, security, or safety regulations."[351] He added: "While this section does not explicitly list employment regulations, that could be fixed through simple clarification in an amended agreement and should not prevent you from pursuing enforcement action."[352]

The Shippers vote also kept the spotlight trained on driver hardship in the rest of the industry, where onerous leases were pushing many further into economic precarity. On January 25, 2015, the *Los Angeles Times* ran a moving profile of Julio Cervantes, who had been a driver for thirteen years, but after the Clean Truck Program, began to struggle to meet the new lease payment on his 2009 Volvo truck, along with "fuel, maintenance, repairs and other costs."[353] As a result, Cervantes burned through $30,000 in savings and ran up $24,500 in debt. In 2013, after moving to a small apartment and with his wife taking on babysitting jobs, Cervantes joined with other drivers to protest their independent-contractor status.[354] NELP's Becky Smith, coauthor of *The Big Rig Overhaul*, laid part of the blame for Cervantes's plight on how companies reacted to the Clean Truck Program, arguing that lease contracts were "a scam to get workers to finance their jobs."³⁵⁵ To Smith, the mounting legal victories showed that "the con game is over—or should be."³⁵⁶

Change to Win's Nick Weiner, though conceding that leases had been bad for port truck drivers, presented a slightly different perspective on the legacy of the Clean Truck Program. Looking back, Weiner thought that clean truck leases helped make issues "more raw" for drivers, constituting a shock to the system: "Drivers were forced to pay for the new trucks [and that] pissed them off," making them "ripe to grieve."³⁵⁷ Yet he was quick to note that the drivers' misclassification did not depend on the program's impact or drivers' lack of truck ownership, but rather on the exercise of company control. Although it was "eye catching" and "egregious" to see "negative paychecks" because of lease payments, from an organizing perspective, there was no distinction between drivers who owned or leased their trucks.³⁵⁸ And Weiner emphasized that misclassification could not be avoided by companies simply saying: "We just have to give trucks to drivers and then they are not misclassified."³⁵⁹

Courts continued to agree with the campaign on this essential point. At the tail end of an already jammed-packed first month of 2015, in another significant legal victory nearly lost in the commotion of campaign activity, a Los Angeles Superior Court judge ruled that the seven Pacer Cartage drivers whose DLSE orders had been appealed the previous year were entitled to over $2 million in damages.³⁶⁰ The judge's opinion emphasized the lack of control Pacer drivers were given through truck lease agreements, which were in English and specified where the drivers could park, prohibited drivers from moving freight for other companies or setting rates, and required drivers to file job applications with Pacer and wear Pacer identification cards.³⁶¹ Parent company XPO Logistics said it would appeal.³⁶² That day, after the decision was issued, lawyers from the Kabateck firm in the federal class action against Pacer filed an amended complaint that included detailed allegations of Pacer's exercise of control over the drivers.³⁶³ Then, in February, the U.S. Supreme Court brought positive closure (from the campaign's perspective) to another contentious suit, denying Pac Anchor Transportation's petition for certiorari and thus letting stand a lower court ruling that the enforcement of California's unfair competition law against trucking companies was not preempted by the FAAAA.³⁶⁴

Around this time, the possibility of another massive work stoppage at the ports reemerged when the ILWU and PMA reached an impasse in their on-again, off-again negotiations over a collective bargaining agreement.³⁶⁵ Bracing for the worst, the Teamsters announced a new plan to support nonunion truckers who stood to lose business if the dockworkers' dispute resulted in a port shutdown.³⁶⁶ On Wednesday, February 18, 2015, the Teamsters touted the establishment of a "rapid-response team" that would be housed at Local 848 and would, with the California Employment Development Department, help "drivers obtain unemployment benefits and other aid such as financial planning advice, food, medical referrals and utility assistance."³⁶⁷ Fortunately, deployment of the team was rendered moot on Friday, when the ILWU and PMA reached a tentative deal, brokered by DOL Secretary Thomas Perez, breaking the stalemate, which in its final phase was prolonged by "the union's demand that a single, low-level dock official in charge of overseeing workplace disputes be fired."³⁶⁸ Despite the breakthrough, and the final ILWU contract that was later approved, the months-long bickering at the ports had produced a cargo backlog that experts predicted would take three months to clear.³⁶⁹

This, of course, was not good news for port officials already stretching to deal with the congestion problem, which was threatening to harm the Los Angeles-Long Beach port complex's reputation as the nation's global trade conduit.³⁷⁰ Officials scrambled to find solutions to now-familiar dilemmas. In March, responding to the turn-time quandary, the Port of Los Angeles put its chassis pool into effect, while Long Beach continued efforts to finalize a similar program.³⁷¹ Compounding the chassis problem was the arrival of a new breed of megaship carrying over 10,000 containers, pushing the limit of port capacity—with shipping giant Maersk poised to order ten ships that could carry an astonishing 20,000 containers.³⁷² In a troubling development, the *Long Beach Press-Telegram* reported that the "increased influx of cargo arriving all at once had led to about two dozen container ships stranded at sea off Long Beach daily, waiting to unload at the ports; long truck lines; clogged terminals; and, weekslong delays in shipments, prompting customers to divert their goods to other ports or ship them by air."³⁷³

In the platitudes of port officialdom, leaders insisted that the solution to congestion was further investment to expand capacity and promote efficiency. With permission to cooperate finally granted by the Federal Maritime Commission, the directors of both ports formed a working group to develop solutions to the congestion problem, which was alleviated to some extent in the short-term by relative labor quiescence.³⁷⁴ By mid-April, with longshoremen back under contract, the number of ships stranded off the port complex fell to six.³⁷⁵ Despite this slight improvement, the Federal Maritime Commission remained concerned about the structural causes of congestion and increasingly sensitive to shipper complaints that they were being unfairly fined for unloading delays that were outside of their control.³⁷⁶ In response, the port directors at Los Angeles and Long Beach

Figure 6.4
Trucks backed up outside the Port of Los Angeles due to congestion. Photo courtesy of Barbara Maynard, Teamsters communication director.

initiated a series of public forums to gain stakeholder feedback on "ways to make operations better along the supply chain."[377] This was followed by the inaugural Southern California Logistics and Supply Chain Summit on April 23 in Pomona, where port officials argued that a $4.5 billion infrastructure project would ease the problem and a Long Beach official pleaded: "We want to win your business back."[378]

The campaign, for its part, was not about to accept a return to business-as-usual at the ports. To the contrary, campaign leaders sought to use official and industry concerns over port congestion to their advantage and by mid-April 2015 planning for a new strike was well underway. As campaign attention zeroed in on the details of driver turnout and the placement of picketing, leaders on the ground were jolted out of their tunnel vision with more good legal news, this time from NLRB Region 21, which was working through its queue of port trucking cases raising the misclassification issue.

The Teamsters' case against hybrid firm Green Fleet was from its inception viewed as the campaign's opening wedge at the NLRB: the logical next step after Toll that would introduce Region 21 to a trucking firm with a subset of independent-contractor drivers and ask for a ruling on traditional ULP grounds requiring the region to accept the premise that such drivers were misclassified employees entitled to NLRA protection. After Judge Gutierrez preliminarily enjoined Green Fleet in November 2014, thus concluding that the Teamsters were likely to succeed on the substance of their claims and ordering the two terminated drivers (Amilcar Cardona and Mateo Mares) back to work, expectations within the campaign of another win were high—particularly following the legal team's performance during the four-week NLRB trial on the merits.

The NLRB administrative law judge presiding over the trial, Jeffrey Wedekind, did not disappoint. On April 9, Judge Wedekind issued a sixty-five-page decision, giving the Teamsters a sweeping victory in which he concluded that Green Fleet had committed roughly fifty separate ULPs and ordered a comprehensive set of remedies: including that Green Fleet "cease and desist" from engaging in a detailed list of antiunion activity, make Cardona and Mares "whole for any loss of earnings and other benefits suffered," and hold a meeting with all drivers, using a Spanish interpreter, to inform them of their labor rights.[379] The decision was significant on its own terms, but also because the judge ruled that *all* Green Fleet independent-contractor drivers—not just the two who had been terminated—were misclassified. Reviewing the legal elements for employee status, Judge Wedekind offered a sharp rebuke to Green Fleet for its treatment of so-called "lease drivers" (i.e., independent contractors):

> In sum, there are some factors of varying significance supporting independent contractor status: the Company did not control the lease drivers' work schedule or how they performed the work; the lease drivers had special skills; and the Company paid the lease drivers somewhat differently than company drivers and did not provide them with fringe benefits or withhold taxes from their checks. However, there are many more factors supporting employee status: the Company controlled what shift and assignments the lease drivers worked and required them to adhere to company policies; the lease drivers were not engaged in a distinct occupation or business; the type of work the lease drivers performed was usually done in the locality under the direction of an employer; the lease drivers' skills were identical to the company drivers' skills; the Company provided the lease drivers with the instrumentalities, tools, and place of work; the lease drivers were employed by the Company for an indefinite period; there were substantial similarities between how the Company paid the lease drivers and the company drivers; the lease drivers' work was a central part of the Company's business; many of the lease drivers did not believe they were independent contractors; the lease drivers and the Company were in the same business; and the lease drivers did not render services to the Company as independent businesses.

> Accordingly, I find that the Company has failed to carry its burden, and that the lease drivers were "employees" protected by the Act during the relevant period.[380]

Green Fleet, having just entered bankruptcy in February 2015, decided not to file an appeal, which meant that the NLRB judge's decision on employee status would serve as precedent for the region.

Green Fleet's bankruptcy—the first by one of the initial trio of targeted trucking firms—set a different kind of precedent that other firms began to follow. Seacon Logix filed for bankruptcy, in February 2015, as did Tradelink Transport the next month. With other trucking firms buckling under the weight of campaign pressure, the moment appeared to arrive for final-stage negotiations with targeted companies over the terms of employee reclassification and a pathway to unionization. As Teamsters Port Division Director Fred Potter recalled, it was during this period that trucking companies, teetering on the brink of insolvency, started "coming to us and saying, 'Listen, here's what we want to do. How do we resolve this?'"[381] In response, Potter and other campaign leaders worked with employers that wanted to develop a "plan and a timeline [to] convert ... misclassified workers into employees" to limit liability, setting the stage for a union contract.[382] In a sign that the campaign's political outreach was starting to have an effect, port officials took the lead in brokering meetings with faltering firms, telling the campaign "we can't have all these strikes."[383] Having succeeded in convincing port officials they had a stake in resolving the misclassification problem, the campaign pressed full steam ahead, insisting that the ports take action "to stop the disruption."[384] In doing so, campaign leaders began to articulate a market-oriented argument that picket lines "hurt commerce," and misclassifying firms engaged in unfair competition—which the ports could police by using their contractual authority to make companies comply with the law.[385]

As if to hammer home the point, on Friday, April 24, 2015, the campaign announced drivers would vote on a new, and bigger, strike.[386] True to its word, three days later, on Monday, April 27, 200 truck drivers walked off the job and set up picket lines at company truck yards. Three of the targets were the now-usual suspects: Pac 9, and the XPO subsidiaries, Pacer Cartage and Harbor Rail Transport. Added to this group was a new member, Intermodal Bridge Transport (IBT), another large all-independent-contractor firm subject to multiple lawsuits,[387] which was also part of a larger logistics conglomerate: China Ocean Shipping Company, or COSCO, a global logistic firm owned by the Chinese government, which itself jointly owned (with SSA Marine) the terminal operating company that controlled the massive Pacific Container Terminal at Pier J in Long Beach (where COSCO vessels docked).[388] In a bit of campaign karma, the Port of Long Beach's effort to expand Pier J to accommodate COSCO's growing fleet had been thwarted by the Natural Resources Defense Council (NRDC) back in 2004; despite the setback, the Long Beach port ultimately succeeded in moving COSCO to the Pacific Container Terminal, where COSCO's flagship port trucking firm, IBT, now faced a strong challenge from the labor movement. Perhaps recalling this history, the Port of Long Beach director, Jon Slangerup, issued a statement that seemed to express a state of denial: "Let's be clear, this is not a strike. We do not expect there to be any adverse impact to port terminals."[389]

However one chose to describe the campaign's April 27th action, no one disputed the notable absence of prior targets TTSI, still in negotiations with the Teamsters, and the family of firms that included QTS, now in bankruptcy, where it sought to protect itself against ongoing DLSE and class action litigation. The conspicuous absence of the campaign's original target, Green Fleet, was quickly explained as it was announced that Green Fleet had entered a "labor peace" agreement with the Teamsters.[390] The Green Fleet agreement was similar to the one that Shippers had struck after its class action settlement, in which the firm committed to reclassifying drivers as employees, while agreeing not to resist union organizing and respect its outcome. As Teamsters lead organizer Carlos Santamaria recalled, on the previous Friday when the campaign announced its intent to strike, Gary Mooney, owner of the bankrupt Green Fleet, called the union local and "said he was giving up."[391] Weiner and Potter told Mooney: "if you really want this, you got to come to the Local [848 office]."[392] Mooney came over the weekend, and in an office strewn with well-worn pickets against Green Fleet, he signed an agreement with the Teamsters.[393] Propelled forward by the prospects of a similar achievement, drivers for IBT, Pacer Cartage, and Harbor Rail Transport carried on picketing for four days before returning to work. Twenty-five drivers from Pac 9—which remained intransigent throughout talks initiated after the November 2014 strike—persisted for an additional day just for good measure, agreeing to go back to work on Friday, May 1.[394]

The following week, the next domino fell. Speaking at a press conference at Dodger Stadium, Mayor Garcetti stood alongside truckers and Teamsters representatives to announce the creation of Eco Flow Transportation, a new "sister" company to TTSI, which was transferring its drivers to Eco Flow while working out legal issues with the Teamsters related to their ongoing labor dispute. Eco Flow, backed by private equity firm Saybrook Capital, entered into a "port solutions agreement" with the Teamsters promising to use only employee drivers (with a pledge to grow from 80 to 500) and agreeing to stay neutral through an expedited NLRB election.[395] In the words of its co-managing partner, Jonathan Rosenthal, Saybrook's model was to invest in "legacy businesses that suffer from legacy thinking."[396] In describing Saybrook's deal with the Teamsters, Rosenthal said that he "went into the lion's den" with labor, management, and ports officials out of a belief that drayage assets were being inefficiently used.[397] He proclaimed that Eco Flow represented "a real transformation, not only for our company but for the industry."[398] In Rosenthal's view, the creation of Eco Flow was a critical test of the idea that "a drayage company with employee drivers can compete in the cutthroat harbor drayage environment in Los Angeles-Long Beach. ..."[399] In an extended interview, Rosenthal emphasized the importance of "legal and liability factors" in the decision to establish Eco Flow:

> Several years ago the best lawyers on the planet were convinced that, if properly structured in accordance with the current law, trucking companies could contract with independent owner/operators and not worry that they might later be determined to be an "employer"—subject to all the costs of being an employer. It has been nearly impossible to predict how the courts will determine who is or is not an employee. It is exceedingly difficult for a company to make investments in new equipment, contract with drivers, upgrade technology, build modern facilities and work to meet growing customer demands without having a reasonable level of certainty that an unexpected liability won't bite you in the butt years down the road. This uncertainty has led us to conclude that for us, the best and most predictable path is to build an employee drayage company.[400]

Rosenthal's view constituted a strong endorsement of the campaign's legal strategy, which culminated two months later when Eco Flow drivers voted in favor of joining the Teamsters (although a collective bargaining agreement would not be executed for another year).

At the Eco Flow press conference in May, Mayor Garcetti called misclassification "not the gripe of a few drivers but a battle cry of a systemic problem that must be addressed."[401] With the deal garnering significant press, the movement against driver misclassification began to gain national attention. On May 28, 2015, the *Wall Street Journal* posted a logistics report online describing how "[s]mall groups of drivers have targeted select employers, demanding full-time employee status and the higher wages and benefits that come with it."[402] Despite favorable coverage, the post highlighted the problem of employee companies achieving scale, quoting ATA director Curtis Whalen diminishing the strikes as "onesies and twosies, it isn't a national trend."[403] The Teamsters, however, had grander designs. As a *Journal of Commerce* article around the same time reported: "According to

Nick Weiner, Teamsters port campaign director, the union's immediate goal is to facilitate a successful, profitable business model in Los Angeles-Long Beach, the largest U.S. port complex, and then to promote the employee-driver model at other major gateways such as New York-New Jersey."[404]

Advancing this ambitious goal as summer approached, the campaign doubled down on its strategy of suing and striking targeted trucking firms, while also wading further into the political arena to press for official action. In early May 2015, two drivers from Los Angeles and Long Beach filed a class action against Carson-based Sterling Express Services, Inc., charging the firm with illegal misclassification of port truck drivers.[405] At the same time, lawyers in the *Talavera* lawsuit against QTS, LACA Express, and Win-Win Logistics initiated a bold legal strategy to keep the pressure on QTS in bankruptcy proceedings, where it was trying to deny payment to QTS drivers in the class action. On May 5, Wage Justice lawyer Matt Sirolli filed a motion to consolidate the class action with the bankruptcy proceeding—effectively asking the bankruptcy court to permit the QTS class action plaintiffs to recover against the assets of Yoo's affiliated companies, LACA and WinWin, in bankruptcy (even though those firms were not formally part of the bankruptcy proceeding).[406] Fighting over that motion, and its implications for class action recovery, consumed the bankruptcy process from that point on—making it clear that QTS was not guaranteed to avoid or diminish driver pay out through bankruptcy and strengthening the class plaintiffs' bargaining power in ongoing settlement negotiations.

On the political front, the campaign's effort to link together the problems of driver misclassification and port congestion in the hope of prompting government action appeared to be bearing fruit. A May news report by the *Long Beach Press-Telegram* recounted a conference at Cal State Long Beach university on the crisis of port congestion, at which Federal Maritime Commission chairman Mario Cordero called port truck drivers "the stepchild of the industry," who did not receive sufficient attention from port officials because of their marginal economic status.[407] On this point, even industry leaders could agree. On May 18, 2015, Weston Lebar, executive director of the Harbor Trucking Association, predicted that small port trucking companies would go out of business if congestion was not alleviated quickly.[408] On May 19, the Los Angeles County Board of Supervisors unanimously approved a motion by Supervisor Mark Ridley-Thomas to consider actions the board could take to aid misclassified port truck drivers. Flexing its muscles, the Board of Supervisors indicated that the county might have authority to suspend licenses and permits for trucking companies that operated in unincorporated areas of L.A. County.[409]

None of this quelled the nerves of shippers concerned about congestion at the Southern California ports. In an alarming trend, the ports continued to lose container traffic to competitors as shippers were increasingly routing their goods through other ports to avoid the congestion and labor unrest at Los Angeles and Long Beach—with some paying a "premium" to ship goods to East Coast ports and transport them farther distances by truck or rail to destinations around the country.[410] Experts stated that the Los Angeles and Long Beach ports' share of U.S. imports was shrinking—down 9 percent between 2002 and 2013—representing a loss of more than twelve thousand jobs and $100 million in state and local taxes.[411] Container ships still coming to Los Angeles and Long Beach continued to grow in size, contributing to congestion—now also exacerbated by increasing shipping industry consolidation. Bound together by agreements to "share vessels and pool cargo to save money," four corporate alliances controlled 80 percent of global fleets, resulting in "immense headaches at port terminals, where longshoremen must sort through containers from multiple carriers before they can be fetched for truckers waiting to pick them up."[412] This chaotic hand off was forcing truckers into longer wait times, which were exacerbated by the continuing shortage of chassis.

In response, the ports were proposing to do what they had always done: grow. On the horizon were plans to replace the Gerald Desmond Bridge, now too low for megaships, and expand two Long Beach terminals as well as the TraPac terminal at the Los Angeles port. These efforts responded to the fact that East Coast ports were also upgrading in anticipation of the impending opening of the wider Panama Canal, which would let larger ships from Asia go directly to the East Coast.[413] Although the Los Angeles and Long Beach ports were losing overall market share to competitors, the absolute number of containers passing through them was nonetheless growing: both ports handled fifteen million cargo containers in 2014, an increase of 4 percent over the previous year.[414] As a result, truck drivers were in demand. Yet that demand exposed ongoing tensions with longshoremen, who in their most recent collective bargaining agreement had negotiated the right to inspect leased chassis, which tended to slow turn times for most port truck drivers.[415] Even the moderate editorial board at the *Long Beach Press-Telegram* expressed concern over how inspection impacted drivers, who lacked a meaningful voice in port governance: "once again it appears the parties were so wrapped up in their own push-and-pull fight, that they didn't give much concern to the very people that they serve and work with."[416]

Reshaping the Industry

As attention increasingly focused on the need for a more stable trucking industry to address the competitive pressures of the ports, and more companies buckled under crushing legal liability, lawmakers finally started to advance concrete policy solutions. One such solution was to provide trucking firms "amnesty": forgiving legal liability in exchange for firms reclassifying drivers as employees. This idea was codified in Assembly Bill 621, proposed in spring 2015 by Assemblyman Roger Hernández, a Democrat from West Covina, who described the proposal as follows:

> Under AB 621, port drayage companies will be provided an opportunity to participate in a limited amnesty program by entering into a consent decree with the labor commissioner. Under the terms of the consent decree, the motor carrier must agree to pay all wages and benefits owed to previously misclassified independent contractors and all taxes owed to the state as a result of such misclassification. In addition, the company must agree to classify any present or future commercial drivers as employees. In exchange, a motor carrier that enters into such a consent decree will be relieved of liability for statutory or civil penalties based on previous misclassification of drivers.[417]

By linking together what the campaign had achieved—holding misclassifying firms financially liable for their pervasive illegal conduct—and what it aspired to—converting the port trucking industry to an employee-driver model—Hernández's bill, which would pass the legislature and gain the governor's approval in October 2015, represented the first indication that a legislative solution to the independent-contractor problem was once again in play.

On the Road: Second-Wave Targets and Labor Law Reform

It was summer 2015 and the campaign was on a roll. Within an intense six-month span, the campaign tallied three targeted trucking companies converting drivers to employees: Shippers Transport Express, which also signed a collective bargaining agreement with the Teamsters, along with Green Fleet Systems and Eco Flow (formerly TTSI), both of which had entered labor peace agreements (with Eco Flow drivers voting for Teamsters representation).[418] The *Wall Street Journal* noted that although employee companies (which also included Toll and Sea-Logix) constituted only 5 percent of the 10,000-driver industry, that figure was double what it had been the previous year.[419] While critics continued to minimize employee companies as "the product of one-off events," the *Journal* quoted Saybrook's Rosenthal, who claimed that the employee-model drayage company was a "test bed for transformation of the industry."[420] The test for the campaign now was to build out this initial beachhead. Of its original trio of targets, only Pac 9 was still holding out. In addition, a number of other firms, all part of larger corporate structures, remained in the campaign's crosshairs: the family-owned network of QTS, LACA Express, and WinWin Logistics; the COSCO-owned IBT; and the XPO firms, Pacer Cartage and Harbor Rail Transport, which had now changed their names, respectively, to XPO Cartage and XPO Port Services (itself sometimes referred to just as XPO Logistics).

Pac 9 continued to be a sought-after prize—even more coveted because of its steadfast refusal to fall into line and the opportunity it afforded to finally present to NLRB Region 21 the specific issue of whether misclassification itself counted as a ULP. On that point of law, the time had come for the campaign to call the question. As Teamsters lawyer Mike Manley recalled, "this is where all the very careful approach to the region paid off. … They viewed us as a serious organizing campaign … and they were clearly of the view that … these folks are misclassified."[421]

But rather than push Region 21 to immediately issue a new complaint against Pac 9 including the misclassification-as-ULP charge, Manley and Gutman Dickinson made the strategic decision to pursue a more indirect route. As Manley recollected, "we didn't expect the region to go with that theory without kicking it upstairs to … the Division of Advice"—the NLRB lawyers who review "cases that present unique or particularly difficult situations" and give legal advice to regional directors.[422] The legal team wanted Region 21 to issue a new complaint, but recognized it was "an untested, untried theory and regions don't like that."[423] Accordingly, Manley and Gutman Dickinson sought to nudge the region to refer the case to the Division of Advice—whose favorable opinion on the misclassification charge would provide strong legal authority for the region to proceed. In this sense, the campaign lawyers were taking a calculated risk: if the regional director refused to issue a complaint including the misclassification charge, the Teamsters could appeal the refusal to the Office of Appeals, but that office was generally reluctant to countermand regional director decisions. Manley believed the campaign's legal "chances were better with the Division of Advice than with the Office of Appeals,"[424] and thus sought to build a record that could be presented for review to the division lawyers.

Toward that end, on June 12, 2015, Manley penned a blunt assessment of the charges against Pac 9 to Region 21 Director Olivia Garcia, outlining numerous traditional ULPs—surveillance, retaliation, and discharge—while for the first time forcefully making the case that Pac 9 had violated section 8(a)(1) of the NLRA by "[m]isclassifying employees as independent contractors, thus denying them the protections of the Act."[425] The crux of

Manley's argument was that, because section 8(a)(1) only protected statutory employees, "misclassification denies workers—in this case, drivers at Pac 9—literally all of the protections of the act. It is difficult to imagine a more dramatic example of interference and coercion."[426] To support this argument, Manley cited a series of NLRB decisions concluding that employers had engaged in ULPs by converting or threatening to convert employees to independent contractors in response to various forms of collective activity.[427] From the campaign's point of view, the letter succeeded in achieving two goals at once: teeing up for the NLRB Division of Advice the fundamental legal question of whether misclassification constituted an independent violation of the NLRA, while laying the ground for another ULP strike.

Following its tried-and-true suing-and-striking template, the campaign had already secured the other essential piece of background legal authority in advance of the impending Pac 9 strike. In December 2014, the DLSE had ordered the company to pay three drivers over $250,000; forty other claims worth $6 million were pending,[428] owing to the driver support provided by LAANE attorney Jean Choi. With the legal foundation cast, the campaign moved to put the strike in motion. On July 21, 2015, Pac 9 drivers announced that they would strike the following day,[429] when they launched an action that ran with the precision of a well-oiled machine, setting up stationary pickets at Pac 9's truck yard and sending ambulatory pickets to the ports. Unable to resist the pun, campaign spokesperson Barbara Maynard announced, "Our goal is to impact this company so that they actually turn their ship around."[430] Thomas Lenz, Pac 9's lawyer, undoubtedly not amused, acknowledged that there was "a fundamental dispute about whether the drivers are employees."[431] The campaign asserted that drivers would walk off "indefinitely."[432]

This time around, the campaign could rely on newly energized political and industry allies. A week into the strike, Long Beach's Tidelands and Harbor Committee called a meeting to hear opposing views of the Teamsters and Harbor Trucking Association on the conflict and to seek solutions.[433] At that meeting, City Council Member Suja Lowenthal denounced driver working conditions as "modern-day indentured servitude."[434] Over ten drivers and labor organizers commented at the meeting, joined by the general manager of recently unionized Shippers Transport Express, Kevin Baddeley, who offered full-throated support for employee conversion: "It's doable. ... We need a fair playing field where every company has to play by the same rules."[435] To force companies to play fairly, in September 2015, as the strike wore on, the California state legislature passed S.B. 588, which empowered

Figure 6.5
Striking drivers at the Port of Los Angeles (left) and Intermodal Bridge Transport (right). Photo courtesy of Barbara Maynard, Teamsters communication director.

the DLSE to put a lien on the property of any employer that had committed wage theft, thereby making it easier to enforce judgments against employers who evaded them by shifting assets among companies and stalling.[436] The bill (which was signed into law in October) was intended to apply to independent-contractor port trucking firms held liable for misclassifying their drivers.

Despite this activity, as the Pac 9 strike entered its tenth week of picketing at the company's Carson truck yard, there appeared to be no resolution in sight.[437] Seeking to apply more pressure on industry and port officials to act, campaign leaders decided to expand mobilizing efforts against more key targets. On Monday, October 26, 2015, a group of drivers from XPO Logistics joined Pac 9 drivers in a ULP strike, arguing that they had been retaliated against for filing the class action against XPO.[438] The *Long Beach Press-Telegram* reported that only a small group picketed the port, causing "little or no impact."[439] The Teamsters responded by intensifying the impact, setting up pickets on Tuesday, October 27, against IBT, whose drivers had submitted a petition to the company for employee recognition and Teamsters representation. As Potter described the petition, the IBT drivers "demanded a dignified and safe working environment; demanded that the company immediately repair all unsafe trucks and ensure every truck has a functioning Air Conditioning unit; and demanded access to a dignified break room with clean drinking water, proper restrooms, and a place where we can prepare their meals and rest during their long days at work."[440] To clear the legal way for the IBT action, Teamsters outside counsel Julie Gutman Dickinson had filed an initial charging letter against IBT in August with Region 21, accusing the company of a variety of traditional ULPs, though not including misclassification on its own terms.[441]

In a significant escalation—a sign that it was time to apply maximum pressure—on Tuesday, the Teamsters also mobilized Amazon warehouse workers hired by California Cartage (Cal Cartage), who had engaged in a short strike in September and sued the company on the ground that its city lease in Wilmington on Port of Los Angeles property required the company to pay workers a living wage under local law.[442] In addition to incorporating a new dimension of cross-sectoral supply chain organizing, which one labor expert opined could provide "considerably more leverage," what was most significant about the move was that—on top of running a warehouse staffed with temporary workers—Cal Cartage was also the largest trucking company at the ports: the undisputed market leader with 965 registered drivers at Los Angeles and Long Beach (dwarfing other port drayage companies), as well as extensive operations nationwide.[443] It had been unionized in the 1980s and was considered "the biggest player and the biggest fight"—a company at the epicenter of the supply chain where organizing success could transform the entire industry.[444]

Precisely because of its scale and diversity, Cal Cartage presented multiple pressure points. The Cal Cartage warehouse property sat on the site of the proposed $500 million intermodal rail yard, the Southern California International Gateway (or SCIG), which was subject to an environmental lawsuit brought by NRDC seeking to block its construction.[445] Arguing that Los Angeles failed to incorporate proper mitigation measures under the California Environmental Quality Act (CEQA), the lawsuit brought back the issue of air pollution and its relation to ongoing labor organizing. Speaking about the suit, NRDC attorney Morgan Wyenn alluded to this point, asserting that the SCIG would "make air pollution worse, especially for the low-income, communities of color that live, work and go to school right next to the SCIG site."[446] As the NDRC suit highlighted, Cal Cartage presented a fat legal target. Its subsidiaries were also subject to a class action (brought by Bramson, Plutzik & Birkhaeuser), another suit on behalf of twenty drivers alleging over $6 million in damages, and numerous DLSE claims.[447] In addition, Cal Cartage was on the radar of state and federal officials. When a driver from a Cal Cartage affiliate was killed in an accident after unlatching his trailer,[448] the California Division of Occupational Safety and Health launched an investigation,[449] while federal authorities fined the Cal Cartage warehouse $20,000 for numerous safety violations, including dangerous forklifts and the lack of protective gear.[450]

Underscoring the enormous stakes, on Tuesday, October 27, drivers and warehouse workers were joined by Teamsters President James Hoffa, who offered this message: "Warehouse is a tremendous area, more opportunity. ... We'll be doing Cal Cartage—that's a beginning. We'll find other targets."[451] Revving up the assembled crowd, Hoffa promised: "You have the support of the 1.4 million Teamster members. We will bring justice to port truck drivers and warehouse workers nationwide!"[452] Hoffa's appearance was not simply an effort to invigorate the strikers, which included Pac 9 drivers—some of whom had been on and off the picket lines since July—it was clear message to industry and port leaders that after the nadir period of the Clean Truck Program, the Teamsters were back: all in. To reinforce that point, on Thursday, October 29, drivers from Gold Point Transportation joined the picket lines in a ULP strike against the company. The next day, drivers from Gold Point, XPO Logistics, and IBT, along with the Cal Cartage workers, took down their pickets—although Pac 9 drivers continued striking with no end in sight.

As striking continued, legal authority kept piling up against Pac 9. On December 14, 2015, in a 299-page opinion, the DLSE ruled in favor of thirty-eight Pac 9 drivers (who had been helped by the campaign's legal clinic), ordering the company to pay nearly $7 million in damages for misclassification, with an average individual award of approximately $180,000.[453] A week later, the parties in the *Castro* class action against Pac 9 filed a motion with the court to approve a settlement agreement to resolve the case.[454]

Another breakthrough came at the same moment, when the NLRB's Division of Advice finally gave its answer to the question of whether misclassification itself constituted a ULP. On December 18, 2015, Barry Kearney, associate general counsel in the NLRB's Division of Advice, issued a strongly worded opinion letter in the Pac 9 case: "We ... conclude that in the circumstances here, where the Employer told its drivers that they were independent contractors and had no right to form a union but treated them as employees in virtually every respect, the Employer's misclassification of its drivers as independent contractors interfered with and restrained the drivers in their exercise of Section 7 rights, in violation of Section 8(a)(1)."[455] Adopting the crux of Manley's reasoning from his June 2015 charging letter to the NLRB, the opinion stressed the fact that Pac 9 "continued to insist in its communications with drivers that they are independent contractors, even after the Region determined that the drivers are statutory employees."[456] The opinion went on to say that the "misclassification conveys that unionization would be futile," while concluding that "in light of the Employer's extensive control over its drivers' day-to-day operations and its awareness that the Region already determined its drivers are statutory employees, the Employer's continued insistence to its drivers that they are independent contractors is akin to a misstatement of law that reasonably insinuates adverse consequences for employees' continued Section 7 activity."[457]

Following the Division of Advice opinion, on January 28, 2016, Region 21 issued a complaint against Pac 9—the first time an NLRB complaint had ever adopted the misclassification-as-ULP theory. It was a remarkable legal victory, though esoteric enough to be fully appreciated only by those labor lawyers whose worlds revolved around NLRB advice memoranda like inside baseball. And while the *Castro* settlement deal continued to be hammered out in the Pac 9 class action suit, Pac 9 drivers remained on strike.[458]

Pac 9 was not alone in its legal trouble. In the campaign's accounting of DLSE claims from 2015, there had been awards in favor of over 150 drivers for more than $22 million. On top of the Pac 9 award from December, the DLSE had ruled in favor of fifty Fargo drivers ($8.6 million), five Pacer Cartage drivers ($950,000), one Performance Team Freight System driver ($22,273), nineteen Superior Dispatch drivers ($1,811,235), thirteen Tradelink drivers ($1,523,037), eight Western Freight drivers ($235,553), and twenty WinWin Logistics drivers ($3.6 million). In addition, it appeared as if the legal actions were wearing some companies down, suggested by the fact that three port trucking firms—Pac 9, Fargo Trucking, and WinWin Logistics—had all declined to appeal DLSE decisions against them.[459]

The year 2015 also saw the campaign expand its online presence as a way to strengthen its communications efforts and diversify its advocacy arsenal. LAANE's Sheheryar Kaoosji took the lead in enhancing the campaign's web profile, launching a blog, *Port Innovations*, that offered extensive information about port trucking firms, driver strikes, litigation activity, and legislative responses to the misclassification problem.[460] Kaoosji also packaged the information into digestible chunks that were disseminated to interested parties through a listserv. From the campaign standpoint, the *Port Innovations* format was a way to easily communicate campaign progress, while also creating documentary evidence of industry practices that could be mobilized by campaign supporters—and sympathetic lawmakers—to address the misclassification problem.

With the force of an awakening giant, Sacramento was starting to take serious note. As of January 1, 2016, several new pieces of legislation went into effect enhancing protections for port truckers. In addition to S.B.

588—the bill authorizing the Labor Commissioner to impose liens against an employer's real or personal property for failure to pay judgments—the state legislature had also passed laws authorizing the DLSE to enforce labor code requirements that employers reimburse employees for all out-of-pocket expenses,[461] tightening rules to require compensation for "nonproductive time" such as waiting for cargo,[462] and expanding protection against employer retaliation and discrimination.[463] Most significantly, Hernández's A.B. 621 went into effect, creating a one-year "amnesty" program for trucking companies that would entitle them to relief from civil penalties for misclassification in exchange for converting their drivers to employees.[464]

Capitalizing on this success, the campaign also kept up political pressure at the local level. As the Long Beach City Council election approached, the campaign sought to make an issue of driver misclassification. In January 2016, Pac 9 driver Julio Cervantes (the same one profiled a year earlier by the *Los Angeles Times*) penned an op-ed in the *Long Beach Press-Telegram* arguing in favor of creating a local wage enforcement agency: "The people of Long Beach deserve a responsive wage enforcement plan, designed with local community input and an emphasis on using wage enforcement to bring revenue lost to wage theft back to the residents and economy of Long Beach."[465] Two weeks later, at a candidate forum for contenders for Long Beach's 2nd District council seat, sponsored by the Harbor Trucking Association, candidate Jeannine Pearce proclaimed her support for truckers being classified as employees, while her two rivals took positions more supportive of the independent-contractor model.[466]

Maintaining political clout with lawmakers was partly a function of the ongoing power of grassroots action. And thus, with furious intensity, the campaign continued to shift back-and-forth from electoral politics to street-level mobilization. The XPO affiliates—combined, one of the top trucking firms at the ports—appeared vulnerable. By early 2016, XPO Cartage had been held by the DLSE to owe drivers over $3 million, and was subject to three pending lawsuits: two class actions, one state and one federal, both filed by the Kabateck firm, and a mass action filed by the Gomez Law Group. XPO Port Services was facing the state class action from Kabateck and a separate mass action from Gomez.[467] In mid-February, XPO drivers again took to the streets in the hopes of pushing XPO to the negotiating table. They conducted two one-day ULP strikes, "picketing at rail and port terminals where XPO Logistics trucks operate at the City of Industry Union Pacific Rail Yard, at Yusen Terminals Inc. at the Port of Los Angeles, and at other locations backing up cargo and causing hours of delay for hundreds of other drayage operators."[468]

Although XPO for the time being stood firm, other port trucking companies started to crumble. With liability mounting and picketing disrupting service, more bankruptcies rippled through an industry that had already witnessed the insolvency of Green Fleet, which had entered a labor peace agreement with the Teamsters, and Seacon Logix, which had shut down under the weight of a half-million dollars in legal judgments.[469] In February 2016, Hub Group, which had converted to an all-employee fleet sixteen months earlier, announced that it was closing its Southern California facility and bringing in "high-service outside carriers," with CEO David Yeager stating that its costs "have been unsustainable and substantially higher than our outsourced core carriers' costs."[470] Then, in March 2016, TTSI filed for Chapter 11 bankruptcy reorganization, citing $3.5 million in damages and $4 million in legal fees resulting from misclassification claims before the DLSE and state courts.[471] Although eye-catching, the TTSI bankruptcy was more of a formality since the firm had already started transferring some of its business to Eco Flow—the Saybrook Capital-owned firm under contract with the Teamsters.[472] It was in this sense that bankruptcy appeared to become an integral step in the process of employee reclassification as port truck drivers agreed to support bankruptcy reorganization that reduced or eliminated outstanding legal debt in exchange for employee status and labor peace.

At this crucial juncture, the Teamsters' Fred Potter believed that the companies began to see the writing on the wall, recognizing they were "fighting a fight they [couldn't] win."[473] In Potter's view, the campaign had gained a level of credibility. He remembered getting calls from company owners saying, "Apparently my lawyer was wrong" to claim the Teamsters could not strike and the ports had no right to bar truckers.[474] The owners would concede: "Here's the Teamsters ... we don't trust you guys ... and yet we find out you're the ones who have been telling us the truth. ... Now we got this liability because we didn't listen to you and I could lose my company. How do I solve this [and] ... stop the bleeding?"[475] Potter also thought the campaign began to have a "major impact at the port" as "terminal operators responded by saying 'we don't want these trucking companies on our property.'"[476] This added pressure on companies to come to the negotiating table with the Teamsters.

In April 2016, this pressure finally got to the last remaining holdout from the campaign's initial trio of targets—Pac 9—which had been in conversations with the Teamsters since the November 2014 strike. On April 26, 2016, Pac 9 filed for Chapter 11 protection,[477] stating that it was unable to pay the $7 million judgment that the DLSE ordered the previous December.[478]

Pac 9 lawyer Vanessa Haberbush stated that the company would continue to operate during the bankruptcy proceedings, which would give it time to "figure out a debt repayment plan."[479] On May 5, 2016, just nine days after the bankruptcy petition, the Teamsters announced they had reached an agreement with Pac 9, settling an indefinite strike that had begun in July 2015.[480] Pac 9 had repeatedly refused to reclassify drivers after reneging on its March 2014 settlement agreement with the NLRB.[481] But after two years of fighting, Pac 9 was ready to make labor peace. In a statement, the parties announced "a global agreement that allows the company to modernize operations by converting all owner-operators to employee drivers, providing a clear pathway for drivers to consider union representation in a neutral environment."[482] The bankruptcy also impacted the parallel settlement negotiations in the two pending class actions against Pac 9, which were essentially complete but required ultimate court approval that could not be conferred until the bankruptcy plan itself was finalized and formally entered.[483]

Significant on its own terms, the Pac 9 labor peace agreement also represented vindication of the campaign's initial misclassification strategy, which had been to deploy its suing-and-striking approach against incrementally more difficult targets—first, the hybrid firm, Green Fleet, and then the all-independent-contractor firms, Pac 9 and TTSI. With Pac 9 now acceding to a labor peace agreement, the steady approach of ratcheting up economic, legal, and political pressure on the initial trio of independent-contractor port trucking firms had produced unambiguous success in facilitating driver organizing—which was the metric that ultimately mattered. One legal consequence of the Pac 9 deal was that, since the labor peace agreement resolved pending legal claims against the company, the Teamsters' NLRB case against Pac 9 would be dismissed, which meant that the carefully constructed argument in favor of misclassification as a ULP would not be tested in front of a judge in the Pac 9 case. However, deferring that legal reckoning seemed a tradeoff worth making to secure an agreement to permit union organizing at what had been a most intransigent trucking adversary.

The Pac 9 agreement also underscored the economic benefits to trucking firms of reclassifying drivers as employees in exchange for eliminating legal liability—and thus spotlighted the new state "amnesty" law ushered in by A.B. 621, which voided DLSE civil penalties for trucking firms that voluntarily reclassified their drivers and paid back wages. Perfectly on cue, the same day that the Teamsters' Pac 9 agreement was announced, the California Labor Commissioner's office publicized that it would begin accepting applications for A.B. 621's Motor Carrier Employment Amnesty Program.[484] In the press release, Labor Commissioner Julie Su said: "The sheer number of claims filed and wages awarded to misclassified port truck drivers over the last several years demonstrates a significant problem in the industry. ... [W]orker misclassification is a form of wage theft as it denies workers all the rights and benefits of employee status. This amnesty program provides an opportunity for motor carriers to remedy these problems and correct past abuses."[485]

The amnesty approach held out promise, but also highlighted the campaign's Catch-22. Success at forcing a handful of firms to reclassify drivers as employees disadvantaged those firms vis-à-vis competitors who clung onto the independent-contractor model, raising—in the *Wall Street Journal*'s view—fundamental questions "about the future of short-haul trucking at the nation's ports."[486] As one industry economist observed, "Unless everybody else is forced to use the same [full-time] labor, these companies won't be able to raise their prices to pay for it."[487] From the campaign's perspective, initial success was not stable, but rather fundamentally depended on bringing more trucking firms into the employee-driver fold.

To protect themselves against this possibility, industry groups promoted ways for trucking firms to adhere more closely to independent-contractor rules in order to shield against misclassification liability from DLSE claims and class actions. In March, industry-side law firm Marron Lawyers conducted a presentation for the Harbor Trucking Association on "Misclassification 3.0."[488] In it, they warned that "The Threat Is Real," citing the DLSE decision in Pac 9, and stressing that "All Companies Are at Risk," necessitating a "Fundamental Model Change."[489] The main thrust of the presentation was to instruct companies on how to insert arbitration clauses into independent-contractor agreements to prevent driver claims coming before the California Labor Commissioner and courts, where there were "stacked odds."[490] Such arbitration agreements, the presentation noted, "Must have class action waiver" which "Minimizes risk of out of control Labor Commissioner and Class Action."[491] In order to assure that such agreements would be upheld, the presentation stressed that "many judges now require you to prove your drivers are ICs [independent contractors] not employees—*before* they will order you to arbitration," and thus companies needed to engage in "fundamental subhaul, lease and operational changes."[492] The presentation reviewed recent cases to provide a "road map" for how drivers were seeking to avoid arbitration provisions, which led to the following "Lessons Learned":[493]

- Cultivate Relationships with Favorable Drivers.
- Be Prepared to Support Motions to Compel Arbitration with Driver Declarations.
- Avoid Harsh Terms in Your Contracts.
- Provide Spanish Translations.
- Comply with the Truth in Leasing Regulations.
- Pay ... Arbitrator Fees.
- Get out of the Leasing Business.

To achieve the last recommendation on the list, the presentation urged companies to "Move away from leasing tractors to your own drivers," and promoted a "Co-Operative Alternative Model" in which trucking companies would reorganize as a cooperative business, selling ownership interests to drivers, thereby negating their employee status; the cooperative would then contract with a management company controlled by the original owners, which would continue to provide "Services similar to the traditional business model," thereby constituting "a real profit center for former ownership."[494]

With both sides to the misclassification dispute seeking relief on the higher ground of law—and with the first-wave trucking firm targets of Green Fleet, TTSI, and Pac 9 all locked into agreements with the Teamsters—the Justice for Port Drivers campaign redoubled its mobilization effort against second-wave targets. First up was IBT. And the first order of business was strengthening the drivers' case in front of NLRB Region 21 as a prelude to another ULP strike. Now that the Pac 9 case had been resolved, campaign lawyers needed to revive their plan to achieve a definitive ruling on the question of whether misclassification constituted a violation of NLRA section 8(a). Toward that end, on March 15, 2016, the Teamsters filed an amended ULP charge with Region 21—incorporating misclassification as an independent ULP.[495] A week later, with misclassification again formally before the region, the campaign effort to achieve its legal recognition as a ULP received a significant boost when NLRB General Counsel Richard Griffin, Jr. issued a memo on "Mandatory Submissions to the Division of Advice" outlining issues that were "of particular interest and would benefit from centralized consideration."[496] Among those "matters that involve General Counsel initiatives and/or priority areas of the law and labor policy," were: "Cases involving the question of whether the misclassification of employees as independent contractors violates Section 8(a)(1)."[497] Although such mandatory advice submission memoranda were not atypical for the general counsel, this one sent a clear signal that he believed misclassification was a cutting-edge labor law issue on which the regional offices should take a consistent position as the issue wound its way to the NLRB for ultimate consideration. Furthermore, since the Division of Advice had already endorsed misclassification as a ULP in the Pac 9 case, it seemed that the general counsel's intent in developing a consistent position on the issue was to ensure the strongest legal case for misclassification in front of the full board.

In the meantime, with ULP charges filed with Region 21, the campaign was prepared to launch a strike against IBT, which it did on April 20, 2016, when IBT drivers set up a one-day picket at several port terminals including COSCO, IBT's parent company.[498] While on the picket lines, drivers learned that Region 21 had issued a complaint against IBT,[499] in which it included, in addition to traditional ULPs for threats and retaliation, a claim that since "at least March 24, 2015, Respondent has misclassified its employee-drivers as independent contractors, thereby inhibiting them from engaging in Section 7 activity and depriving them of the protections of the Act."[500] A hearing date was set for July 2016, marking the first time the misclassification-as-ULP theory would be tested in court. Julie Gutman Dickinson exuded optimism: "The Complaint issued by the NLRB Regional Director represents a determination that misclassifying drivers in and of itself violates the NLRA. ... The complaint will lead to an historic trial where for the first time, a Judge will determine whether the act of misclassifying drivers in and of itself violates the National Labor Relations Act."[501] In a *Port Innovations* blog post a month later, Kaoosji subtly reiterated the campaign's offer to the trucking industry: "Port drayage companies continue to face governmental scrutiny of driver misclassification. Forward thinking companies are weighing the risks of continuing on a path ripe with litigation that may result in filing for bankruptcy protection versus converting to a risk-free asset-based model that can provide more control and consistent customer service."[502]

Seeking to nudge companies in the right direction, the campaign kept up the picketing pressure. On May 31, IBT truck drivers and Cal Cartage warehouse workers struck for nearly a week,[503] picketing at both ports in protest of their misclassification as independent contractors; other drivers participating in the strike were from the XPO companies—XPO (formerly Pacer) Cartage and XPO Port Services (formerly Harbor Rail Transport)—and from K&R Transportation, a trucking subsidiary of Cal Cartage.[504] Despite the showing of solidarity, there were signs of strain as IBT drivers crossed the picket lines to return to work on Monday, June 6, prompting the Teamsters' Barbara Maynard to explain: "The drivers could only stay out for so long from a financial perspective, they still had to pay the companies for the leases on their trucks. ... At one point we had 20 picket lines going on simultaneously. That was a very big effort."[505]

To further buttress that effort, in July 2016, Region 21 issued a barrage of complaints against port trucking firms. In addition to targeting misclassification, the complaints focused attention on the precise issue that industry lawyers had touted only a few months earlier: the inclusion of mandatory arbitration clauses designed to keep misclassification disputes out of court. On July 5, Region 21 issued an amended consolidated complaint against IBT, which included new allegations of the harassment, suspension, and discharge of lead union organizer, Eddie Osoy,[506] who was quickly rehired after Gutman Dickinson threatened to seek a 10(j) injunction. On July 14, the region issued a similar complaint against XPO Cartage, which in addition to claims of harassment, retaliation, and misclassification, argued that by making drivers sign agreements forcing them "to resolve disputes through individual arbitration proceedings and relinquish any rights they have to resolve disputes through collective action," the company was precluding "employees from engaging in conduct protected by Section 7 of the Act."[507] Another NLRB complaint was issued against XPO Port Services nearly two weeks later, again targeting mandatory driver arbitration clauses as ULPs, this time including a lengthy excerpt from the company's "Vehicle Lease and Independent Contractor Hauling Agreement," "requiring employees to resolve employment-related disputes exclusively through individual arbitration proceedings and relinquish any rights they have to resolve disputes through collective or class action."[508]

Summer brought more legal success for port drivers—but also revealed some limits to that success, while industry opponents began seizing the legal offensive. There were two major class action settlements against central campaign targets, which brought drivers monetary damages but failed to yield organizing dividends. In June 2016, Pacer Cartage reached a preliminary settlement in its pending class action, agreeing to pay $4.25 million to misclassified drivers over a period stretching from August 2009 to April 2016 (the court approved a modified agreement with a nearly $2.7 million pay out to drivers in October).[509] Despite the large number, the settlement in that case, which had been brought by the Kabateck firm without campaign involvement, laid bare the tension in the campaign's strategy of seeking to benefit from the economic pressure class actions afforded against targets without letting lawsuits displace organizing objectives. In the Pacer Cartage suit, that tension erupted when a group of drivers involved with the organizing campaign—which included the lead plaintiff in the class action—objected to the proposed settlement, which did nothing to promote reclassification, and ultimately decided to opt out.[510]

The one case brought by campaign-affiliated lawyers produced another impressive pay out—although it also lacked an organizing pay off. In July 2016, nearly four hundred drivers for QTS, Inc., LACA Express, and Win-Win Logistics (the Yoo family firms) agreed to a $5 million class-action settlement.[511] That case had been brought by the Wage Justice Center with the collaboration of Sanjukta Paul, continuing after she left LAANE's port drivers legal clinic, and the Asian American Center for Advancing Justice–Los Angeles. Though undoubtedly a victory, it was presented by the campaign as yet another step in the ongoing fight against misclassification.[512] Barbara Maynard from the Teamsters offered a cool assessment: "It doesn't change anything. ... The clock starts again the minute the settlement is over. The misclassification is continuing."[513] The *Los Angeles Times* viewed the settlement with more skepticism, stating: "It is unclear ... whether the case will prompt a radical transformation in the trucking business, which has clung to a contractor model despite an onslaught of legal challenges brought by drivers."[514]

What did seem increasingly clear was that the industry was no longer going to simply permit the campaign to dictate the terms of legal engagement. Instead, port trucking companies sought to seize the upper hand by initiating affirmative lawsuits. In an egregious example, a week after the QTS settlement was announced, the California Trucking Association (CTA) filed suit against Labor Commissioner Julie Su, attempting to shut down the DLSE's entire system of misclassification enforcement by claiming (in an argument that echoed the failed Pac Anchor challenge to the DLSE) that the Labor Commissioner's application of state employment law against trucking firms was preempted by the FAAAA.[515] A district court judge in January 2017 dismissed the claim,[516] but the CTA appealed. And, as if to prove that its litigation strategy was just getting started, the CTA promptly filed another state court suit against Commissioner Su alleging that the DLSE, by treating drivers as employees, was engaged in a coordinated effort to undermine their independent-contractor status in violation of trucking companies' due process rights.[517]

As this counteroffensive suggested, the threat of liability remained a central concern for port trucking companies in their cost analysis of the tradeoffs of reclassification.[518] No longer wanting to face that threat, Gold Point, which had withstood one strike and was still facing an ongoing class action, agreed—in the campaign's last major organizing victory of the year—to a labor peace agreement with the Teamsters on October 13, 2016.[519] Seeking to make more progress at ground level, on October 25, truck drivers from IBT and K&R Transportation LLC (the Cal Cartage subsidiary that was also a U.S. Department of Defense contractor) launched a two-day strike to coincide with the rollout of President Obama's Fair Pay & Safe Workplaces Executive Order, which required federal contractors to report labor violations for use in future contracting decisions.[520] The IBT and K&R drivers were joined by others from California Cartage Express and Container Freight, who coordinated with law enforcement and company and terminal management to set up pickets at the company facilities and the ports.[521] Yet, in an ominous sign of industry's new legal militancy, the strike was immediately met with litigation, as lawyers from the Law Office of Jewels Jin—claiming to represent roughly thirty independent drivers—filed suit against terminal operators and the Teamsters alleging that the strike constituted a civil conspiracy to interfere with their business operations.[522]

Amid this cacophony of competing legal claims, it was perhaps appropriate that the year ended in court with the much-anticipated IBT case in the NLRB commencing on August 22, 2016. In a grueling trial that spanned nearly four months, Teamsters counsel Julie Gutman Dickinson and her colleagues put fourteen drivers on the witness stand before NLRB administrative law judge Dickie Montemayor to testify in support of the Teamsters' central argument: that IBT "has created and perpetuated a fiction that it has been profiting from since 2010—the fiction that its drivers are independent businesses making all their own business decisions and merely providing services to IBT."[523] As Judge Montemayor lowered the final gavel on December 7, sounding the trial's close, he sat poised to make new law, ruling on an issue of first impression at the heart of the Justice for Port Drivers campaign: whether misclassification by itself constituted an unlawful labor practice under the terms of the NLRA.

Nearly four years after campaign leaders had conceived of a legal strategy to resolve this seminal question—hoping against hope for a favorable resolution throughout myriad twists and turns—it was finally possible to imagine an outcome in which an arm of the nation's most powerful labor agency would actually side with some of the least powerful workers in the contemporary economy. What campaign leaders probably could not have imagined was the radically transformed political context—and with it, the imminent restructuring of the NLRB—in which Judge Montemayor's ruling, still pending as of August 2017, would ultimately be handed down.

Ending at the Beginning: Toward a Tipping Point and Labor Peace

As 2016 drew to a close—and the era of President Donald Trump began—there was only one real certainty: the Justice for Port Drivers campaign was not going to end. Indeed, if the history of mobilization at the port over the previous two decades was any indication, the fight would continue on indefinitely, changing phases and shifting ground. But the fundamental clash over the nature of work and the role of the ports in the Southern California economy would be a constant—though now playing out in a national labor law system turned hostile toward organized labor once again.

Looking back from this pivotal point, the campaign strategy of suing and striking could claim significant success (outlined in table 6.1)—even as it faced new challenges going forward. On the suing side, the campaign and its allies have accumulated an unbroken string of legal victories on claims within the three institutional spaces targeted as sites to challenge driver misclassification: the assertion of ULP charges at the NLRB, individual wage claims at the DLSE, and class actions in federal and state courts. In NLRB Region 21, campaign lawyers won a ULP case against the hybrid firm Green Fleet in which an NLRB judge found drivers to be misclassified. In the Pac 9 case, campaign lawyers succeeded in convincing the NLRB Division of Advice to issue a path-breaking legal opinion that Pac 9's misclassification of its drivers constituted an independent ULP, leading the region to authorize the first-ever complaint charging a company with violating section 8(a) of the NLRA by misclassifying employees as independent contractors. And in IBT, campaign lawyers presented the misclassification as ULP claim at trial after another complaint was issued on that ground.

At the DLSE, to date, campaign records indicate that there have been approximately eight hundred individual wage claims filed by misclassified drivers, resulting in more than $35 million in judgments.[524] Approximately one hundred of these claims have been filed with support by the Justice for Port Drivers campaign. Of the more than twenty class action lawsuits filed on behalf of port drivers in California state and federal courts alleging misclassification and unfair competition,[525] a partial list of outcomes includes the 2014 settlement in the Shippers Transport case, paving the way for a collective bargaining agreement with the Teamsters; the $2.7 million settlement with Pacer Cartage in 2016; and the 2016 settlement for $5 million between drivers and QTS, LACA Express, and WinWin Logistics (with which the campaign has had periodic conversations about the possibility of labor peace but has not reached any agreement).

Striking has also yielded unprecedented, though still limited, success. As of August 2017, the campaign has seen drivers vote for Teamsters representation and sign collective bargaining agreements at three port trucking firms (Toll, Eco Flow, and Shippers).[526] It has secured labor peace agreements at three other companies (Green Fleet, Pac 9, and Gold Point

Table 6.1
Campaign Results as of January 1, 2017

Company	Model	Special Features	NLRB	DLSE	Private Litigation	Strikes	Bankruptcy	Outcome
Toll Group	100% employee (EE) drivers	U.S. affiliate of unionized Australian logistics company	traditional ULPs filed	n/a	n/a	1	n/a	collective bargaining agreement (Dec. 2012)
Green Fleet	hybrid firm with minority independent-contractor (IC) drivers	first target with some IC drivers	traditional ULPs filed complaint issued (June 2014) favorable ALJ ruling (Apr. 2015)	$280,000 for 4 drivers	n/a	4	Feb. 2015	labor peace agreement (Apr. 2015)
Pac 9	100% IC drivers	top 10 port trucking company (> 150 drivers) = first target with 100% IC drivers	first charge claiming misclassification as ULP Division of Advice agrees (Dec. 2015) complaint issued with misclassification as ULP (Jan. 2016)	$7.15 million for 42 drivers	2 state class actions settled (2017)	7	Apr. 2016	labor peace agreement (May 2016)
TTSI-Eco Flow	100% IC drivers	owned by private equity firm Saybrook Capital	traditional ULPs filed	$1.25 million for 17 drivers	2 state class actions settled (2013, 2015)	3	TTSI announced transfer of assets to Eco Flow (May 2015) TTSI bankruptcy (Mar. 2016)	collective bargaining agreement (July 2016)
Shippers Transport	100% IC drivers	subsidiary of Carrix-SSA Marine logistics firm, which owns SSA Marine Terminal at POLB	n/a	n/a	federal and state class actions settled (2015) DOL suit settled (2014)	0 (yard picketed in separate 2013 strike)	n/a	collective bargaining agreement (Feb. 2015)

Company	Classification	Description	ULP/Charge	Settlement	Class Actions	Count	Other	NLRB Status
XPO Cartage (f/k/a Pacer Cartage)	100% IC	subsidiary of XPO Logistics	misclassification as ULP charge filed complaint issued with misclassification as ULP (July 2014) ALJ trial (July 2017)	$3.15 million for 12 drivers	federal class action settled (2016) state mass action pending	5	n/a	NLRB decision pending
XPO Port Services (f/k/a Harbor Rail Transport)	100% IC	subsidiary of XPO Logistics	traditional ULPs filed complaint issued (July 2014) ALJ trial (Oct. 2017)	n/a	state mass action pending state private attorney general act lawsuit pending	5	n/a	NLRB decision pending
Intermodal Bridge Transport	100% IC	top 10 port trucking company, subsidiary of COSCO, global logistics firm that co-owns terminal operator of Pacific Container Terminal at POLB	misclassification as ULP filed as amended charge complaint issued with misclassification as ULP (Apr. 2016) ALJ trial (Aug. 2016)	n/a	state class action pending state mass action pending	5	n/a	NLRB decision pending
QTS, WinWin Logistics, LACA Express	100% IC	network of firms owned by Yoo family	n/a	$4 million for 26 drivers	class action settled (2016)	1	QTS bankruptcy (Aug. 2014)	n/a
Gold Point	100% IC	market leading firm, owned by Peter Kim, who also owned affiliate Harbor Express	traditional ULPs filed	n/a	state class action settled (2017) state class action pending state private attorney general act lawsuit pending	1	n/a	labor peace agreement (Oct. 2016)

Transportation),[527] where conversations to reclassify drivers and lay the groundwork for a union vote are ongoing, although progress with Green Fleet and Pac 9 has been slowed by the companies' bankruptcies.[528] Through striking, drivers have been mobilized and continue to be militant. As Gutman Dickinson reflected, the most rewarding aspect of the campaign thus far has been to witness how "workers feel empowered to understand misclassification" and challenge "wage theft at work." In addition, campaign organizers have also systematized their striking model, securing agreement from terminal operators on protocols for how strikes should be handled that permit the campaign to exert pressure without operators grinding to a halt. As Nick Weiner put it: "It's become a thing."[529] This "thing" has drawn state and local political attention to the problem of misclassification resulting in several new state laws—most significantly the amnesty program created by A.B. 621—and galvanizing city and port officials in Los Angeles and Long Beach to initiate serious discussions over possible solutions. Within this evolving legal landscape, the campaign has been able to develop a process in which independent-contractor firms can effectively convert to an employee model, and as of August 2017 there were approximately thirty-five companies at the ports with employee drivers.[530] With these firms successfully reclassifying their drivers, the Teamsters could more credibly assert their mantra: "You can in fact do business with us."[531]

As for the next steps, the picture is complicated, with several opportunities and challenges confronting the campaign. Its leaders remain focused on pushing forward the misclassification strategy, especially as trucking companies are attempting to legally reformulate relations with truck drivers to reinforce their contractor status. The misclassification argument has hinged on the degree of control that the companies exert, which has been codified in contract provisions preventing drivers from operating equipment for other trucking firms, imposing detailed inspection and reporting requirements, and setting nonnegotiable rates. LAANE's Sheheryar Kaoosji reports that in addition to inserting arbitration clauses in leases in an effort to block legal action in courts or the DLSE, employers have tried to require drivers to establish themselves as limited liability companies to enhance the appearance of arm's-length business contracting.[532] "They are telling people to act independently—kind of a contradiction."[533]

Yet as the campaign model of suing and striking moves ahead, leaders are still "looking for that tipping point," where the industry will start to move to an employee model en masse.[534] To reach that point, the campaign continues to engage multiple stakeholders in the hope of defining a new master legal solution to the fundamental problem of illegal driver misclassification. As Kaoosji reflected, "the industry has to decide whether they are going to move over to an employee model,"[535] begging the question: Could an employee model reach scale and be sustainable? "We are not sure," Kaoosji said, although he believes that companies "need employees to do business consistently," particularly as megaships carry hundreds of containers from different retailers, requiring significant control and coordination to ensure that containers make it to disparate retail destinations.[536] However, there have to be enough companies that choose the employee model to level the playing field.

To help achieve this leveling, the campaign remains committed to negotiating with local political officials to resolve the disruptive impact of driver suing and striking, while also trying to enlist new industry partners to help forge a sustainable path forward. Engaging retailers is part of this strategy, since they pay the shipping rates that are passed on to trucking companies. In the campaign's view, big retailers could change the industry very quickly, since the cost of paying trucking companies to move to an employee model would be "a rounding error" for retail giants like Walmart or Target.[537] "Power is obviously at the top."[538] To pressure retailers, the campaign has started engaging brands on social media and trying to apply codes of conduct on responsible contracting to retailers' interaction with trucking firms as a way to "get to scale and engage the biggest players and to engage the public."[539]

Given the economic consequences of ongoing disruption, a critical problem confronting the ports is how to protect their proprietary interest in efficient business operations. The Justice for Port Drivers campaign continues to make its case to the ports that they should exercise their power to bar companies found liable for misclassification under existing concessions that require trucking companies to be in good legal standing. Because the ports have already effectively been barring struck company trucks to avoid the disruption of ambulatory picketing, the campaign has used this fact to its legal advantage by telling the ports: "You *are* allowed to bar struck companies from the port. You do it all the time."[540] In addition, there may be a political opportunity to have the ports incorporate labor peace provisions into concession agreements with trucking companies—contractually requiring companies to remain neutral in relation to any union organizing—thus reformulating the model developed in the Clean Truck Program, only now on a more solid legal basis. Teamsters organizer Carlos Santamaria alluded to this possibility: "We have created a militant base of truckers at the port and they are not going to stop. ... That slows down productivity ... and puts a lot of economic pressure" on trucking companies, terminals, and the ports.[541] In Santamaria's view, the port's "best interest is to help us get labor peace."[542]

From this vantage point, the campaign's success in suing and striking has led it back to where it started: advocating for local policy to help resolve the independent-contractor problem. Yet the goal of this advocacy looks different in important ways. In the original challenge by the ATA to the Los Angeles Clean Truck Program, the Ninth Circuit's objection to the employee provision was that it was unrelated to the port's propriety interest as a market participant in promoting growth by stopping the constant disruption of environmental lawsuits. Although the current argument for labor peace at the ports contains echoes of the Clean Truck Program's market participation claim, it seeks to avoid the Ninth Circuit's critique by now resting that claim squarely on a labor, not environmental, foundation: promoting operational efficiency by stopping the disruption caused by labor strife. Whereas the Clean Truck Program was legally justified on the ground that environmental litigation constituted a business threat that the ports could resolve through their proprietary power, now the argument is that union striking constitutes the same type of business threat to the free flow of goods at the ports—which could be resolved through the same type of market participation (i.e., changing the terms of concession agreements), thus avoiding federal preemption. Moreover, the goal of labor peace at the ports now would look significantly different from the old idea of employee conversion. Although the ports could mandate labor peace provisions in trucking concessions, such provisions would only require labor neutrality—that is, noninterference in driver organizing—and would not mandate any change in employment status. This could help nudge the industry forward by requiring companies, in order to access the port, to ensure that there would be no strikes by having agreements with labor groups seeking to represent company drivers.[543]

In this sense, any labor peace approach, predicated on the ports' economic interest in mitigating disruption, would be built on a stronger legal foundation. Yet asking for labor peace as a solution would continue to depend on the campaign's ability to successfully organize misclassified workers.[544] And this clearly has been complicated by the election of President Trump, who has moved to restore an antiunion majority on the NLRB,[545] imperiling the careful work that has been done to create the opportunity for misclassification organizing in the Los Angeles and Long Beach port trucking industry. In this new context, the campaign will be unlikely to create precedent establishing misclassification as a ULP at the full board, though the historic autonomy of regional offices may still mean that there are significant local victories to be won. In addition, to the extent that any modification of the ports' concession agreements through local policy reform might reprise the preemption litigation that doomed the original effort at employee conversion, the election of Trump has solidified a conservative majority on the Supreme Court that will be unsympathetic, if not hostile, to labor claims.

Despite this uncertainty, the fundamental terrain at the port complex has changed dramatically since the launch of the Justice for Port Drivers campaign in 2013, sustaining campaign optimism even in the face of yet another unexpected political setback in the presidential election. In the impassioned words of Julie Gutman Dickinson: the "drivers are going to continue to rise until misclassification is dead. We'll take to the streets. We'll take to the courts. … We are here to stay."

Parallel Path: The Persistent Problem of Port Pollution

The resurgence of organized labor—the "blue" side of the original Coalition for Clean and Safe Ports—as a force in the port trucking industry after the Clean Truck Program invites final reflection on the parallel trajectory of its "green" partner: the environmental movement. As recounted in chapter 5, the essential story of the Campaign for Clean Trucks was one of environmental success and labor setback. With the passage of the Clean Truck Program in 2008, environmentalists accomplished an enormously impressive transformation: changing port law to produce a "green fleet." However, as industry litigation took away employment status from the drivers, the labor side of the coalition was left with little to show for all of its investment and hard work.

Following that defining moment, the tight coordination that had characterized the Campaign for Clean Trucks in its policy reform phase gave way to looser interaction and, on the part of labor, an internally driven focus on building the Justice for Port Drivers campaign on a new legal footing—misclassification—that would support robust striking and, ultimately, union organizing. As suggested above, this distancing was partly the inevitable after-effect of completing a hard-fought campaign around a clearly defined goal and partly a deliberate strategic decision by organized labor to rebuild the misclassification strategy upon an unambiguous labor foundation (borrowing from the legal template used in the Clean Truck Program, while environmentalists would, by design, recede into the background). After the ATA litigation, the coalition became inactive and coordination between labor and environmentalists around port issues was more ad hoc. Yet during this time—as the increasing focus on megaships, congestion, and growth at the ports highlighted—environmental concerns remained salient and environmental lawyers and activists stayed involved in advocating to reduce port pollution. Further, as environmental problems persisted and intertwined with labor issues, there were new opportunities for blue-green activity, though in more episodic fashion, around the goal of enhancing efficiency through consolidating and rationalizing port trucking.

From an environmental standpoint, the post-Clean Truck Program era has been defined by heightened expectations clashing with business-as-usual at the ports. The expectations grew from expressions of real commitment by local political leadership, codified in the Clean Truck Program at both ports, to seriously address port pollution in a genuine effort to go green. Yet this commitment has run headlong into the ports' implacable growth imperative, fueled by familiar forces of global competition and novel challenges posed by industry consolidation—interwoven with an effort by both ports to reduce dependence on what has become a less reliable trucking industry.[546]

In the years immediately after passage of the Clean Truck Programs in Los Angeles and Long Beach, diesel emissions from trucks declined dramatically, with some pollutants reduced by 90 percent,[547] making the San Pedro port complex one of the cleanest in the world. This success sustained momentum for further reducing air pollution,[548] and advancing new green initiatives at the ports,[549] though the challenges of inter-port competition, congestion, and community participation remained. Long Beach continued to gain ground in the competition for cargo against its more labor-friendly Los Angeles counterpart.[550] Although container volume at both ports dipped after the Great Recession, it jumped back up in 2010 and by 2015 had reached levels close to its pre-recession peak—with Los Angeles holding a slight edge.[551] The arrival of megaships raised the stakes at both ports, focusing attention once again on the ever-expanding need for space: to dock the larger vessels and to load and unload the increased number of containers. The issue, as always, was who would be in charge of designing expansion plans and whose interests would be served.

On the Los Angeles port side, the leadership would not include key architects of the Clean Truck Program. Amid early efforts to revise and thus revive elements of the concession plan struck down by the Supreme Court, the Port of Los Angeles lost its executive director, Geraldine Knatz, and director of operations, John Holmes.[552] In a sign that the port was no longer as interested in community participation in future planning, the Los Angeles harbor commission voted in 2013 to get rid of the Port Community Advisory Committee—the group of environmental activists, community members, labor leaders, and businesses, created back in 2002 at the height of the China Shipping litigation to keep local residents informed and hold the port accountable.[553] This occurred just as the port was announcing $400 million in expansion projects that threatened further community and environmental impacts.[554] Yet even these projects did not appear to sate the appetite of local officials for growth. By 2015, the *Long Beach Press-Telegram* editorial staff was arguing that new investments should go beyond port-side infrastructure improvements to expand freeways to accommodate the influx of "toys, TVs and other goods" coming into the ports.[555]

It was in response to this familiar growth ratchet—and the derogation of community input it seemed to necessarily entail—that environmental issues returned to the fore.[556] In what must have felt like déjà vu to environmental and community activists, the post-Clean Truck Program world at the ports revolved around competing proposals to expand infrastructure facilities to alleviate congestion and deal with more cargo. In September 2011, the Port of Long Beach circulated its draft environmental impact review (EIR) of a new cargo terminal on 160-acres of land on the inner harbor of Terminal Island to be called Pier S.[557] This proposal came against the backdrop of fights over two proposed rail yards near the port:[558] the Port of Los Angeles's construction of the SCIG project, to be operated by the Burlington Northern Santa Fe (BNSF) railroad in Wilmington;[559] and the expansion of the Intermodal Container Transfer Facility (ICTF), jointly owned by the ports and operated by Union Pacific five miles to the north (directly adjacent to the SCIG site). Community opposition to the ICTF expansion was immediate and fierce, causing a withdrawal of the plan for further development and review.

For NRDC, the group that had anchored the green side of the trucking campaign's blue-green coalition, the focus would not fully shift back to these port battles until after the final, exhausting phase of clean truck litigation was over at the Supreme Court in 2013. NRDC attorney Morgan Wyenn reflected that "once the dust settled after the ATA case," everyone took a collective step back and said: "We know now what the Clean Truck Program is," so "what are the other capital improvement projects or parts of the Clean Air Action Plan ... [to] push forward?"[560] Three major areas of advocacy quickly emerged: (1) challenging efforts to expand near-dock rail; (2) lobbying for stronger state-level regulation of the ports; and (3) shaping the renewal of the Los Angeles-Long Beach port complex's Clean Air Action Plan.

Near-Dock Rail

The first area of environmental advocacy has revolved around the proposed near-dock rail projects: the SCIG and ICTF expansion. Leading this effort at NRDC has been Morgan Wyenn, a 2008 graduate of Lewis & Clark Law School's Environmental, Natural Resources, and Energy Law program, who joined NRDC's Climate and Clean Air unit after fighting oil and gas exploration at the Southern Utah Wilderness Alliance.[561]

As the ICTF expansion has been put on hold, the $500 million SCIG project—a near-dock rail yard to be operated just south of the current ICTF four miles from the ports adjacent to a Long Beach residential community packed with homes, schools, and a homeless services organization—has received the most attention. As Wyenn described, since 2007 the port has oscillated between two narratives: one is "this is a great green project so good for the environment," while the other is "this is going to expand operations so good for business"—flipping "between those two message points."[562] From the perspective of community members, the project represents a dangerous step backward. The Port of Los Angeles first released a draft EIR in September 2011 (the same month Long Beach released its Pier S draft), prompting a furious community response, which came together around a new coalition, Stop the SCIG!, whose members included battle-worn groups from the Green L.A. Port Working Group: the Coalition for a Safe Environment, Coalition for Clean Air, Communities for a Better Environment, East Yard Communities for Environmental Justice, End Oil/Communities for Clean Ports, Greater Long Beach Interfaith Community Organization, and West Long Beach Association.[563] In addition to raising concerns about community impact, a central coalition objection has centered on project need. In a letter to the Los Angeles port after the release of the draft EIR, the coalition argued: "The cargo forecast used by port planners appears to be based on economic assumptions from before the recent recession and now appears extremely inflated."[564] As Wyenn put it, after the recession, with container volume in decline, the urgency of near-dock rail expansion diminished, strengthening the coalition's argument that "you really don't need the SCIG anymore."[565] Questions about project need were reinforced by Long Beach's 2015 completion of a $93 million on-dock rail project, known as the "Green Port Gateway."[566] Wyenn contrasted the environmental impact of this on-dock project with the SCIG, which "would be located close to several schools and residential neighborhoods, and would increase toxic air pollution for nearby communities that already face high levels of air pollution."[567]

Doing what they do best, advocates have taken the SCIG fight to the streets and to court. Because BNSF is part of Warren Buffett's Berkshire Hathaway empire, Stop the SCIG! coalition members produced a video imploring Buffet "to learn more about the proposal and how it will harm the nearby community."[568] After release of the second recirculated draft EIR in September 2012, coalition members spoke out against the project at a public comment hearing on October 18. When the port refused to have a second meeting, Long Beach City Council Member James Johnson, whose West Long Beach district abutted the proposed yard, held his own gathering in front of an overflowing audience, which included NRDC's Wyenn, who gave three minutes of testimony on the negative health impacts.[569] Despite these protests, the port voted to approve the project in March 2013, prompting a June lawsuit by NRDC on behalf of its clients, East Yard Communities, Coalition for Clean Air, the Villages of Cabrillo, and two individuals who live near the proposed site.[570] The NRDC suit was combined with other actions challenging the SCIG, including those brought by the Long Beach Unified School District, with schools adjacent to the proposed rail yard, and Cal Cartage, with a warehouse on what would be the SCIG site (which, as discussed earlier, was the subject of ongoing Teamsters organizing).[571] On July 26, 2016, after three years of legal battles, the Contra Costa County Superior Court held that the Los Angeles port and BNSF violated state law by "failing to adequately assess the environmental impacts of the proposed Southern California International Gateway."[572] The court specifically concluded that the insufficient analysis of the cumulative environmental effects of the rail yard and its operations were in violation of CEQA, and that BNSF's EIR lacked analysis sufficient to inform the public and allow for meaningful rulemaking.[573] A key problem was BNSF's purported bait and switch: although it had argued the SCIG would reduce trips to BNSF's Commerce truck yard, according to the *Los Angeles Times*, "it appeared that the new yard would supplement, not replace, the old yard, essentially doubling BNSF's capacity to move cargo to and from the port."[574] The case is on appeal.[575]

Although the Stop the SCIG! coalition has operated at a distance from labor, there have been uncomfortable points of intersection around the lawsuit. Cal Cartage, one of the Teamsters' prime targets in the ongoing misclassification campaign, is on NRDC's side of the lawsuit to block the SCIG (though for different reasons), creating a "sensitive alliance" between NRDC and Cal Cartage at a moment when the Teamsters are seeking to organize against the company.[576] In addition, although the Teamsters themselves have stayed neutral, the building trades, which stand to gain jobs from the SCIG, have been in strong support, rallied by BNSF at public hearings to tout the job-creation benefits of the proposal.

State-Level Regulation

The second main area of post-Clean Truck Program environmental advocacy has focused on state-level efforts to further reduce pollution at the ports and in the broader freight transportation system. Spearheaded by the statewide California Cleaner Freight Coalition, this effort has promoted "eco-friendly alternatives" to diesel trucks to move goods, including more "plug-in hybrid, fuel cell and electrified freight shuttle."[577] Advocates have urged the South Coast Air Quality Management District (SCAQMD) to take a leadership role in regulating port pollution, promoting changes like the so-called "backstop rule," which would "require the ports to reduce their emissions to the levels the ports have already promised to reach via their Clean Air Action Plan"[578]—effectively taking regulatory oversight away from the ports themselves. In addition, NRDC has asked the California Air Resources Board (CARB) to take a greater role in regulating port trucking emissions.[579] These requests have fallen upon more sympathetic official ears. Since 2014, California state and regional air quality authorities have been in "unanimous consent" that the state has to move to "zero emissions" at the ports to meet state air quality and climate goals.[580] To that effect, Governor Brown has issued a supportive statement on port emission reduction, while the SCAQMD and CARB have incorporated zero emissions into their planning documents.

The drive for state regulation has grown stronger amid sensational revelations of local port enforcement failures. While NRDC has been pushing the ports to do more to enforce basic truck maintenance under the Clean Truck Program,[581] a series of revelations of noncompliance with environmental rules at the Port of Los Angeles has increased the sense of urgency. In late 2015, the port came under fire for failing to enforce court-mandated air pollution measures against China Shipping.[582] In an explosive front-page story, the *Los Angeles Times* reported that, beginning in 2009, then-Port of Los Angeles Executive Director Geraldine Knatz waived the requirement that China Shipping connect at least 70 percent of its ships to shore-based electricity, or "Alternative Marine Power," instead of allowing the diesel-fueled vessels to idle off shore.[583] Other mandates that China Shipping was permitted to violate included those requiring the company to "deploy cleaner natural-gas fueled trucks, less-polluting yard equipment and to slow down ships as they approach to cut emissions of diesel particulate matter. ..."⁵⁸⁴ In her response, Congresswoman Janice Hahn called for reinstatement of the Port Community Advisory Committee,⁵⁸⁵ which thus far has remained defunct.

In February 2016, investigators uncovered a similar pattern of violations at the TraPac terminal, though the violations were not nearly as severe. The TraPac terminal was in violation of only three out of fifty-two environmental standards, as compared to eleven at China Shipping.⁵⁸⁶ In March, the Port of Los Angeles was again under fire when it was revealed that it had paid China Shipping $5 million in 2005 to upgrade seventeen vessels to meet more stringent environmental standards (mandated under the China Shipping litigation settlement) and then secretly waived those standards four years later after the global recession.⁵⁸⁷ Although the port had already spent the money to retrofit China Shipping vessels so that they could plug into electrical shore-side power, China Shipping decided "to use the upgraded vessels in other parts of the world where they might be more profitable because of features they had unrelated to shore power."⁵⁸⁸ As a result of these episodes of behind-the-scenes agreements to undo hard-won environmental gains, current Executive Director Gene Seroka has been faced with the challenge of regaining community trust while reinforcing compliance with air standards,⁵⁸⁹ against the backdrop of weakened faith in local government to meet its public health and environmental obligations.⁵⁹⁰

This shaken faith has been one of the central reasons environmental advocates have redoubled their efforts to promote stronger statewide oversight of the freight transportation industry. Toward this end, advocates have worked with leaders of state environmental, transportation, and energy departments on the governor's 2015 executive order to draft California's Sustainable Freight Action Plan, an integrated framework that "aims to make the freight-moving industry run cleaner, without sacrificing its competitive edge."⁵⁹¹ The Plan's key goals include: deploying "100,000 trucks, trains and cargo-moving machines fueled by cleaner fuels or electricity by 2030," while creating "incentive programs" and a "sustainable-freight 'think tank'" to encourage more green technology focusing on making the industry generally more efficient.⁵⁹² Former Campaign for Clean Trucks attorney Adrian Martinez, now with EarthJustice, said that the freight plan will "allow this industry to coexist with communities living in close proximity to it."⁵⁹³ As this initiative underscores, in the wake of disappointing local enforcement efforts to green the ports, environmental advocates have sought to leverage state regulatory frameworks to combat port-related pollution—while still carrying on the fight at the local level to make the ports accountable to the air quality concerns of the communities they adjoin and the broader region they serve.

Clean Air Action Plan 3.0

The local fight to strengthen the ports' environmental accountability has centered on impending revisions to the seminal Clean Air Action Plan (CAAP)—the original regulatory vehicle for the Clean Truck Program, first created in 2006 to provide a comprehensive plan to reduce emissions at the Los Angeles and Long Beach ports, which is up for renewal.[594] According to NRDC's Wyenn, this renewal process, which will create the third version of CAAP, poses a fundamental question to the ports: "What is the next chapter of their leadership?"[595]

Despite Los Angeles's failure to enforce aspects of the China Shipping agreement, the ports have taken significant steps over the past five years to further reduce emissions—steps that environmental advocates seek to build upon in strengthening CAAP. In 2012, both ports agreed to implement incentive programs to encourage the adoption of cleaner fuel technology, providing financial awards to companies to upgrade ship engines and deploy other clean devices, like on-dock electrical plug in.[596] Long Beach, in its quest to become a "zero-emissions" port has already increased the use of electrical shore-side ship plug in (although it is still not required for all ships), while also experimenting with "sock on a stack" technology that places "a large bonnet over all of the exhaust stacks" of docked ships that run their engines.[597]

In Los Angeles, Mayor Eric Garcetti announced the 2016 formation of a freight advisory committee dedicated to reducing emissions by transitioning the local shipping industry "from diesel truck and ships to zero and low-emissions alternatives."[598] Seeking to complement Governor Brown's statewide Sustainable Freight Action Plan, the local freight advisory committee is specifically tasked with helping the Port of Los Angeles increase "the percentage of goods-movement trips that use zero emissions to 15 percent by 2025."[599] One step in that direction already underway is construction of a near-zero emissions terminal at the Los Angeles port, known as the Pasha Green Omni Terminal Demonstration Project, which will aim to cut pollution by operating on a massive solar panel system and using electric-powered machinery.[600] Funded by revenues generated by the state's cap-and-trade program, along with private funds from the Pasha terminal operating company itself, the Pasha project has been touted as "the showcase of green technology for the shipping world and ... one that California and the nation is likely to keep their eyes on."[601] Although completion of the cutting-edge terminal is far off, it highlights ongoing efforts by both ports to maintain their lead in the U.S. shipping industry, while developing a more environmentally sustainable freight management system.[602]

It is precisely the drive toward greater sustainability that has galvanized environmental advocates around the CAAP update, which has once again spotlighted the ongoing problem of dirty drayage trucks. In a decisive move to eliminate diesel trucks from the ports, in November 2016, the discussion draft of the CAAP update required all trucks entering the ports to meet the EPA's 2010 "near-zero" emission standard by 2020 (three years ahead of the state's own 2023 deadline), while eventually completing a total phase-out of diesel trucks by requiring all trucks serving the ports to be zero-emissions by 2035.[603] The trucking industry objected to the accelerated timeline for emission standard compliance, successfully lobbying for a controversial provision in a statewide transportation bill, signed into law in April 2017, which now prohibits CARB from requiring trucking companies to retire or retrofit trucks for thirteen years after the release of a new model year vehicle or before the trucks travel 800,000 vehicle miles, whichever is later.[604] The draft CAAP update, circulated in July 2017, cited this new law in watering down its own proposed clean truck framework, stating that "unless and until CARB is able to adopt a new state truck standard required for port drayage trucks, the Ports are unable to follow our previous strategy of advancing a deadline of a new State truck regulation. Instead, in compliance with our jurisdiction and authority, the Ports are proposing a suite of actions to encourage acceleration of new trucks entering the fleet to meet the cleanest standards, including near-zero emissions and zero-emissions."[605] This suite of actions includes offering "financial inducements" to trucking companies for early compliance with the near-zero 2010 EPA standard, while prohibiting new trucks from serving the ports beginning in 2018 that do not meet model year 2014 standards (pre-2010 model year trucks already serving the ports will be permitted to continue doing so until the 2023 state deadline).[606] This proposal, and the entire CAAP update, await final approval.

As CAAP renewal highlights, the current regulatory moment—in which negotiations over clean trucks have again taken center stage in environmental advocacy—creates new opportunities for potential reconnection between environmentalists and labor advocates, whose efforts over the past decade have transformed the port trucking industry. Even as the labor movement seeks to define its own path in contesting port driver misclassification and promoting unionization, labor leaders are quick to draw links between their efforts and those of their long-standing environmental partners to strengthen the case for industry reform. In this vein, LAANE's Sheheryar Kaoosji argues that port driver misclassification and truck pollution are both "products of an atomized port trucking sector" that needs to be systematically reformed: "If the port trucking industry is going to meet increasingly stringent environmental requirements and the demands of megaships regularly calling at the terminals in Los Angeles and Long Beach, it will require an industry that has the financial capacity to invest in cutting edge technology [and] adhere to environmental regulations, while getting a financial return on its investments. Some industry leaders have already called for industry consolidation to meet these demands for improved efficiencies."[607] In this call for change—echoing the original Coalition for Clean and Safe Ports' call for an asset-based trucking industry that internalizes the environmental and economic costs of doing business at the ports—the message of the Campaign for Clean Trucks lives on.

The Long Haul: Los Angeles and Beyond

Nearly a decade after the Clean Truck Program was initially passed, the ports of Los Angeles and Long Beach can claim significant achievements despite ongoing threats. The positive environmental impact of clean truck conversion is undeniable—even if still incomplete. In 2015, the Port of Los Angeles reported that "diesel particulates [were] down an unprecedented 85 percent in the past 10 years and sulfur oxides bordering on total elimination."[608] Yet it is precisely the glaring success of emission reduction that illuminates ongoing structural tensions in the global supply chain, which not only present a constant threat to environmental progress but also reinforce long-standing pressures on surrounding communities and the labor system upon which port logistics fundamentally relies. In this way, although there has been significant change at the ports since the inception of the Campaign for Clean Trucks in 2006, much of it pushed forward by the campaign itself, there are essential continuities—deep interconnections between industry growth, port development, and community impact—that bend the arc of economic and environmental advocacy at America's port full circle. Particularly as the blue-green alliance at the forefront of that advocacy begins a new and uncertain chapter, reflecting on structural port dynamics reveals how ongoing battles over clean trucks and driver misclassification launched in Los Angeles will necessarily be shaped by broader forces—while also creating new templates for mobilization and regulation that spread throughout the rapidly evolving gig economy. This last section therefore offers a preliminary account of

Reshaping the Industry

how short-term prospects at the ports will affect the enduring drive for economic and environmental justice—at America's port and beyond—over the long haul.

In perhaps the most uncompromising storyline in the ports' long history, industry transformation and port congestion continue to assert profound challenges to port competitiveness and community impact—even as those challenges sometimes come in surprising new forms. In a dramatic recent variation on the competition theme, South Korean corporate giant Hanjin Shipping Company filed for bankruptcy protection on August 31, 2016.[609] The bankruptcy sent immediate shockwaves that reverberated across the vast global supply chain: Hanjin was the seventh largest global shipping company, controlling 2.9 percent of the global shipping market,[610] and comprising approximately 8 percent of the U.S. market's transpacific trade volume.[611] At the local level, Hanjin imports accounted for a significant portion of containerized cargo: 4 percent of containerized imports in Los Angeles and roughly 12 percent in Long Beach.[612] The Hanjin bankruptcy thus presented what Congresswoman Janice Hahn proclaimed an "unprecedented global crisis,"[613] with potentially drastic consequences for local retailers awaiting merchandise, as well as local workers—on the docks and on the roads—whose pay depended on Hanjin honoring its outstanding shipping contracts. Because the company was unable to guarantee payment to the ports to unload cargo in the immediate wake of insolvency, the ports stopped accepting Hanjin ships,[614] stranding three off of the coast, where they floated a week after the bankruptcy with millions of dollars of containerized goods and hundreds of Hanjin employees on board.[615] After a few days of tense standoff, the company received bankruptcy protection and a $90 million pledge by its parent company to meet current obligations.[616] As a result, the cargo crisis began to ease,[617] and the first of the stranded vessels began unloading cargo on September 10, 2016.[618]

Although the immediate crisis was averted, Hanjin's collapse was viewed by experts as just the "tip of the iceberg" for the global shipping industry,[619] foreshadowing increased economic instability leading to greater industry consolidation. This consolidation, in turn, seems destined to put ever more pressure on the ports to respond to the economic demands of major shippers for better facilities and lower rates, while also concentrating power in global shipping conglomerates with interests in driving down payments for cargo transportation from the docks to final destinations—once again undercutting the economic position of drayage truck drivers. The roots of the current shipping crisis trace back to the global recession, when shipping companies bullish on the prospects of economic recovery overinvested in large format container ships to carry goods to meet potential demand from the United States that never fully materialized—leaving the United States market oversaturated with partly empty ships carrying freight at reduced rates.[620] Hanjin was the most sensational casualty, although Port of Los Angeles director Gene Seroka emphasizes that oversupply has put downward pressure on the entire industry and predicts new shipping alliances will lead to more market consolidation—and thus greater pressure on the ports and local economy.[621] At the local level, industry consolidation may have a sustained impact on Long Beach, which relied more heavily on Hanjin imports than Los Angeles.[622] Indeed, the ripple effect of the Hanjin collapse was felt acutely in Long Beach for several months, although by February of 2017, that port finally began reporting cargo increases.[623]

As the global shipping market recalibrates after the Hanjin collapse, with fewer carriers traversing the supply chain, the companies left standing continue their development of ever-larger vessels: maximizing container volume per trip in the relentless pursuit of marginal profit and thus reprising the interlocking debate over congestion, port expansion, and community impact.[624] The current flashpoint is the impending arrival of new megaships with the astonishing capacity to carry 18,000 containers,[625] about 2,500 more than the next largest vessel. As carriers fighting for market share have raced to place new megaships online, thereby enabling larger loads with lower fuel costs per unit,[626] they still have struggled to maintain profits due to the lower freight rates paid by cargo owners since the recession.[627] Nonetheless, the pressure for growth symbolized by the megaships has triggered predictable calls for more automation and capital improvements at both ports as they struggle to maintain their central position in the supply chain amid ongoing competition. The reopening of a newly enlarged Panama Canal in the summer of 2016—a shortcut between the Pacific and Atlantic Oceans—presents a newly viable route for megaships from Asia to access East Coast ports. These ports are closer to the lucrative Midwest market and the ability of shippers to access that market without using transshipments from the West Coast via truck and rail could cause some shippers to bypass the Los Angeles and Long Beach ports, thus cutting into their roughly 40 percent share of U.S cargo imports.[628] In a sign the competition is heating up, the Port of New York and New Jersey has deepened its harbors, expanded access to rail lines, and installed cranes.[629] As discussed above, Los Angeles and Long Beach's attempts to respond in kind, with projects like the SCIG, have triggered new cycles of contention around the community and environmental impact of the inexorable port growth ratchet.

This growth ratchet exacerbates existing problems in the drayage trucking industry. As megaships take longer to unload and chassis remain in short supply, trucks must wait longer for their cargo.[630] This is partly a function of sheer volume—more containers take longer to unload—but also a product of complexity: with the cargo of an increasing number of distinct owners loaded onto larger vessels, part of the challenge for drivers is wading through the thicket of containers from different sources to find their loads at the port.[631] The failure of Hanjin contributed to the chaos, as empty containers with no place to go have been left stranded, causing overcrowding and further complicating the pursuit of chassis, many of which remain hitched to the "empties."[632] The *Long Beach Press-Telegram* reported in October 2016 that there were as many as fifteen thousand cargo containers sitting in the Long Beach port alone,[633] requiring (when combined with those in Los Angeles) at least a fifty-acre site to assemble the scattered containers so that their chassis could be placed back in service.

The ports' use of PierPass—a decade-old program that charges cargo owners mitigation fees for day-time shipments that are used to pay terminal operators and longshoremen to unload during off-peak hours—has been both praised and criticized for its response to the congestion problem.[634] Its supporters point to evidence that trucker turn times are dropping, owing to PierPass's shift of 40 percent of container pickup to off-peak hours as well as higher enrollment in an "appointment system" that designates specific pickup times and a more concerted effort by shippers to prepare containers in advance to facilitate pickup.[635] Critics, in contrast, charge that lower reported turn times (below forty minutes on average in May and June 2016) are misleading because they do not include the significant time truck drivers spend outside the ports (some waiting for peak hours to pass because cargo owners do not want to pay mitigation fees).[636] These lengthy truck wait times, coupled with concerns about labor disruption and environmental pollution, have prompted government officials to pursue alternative solutions to port congestion. These include eliminating PierPass and, more ominously for the port drayage industry, moving toward greater reliance on short-haul rail to transport containers to inland warehouses[637]—thus threatening to reduce the port trucking sector at the very moment that labor organizing headway is being made.

Just as growth pressures continue to challenge the movement in Los Angeles and Long Beach to green the ports and rationalize the trucking sector, they also draw attention to national trends and spotlight campaigns to reproduce some of the successes of Southern California elsewhere. These wider movements—at other ports around the country and in other sectors of the economy experiencing analogous patterns of labor precarity—showcase the strength and vibrancy of coordinated national efforts, while also tracing distinct environmental and labor trajectories that the story of blue-green advocacy in Los Angeles and Long Beach has etched in microcosm.

On the environmental side of the ledger, there are now four U.S. ports with Clean Truck Programs (Los Angeles-Long Beach, Seattle-Tacoma, Oakland, and New York-New Jersey). Although Oakland's program, passed in 2009, largely tracks that of Los Angeles and Long Beach,[638] the New York-New Jersey plan has drawn more controversy since that port, run through a joint authority controlled by the state governors,[639] has relied on a patchwork of funding, including federal grants, to replace pre-2007 trucks.[640] While the New York-New Jersey port was able to use nearly $30 million in federal aid and port funds to completely replace more than four hundred of the oldest trucks, it would need another $150 million to replace the entire pre-2007 fleet. Unable to extract additional cargo fees from shippers, the port recently announced it was lifting the 2017 deadline for complete replacement,[641] prompting a New Jersey state legislative proposal that would impose tariffs on the beneficial owners of cargo for trips using old, dirty trucks.[642]

There have been hopeful expressions of environmental advocacy around freight issues at the national level—although such progress, of course, faces an unprecedented and unpredictable threat in the Trump administration. NRDC, building on its state-level regulatory efforts around California freight, has also taken on national leadership, joining forces with core allies from the Campaign for Clean Trucks. The Moving Forward Network has emerged in a prominent role, bringing together a coalition of local environmental and environmental justice organizations, led by Angelo Logan, formerly of East Yard Communities—one of the groups that spearheaded the Los Angeles clean trucks fight. The Network has sought to "link all local struggles" around freight issues into a united front and has worked to press freight justice issues at the EPA, successfully urging state agencies, including the SCAQMD, to file comments supporting ultra-low emission standards for trucks.[643] While the Network represents a mature and energized national movement that brings together historically fractious environmentalists and environmental justice activists, its national advocacy program faces the daunting prospect of confronting an industry-friendly EPA leadership that observers have charged with systematically "dismantling the agency's policies—and even portions of the institution itself."[644] In the short-term, at least, national clean freight advocates seem destined to play defense and protect local victories.

On the labor side, the fight over employee misclassification at the center of the Justice for Port Drivers campaign has also taken on wider national significance. With misclassification not confined to drivers at the Los Angeles and Long Beach port complex, the Teamsters have sought to extend the fight against independent-contractor firms to ports around the country. Toward this end, there has been notable organizing activity at the grassroots alongside high-profile legislative campaigns. On the ground, seeking to build organizing pressure, the Teamsters have launched coordinated actions targeting employee-based affiliates of national firms serving the less-than-truckload market. Some of these companies also own contractor-based trucking subsidiaries in the port drayage sector. In the most prominent example, drivers at XPO Logistics affiliates in suburban Philadelphia and Aurora, Illinois, voted for Teamsters representation in October 2016, joining already unionized XPO drivers in Miami, Florida, Laredo, Texas, and Vernon, California.[645] The Teamsters have also made forays into hostile territory, organizing a 2013 walk-off of port drivers protesting misclassification in Savannah, Georgia—thereby launching the Stand Up for Savannah movement, which has aimed to mobilize drivers at the nation's fourth-largest port in the heart of a deeply conservative state. Although that movement has met with stiff resistance and limited success, efforts in labor-friendly states have notched important policy wins. In a legislative move to clamp down on misclassification, New York passed a state law in January 2014 creating a strong presumption that drivers for commercial goods carriers are statutory employees.[646] That same year, the New Jersey legislature took up a similar measure—provoking a furious counter-attack from industry lobbyists in the hope of maintaining their business model—which led to a compromise bill coming out of a key legislative committee in the summer of 2016 that the Teamsters roundly condemned as simply authorizing the industry to continue misclassification.[647]

This acrimony has bled into the growing national conversation around independent-contractor misclassification, which has increasingly centered on the highly visible application-based, on-demand—or "gig"—economy.[648] In July 2015, NELP released a report on misclassification across the United States, which it called a "persistent problem in many of our economy's growth industries," concluding that misclassification imposes significant social costs: "denying [employees] the protection of workplace laws, robbing unemployment insurance and workers' compensation funds of billions of much-needed dollars, and reducing federal, state and local tax withholding and revenues, while saving as much as 30% of payroll and related taxes otherwise paid for 'employees.'"[649]

It is against this backdrop that iconic gig economy companies have come under pressure for their labor contracting practices. Class action lawsuits continue to mount against the controversial ride-sharing app company Uber, filed by drivers the company insists are independent contractors. In California, a $100-million class action settlement was overturned in 2016 as "not fair, adequate, and reasonable" by a federal judge, who suggested the misclassification case was worth many times more; in July 2017, a court in North Carolina gave conditional certification to a class of roughly eighteen thousand drivers who had opted out of mandatory arbitration clauses in their contracts.[650] As drivers in private lawsuits assert Uber's misclassification entitles them to recover back pay for the company's failure to meet minimum wage and overtime standards, drivers are also bringing enforcement actions in state agencies to recover public employment benefits. In this regard, the New York State Department of Labor recently ruled that two drivers are eligible for unemployment payments, finding the drivers to be employees rather than independent contractors.[651] California's Employment Development Department has similarly awarded Uber drivers unemployment benefits.[652]

In the trucking industry, the online retail giant, Amazon, Inc., with its vast network of independent-contractor drivers delivering goods to every corner of the country, and FedEx, a leading home delivery service that also operates through independent-contractor firms, are fending off misclassification legal challenges. Amazon, which has built an economic empire on contract labor, is subject to a federal class action suit brought in 2016 by delivery drivers seeking "back wages, overtime pay and compensation for fuel, car maintenance and other expenses"[653]—claims echoing those made by truck drivers in the myriad class actions against misclassifying firms at the Los Angeles and Long Beach ports. Amazon—an important player at the ports because of its massive warehouse system—is a critical bellwether in the misclassification fight, and is aggressively growing its control over the supply chain through the purchase of truck trailers, leasing of Boeing 767 jets for air cargo, and receipt of a Federal Maritime Commission license to operate as an ocean freight forwarder.[654] Like Amazon, FedEx has resisted unionization efforts, in part by contracting directly with drivers—a practice it continued until 2011, when it began entering delivery contracts with trucking firms that, in turn, hired or contracted with drivers. As a result of its pre-2011 practice, FedEx faced nationwide litigation resulting in a $240 million class action settlement with drivers in twenty states for misclassification (this followed a $238 million settlement in 2015 with drivers in California alone).[655]

As these legal challenges to misclassification reveal, a central fault line of the new economy runs between the domain of formal employment—with its greater benefits and security—and that of independent contracting—with its increased instability and risk.[656] With the most powerful corporations in the world vying to dominate the markets for online retail and on-demand service, the push toward labor "independence"—with the attendant corporate benefits of lower costs, looser regulation, and no unions—is moving ahead at a frenzied pace: foreshadowing a future in which independent contract labor is the norm, employment the exception. While the FedEx and Uber misclassification settlements suggest that litigation is still an effective tool to resist corporate efforts to shed the regulatory strictures that come with employment status, they also throw into stark relief law's limits: capable of bringing financial redress to workers wronged by corporate misconduct, but unable to force structural changes that staunch the flow of workers into the more precarious status of contractor. It is in this sense that the national battles over misclassification, which have resulted in significant monetary victories but have not strengthened worker organizing, render all the more impressive what has been achieved at the Los Angeles and Long Beach ports. There, the Justice for Port Drivers campaign of suing and striking has left an enduring impact on the port trucking industry, resulting in precisely the worker mobilization and unionization that have eluded the labor movement on the national stage. For this reason, the struggle against misclassification at America's port stands out as a story of how much it takes for low-wage workers to successfully stand up for their collective rights—and makes the outcome of that local struggle so crucial as a test of what might be achieved in America's uncertain future.

7 Assessing the Movement: Impact and Implications

Since the birth of the modern environmental movement in the 1960s, the relationship between labor and environmental activists has often been defined by conflict rather than collaboration. Although coalition building between the two movements is not new,[1] the pursuit of industry regulation by environmentalists, empowered with new legal tools in the 1970s and endowed with growing political clout, has been perceived by many within a beleaguered labor movement, itself experiencing a dramatic decline in private-sector union density and political influence, as a threat to jobs and membership. Yet within American politics, there are no other movements on the political left with the same scope and power to mobilize for local change and shape national policy. Stories of how they mesh are therefore important not simply as micro-histories of political resistance but as windows into deeper dynamics of collaboration revealing ways that social movements and interest groups in American politics find common ground in the quest to produce sustainable policy reform and redistribute power to those marginalized from the democratic process.

The story presented in this book—of the blue-green challenge to economic and environmental degradation at America's port—is of particular significance: not only for what it accomplished but, just as importantly, for what it left unfinished. The struggle against the linked problems of air pollution, caused by dirty diesel trucks streaming in and out of an ever-expanding port complex, and economic precarity, experienced by drivers of those trucks buckling under the strain of delivering global trade, joins together two of America's most pressing social problems—environmental damage and economic inequality—at one of its most vital commercial locations: the Los Angeles-Long Beach port complex through which over $1 billion in cargo flows each day. The histories the book has traced, critical on their own terms, become more complicated and powerful as they intersect and combine to show how residents of the port communities of

Wilmington and San Pedro asserted standing to challenge the deterioration of their neighborhoods and exclusion from political decision making; how environmentalists mobilized law in courts and policy-making processes to check the powerful growth imperative of global trade; and how the labor movement saw in the environmental challenge to port growth an opportunity to redress the fundamental problem of drivers' independent-contractor status that had stymied organizing since the 1980s. It is these actors' collaborative achievements—empowered communities with an ongoing voice in port governance, cleaner trucks at both ports, a political commitment to zero-emissions, and mobilized drivers who have won labor agreements in a trucking industry without any for more than three decades—that make the port struggle so riveting and consequential. But it is also in the ongoing challenges—of environmental regulatory implementation and industry resistance to driver reclassification and unionization—that one gains a deeper appreciation of the difficulty of sustaining cross-sectoral movement coalitions over time, particularly when the benefits of coordinated action are unequally distributed, as they have been at the ports. The pursuit of a sustained blue-green alliance in the face of forces pulling each movement back into familiar silos is perhaps one of the most enduring legacies of the port trucking struggle.

Assessing the legacy of complex social movement challenges like the one at the Los Angeles and Long Beach ports is, as movement scholars have long pointed out, an exceedingly difficult task, complicated by disagreement over what outcomes matter and how to measure them[2]—particularly when, as here, the challenge is ongoing and thus judgment is rendered at an arbitrary point. There are also legitimate concerns about how much one can extrapolate from events that transpire in a particular time and place to draw broader conclusions. Yet paradigmatic cases—those, like the port trucking struggle, which disrupt powerful systems in ways that gain wider recognition as symbols of innovation in social struggle—are important to dissect precisely because they become reference points in larger debates about what type of social reform is possible and how best to achieve it. In a city, Los Angeles, that is frequently identified as one of the most important in the nation for leading-edge labor and environmental justice activism,[3] the movement to transform the port trucking industry has been one of the biggest and most ambitious efforts over the past twenty years. Although Los Angeles is a unique place and port trucking a unique industry, the story of what happens when global forces driving economic change clash with local forces resisting that change—and how that clash shapes the meaning of economic and environmental justice where it occurs—offers lessons that

Assessing the Movement

transcend the space. This last chapter is an attempt to crystallize some of those lessons: drawing upon materials presented in the preceding chapters to analyze what happened and assess what worked, while generalizing from the specific context of port struggle to think in a grounded way about theoretical implications for understanding the complex relationship between social movements, legal mobilization, and local reform.

This book began by asking how law, understood both as a tool and a target of social movement activism, has influenced the drive for justice at the ports—and what it has achieved. It has explored this question by offering a close historical account, viewed through the lens of a social movement campaign navigating complex and ever-shifting legal and political challenges. In this sense, it has presented a partial view emanating from the perspective of the main campaign actors who imagined change was possible and endeavored to will it so. Yet the very partiality of this perspective enables a valuable type of appraisal—one which judges the struggle in relation to what its main architects understood themselves to be pursuing, even when those understandings may have been in tension or turned out to be wrong.

The book ends, therefore, by considering the larger contributions of the ports case study. It begins by reflecting back on how law shaped the strategy of port struggle—in its foundational Campaign for Clean Trucks and Justice for Port Drivers phases—and how this strategy impacted outcomes, both in terms of policy reform and industry change. On the strategy side, the chapter explores how legal rules governing port operations and driver status contributed to the central problems the campaign sought to redress—pollution linked to misclassification—and the advocacy approaches used to do so, which changed over time based on assessments of legal and political opportunity. Turning to outcomes, it discusses how each movement's initial legal endowments shaped different paths to reform—and how industry opponents were able to exploit those differences to produce divergent results.

This final chapter then returns to the central theoretical frameworks in which the port study is situated—on social movement coalition building, legal mobilization, and local policy making—to explore the role of law in three significant and underappreciated social movement dynamics.

First, it examines how law simultaneously creates conditions of possibility for cross-sectoral coalition building, while challenging the cohesion and sustainability of a coalition's mutual effort. In the port struggle, strategic congruence, alongside common legal analysis, brought the labor and environmental movements together in a challenge that leveraged the comparative advocacy advantages of coalition partners. That the challenge ultimately yielded different benefits for labor and environmental stakeholders highlights the unique nature of the legal obstacles they faced and the difficulties movement leaders confronted in trying to represent diverse constituencies within the blue-green coalition.

Second, the story of struggle around the ports spotlights dimensions of legal mobilization by movements and the connected role of movement lawyering that have received little attention in the law and social science literature. In contrast to the dominant scholarly focus on the benefits and costs of impact litigation as a social movement strategy,[4] the port struggle reveals essential non-litigation roles for lawyers in social movement campaigns to change public policy and industry practice, and distinctive methods of leveraging the litigation that does occur.

Third, this chapter examines how the history of blue-green mobilization at the ports contributes to understandings of the city as a place for progressive legal reform generally,[5] and local labor law making in particular.[6] It does so by supplementing scholarly accounts of social movement-led "law making from below" to spotlight the inherent connection between local mobilization and federal law that shapes local legal strategy around avoiding preemption and continuously pulls local efforts back into federal forums to defend and build upon gains.[7] In this sense, the relation of local law making to federal law is less complementary than it is contestatory—part of an ongoing struggle to create space for the exercise of power.

The chapter closes with final reflections on broader themes that emerge from, but ultimately transcend, the port story: the crucial role of professional expertise in grassroots challenges to entrenched power; the importance of tenacity in social movement leadership; and, most foundationally, the transformative potential of courage—exhibited by port drivers and community members with much to lose, who stood firm against reprisals of power to lead the way along the long road to justice at the ports.

Analyzing Strategies and Evaluating Outcomes

A core focus of this book has been on tracing the emergence, intertwining, and joint mobilization of social movements using a coordinated set of legal and political tools to force private industry and government actors to be more responsive to those they impact. Each of these social movements—labor and environmental—is directed toward sustained engagements with powerful adversaries to bring about social change: working both inside the system of institutional politics, through legal action and lobbying for policy reform, and outside the system, through protest, media campaigns, and community education. In each movement, the central organizations involved—the Teamsters, LAANE, and NRDC—defy conventional categorization. They are professionalized, but have leaders who care deeply about community engagement and support; they pursue "inside game" strategies, thus resembling traditional interest groups, while coordinating with "outside game" efforts to mobilize dissent and pressure decision makers. Yet, despite the similarities of movement organizations, the movements themselves are differently situated in American law and politics. The labor movement comes out of a history of ground-level mobilization and lays claim to represent a majority of Americans—workers—through collective action governed by a legal framework achieved at the zenith of its now-diminished power. The environmental movement, in contrast, emerges from a tradition of using legal action in court to prevent environmental harm, mobilizing environmental laws of more recent vintage to overcome the collective action problem associated with representing a diffuse constituency of Americans who care about environmental protection but are not readily organizable in specific locations. The structural features of each movement influenced the role they played in the port struggle and what they have been able to accomplish.

In addition, the labor and environmental movements both have agendas that stretch far beyond efforts to reform the ports. Thus, a key question the book has explored is how these larger, distinctive social movements converge at a particular moment around a particular issue—in this case, the economic and environmental costs of port trucking—and how that convergence produces a process of coordinated planning and execution in the form of a *campaign*: a series of interconnected operations intended to achieve a specific objective. From a campaign perspective, the port struggle illuminates how distinct organizations and their leaders come together at the local level with affected community members to design and implement *strategy*: a set of conscious choices made by movement actors that generally target particular decision makers, identify resources and pressure points, and proceed through sequential steps toward a predefined goal.[8] Movement leaders then make decisions about the discrete means, or *tactics*, used to advance strategy, such as litigation, lobbying, and protest.[9]

One of the unique contributions of this account has been to show how such tactics were deliberately coordinated by movement leaders and executed according to an overarching strategy designed to maximize their combined power in local politics, while highlighting the role that law played at each stage of strategic development: driving the blue and the green coalition members toward mutual, though ultimately distinctive, ends. This section looks more carefully at how law informed campaign strategy—shaping the coalition's efforts to define problems and solutions, and turn weakness into strength—and how campaign outcomes varied in relation to each coalition partner's initial legal endowments and the divergent response of industry opponents.

Turning Problems into Solutions

Scholars have treated the role of law in defining social movement problems and solutions in relation to the concepts of "framing" and "political opportunity." Framing studies have generally emphasized the productive role of social movement actors in developing "emergent action-oriented sets of beliefs and meanings that inspire and legitimate social movement activities and campaigns."[10] Framing, in this sense, is a key bridge between the individual experience of grievance and collective action. By defining grievances as part of broader social problems that can be redressed, framing helps to shape collective movement identities and motivate action.[11] Legal scholars have suggested that law itself provides principles of rights and justice that frame how social movements mobilize constituents and how lawyers make persuasive arguments on behalf of movement causes in and out of court.[12] Political opportunity theory, in contrast, posits that movement mobilization is shaped by the opportunities and challenges embedded in the pre-existing political environment—comprised of political officials, media outlets, financial patrons, local organizations, and mobilized opposition.[13] Movement scholars have argued that law is an essential part of the opportunity structure: with courts in certain periods opening to social movement claims in ways that invite movement-advancing litigation,[14] while at other times erecting legal barriers to be overcome or changed.[15]

This book's study of the port struggle, while validating central insights of the framing and opportunity research, moves beyond it to spotlight a distinctive, and equally critical, role of law in social movements. Specifically, the history of blue-green mobilization around port trucking shows how *legal rules themselves actively contribute to the creation of relevant social problems*—air pollution, community congestion, and labor precarity—in ways that make *the modification of legal rules a central target for social movement reform*. This perspective on the role of law, in which it fundamentally contributes to how movements define problems and solutions, helps to explain the peculiar pathways of movement mobilization in the Campaign for Clean Trucks, which pursued local policy reform to achieve clean truck and employee conversion, and the Justice for Port Drivers campaign, which has attacked driver misclassification and promoted unionization through "suing and striking."

The problem targeted in the Campaign for Clean Trucks had its origins in the conservative push for trade liberalization and industry deregulation decades before. Beginning in the 1970s, legal rules promoting free trade transformed the ports from engines of regional growth to conduits of globalization, imposing negative externalities on surrounding communities in the form of pollution and incompatible development. Deregulation facilitated the explosion of trade by reducing the cost of transport, while also producing the specific problem of fragmentation in the drayage trucking industry and the shift toward independent-contractor drivers. Since that fragmentation occurred, starting in the 1980s, the independent-contractor relation has been the crucial legal barrier to unionization, since independent contractors receive no labor protection and are legally proscribed from organizing under antitrust law. Therefore, from the perspective of organized labor looking at port trucking at century's end, weak federal labor law interacted with federal deregulation and antitrust law to put port drivers in a legal box: economically marginalized but unable to organize to improve their dismal working conditions. These working conditions, in turn, interacted with port expansion to produce pollution. As the size of the drayage industry grew in relation to imports, independent-contractor drivers unable to afford to maintain and upgrade their vehicles to current environmental standards became stuck with aging diesel trucks that were central sources of port air pollution. In the coalition's terms, the port was where old, dirty diesel trucks "went to die." In Los Angeles and Long Beach, pollution grew consistently worse with the arrival of ever-larger ocean carriers, which strained port capacity and continuously motivated official calls for more growth. As a result, city government officials in the early 2000s promoted port expansion as an engine of job creation, but lacked the legal tools and political will to directly regulate labor and environmental externalities.

For the gathering Coalition for Clean and Safe Ports, crafting a solution to the economic and environmental impacts of port trucking required identifying the institutional space, political power, and legal strategy to address the underlying legal problems. Coalition partners approached these problems from distinct points of view. For organized labor, finding a way to restructure the industry on the legal foundation of employment was the central aim. The Teamsters, in particular, had devoted significant internal resources to experimenting with possible legal solutions to the independent-contractor problem, which included supporting recognition strikes and advancing state legislation to permit independent-contractor organizing in the late 1990s and early 2000s. But none of these early efforts gained political traction and fizzled out.

The labor movement kept a hopeful eye on port trucking as a potential organizing target, not because of its size (it was a relatively small proportion of the overall trucking industry), but because of its strategic importance in the logistics supply chain—with the ports viewed as a key "choke point" to exert organizing pressure on warehouses and big retailers like Walmart. It was in the context of this broader strategy that labor movement leaders conceived the idea of leveraging the ports' contracting power to change driver status and transform drayage labor relations. The specific idea of embedding employee conversion in port contracts with trucking firms had roots in labor campaigns to promote living wage laws and attach labor-friendly requirements to city concession agreements at publicly owned airports and development agreements relying on public subsidies. The legal foundation of many of these efforts was the doctrine of market participation, which justified a local government's proprietary action to enact labor-enhancing regulation without running afoul of federal preemption. For organized labor facing the independent-contractor problem in port trucking, the legal calculus thus converged around a local policy campaign to use the ports' contracting power to reclassify port drivers as employees.

For environmentalists, the ports presented a different legal problem, but one that ultimately was seen as connected enough to the independent-contractor status of drivers (which caused driver dependence on old, cheap diesel trucks) to warrant a joint solution. By the early 2000s, environmentalists began drawing a link between port air pollution and drayage trucking, noting that idling trucks queuing up for port entry were spewing diesel fumes and that port terminal operators and company owners, because they did not bear the costs of long wait times, had no incentive to take remedial actions. As the Sierra Club's Tom Politeo recalled, environmentalists began arguing to port officials that port truck drivers "should be paid by the clock and not by the can"—and thus the idea of reforming drayage trucking as a solution to the twin port problems of economic and environmental injustice was born of "multiple inventors."

As the labor and environmental movements converged around a common analysis of the problem to be addressed—a dysfunctional drayage trucking market producing economic precarity and air pollution—their formulation of a solution necessarily came to be shaped by the legal possibilities for local redress. The ports of Los Angeles and Long Beach were targeted together by the coalition as a strategic site precisely because they were city entities with authority to enact policy to address truck emissions and driver status. To achieve this result, the policy had to be crafted to take advantage of local government power while avoiding federal preemption under the Federal Aviation Administration Authorization Act, the FAAAA, which prohibits localities from enacting any law "related to a price, route, or service of any motor carrier. ..."[16]

The result was a policy solution to port trucking nested in local government law that provided the crucial legal fix to the independent-contractor and pollution problems—but also built in legal vulnerabilities. The fix, enacted in the Los Angeles Clean Truck Program in 2008, was to modify the port's concession agreement—the legal contract between the port as a city entity and trucking firms seeking to enter its property—to require a *double conversion*: of *trucks* from dirty diesel to clean fuel and of *drivers* from independent contractors to employees. Under the program, the environmental movement achieved one of its long-standing goals—emission reductions in the drayage trucking fleet as a step toward "greening" the port—while the Teamsters won employee status for drivers that enabled potential unionization. It was a mutually beneficial solution: by taking maintenance out of the drivers' hands and making trucking companies internalize the costs, the program promised to create a sustainable foundation for clean trucking and driver prosperity over time.

To do so, the Clean Truck Program required trucking firm compliance with operational standards such as off-street parking and placard posting, and mandated truck and employee conversion as a way to solve the port's central proprietary problem: its inability to achieve necessary growth due to legal fights over environmental pollution caused by diesel trucks. Failure to comply on the part of trucking firms subjected them to penalties—including contract termination and port disbarment. The program hinged on enforcement by terminal operators, which were prohibited from admitting noncompliant trucks on the threat of port-imposed criminal sanctions. These features of the policy solution succeeded in leveraging local government power to address the structural legal problems at the heart of port trucking, but in so doing also provided the basis for the program's legal demise in the American Trucking Associations (ATA) preemption lawsuit. In that suit, the program's legal assets—welding together truck and employee conversion, and enforcing concession agreements through the port's police power—became legal liabilities. The Ninth Circuit ruled that employee conversion was too tenuously connected to environmental remediation to qualify as permissible market participation, while the Supreme Court invalidated the concession plan's off-street parking and placard provisions on the ground that their enforcement via criminal sanctions constituted a regulatory, not proprietary, act.

For organized labor, the failure of employee conversion in the Clean Truck Program required reframing the legal problem in terms that could mobilize a different set of actors and resources. "Misclassification" became the organizing frame that defined the legal problem in the Justice for Port Drivers campaign—and pointed toward a new set of legal solutions. Misclassification, of course, was also the underlying problem targeted by employee conversion in the Clean Truck Program; but the local policy campaign to promote clean trucks turned on suppressing the labor law dimension of fighting misclassification (especially its relation to unionization) in favor of stressing its connection to combatting pollution. When the Clean Truck Program failed to achieve employee status for drivers, labor leaders in the Justice for Port Drivers campaign decided to reframe the industry problem explicitly in terms of misclassification and to amplify the labor law consequences of that reframing.

Drawing attention to pervasive drayage industry misclassification in Los Angeles and Long Beach was deemed necessary to advance the campaign's new strategy: challenging independent-contractor firms at the California Department of Labor Standards Enforcement (DLSE) and in court; using the precedent from successful cases in those venues to pursue federal labor protection for misclassified drivers at the National Labor Relations Board (NLRB); leveraging that federal protection to launch driver strikes against targeted trucking firms, which along with litigation liability, would pressure firms to negotiate union contracts; and building political power from this suing-and-striking model to bring port and city officials back to the negotiating table. This strategy hinged on defining the problem in terms of misclassification in order to draw crucial legal enforcement resources from sympathetic government labor officials, while building a pipeline of cases against campaign targets at the DLSE and in court. Redefining the port trucking problem more explicitly around misclassification thus facilitated the campaign's pivot away from the local policy arena toward state and federal administrative agencies, where campaign leaders believed they could build favorable misclassification precedent that would buttress their organizing campaign and impose systemic costs. In this way, the campaign sought to leverage the power of federal and state enforcement to rebuild the political capital needed to ultimately return to the ports to make the case for a new policy solution to the independent-contractor problem.

Turning Weakness into Strength

How the campaign defined problems and solutions influenced not only where it mobilized law, but also how it did so—shaping the nature of legal advocacy it deployed and its relation to other tactics. In studies of law and social movements, legal advocacy is often treated, at best, as complementary to movement mobilization if confined to a limited role and, at worst, counterproductive.[17] The port struggle reveals a more sustained and interconnected relationship between legal and political action. First, law was always an intrinsic part of overall campaign strategy—neither a centerpiece nor an afterthought—its role shaped by the specific campaign goals pursued. Campaign planning consistently integrated advocacy tactics in ways designed to maximize their collective impact. Second, because the Campaign for Clean Trucks and the Justice for Port Drivers campaign each focused on different types of legal solutions to address driver misclassification, at each phase, campaign leaders had to devise new ways of converting an initial legal weakness—the rules governing independent contractors—into strength. Because it was launched from a position of relative disadvantage, this project necessarily required campaign leaders to combine legal concepts and advocacy tactics that had previously been viewed as distinct: linking environmental litigation to labor law reform, employee reclassification to emission reduction, port concession agreements to environmental and labor law making, suing to striking, and misclassification to labor peace.

Throughout the port struggle, leaders integrated legal and other modes of advocacy in campaign strategies that were both proactive and responsive to unplanned events. Port advocacy thus evolved through an adaptive process in relation to the development and revision of campaign goals and targets—a process that hinged on converting legal liabilities into opportunities for change. Throughout this process, campaign leaders on both the environmental and labor sides had to invest in research and planning in order to be nimble enough to seize opportunities when they were presented.

The formative period leading to the launch of the Campaign for Clean Trucks provides an important illustration of this adaptive process. On the organized labor side, the Teamsters began to develop a strategic approach to the nation's ports in 2000, when the union announced a nationwide port trucker initiative focused on promoting policy solutions to address driver misclassification. The initiative, housed within the Teamsters Port Division, responded to a string of failed efforts in the 1990s: independent organizing by the Communications Workers of America; an attempt to create a model trucking company, Transport Maritime Association, that would hire employee drivers; and class actions like the California workers' compensation case, *Albillo v. Intermodal Container Services*, which succeeded in gaining compensation for misclassified workers but not reclassification itself. When Mike Manley was hired by the Teamsters general counsel office in 2004, he and Port Division Assistant Director Ron Carver researched how to leverage the ports' legal authority to reclassify drivers as employees, but did not get "a lot of traction ... in terms of resources" from national labor leaders until the creation of Change to Win in 2005 and the launch of its Strategic Organizing Center. It was there that national campaigns organizer, Nick Weiner, and ports campaign director, John Canham-Clyne, developed a national plan focused on leveraging "potential hooks" for organizing port drivers at five ports in friendly political jurisdictions, including the port complex at Los Angeles-Long Beach. That plan enlisted the help of labor lawyers from progressive unions, like the Service Employees International Union, and well-known labor law firms, like Davis Cowell & Bowe in San Francisco, to develop a legal framework for using the ports' concession authority to convert drivers into employees. Thus, by 2006, the national labor movement had a working theory to advance a concession-based model of employee conversion and union organizing at the Los Angeles and Long Beach ports—but lacked a specific political strategy that identified concrete levers of power or that made any connections to environmental issues.

It was against this backdrop that the opportunity to advance a serious local campaign to achieve employee status for drivers came in an unforeseen package: an environmental legal challenge to port expansion that laid the groundwork for a sweeping environmental overhaul at the ports. This challenge itself was created by sustained community push-back against port growth, amplified by the Los Angeles port's 2001 announcement that it was expanding into adjacent communities to accommodate a massive new China Shipping terminal. In response, NRDC sued on behalf of resident groups under the California Environmental Quality Act (CEQA), winning a 2002 state court ruling that the port had not complied with its duty to identify and mitigate environmental harm resulting from the China Shipping project. As a result, the port was ordered to conduct a new Environmental Impact Report (EIR) and was enjoined from completing an expansion that was already 90 percent finished.

The legal challenge to the China Shipping project—strengthened by the simultaneous effort by San Pedro to secede from the city of Los Angeles—extracted political concessions from local politicians: James Hahn's election as Los Angeles mayor, accompanied by his sister Janice Hahn's election to city council, were built on promises to community groups to mitigate the port's impact. At the height of the China Shipping dispute, Mayor Hahn committed his administration to a policy of "no net increase"—promising to cap port emissions at 2001 levels—which sparked a policy dialogue that led to the 2004 creation of a No Net Increase Task Force, comprised of the NRDC lawyer who had brought the *China Shipping* lawsuit, Gail Ruderman Feuer, along with labor and community representatives who would become central to the Campaign for Clean Trucks. The work of this Task Force laid the groundwork for the next Los Angeles mayor, Antonio Villaraigosa, to initiate discussions in 2005 between the Los Angeles and Long Beach ports around a comprehensive environmental policy, the Clean Air Action Plan (CAAP). It was CAAP, which introduced standards for reducing emissions by diesel trucks, that provided the policy framework and political leverage for the labor movement to finally translate employee conversion from theoretical concept to concrete reform—embedded in a local policy designed to promote environmental remediation.

The pursuit of this end shaped the legal and political advocacy that the campaign pursued. To achieve the Clean Truck Program, campaign leaders combined an outside game of protest and public relations with an inside game focused on alliance formation and lobbying within government policy-making arenas. In Los Angeles, pressure by labor and environmental activists locked Mayor Villaraigosa into a process of genuine port reform, which he strategically advanced through his appointments to the harbor commission and his selection of the port's executive director, Geraldine Knatz, former managing director of the Port of Long Beach, who was a supporter of "greening and growing" and seen as well positioned to build bridges between the Los Angeles and Long Beach ports. With these personnel in place, CAAP was launched to advance green growth through coordinated action at both ports, but without specific attention to implementing employee conversion for port truck drivers.

Coalition advocacy followed a carefully designed script. Because CAAP required its own EIR for approval, it gave the coalition, led by LAANE's Jon Zerolnick along with campaign lawyers, the Teamsters' Manley and NRDC's Adrian Martinez, a formal opportunity to weigh in on its terms. The coalition thus submitted a comment letter that made the case for a "long-term solution" in which the ports would implement the concession model developed by Change to Win—with trucking firms required to comply with "clear standards" to be admitted to the ports, including the promise of "employee status for drivers" and "labor peace." After the CAAP framework was approved by both ports in 2006, the coalition initiated an internal policy process to develop an implementation plan for truck and employee conversion, conducted in coordination with external political and legal advocacy.

It began in Los Angeles with a high-level contact by Teamsters President James Hoffa and a commitment by Mayor Villaraigosa to support the campaign in November 2006. With local official support thus secured, the campaign proceeded through a series of negotiations among industry, labor, environmental, and community stakeholders. Inside pressure on local decision makers was brought to bear by the unions; outside pressure was exerted by community groups as well as ongoing legal challenges by NRDC, most notably around the TraPac terminal expansion at the Los Angeles port. This advocacy created the conditions for policy success, which the mayor's staff reinforced through key decisions: placing the Los Angeles port's Deputy Executive Director of Operations John Holmes—a credible insider supportive of the program with deep operational knowledge—in charge of concession plan development and commissioning the Boston Consulting Group to reanalyze the industry impact of employee conversion—which was essential to convince port officials and city policy makers of the economic viability of the program in response to an earlier economic report raising red flags.

After the ATA litigation invalidated the Clean Truck Program's employee provision, the goals of the labor movement shifted in ways that reshaped its strategy and tactics. The pivot away from local policy reform toward misclassification enforcement required a new power analysis informing a revised advocacy strategy to rebuild legal and political strength. The ascension of prolabor leadership at key enforcement agencies at the federal and state levels provided a transformative opportunity. At the federal Department of Labor (DOL), Obama appointees were supportive of targeting misclassification throughout the economy and agreed to allocate specific resources to port trucking, filing a 2012 lawsuit against Shippers Transport Express seeking reclassification at the Los Angeles and Long Beach ports. In California, the 2010 appointment of Julie Su as state labor commissioner at the DLSE also changed the legal calculus. Su was a prominent workers' rights lawyer whose pioneering litigation and policy work around garment subcontracting connected labor abuse at the production level to retailers with industry power—and thus focused legal attention on abuse at the heart of contracting relationships like those characterizing the drayage trucking industry. Although her role required evenhanded enforcement of state employment law, she could appreciate the systemic barriers to enforcement and use her discretion to allocate resources to spotlight misclassification, which she did by prosecuting notable port trucking claims, like Seacon Logix, and permitting the adjudication of multiple cases against the same company en masse.

Having an effective DLSE enabled a strong play at the NLRB, which was also undergoing a transformation under President Obama that created new possibilities to organize misclassified drivers. The campaign's advocacy centerpiece—the suing-and-striking model—aimed to leverage litigation to win tangible legal victories with economic bite that would contribute to a striking strategy through the NLRB to mobilize collective action by drivers without fear of reprisal. The resolution of the logjam at the NLRB—broken after the *Noel Canning v. NLRB* case limited President Obama's use of the recess appointment power and Senator Harry Reid threatened to eliminate the filibuster if Senate Republicans continued to block NLRB appointments—ushered in a sea change that was an essential foundation for striking success.[18] With a functioning prolabor NLRB and sympathetic regional directors and staff attorneys, the Justice for Port Drivers campaign in 2013 could begin to contemplate a strategy that only a few years before would have seemed unthinkable: directly organizing independent-contractor drivers as misclassified workers whose status in itself constituted an unfair labor practice (ULP) under the National Labor Relations Act (NLRA).[19]

By turning legal weakness into potential strength, suing and striking was innovative on two fronts. First, with respect to the DLSE, the campaign leaders' decision to help drivers file individual claims against target companies sought to transform a system long viewed as inhospitable to individual worker challenges to their employers—to say nothing of system-wide reform efforts like the misclassification campaign. Historically, the problem with the labor commission wage-and-hour adjudicative process was that it permitted employers to undercut collective action by individualizing grievances through bureaucratic procedures that placed significant barriers in front of claimants and thus incentivized low-ball settlement. Claimants, often appearing without lawyers, confronted companies with more resources and sophistication in the process, able to "play for rules" by settling strong claims and thus keeping them out of the appellate process where adjudication on the merits might risk unfavorable precedent.[20] In this regard, earlier studies of DLSE outcomes had shown that workers received far less than they were owed—and even then confronted significant barriers in collecting damages.[21]

The campaign's mobilization of collective driver claims against targeted trucking firms in the DLSE turned all of these disadvantages on their head. The campaign exploited the fact that drivers in the wage claim process, with streamlined timelines and less formal rules, benefited from legal representation but did not require it. Nonlawyer campaign staff, particularly Teamsters organizers, could therefore help educate and advise drivers in the preparation of their wage claims—allowing them to reach more drivers and build the legal foundation for high-volume claims. Campaign staff could then directly assist drivers in the DLSE process when it was effective to do so, while also connecting drivers to lawyers—like solo employment lawyer Stephen Glick and attorneys from progressive labor and civil rights firms such as Gilbert & Sackman, Hadsell & Stormer, and Traber & Voorhees—to add legal firepower in support of individual cases when doing so could make a difference. In this way, the campaign was able to allocate some of the DLSE fact development and driver relations work to nonlawyers, thus reserving lawyer time for arguments and appeals in which legal expertise was crucial.

This approach turned on mobilizing large numbers of claims against targeted companies, increasing liability exposure and building economic pressure, while simultaneously wining favorable legal rulings from the DLSE (and from courts on appeal) that would add weight to the campaign's central misclassification argument—since in order to succeed in the DLSE process the labor commissioner had to decide as a threshold matter that port truck drivers were in fact misclassified employees. With employee status established as a legal fact in the DLSE, the campaign could crossover to the NLRB, where it would have a stronger basis to file ULP charges predicated on the theory that drivers were statutory employees permitted to engage in protected striking. Within the DLSE itself, claims from multiple drivers were deliberately filed together to create a collective action despite the formal individualism of the claims processing. And campaign lawyers found an effective legal hook, deductions for business expenses, which exposed companies to significant damages that far exceeded what would have been available for routine wage-and-hour violations—and which was so unfamiliar to the DLSE that it required campaign staff to create a new form for drivers to document business expense claims.

The role of organizers in the wage claim process also avoided thorny ethical problems by making them part of the legal team covered by attorney-client privilege, thereby allowing them to have sustained conversations with drivers about the option of potentially trading off DLSE damages for agreements by companies to come to the bargaining table and commit to labor peace. In addition, once the DLSE strategy gained visibility and won early success, private plaintiffs' lawyers—particularly Joe Sayas, Matthew Hayes, and the class action firm Kabateck Brown Kellner—started bringing class and mass actions against major trucking firms exposing them to even larger, and potentially ongoing, liability that further ratcheted up economic pressure and added to the weight of favorable misclassification legal precedent.

The Justice for Port Drivers campaign's second innovation was turning liability into strength at the NLRB in ways that produced real organizing power on the ground. The critical legal move was to reject the conventional approach of seeking driver reclassification prior to striking—either as a legislative strategy (as in the Clean Truck Program) or a legal strategy (as in first-wave misclassification lawsuits). Instead, the Teamsters' lawyers tested a new theory: bringing ULP charges directly asserting that drivers were misclassified employees with the right to strike. This was risky for drivers and the campaign more broadly. Any finding by the NLRB that drivers were not misclassified would mean, as unprotected contractors, they could be summarily dismissed, which could destroy the drivers' livelihoods and effectively end the campaign.

To avoid this problem, the campaign's test-case strategy sought to educate staff at the NLRB Region 21 office (covering the ports) about the driver misclassification problem and then to nudge the region, and possibly the full board, to issue a legal opinion affirming that misclassification constituted a ULP. This strategy proceeded by introducing the region to port trucking in a familiar legal context: as part of the Teamsters' traditional unionization campaign against the 100-percent employee firm, Toll, which entered a collective bargaining agreement in 2012. The campaign's next step was to bring ULP charges against a firm, Green Fleet, that had a minority of independent-contractor drivers, thus demonstrating to Region 21 the functional equivalence between employee and independent-contractor drivers as a prelude to making the argument for their legal equivalence under the NLRA. With Region 21 exposed to misclassification and the Teamsters' unionization drive, the campaign next built a pipeline of 100-percent independent-contractor firms to strike (Pac 9, TTSI, XPO, and IBT). In each case, the campaign started by filing traditional ULPs (e.g., for retaliation and threats), asking Region 21 to rule in the drivers' favor on those claims—which required a foundational finding that they were employees with labor law rights in the first instance. This initially happened in February 2014, when the Region 21 director notified Pac 9 of her intent to issue a complaint against the company. With this strong signal of legal support in hand, the campaign took the next step in building the legal basis for a direct misclassification-as-ULP claim: seeking a legal opinion in the Pac 9 case on the misclassification issue from the NLRB Division of Advice, charged with giving guidance to regional directors on novel issues of substantial importance. That opinion, issued in 2015, concluded that "the Employer's misclassification of its drivers as independent contractors interfered with and restrained the drivers in their exercise of [NLRA] Section 7 rights." On the heels of the Division of Advice's opinion, Region 21 issued an amended complaint against Pac 9, the first ever to include the misclassification-as-ULP theory. That complaint, combined with a March 2016 NLRB general counsel memo requiring "Mandatory Submissions to the Division of Advice" of misclassification claims, was the crucial legal authority the campaign needed to advance aggressively against its all-independent-contractor trucking firm targets.

As a tactical complement to asserting misclassification to authorize ULP striking by drivers, the campaign also tested the legal boundaries of ambulatory picketing, in which strikers followed struck company trucks to the places they traveled, gradually moving from trucking yards to port terminals in order to demonstrate to industry, port, and city officials the campaign's power to disrupt operations. During the campaign's first two strikes (against Green Fleet in August 2013, and Green Fleet, Pac 9, and American Logistics in November 2013), strikers refrained from picketing at the ports, choosing instead to limit disruption to struck company truck yards. During the November 2013 strike, the Teamsters set up the first ambulatory pickets at the Shippers Transport Express storage yard, where struck company trucks went to pick up and drop off containers for ships at the terminal operated by SSA Marine, Shippers' parent company. Although Shippers called the police to shut down the pickets, the police respected the picket line under legal authority the campaign had shared in advance with local law enforcement. When the Shippers picketing created, in Nick Weiner's terms, "complete mayhem," campaign leaders realized the power they held and starting in the April 2014 strike against the initial trio of major targets—Green Fleet, Pac 9, and TTSI—the campaign extended pickets to port terminals and retail warehouses.

As picketing at the ports grew more frequent and lengthy during the wave of 2015 strikes, the Justice for Port Drivers campaign was able to reestablish political credibility tarnished during the failure of employee conversion in the Clean Trucks Program. That credibility translated into direct political gains, particularly California Assembly Bill 621, which offered "amnesty" from DLSE civil penalties to trucking firms that agreed to reclassify drivers as employees. In addition, the campaign's demonstrated power to disrupt rebuilt political capital enough to motivate city and port officials to become more actively involved in brokering negotiations between the Teamsters and trucking firm targets, helping usher in deals with firms like

TTSI/Eco Flow and reopening conversations about local policy solutions to the misclassification problem.

Success as a Function of Legal Endowment

Because the labor and environmental movements, as a matter of strategy, mobilized around different legal strengths, it was inevitable that those strengths defined the distinctive boundaries of what outcomes each was able to achieve—inviting appraisal of how law affected what worked, what did not, and why. Assessing the relationship between legal mobilization and social movement outcomes has long been one of the most contested areas of empirical inquiry. Scholars have generally focused on the achievement of two types of results: (1) policy reform that advances a social movement's cause, issued either by a court (such as in *Brown v. Board of Education*) or codified through the legislative or administrative process (such as the Voting Rights Act of 1965 or, more recently, the Deferred Action for Childhood Arrivals program created by executive order); and (2) deeper changes in official practices (i.e., the implementation of policy reform), public attitudes, political consciousness, or the distribution of political power or socially valued goods.[22] There is a large body of empirical research on the factors that promote achievement of each set of outcomes and the connection between the two—the classic relationship between "law on the books" and "law in action."[23] Within this literature, there is no consensus on the meaning of "success" and the appropriate baseline against which to measure it.

For any challenge to entrenched power, what is achievable is a function of where a movement begins in terms of its available resources, political advantages, and public support.[24] As the port struggle reveals, it is also a function of initial legal endowments: *What a movement can win both in terms of policy reform and on-the-ground transformation depends profoundly on the strength of the legal tools at hand and the legal barriers to be overcome.* Evaluating outcomes in relation to legal starting points offers an important perspective from which to judge success for the labor and environmental movements at the ports—and to explain why each movement's outcomes diverged when and how they did.

In terms of policy impact and industry reform, one can trace distinct blue and green trajectories through the Campaign for Clean Trucks and Justice for Port Drivers campaign. The seminal policy achievement of the Campaign for Clean Trucks produced mixed results for its labor and environmental coalition partners—a function of its initial legal architecture. The Clean Truck Programs in Los Angeles and Long Beach contained two essential components. The first, adopted in identical form by both ports, required trucking companies to convert their fleets to clean emission vehicles through a progressive ban on old-model year trucks. The financial impact of that ban was lessened by a set of incentives passed shortly thereafter. Each port imposed a $35 per container Clean Truck Fee on shippers and used the revenue generated therefrom to subsidize the "replacement and retrofit" of noncompliant trucks. From this point, the ports diverged on the second essential component with the Port of Los Angeles passing a concession plan in 2008 that mandated employee status for drivers, while the Port of Long Beach's concession plan rejected employee status in favor of simply requiring trucking firms to ensure compliance with existing law. Thus, as enacted, the ports' Clean Truck Programs contained different results for the coalition's environmental and labor partners: environmentalists won publicly subsidized clean truck conversion at both ports as organized labor secured employee conversion only in Los Angeles.

Yet even that more limited victory by labor was short-lived as employee conversion was struck down in the Ninth Circuit and other features of the Los Angeles concession plan were held to be preempted by the Supreme Court. The one silver lining of the Supreme Court's decision was that it did not erode the underlying doctrinal basis for local government authority to address port trucking through the market participant exception to federal preemption—leaving the door cracked open for a new concession policy on more solid legal grounds. But that was cold comfort to the campaign lawyers and activists, particularly those on the labor side, who had worked so hard to get the Clean Truck Program passed in the first instance. It was also a blow to officials at the Los Angeles port, who viewed themselves as caught in a bind: accountable for pollution and other negative externalities, but without complete authority to redress them. Most profoundly, the litigation outcome was a serious setback for port truck drivers, who found themselves additionally burdened: obliged to acquire (mostly through leasing) and maintain new, more expensive low-emission trucks, yet still in the degraded economic position of independent contractors without the financial means to do so.

The Justice for Port Drivers campaign challenged that degradation, tapping into driver discontent over onerous lease terms and increased expenses to successfully mobilize misclassification strikes and bring some industry leading firms to the bargaining table. As this misclassification campaign has proceeded, environmental groups, particularly NRDC, have focused primary attention on challenging the environmental impacts of further port expansion (like the proposed Southern California International Gateway, or

Assessing the Movement

SCIG, near-dock rail project), while keeping pressure on both ports to implement existing clean truck commitments and enhance emission standards going forward. From this perspective, one can characterize the current state of blue-green mobilization at the ports as one in which environmentalists are primarily fighting over implementation after winning significant policy reform, while labor is still fighting to achieve systemic policy reform in the first instance, building toward that outcome through a company-by-company misclassification approach that has yielded some success, though is still a work in progress.

How one judges these disparate outcomes for the labor and environmental movements at the ports is a function of their relation to each movement's legal endowments. Because of the exclusion of independent contractors from the protections of federal labor law, the labor movement from the outset of the campaign was cast in the difficult position of having to use alternative legal regimes as proxies to leverage stronger labor rights. The Campaign for Clean Trucks in particular was an instance in which the Teamsters, lacking legal power because of independent-contractor rules, sought to build on the power that NRDC litigation created to change the union's structural bargaining position—using environmental law as a springboard to surmount the legal barrier to organizing independent contractors. However, once labor movement leaders decided to take advantage of the opportunity that environmental law afforded, they simultaneously became bound by countervailing legal constraints: tethering labor goals to environmental policy in a way that ultimately undermined the market participation argument for employee conversion.

For the environmental coalition partners, in contrast, litigation was a strong tool in the port context to advance emission reduction objectives. The NRDC challenge to China Shipping changed the power dynamic of port growth by demanding greater accountability in port expansion plans. Environmental review, while affecting process not outcomes, could prove costly if the ports had to redo EIRs for failing to consider and mitigate significant impacts. At the ports, environmental litigation had the power to force environmental reform in a context in which two factors were present: first, the polluting industry was fixed in place based on massive up-front infrastructure investment and, second, there was intense interregional competition for the industry's service. Because the ports were geographically bound, but could potentially lose business to other regional competitors, delay and uncertainty were potent bargaining chips. Since litigation could impede port expansion plans and shippers could reroute cargo to other West Coast ports (or eventually through the Panama Canal), the Los Angeles and Long

Beach ports had a strong incentive to mitigate uncertainty to maintain profitability. This incentive gave NRDC negotiating power through lawsuits to enforce environmental compliance. In addition, because environmental review of port development ultimately had to be approved by city council, it provided a means for exerting political influence in that body.

It was political influence that organized labor brought to bear in passing the Clean Truck Program, though on terms that could not ultimately overcome its structural legal deficits. To win the concession plan, labor's argument for driver employee status was firmly connected to the idea of environmental sustainability: without employee drivers, a short-term incentive program might produce clean truck conversion, but over time drivers would be unable to maintain truck quality, thus necessitating another round of public funding. That argument proved compelling as a way of advancing the Clean Truck Program through political channels in Los Angeles. However, it was legally insufficient to survive preemption. The Ninth Circuit characterized employee conversion as an effort to "impact third party behavior unrelated to the performance of the concessionaire's obligations to the Port"[25]—in other words, an attempt to change trucking company labor practices to enable unionization. The court thus did not buy the argument that employee conversion was about environmental sustainability first and unionization second (if at all). As a result, the environmental components of the Los Angeles Clean Truck Program were left standing, while the labor and community provisions were gutted.

This outcome correlated to ex ante legal strength: because environmental law was the most potent weapon in thwarting port growth, mitigating environmental concerns was ultimately viewed by the court as most central to the ports' role as market actors. Conversely, requiring employee conversion (and thereby strengthening labor law) was not seen as market participation, despite the arguments connecting conversion to long-term environmental sustainability. In this way, labor law making was doubly disadvantaged by its relationship to federalism: the weakness of the NLRA regime pushed labor law down to the local level, where that same weakness made it too insignificant to count toward legitimate city market participation—at least in relation to a Clean Truck Program designed around environmental redress.

From this perspective, the policy outcome in the Campaign for Clean Trucks was a product of the deeply uneven playing field on which organized labor sought to advance. The campaign itself spotlighted the high threshold barrier to effective local labor activism in low-wage sectors defined by contracting: specifically, the legal predicate for unionization—employee

Assessing the Movement

status—had to be in place before local legal strategies to facilitate it could work. As labor scholars have pointed out, in industries in which statutory employees already exist, local governments may bargain with employers for labor neutrality agreements in exchange for public benefits in order to facilitate union organizing without running afoul of labor preemption.[26] However, the Campaign for Clean Trucks underscored the challenge of getting to this step in low-wage industries characterized by pervasive independent contracting and thus outside the formal purview of labor law. In such industries, where baseline employment conditions are not established, federal preemption on nonlabor law grounds (i.e., deregulatory transportation rules codified in the FAAAA) can preclude local legal reforms to create employment relationships and thus get to a locally mediated bargaining process in the first place.

As this suggests, the structure of low-wage industries defined by independent contracting creates a formidable double barrier to labor organizing since unions must *first* overcome the legal challenge of transforming the employment status of workers in the industry in order to even create the *possibility* for unionization—which still has to be fought for and won. Because this weak starting position is a product of federal law—and organized labor is too politically disfavored to change that law outright—the labor movement has sought to anchor reform efforts in receptive local government law processes. But doing so subjects labor efforts to the risk of *non*labor-law preemption—underscoring the movement's deeply disadvantaged position in pursuing unionization. Thus, it is *both* the weakness of labor law *and* the strength of nonlabor law (antitrust and deregulation) that converge to erect significant challenges to organizing industries—like port trucking—in which workers are assigned the label of "owner-operators" but live a reality indistinguishable from that of low-wage employees. In such industries, even if (against the odds) efforts to change local policy to reclassify workers as employees manage to succeed, they do so merely by restoring workers back to the baseline of being *organizable*—aligning their legal label with their lived reality—and thus removing extant legal barriers that now preclude even the possibility of demanding better conditions.

The Justice for Port Drivers' misclassification campaign has endeavored to get out of this difficult box by bypassing the need to establish employee status *before* initiating collective action. Thus far, this strategy has proven legally potent: resulting in $35 million in DLSE awards against port trucking firms and extending protection for misclassified drivers in NLRB Region 21—with important cases still pending against XPO, IBT, and the Yoo family firms (QTS, WinWin Logistics, and LACA Express). This legal success has resulted in tangible union advances: collective bargaining agreements at Toll (2012), Eco Flow (2016), and Shippers Transport Express (2015); and labor peace agreements at Green Fleet (2015), Pac 9 (2016), and Gold Point (2016). Significant as this is, there have been ongoing challenges in converting labor peace into collective bargaining agreements. Most prominently, although Pac 9 struck a labor peace agreement with the Teamsters in May 2016, the company remains in bankruptcy, where uncertainty over its future has produced a frustrating legal limbo that has hindered progress on organizing.

In addition to tangible organizing outcomes, the campaign's misclassification strategy, by combining suing and striking to disrupt port options, also builds a legal foundation for a potential labor peace policy at the ports that responds to the legal deficits that undermined the Clean Truck Program. It does so by seeking to convert the legal vulnerability of employee conversion as a mechanism to address *environmental harm*, and thereby reduce costly environmental litigation, into a legal asset as a way to avoid costly *labor strife*. In this sense, the misclassification strategy is geared toward surmounting the preemption hurdle by repositioning employee status for drivers within the space for market participation that the Ninth Circuit and Supreme Court left open in the ATA case. By firmly linking driver misclassification to the problem of labor unrest, any new labor peace policy would thus seek to avoid the legal weakness at the heart of the campaign's loss of employee conversion in the ATA litigation: that it was disconnected from the port's interest as a market participant in environmental peace. Whether this strategy can withstand a new presidential administration committed to probusiness policies and a newly fortified conservative majority on the Supreme Court is the critical question now confronting the campaign—one that underscores the ongoing vulnerability of organized labor in American politics.

Progress as a Function of Legal Opposition

Where the protagonists of port struggle ended up was not simply a function of how law structured legal endowments—but how industry opponents mobilized them to contest movement progress. Social movement challenges by definition oppose existing arrangements of power. The success of those challenges is thus affected by how strong and skillful are the opponents, who advance their own legal and political strategies in ongoing cycles of contention.[27] Countermobilization, or the threat thereof, shapes social movement strategy by forcing movements to stake out legal positions in relation to the arguments being made by adversaries.[28] Accordingly,

Assessing the Movement

the legal advantages that accrue to those adversaries, and how they choose to deploy them, affect the chances for movement success and therefore how one evaluates movement outcomes. In the port context, the trucking industry's structural legal position vis-à-vis the labor and environmental movements permitted it to mobilize different legal strategies to impose differential costs on movement coalition partners: exploiting organized labor's disfavored legal position to more aggressively resist employee conversion and unionization, which industry viewed as existential threats.

For port trucking firms in Los Angeles and Long Beach, legal conditions at the outset of the Campaign for Clean Trucks were the mirror image of those confronted by organized labor—and thus placed industry in a favorable starting position. The legal playing field that disadvantaged labor mobilization empowered industry countermobilization. Just as the structure of local government law and its relation to preemption placed limits on coalition efforts to reform port trucking, it facilitated industry resistance and, ultimately, industry success in overturning key elements of the Los Angeles concession plan. Industry began from a position of strength—seeking to protect the legal and economic status quo against change. Its strategy from the outset was to defend against structural market reform and thus keep labor in the posture of litigating individual misclassification suits, which the industry could defend in a war of attrition.

Politically and legally, industry sought to sow division in the Campaign for Clean Trucks where the coalition attempted to build unity. When industry could not stop the Clean Truck Program in Los Angeles, it ramped up lobbying pressure on Long Beach and succeeded in defeating employee conversion there. The Los Angeles-Long Beach split paid political and legal dividends. Politically, it heightened inter-union division, further undercutting the support of already unionized longshoremen by increasing the risk that the Los Angeles program would divert cargo to neighboring Long Beach—or other West Coast ports. The split also posed a challenge to Los Angeles officials who confronted the prospect of lost economic benefit to the city and the real chance of political failure. This concerned Los Angeles city officials, who feared an outcome in which the blue-green goals of the Clean Truck Program would be only partly realized. If employee conversion ultimately failed in court, there could be environmental victory and labor failure. Although the administration fought against this outcome, the possibility was embedded in the structure of the policy, which contained legally severable provisions allowing the overall plan to remain in effect even if individual provisions were invalidated in court. In the end, the ATA exploited this structure to focus its attack on employee conversion while professing support for clean trucks. The split between the two ports therefore produced legal benefits for the ATA: sharpening the legal focus on employee conversion (and its connection to unionization), facilitating the ultimate legal settlement with Long Beach (thus allowing the ATA to direct its full resources to defeating the Los Angeles plan), and adding weight to the ATA's preemption claim (by revealing the danger of inconsistent port regulation).

In similar fashion, industry also aimed to weaken the ties binding the labor and environmental coalition partners—using litigation to promote the breach it could not achieve during the policy campaign. This divide-and-conquer strategy was enabled by the industry's favorable legal starting point. Labor and environmental partners had to overcome substantial legal uncertainty weighing against their core argument that local government had authority to enact joint clean truck and employee conversion as a coherent—and intrinsically connected—market participation strategy. Recognizing this vulnerability, industry pushed against the weakest part of the coalition's argument: that employee conversion was central to the port's proprietary interests in mitigating environmental harm. And industry skillfully pitted the coalition partners' ultimate interests against each other. Although the labor and environmental movements stood firm in their commitment to the integrity of the Los Angeles Clean Truck Program as an integrated package of reforms, their constituencies ultimately had distinct aims. Environmentalists wanted emission reductions while organized labor sought unionization. The industry plan was to drive a wedge by effectively conceding on emissions in order to thwart unionization. This plan was advanced through the industry's litigation decision not to challenge the progressive dirty truck ban and its public relations strategy repeatedly linking the concession plan to the Teamsters. Although NRDC lawyers steadfastly defended the environmental-labor linkage, they could not convince courts to agree on the connection between employee conversion and environmental sustainability. In the end, whereas NRDC litigation brought the coalition together, ATA litigation succeeded in dividing the coalition's achievement apart.

As the blue and green partners moved onto separate tracks after the ATA litigation, countermobilization persisted, but in different forms. On the environmental side, the most significant opposition has come from shipping lines, which cut side deals with the ports to defer or turn a blind eye to ongoing noncompliance with green mandates from CAAP and the 2003 *China Shipping* settlement—sensationally revealed in the 2015 *Los Angeles Times* exposé of the Los Angeles port's decision to suspend China Shipping

Assessing the Movement

requirements after the Great Recession. Resistance has also come from the ports themselves whose continued investment in expansion projects like the SCIG have reopened old community wounds.

In response to labor's Justice for Port Drivers campaign, trucking firms have been able to fall back on the familiar strategy of aggressively defending against lawsuits, which consume time and resources, in an effort to wear the campaign down. In addition, as the campaign has moved beyond individual and class litigation toward an integrated suing-and-striking model that leverages public enforcement resources, industry opponents have initiated strategies to protect themselves against economic liability, while undercutting the campaign's newly won legal power. This opposition strategy has taken three forms: attempts to revise lease agreements to legally solidify drivers' independent-contractor status; the use of bankruptcy to protect against litigation exposure; and affirmative litigation against government agencies and campaign organizations to thwart suing-and-striking efforts.

Trucking lease terms, in which drivers are formally treated as independent owners but required to follow detailed rules, such as those prohibiting drivers from moving freight for other companies, have provided documentary evidence of company "control" used by courts and the DLSE to establish misclassification. In response, industry lawyers have sought to revise leases in two crucial ways to protect companies from misclassification liability. First, lawyers such as those at the industry-side Marron law firm have promoted "Fundamental Model Change" to help companies avoid ensnaring themselves in the employee trap. To do so, they have advanced a "Co-Operative Alternative Model," in which drivers are sold ownership interests in a cooperative trucking firm, negating their claim to employee status, while the firm continues to be managed by owner representatives. In this way, industry lawyers seek reorganization that slots drivers into the formal category of "owner" to undermine the legal argument for misclassification. Similarly, trucking firm owners have asked some drivers to incorporate themselves as legal entities, such as limited liability companies (LLCs), also to permit firm owners to argue against misclassification on the ground that LLCs cannot be legal employees. Furthermore, companies have increasingly inserted class action waivers and arbitration clauses in new lease agreements to block drivers from pursuing class actions in court or individual actions at the DLSE. The effectiveness of the arbitration clauses, in particular, hinges on establishing that the drivers are independent contractors, since the California Supreme Court has held certain types of mandatory employee arbitration provisions to be unconscionable.[29] Arbitration clauses have become a point of contention between the campaign and target companies in NLRB Region 21, which has issued complaints, as in the 2016 XPO cases, challenging such clauses as ULPs.

For companies already facing significant misclassification liability and thus unable to take advantage of this type of prospective contractual protection, an alternative response to suing and striking has been to file for Chapter 11 bankruptcy. This is the path followed by Green Fleet, Pac 9, TTSI, QTS, and Seacon Logix. Chapter 11 bankruptcy provides firms with some tools to resist litigation and mitigate legal liability, while reorganizing operations and financing under court supervision in ways that permit firms to carry on their business (i.e., it does not lead to dissolution). Specifically, a bankruptcy filing triggers an automatic stay of all litigation against the debtor company, which prevents plaintiff-drivers from making demands for payment or from enforcing judgments already issued until the bankruptcy is resolved. In addition, because plaintiffs in employment litigation are considered unsecured creditors, they are forced to have their claims compromised as part of negotiated settlements with other creditors. In this way, bankruptcy gives beleaguered trucking companies tools to protect themselves against misclassification liability. However, bankruptcy cannot preclude future litigation once a company emerges from bankruptcy and thus also empowers drivers to participate in structuring a resolution that includes a commitment by the debtor company to reclassify drivers to avoid prospective lawsuits. In the ports campaign, three bankrupt firms—Green Fleet, Pac 9, and TTSI/Eco Flow—have come out of reorganization adopting an employee model in exchange for a global settlement of outstanding liability. QTS, in contrast, attempted to use bankruptcy more aggressively to limit its liability in the *Talavera v. QTS* class action by shifting assets to affiliated firms owned by the Yoo family, LACA Express and WinWin Logistics, which remained solvent. In a significant legal victory for the campaign, class counsel succeeded in resisting this maneuver by convincing the bankruptcy court to consolidate the bankruptcy proceedings and class action, thereby permitting plaintiffs to recover against the assets of the combined Yoo family entities.

In addition to these protective strategies, industry actors have taken affirmative legal steps to undercut the power of campaign suing and striking. In one bold effort, companies have filed lawsuits seeking to preclude state enforcement efforts under a theory of FAAAA preemption. This strategy has been used in a number of cases: by Pac Anchor to challenge the California attorney general's 2008 lawsuit against it on unfair competition grounds (a claim rejected by the California Supreme Court); by Shippers Transport Express in the 2012 *Taylor v. Shippers Transport Express, Inc.* class

Assessing the Movement

action in a motion for summary judgment on plaintiffs' state employment claims (denied by the district court); and most recently (and radically) in a suit filed by the California Trucking Association arguing that the entire system of DLSE wage claim adjudication is preempted. There are also growing examples of industry actors using litigation to chill strikes. In response to the 2016 strike against IBT and K&R Transportation, the Law Office of Jewels Jin filed suit on behalf of a group of independent-contractor drivers against terminal operators and the Teamsters for conspiracy to interfere with business operations. Taken together, these efforts augur a mounting legal counteroffensive by the trucking industry to undermine the advantages the campaign has built through DLSE claim making and ambulatory picketing at the ports. They also reflect industry's different treatment of the environmental and labor movements. While environmental subsidies and the promise of greater energy efficiency have made it politically and economically possible to contemplate green initiatives, corporate resistance to organized labor—perceived as presenting zero-sum tradeoffs between worker wages and corporate profits—has remained implacable.

Legal Mobilization in Cross-Sectoral Movements for Local Change

Although the context of the blue-green challenge to port trucking in Los Angeles and Long Beach, and the particular way it has played out, are unique, this study of its history illuminates structural dynamics and patterns of practice that contain higher-order lessons for other movements and for the local governments that respond to their regulatory demands. This section explores those lessons in relation to the three main theoretical strands that have run through the study—on coalition building, lawyering, and labor regulation—asking how the peculiar nature of port struggle deepens understanding of legal mobilization in cross-sectoral movements for local change.

Coalitions in Local Politics

Coalitions are formed as combinations of preexisting organizations that come together to advance a policy reform or organizing goal, usually within a discrete time frame. Social movement scholars have analyzed coalitions primarily in terms of the tradeoff between, on the one hand, the benefits that accrue from coalitions able to mobilize new resources and allocate them efficiently to coalition partners based on a competence-based division of labor, and on the other hand, the costs imposed on movements when coalitions become professionalized and oriented around conventional processes of institutional change that undermine their power to disrupt.[30] Legal scholars, in turn, have thought of coalitions, when at all, primarily in terms of conflicts of interest and client accountability. For a lawyer representing a coalition of multiple organizations, to whom is the lawyer's loyalty owed and how are disagreements managed?[31]

Revisiting blue-green coalition politics at the ports deepens each of these theoretical perspectives in novel ways. First, with regard to coalition formation and the tradeoffs it presents, the port story shows how law can create targeted opportunities for cross-sectoral coalitions to come together around projects of mutual gain, while also presenting challenges to sustaining such coalitions over time. During the Campaign for Clean Trucks, law was a glue that brought the labor and environmental movements together in joint struggle, enabling a division of advocacy based on comparative advantage. Yet, over time, as legal compatibility eroded in the face of countermobilization, law was also a wedge that split coalition partners onto separate advocacy tracks. As that split underscored, efforts by campaign leaders to represent different interests in the blue-green coalition were complicated across the inter-movement partnership and within each movement itself. The challenges leaders faced balancing obligations to labor and environmental stakeholders, and defining how the labor movement's broader strategic interests should relate to its obligations to port drivers, reveal the fundamental difficulty of finding zones of interest convergence among diverse constituencies and staying accountable to those constituencies throughout the contested process of social movement mobilization.

Convergence and Divergence In the Campaign for Clean Trucks, complementary legal and political needs coalesced around a common legal analysis of the preemption doctrine to draw the blue-green coalition together. The campaign offered a moment of labor and environmental interest convergence: the legal framing of the concession plan as a win-win for the labor and environmental movements—simultaneously addressing the twin problems of air pollution and low-wage work at the ports—was the crucial foundation for a coalition building process in which each side had something to offer the other.

Labor leaders saw the environmentalists as bringing legal power and political appeal. Framing the campaign as primarily about cleaning the environment was an important strategic move that provided political and legal benefits for labor: minimizing the salience of unionization in an industry context hostile to that aim, while also emphasizing the threat of

Assessing the Movement

litigation if the campaign failed and thus crystallizing the market participation argument against preemption. The employee provision of the Los Angeles Clean Truck Program was presented as a way to strengthen the green growth project through enhancing job quality for drivers, not to permit an independent mechanism for unionization, an issue on which NRDC lawyers took no official position.

For their part, environmentalists appreciated the local power of the labor movement to move policy and mobilize on the ground, and were eager for an inter-movement alliance that would alleviate tensions over development, which labor unions (particularly the longshoremen) often promoted for the job benefits, but which environmentalists opposed on pollution grounds. In fact, it was during the first meeting in 2006 between Sierra Club director, Carl Pope, and Change to Win leaders to build the national Blue-Green Alliance that the seeds of the Campaign for Clean Trucks were sewn. Pope's mention of the *China Shipping* litigation and CAAP, though "just happenstance," was the spark that caused Change to Win's Nick Weiner to "connect the dots" and see the potential benefits to labor of a local coalition with environmentalists in Los Angeles to address port trucking. What labor gave to environmentalists was also critical. Although NRDC lawyers could threaten additional port slowdowns, they needed proactive policy reform that would guide future port development toward sustainable green goals. CAAP promised that and, along with structures like the Port Community Advisory Committee and No Net Increase Task Force, provided critical institutional frameworks connecting activists and putting pressure on local political officials to come up with genuine reform. But the details of CAAP were vague and environmentalists' influence in city hall limited. Folding employee conversion into the Clean Truck Program thus created the opportunity for stronger environmental policy—leveraging organized labor's interest in driver status for a more aggressive attack on air pollution and comprehensive drayage fleet conversion. Although there were important local environmental and community organizations working on port trucking, none had the staff and political connections to mount a decisive policy campaign on their own. Change to Win's entry on the local scene and collaboration with LAANE altered that equation. Change to Win provided money and staff to craft a concession model linking air pollution to employee status and to mobilize the formidable local labor movement political machine. Supported by Change to Win funding and expertise, LAANE became the organizational control center of the campaign—able to draw upon a store of local contacts and internal experience built upon two decades of policy work in Los Angeles.

A shared legal interest in supporting local power to craft progressive policy reform within the crevices of federal preemption doctrine also cemented the blue-green partnership at the heart of the Coalition for Clean and Safe Ports. Because the political climate was inhospitable to progressive policy change at the federal level, even during the Obama presidency, the labor and environmental movements promoted state and local policy experiments to advance goals that could not be achieved nationwide. In the early 2000s, on the cusp of the Campaign for Clean Trucks, there was limited legal precedent guiding local lawmakers on the scope of what they could do legislatively to affect labor and environmental regulation in ways that departed from federal standards. Although the Supreme Court had held federal labor law did not preempt active city participation in supporting project labor agreements (pre-hire collective bargaining agreements between developers and unions),[32] the court had given ambiguous signals about whether local governments could regulate emission standards under the market participant exception to Clean Air Act preemption,[33] and had not weighed in on the scope of FAAAA preemption of local action around port trucking. In this murky legal space, environmental and labor lawyers had incentives to interpret the doctrine of market participation broadly to support local policy, like the Clean Truck Program, that sought to enhance labor and environmental standards. This common doctrinal analysis was a key factor that solidified the coalition. By developing this common analysis, Teamsters and NRDC lawyers played an important brokerage role, helping to bridge gaps in social movement networks and thereby link together labor and environmental stakeholders in order to facilitate campaign development.[34]

Ultimately, however, the legal and political interests of the blue-green alliance were not completely aligned. The legal tensions that were baked into the coalition at the outset emerged as fissures as the policy campaign resulted in mixed success. Environmentalists could claim victory with a program that banned dirty trucks and promoted the acquisition of new clean vehicles, while labor needed employee conversion to win. The linked package of Clean Truck Program reforms—the dirty truck ban, financial incentives for clean truck conversion, and a concession plan with employee conversion—constituted a political compromise that deftly advanced all stakeholder interests. However, when that compromise was pulled apart in court, the tight legal bond holding the coalition together was also weakened.

After this occurred, each side of the coalition responded by pursuing parallel agendas. Labor leaders from the Teamsters and Change to Win did not want to see their massive investment of political and financial capital produce no gain, but understood that the path forward had to proceed outside of the blue-green alliance—at least in the short term. This decision was the product of political expedience and legal calculation. Politically, the Justice for Port Drivers' misclassification strategy focused singularly on mobilizing traditional employment and labor law to protect the rights of misclassified workers, which caused the coalition's labor partners to break away from their environmental counterparts. Legally, the Ninth Circuit's decision in the ATA case made clear that for organized labor to pass local policy requiring employee reclassification, it would have to detach its rationale from environmental remediation and instead make an independent case for avoiding preemption on the basis of the ports' proprietary interest in reducing the costs of striking.

On the environmental side, the post-Clean Truck Program calculus was quite different. Empowered with a new regulatory tool—clean truck conversion at both ports—the goals of environmental partners shifted toward enforcement. In the face of competitive pressures to retain big shipping lines, the ports' own disinclination to press shippers and terminal operators hard on green mandates motivated environmental efforts to bring state regulatory agency resources to bear on the enforcement of port air-quality standards. This has been seen in the California Cleaner Freight Coalition's attempts to shift port regulatory oversight to the South Coast Air Quality Management District and California Air Resources Board. In a move that paralleled labor's own pivot toward sympathetic agencies, NRDC and its environmental justice partners have promoted near-zero-emissions policies at the ports as part of state and national climate commitments, while maintaining the pressure on local authorities to completely eliminate diesel trucks from the ports through the update to CAAP. In addition, NRDC's pending lawsuit against the SCIG near-dock rail project highlights that organization's ongoing commitment to using the courts to disrupt port development when needed to protect air quality and local communities. That lawsuit, in which NRDC is joined in litigation against SCIG by trucking firm Cal Cartage, whose warehouse is on the proposed SGIG site—and which remains a central target in the Justice for Port Drivers misclassification campaign—also underscores the legal divisions that have pulled the coalition apart since the Clean Truck Program.

Representation and Accountability The strange-bedfellows aspect of the SCIG litigation—with NRDC aligned with the Teamsters' bitter foe—illustrates another challenge of coalition politics over time: the changing nature of representation and the complexity of constituent accountability in ongoing cycles of social movement conflict. The concept of accountability holds that leaders who purport to speak on behalf of a social movement constituency, particularly one that is structurally disadvantaged, have to answer for decisions to those they represent. This concept is fundamental to the way that legal scholars think about the tradeoffs of social movement lawyering, but has not figured prominently in social movement scholarship more generally, despite a growing interest in leadership. Legal scholars, focused on lawyers' fiduciary obligations to clients, have expressed concern over what legal ethicist David Luban called the "double agent" problem: that lawyers committed to advancing particular causes could have their loyalty divided between what was best for the client and what ultimately served the cause.[35] Harvard Law School professor Derrick Bell famously made this controversial charge against NAACP lawyers advancing school desegregation after *Brown*, arguing that they were "serving two masters": beholden to their own (and their elite funders') "integration ideals" at the expense of their clients, black parents who wanted a quality education for their children, whether integrated or not.[36] In Bell's powerful words: "Idealism, though perhaps rarer than greed, is harder to control."[37]

The port struggle broadens Bell's perspective by spotlighting *the challenge of accountability among social movement leaders*—lawyers and nonlawyers alike. Although lawyers have special obligations to clients, the problem of serving multiple masters is inherent in contested social movement campaigns—made even more difficult when leaders seek to forge unity among competing interests in cross-sectoral coalitions. Looking back at the Campaign for Clean Trucks and the Justice for Port Drivers campaign shows how leaders struggled to bridge differences, promote stakeholder participation, and make critical decisions with serious consequences for drivers; and how, even with the best intentions, there was never an easy way to make uncomplicated, strictly accountable choices to any specific constituency with monolithic interests. For labor leaders seeking a vehicle for organizing supply chain workers, their approach was to establish good processes and exercise judgment in ways that promoted what they viewed as the best interests of drivers in Los Angeles and Long Beach, while also advancing the larger goals of the national labor movement. Considering how well they did so, and what consequences flowed from their choices, is a critical dimension of outcome evaluation.

The coalition that would guide the Campaign for Clean Trucks was forged at the intersection of top-down planning and bottom-up resistance, and reflected a genuine effort to promote inclusion and participation among a broad range of stakeholders affected by the port trucking industry. From the top-down, Teamsters and Change to Win leaders developed a campaign plan, reached out to LAANE to host the local campaign staff, and hired the LAANE port team (Jon Zerolnick and Patricia Castellanos), who then undertook extensive outreach to local union leaders, NRDC lawyers and representatives from other key environmental justice groups, like East Yard Communities for Environmental Justice, as well as public health groups, like the Long Beach Alliance for Children with Asthma. Some of these environmental and public health organizations were already involved in bottom-up advocacy—led by the Sierra Club, East Yard Communities, and the Coalition for a Safe Environment—which had sought a larger voice in harbor commission governance and came together in formal networks, like the Green LA Port Working Group, created after Mayor Villaraigosa's election to develop plans for port environmental cleanup and CAAP development.

In this way, the Coalition for Clean and Safe Ports brought together national labor leaders focused on the independent-contractor problem with local organizations fighting against the environmental and community impact of port expansion. As it formally convened in 2006, the coalition operated as a group of representatives from these diverse groups. The coalition did not formally include drivers themselves, whose interests were represented by the Teamsters. Decision making processes were established in democratic fashion: major decisions required a supermajority of all coalition members (though were typically by consensus); policy development on issues like truck routes and parking was assigned to subcommittees staffed by members with relevant expertise; time-sensitive strategic issues were delegated to an elected steering committee, whose members thought of themselves as representing the campaign, even as they also worked to ensure that their particular group interests were heard. Some coalition members acknowledged that organized labor, because of its political clout, and LAANE, because of its full-time staffing, had more power to set the agenda and influence strategy. As one coalition member put it, LAANE was "really in the driver's seat." This power sharing arrangement worked well as the interests of labor, environmental, and community groups aligned around the finely wrought package of changes in the Clean Truck Program. When that package unraveled in court, however, it focused attention on how the loss was distributed—and particularly whether the decision-making process adequately addressed the potential for harm to drivers who, without employee conversion, were saddled with the costs of clean trucks without the benefits of employee status.

The problem, from labor's perspective, was how to evaluate the legal viability of the Clean Truck Program in relation to the ambiguous boundaries of preemption—and how to pursue the program in a way that served the interests of port drivers while also advancing the labor movement's broader ambitions for organizing supply chain workers and rebuilding its national power to promote workers' rights. The decision to advance employee conversion as part of an integrated Clean Truck Program, although resting upon legal analysis, was ultimately one made by top campaign leadership, which included key nonlawyer members of the campaign team from the Teamsters, Change to Win, and LAANE. The crucial legal prediction centered on the risk of litigation failure—specifically, the risk that the program in general and the employee conversion piece in particular would be struck down. Movement lawyers from the Teamsters and NRDC believed that the ports would win a FAAAA challenge to the Clean Truck Program at the Ninth Circuit and that the Supreme Court would not grant certiorari. This prediction turned out to be wrong: only the concession plan was challenged, the Ninth Circuit struck down the critical employee conversion provision, and the Supreme Court decided to consider, and reject, remaining concession provisions.

At one level, this outcome, while disappointing for the campaign, is analytically mundane. Legal predictions turn out to be wrong all the time, in part because law is indeterminate and judges have significant discretion, in part because lawyers and their clients are fallible. However, the nature and process of legal analysis in relation to the Clean Truck Program was itself not mundane. As part of a social movement-led local policy-making campaign, the movement lawyers drafting the preemption analysis were not representing drivers directly, but rather were representing their own organizational interests. And the stakes were high: the legal judgment about preemption affected passage of a significant social policy affecting a critical national industry. Although legal judgments from government lawyers routinely precede the enactment of significant policies, a unique feature of the Campaign for Clean Trucks was the extent to which movement-generated doctrinal interpretation informed internal government legal analysis and political decision making. Understanding the contours of that decision-making process and how it misjudged the legal outcome, then, presents the accountability question most directly: Given the risk of error, was analysis and decision making sufficiently accountable to the group upon whom error would fall most heavily?

Coalition leaders knew that there was a small, but nontrivial, risk that if clean truck conversion passed without employee conversion, drivers could be left worse off: remaining independent contractors forced to bear the cost of acquiring and maintaining more expensive clean trucks. In a context in which the prize was so big (trucking industry transformation with the possibility of organizing the entire logistics supply chain) and the legal analysis suggested that the risk of losing was so small, coalition leaders decided to pursue the policy in the face of risk. Because drivers were not independent members of the coalition, there was no mechanism for them to formally weigh in on the risk, though it is reasonable to believe that those who were active in the campaign would have agreed with the coalition's analysis and thus supported the policy. On the ground, organizers promoted the promised benefits of the Clean Truck Program to drivers without engaging in deep analysis of potential costs. This was, undoubtedly, because the campaign was so invested in playing to win—and believed winning was likely—that it did not want to dwell on the complexities of what it would mean to lose.

Yet, even with activist-driver support of the plan, the accountability problem in the Campaign for Clean Trucks was a thorny one. The activist drivers were only a fraction of the total port trucker population and could not claim representative status (and, indeed, there were challenger groups, like the National Port Drivers Association, which purported to represent thousands of owner-operators who wanted to maintain that status). Without organizing and polling all drivers, coalition leaders had to rely on the support of those drivers most committed to the clean truck project.

Moreover, the nature of the coalition itself—while strengthening its claims to broad stakeholder representation—complicated its members' representative roles. Environmentalists were committed to emission reduction and not in a position of direct accountability to the drivers. Although the Teamsters' constituency, broadly defined, included port truck drivers, the drivers were not union members—only potential ones. The Teamsters and their allies at Change to Win thus had to consider the impact of a successful Clean Truck Program not just on port truck drivers, but the union movement more generally. And this was a judgment ultimately made by labor movement leaders, *not* the movement lawyers, who were in a more conventional position of general counsel to the unions—thus shifting the crucial representational choices to the client level. In the end, labor movement leaders decided that the campaign was strategically important enough (with significant potential benefits to a much wider universe of workers), and the downside limited enough, that it was worth the risk—that to change the equation for organized labor in the United States, it was necessary to seize the once-in-a-generation chance to go big at the ports. Once the political decision was made to pursue this chance in tandem with clean truck conversion, the potential conflict with driver interests was unavoidable: If the total package succeeded, drivers would benefit, but if employee status failed, the largest cost would fall on the constituency members with the most to lose and the weakest voice in movement decision making. That it ended precisely so underscores the unequally distributed risks of movement ambition and the challenge of ever neatly resolving the dilemma of "serving two masters" in complex and high-stakes campaigns for transformative social change.

Had organized labor's gamble for employee conversion paid off, its claim to represent port truck drivers would have been put to the ultimate test in the unionization drive that ensued. Although not exactly what labor leaders had envisioned, the Justice for Port Drivers misclassification campaign has nonetheless sought to make the case for representation through its suing-and-striking strategy. While campaign leaders have not had to directly navigate the complexity of blue-green collaboration, they have continued to confront the dual challenge of representing the interests of disaffected drivers and the broader labor movement. As this strategy has evolved, there have been points of convergence and ongoing tensions. Interest convergence has been strengthened through the campaign's clinic-based organizing. Many aggrieved drivers, "rubbed raw" by mistreatment, have arrived at the campaign's legal clinic already activated to pursue change. In a process that echoes successful worker center advocacy in other contexts,[38] campaign organizers have counseled drivers on their entitlement to legal recovery for misclassification, while also educating them on the connection between the pursuit of monetary redress for past wrongs and the campaign's efforts to change the system going forward. While campaign staff has helped drivers through the initial process of assembling and presenting their claims to the DLSE, in the words of campaign organizer, Carlos Santamaria, the campaign has also emphasized that wage claims are leverage "to negotiate things that are worth more than money." Thus, when lawyers sympathetic to the campaign have agreed to help drivers in the DLSE process and on appeal in court, they have taken on driver clients who have a clear understanding of how their wage claims relate to the organizing goals already established.

This process of driver outreach and representation raises issues that have been well-rehearsed in legal scholarship: particularly, the possibility of coercion both in terms of pressuring drivers to accept organizing goals as a quid pro quo for legal assistance and in terms of steering them toward

Assessing the Movement

settlements in which they are asked to sacrifice more money than they might be inclined to in the short term for the uncertain promise of longer-term economic benefits resulting from unionization. Yet, at least with respect to the pursuit of major campaign targets, driver involvement has been informed and empowered, resulting in settlements—in the cases of Green Fleet, Pac 9, and TTSI/Eco Flow—that strike a balance between drivers' personal interests in recovering for past harms and the future economic benefits of labor peace (Green Fleet and Pac 9) and collective bargaining (Eco Flow) agreements. Moreover, the campaign's striking strategy, which has won legal protection for misclassified drivers under labor law, has demonstrated to drivers that campaign leaders are taking care to minimize the risks of collective action, which has helped them accrue tangible economic benefits and political power.

The interface between campaign organizing and large-scale private lawsuits against campaign targets has been more jagged. Those suits, brought almost entirely by private plaintiffs' lawyers—some sympathetic to the campaign's goals, others singularly focused on securing drivers' monetary damages—have created tensions at the border of litigation settlement and union organizing. There have been three types of lawsuits so far. First, from the campaign's perspective, the most positive case—Shippers Transport Express—is also the most unique. With Shippers, there were two parallel and ultimately intersecting lawsuits: one brought by the DOL seeking injunctive relief in the form of employee reclassification and the other, the *Taylor* class action, brought by small firm lawyers Matthew Hayes and Joe Sayas, who campaign leaders viewed as sympathetic to campaign goals. Because the two separate suits allowed lawyers to simultaneously press for two separate remedies—reclassification and damages—without either legal team having to worry about Shippers seeking to directly trade one against the other, the suits could work in tandem to advance both remedies without internal conflicts. Particularly since the DOL was not seeking to recover fees, it was under no pressure to ensure a monetary recovery. Damages could be the singular focus of the *Taylor* class action suit. This compatibility put significant pressure on Shippers, which acceded to a settlement with the DOL in which it agreed to reclassify drivers in 2014, paving the way for a collective bargaining agreement in February 2015 and an $11 million class action settlement four months later.

This outcome was in contrast to what occurred in the second type of lawsuit—the only case directly sponsored by the campaign, *Talavera v. QTS*, brought by a group of lawyers led by the nonprofit Wage Justice Center. Despite the campaign's effort to bridge legal and organizing strategy through formal involvement in the lawsuit, that case became bogged down by QTS's bankruptcy and, although drivers have won nearly $4 million in DLSE awards against Yoo family trucking firms and ultimately negotiated a $5 million settlement in the *Talavera* class action against QTS, the campaign has conducted only one strike and made little progress in negotiations.

The third type of private lawsuit is the most common—and most ambiguous from an accountability perspective. In it, unaffiliated private firm lawyers bring large-scale cases against campaign targets (sometimes more than one case against a single trucking firm), which proceed largely on a separate track, without much, if any, communication between campaign staff and private counsel. In two such cases, there have been successful legal settlements and union agreements. In Pac 9, parties to the class action brought by the Kabateck firm agreed to settle in late 2015, which was followed by Pac 9's bankruptcy and then a labor peace agreement in 2016. In the 2012 Gold Point class action filed by James Hawkins, lawyers moved for approval of a settlement in 2017, after Gold Point had agreed to labor peace. Other cases have produced mixed results. There have been several competing actions filed against XPO affiliates by the Kabateck firm, the Gomez Law Group, and Harris & Ruble—only one of which (*Molina v. Pacer Cartage*, a 2013 class action brought by the Kabateck firm) has reached settlement. Not only have there been no labor agreements with the XPO firms, there has been open tension between the campaign and lawyers in the XPO cases—highlighted when some of the drivers in *Molina*, including the lead plaintiff, opted out of the class action over disagreements about the lawyers' failure to include reclassification in the settlement proposal. In other signs that plaintiff-side lawyers are pursuing monetary recovery disconnected from the organizing campaign, the Gomez Law Group has brought mass action suits on top of already-filed class actions against XPO Cartage, IBT, and Gold Point—the latter after the labor peace agreement with the Teamsters was already signed.

As these complex cases highlight, the labor movement's claim to represent workers with tangible interests and sometimes live legal cases distinguishes it strongly from the environmental movement, in which lawyers at places like NRDC are able to bring cases on behalf of their organizational members and partners in defending the environment. During its involvement in the Coalition for Clean and Safe Ports, NRDC lawyers represented its own members, not the coalition directly. Similarly, when NRDC intervened in the ATA litigation over the Clean Truck Program, it did so on behalf of its members' interests, not the coalition's. Although, in cases like *China Shipping*, real people's lives were at risk because of port pollution, those people were not seeking parallel monetary redress and to the extent they were mobilizing against port expansion, the lawsuit—by virtue of the power afforded under environmental law to delay development—served to strengthen organizing efforts. As such, environmental litigation did not create the same sort of client-constituency conflicts as did litigation on the labor side. And to the extent that environmental justice advocates have historically criticized the failure of elite, predominantly white environmental lawyers to protect low-income communities of color against disproportionate environmental impacts,[39] the ports campaign offered an opportunity for solidarity and even redemption. Environmentalists and environmental justice activists could rally together against a common foe, building the foundation for sustained collaboration around green ports locally and clean freight nationally.

Lawyering in Local Movements
The concept of legal mobilization in social movements has received significant treatment over the past decade, as scholars have provided diverse accounts of how lawyers and activists use law in struggles for social change.[40] Legal mobilization theory treats law as multifaceted and contested: not simply what judges say in court opinions, but a set of ideas about what is just that can be used by lawyers, activists, and people on the ground to challenge unfairness and mobilize around collective aims.[41] Litigation is the most recognizable form of legal mobilization, but not always the most effective, since it can divert resources from more powerful political action or narrow collective demands around concepts of individual rights.[42] Yet, in some cases, litigation is essential to hold more powerful actors to account, impose costs on adversaries, or publicize wrongdoing; rights can also be used by social movement actors as critical rhetorical tools to motivate individuals to take collective action around common mistreatment.[43]

Lawyers are useful but not always essential to mobilize law. Nonetheless, particularly in the context of large-scale social movement efforts, like the ports campaign, lawyers tend to figure in prominent roles. Whether they do more good than harm has been a key focus of law and social science scholarship, which has concentrated on the accountability problem raised by Bell, earlier, as well as concerns about social movement demobilization, cooptation, and backlash.[44] In response to these concerns, recent legal scholarship has explored the role of "movement lawyers," who are deeply embedded in social movement contexts, where they collaborate closely with nonlawyer activists to coordinate legal and political strategy on behalf of mobilized groups in a way that is designed to ensure accountability and enhance their power to make change.⁴⁵ Within this literature, stories of movement lawyers deploying multidimensional strategies for groups advancing marriage equality, workers' rights, and immigrants' rights (among others issues) have received considerable attention.⁴⁶ Yet most of these stories continue to foreground litigation-centered models of legal mobilization and emphasize policy reform battles at the national level. The history of struggle at the ports contributes to this conversation by providing a unique perspective, over a sustained period of social movement contention, on how the nature of cross-sectoral movement mobilization at the local level shapes the role of lawyers—on the inside and outside of movements—and their use of law.

In the port struggle, movement lawyers with deep experience and strong ideological commitments to their respective causes provided strategic guidance throughout, while lawyers outside the labor and environmental movements played more discrete roles by virtue of their specific expertise or access to resources. The approach of movement lawyers in the campaign—particularly those from the Teamsters and NRDC—was influenced as much by practice background and professional ideology as it was by the specific local context of port struggle. Teamsters' lawyers Mike Manley and Julie Gutman Dickinson thought of themselves as active problem solvers dedicated to using their legal expertise to promote labor organizing and worker empowerment first and foremost. Gutman Dickinson explicitly understood her professional identity in terms of "lawyering for social change," in which law was a tool used in combination with grassroots organizing, research, and policy work in "looking at the bigger problem and analyzing how to solve it." Similarly, NRDC's Morgan Wyenn viewed her role in broad terms, stressing that "policy and regulatory advocacy is another big tool that we use a lot"—"not always just litigation."

Although professional ideology shaped the background choices that movement lawyers made about who to represent and how, the specific nature of their advocacy in the port struggle was influenced by the local context in which it was undertaken. What this meant varied over time and space, but the book's account of the campaign's arc reveals a trajectory of legal mobilization that tracks the movement's effort to create local policy, defend it from attack, and then reformulate strategy to challenge misclassification head on. Retracing this trajectory illuminates important, but often hidden, roles for movement lawyers as campaign strategists, behind-the-scenes legal technicians, and advocates for policy reform. Historical and contemporary accounts of movement lawyering focused on litigation fail to explore many of these roles, and when they do, the roles are not linked together in relation to an overarching program of local movement mobilization. Responding to this gap, this section offers a framework for understanding what precisely it is that lawyers do "for, and to" social movements that seek progressive local policy reform to improve conditions for marginalized community members.[47] It spotlights five essential roles—disruption, validation, protection, authorization, and inspiration—that are described as follows.

Disruption Throughout the "Hundred Years' War" between the ports and communities challenging its expansion, there was a steady-state equilibrium. The ports, supported by cities eager for revenue and corporations eager for profit, would advance capital projects that incurred progressively on adjacent communities, which were treated as adjuncts of the ports themselves. Those communities would complain, attend port meetings, even threaten to secede, largely to no avail. The *China Shipping* lawsuit disrupted this equilibrium by asserting the real possibility that future development would become embroiled in lengthy legal dispute. This disruption produced a scramble for a new equilibrium that environmentalists would support, which led to the creation of the policy process that resulted in the Clean Truck Program.

Disruption in this sense echoes the concept of "destabilization rights" pioneered by Roberto Unger and applied by Charles Sabel and William Simon to the public law litigation context.[48] The idea in both cases is that the status quo ante is no longer tenable because of legal intervention and new arrangements are necessary to reestablish functional institutions. As Sabel and Simon describe it, destabilization can draw greater attention to social problems, lead to more open systems amenable to policy experimentation, and encourage a broader range of stakeholder participation since the "balance of power" shifts toward challenger groups.[49] Upsetting the status quo in this way may produce an opportunity for innovative solutions, less rooted in old regulatory paradigms and more flexible in adapting to change.

The response to *China Shipping* in the ports case follows this model, but also departs from it in important ways. The disruptive effect of *China Shipping* was essential to give leverage to environmentalists and community residents to negotiate emission standards from a stronger position with both ports. CAAP itself reflected values of policy experimentalism—particularly stakeholder participation and flexible rule making with built-in opportunities to revisit and revise—but it also created a framework in which impacted social movements could exercise power. Within the initial CAAP platform, designed to be a "living document," the emission control framework for "Heavy-Duty Vehicles" like port trucks did not specify enforceable remedies but did identify drayage trucks as a significant source of air pollution—asserting an aspirational goal of replacing or upgrading port trucks to meet lower-emission standards and outlining several different plans to achieve it. By asking port staff to develop an "implementation plan" by a fixed date, CAAP established no firm rules, but created the bargaining framework within which the campaign elaborated and advocated for the Clean Truck Program. The key point is that the disruptive force of litigation enabled the blue-green coalition not only to win a flexible policy in CAAP, but also to mobilize other forms of power within the negotiation framework set in motion by that policy to achieve deeper change built on hard law enforcement mechanisms: mandatory fleet and employee conversion imposed on penalty of port exclusion.

Validation What movement lawyers do to analyze proposed legislation in light of existing legal standards and how they wield legal interpretations as resources for advancing policy in the face of resistance by law makers or career government attorneys is a critical dimension of mobilizing law in local campaigns. In the Campaign for Clean Trucks' policy development phase, movement lawyers from the coalition's labor and environmental partners circulated and defended legal opinions on the market participant doctrine in order to minimize concerns over the well-known risk of FAAAA preemption. This advocacy occurred far from the glare of the courtroom in the more prosaic—but no less critical—process of administrative review through meetings with elected officials, city attorneys, and port staff. In this process, city officials ultimately had to sign off on legislation that they were confident passed relevant legal standards. These officials cared both as a matter of principle and because they did not want to be on the hook for provoking, and potentially losing, expensive litigation when strong arguments did not support their positions.

Although movement leaders and city officials understood that a litigation challenge to the Clean Truck Program would ensue, the lawyers' preemption analysis was critical to providing those officials with some degree of confidence in the outcome in the face of doctrinal uncertainty. In general, legal opinions generated by lawyers invested in a movement may be discounted as biased in favor of the lawyers' own reform position. Yet, in policy campaigns, movement-side legal opinions must have enough credibility to be accepted by lawyers on the inside of the political decision-making process. In this regard, the Teamsters and NRDC's preemption analysis was also vetted by Los Angeles port counsel, as well as the mayor's general counsel—which enhanced opportunity for independent review. The presence of Teamsters and NRDC lawyers in this vetting process meant that government lawyers were accountable to a wider audience for their own internal judgments. In the end, movement and government lawyers converged around the same ultimate conclusion: The Clean Truck Program was a legally defensible exercise of local power under the market participant exception to preemption. Legal review acknowledged the risk of litigation failure, but ultimately concluded that it was a risk worth taking.

From the point of view of port counsel (both inside counsel at the city attorney's office and outside counsel at Kaye Scholer), one might understand legal support of the Clean Truck Program in terms of presenting a client with the strongest defense of its proposed course of action. When port lawyers evaluated the Clean Truck Program, there were strong political currents already in its favor, and therefore port lawyers may have viewed their job as advancing the wishes of their client, the harbor commission. (The mayor's general counsel played a similar role relative to his client.) The fact that campaign and government lawyers all understood that the ATA would sue the ports *no matter what their programs required* strengthened political resolve, at least in Los Angeles, to press for the most ambitious set of policy changes.

Legal review was also intimately connected to an assessment of political risk by public officials. For the Los Angeles mayor's office, the political upside of the Clean Truck Program was transformative policy that made two important constituencies happy—and advanced goals that Mayor Villaraigosa believed in. On the other hand, losing the entire program in court would have been another political failure (following the mayor's defeated bid to take over the Los Angeles school district). The political compromise embodied in the Clean Truck Program's severability provision addressed this risk by fusing together reforms, truck and employee conversion, that were legally independent of one another. This meant that in the event employee conversion was struck down, the environmental provisions could survive—allowing the mayor to claim a major environmental win, while leaving the Teamsters (though not the truck drivers) no worse relative to where they started.

Protection Once policy reform is enacted, lawyers must defend it from attack. This an embedded feature of any policy reform campaign, evidenced in litigation against President Obama's health care policy and his executive orders on immigration, overtime law, and climate regulation. This protective lawyering function is intrinsically connected to the lawyer's role in providing ex ante policy validation, which makes a prediction about what legal challenges will occur and whether they will succeed. Thus, the protective element of lawyering for local policy reform—defensive strategies to convince judges that validation arguments are correct—is the ex post counterpart to the ground work that lawyers do in advance of policy enactment. Although defending policy requires the entity that enacts it—in this case, the city—to use its own counsel, movement lawyers play ancillary, though important, roles.

NRDC played such a role in the ATA litigation through its "joint defense agreement" with port counsel. During that litigation, Kaye Scholer partner Steven Rosenthal understood his role as outside port counsel in conventional client-centered terms—defending his client's policy from preemption attack—yet saw the benefits of coordinating legal strategy with NRDC as intervenors and dividing up advocacy from trial through the Supreme Court to best present the city's market participation argument and defend the program. At trial and in the Ninth Circuit, NRDC lawyers took the lead in arguing in favor of employee conversion as a way to advance the port's proprietary interest in environmental compliance since they were deemed to have no essential political stake in that position. The division of labor between port lawyers and NRDC persisted through the Supreme Court phase, where NRDC was tasked with stressing the positive environmental and public health impacts of Los Angeles's concession plan. During the ATA litigation, the Teamsters' legal team largely faded into the background, particularly when the employee conversion issue was still alive. Nonetheless, Teamsters' counsel remained active behind the scenes in helping NRDC frame its preemption arguments given the union's ongoing interest in protecting local authority to enact progressive legislation through the market participant exception.

Authorization One of the most significant and least appreciated elements of legal mobilization in the Justice for Port Drivers campaign was the way in which Teamsters' lawyers built the legal authority for misclassification strikes protected under labor law. This effort was crucial to *enabling driver collective action in the first instance*—reinterpreting labor rights to create the conditions for organizing success. Doing so required seeking the appropriate legal authorization from the relevant government agency, the NLRB, to assure drivers that any striking activity to challenge their treatment would be legally protected—as would their jobs. The pursuit of legal authorization for collective action resonates with other types of lawyering activity, particularly, First Amendment advocacy to protect the right to protest. But the campaign lawyers' under-the-radar advocacy in the NLRB went one step

Assessing the Movement 339

beyond this: fashioning an entirely new legal foundation for bringing the protections of labor law to a group of misclassified workers formally outside its scope.

In this way, advocacy in the NLRB to gain authorization for misclassification strikes was essential to the Justice for Port Drivers' suing-and-striking model. Such authorization work required Teamsters' lawyers to develop internal legal analyses marshalling precedent and making creative arguments supporting their central claim that misclassified drivers were entitled to the protections of the NLRA. To build a test case around this claim, lawyers incrementally advanced their misclassification arguments in NLRB Region 21: through one-on-one conversations with region staff, in ULP charges against campaign targets, and ultimately in formal hearings in front of NLRB judges. The lawyers' ultimate goal was to reclaim the arcane striking rules as a tool to benefit drivers misclassified as independent contractors, rather than a barrier to their collective action, by convincing the region to undertake two critical legal changes: (1) extending the meaning of who counted as statutory employees under the NLRA to include workers misclassified as independent contractors, and (2) extending the meaning of what counted as a ULP to encompass the practice of misclassification by trucking firms.

The lawyers' success in both regards had two major campaign consequences. First, widening the scope of statutory employees with NLRA section 7 rights to include port drivers permitted the Teamsters to run traditional organizing campaigns against key trucking firm targets, using strikes as crucial levers of economic pressure. Defining drivers as employees meant that if they were subject to traditional ULPs, like employer retaliation or threats, drivers could file charges with Region 21 and walk off the job in protest while remaining legally protected against permanent replacement. The campaign's success in expanding the scope of ULPs to include misclassification—convincing the region, as the Teamsters did in the Pac 9 case, that "misclassification denies workers ... literally all of the protections of the Act"—was also crucial to reducing barriers to striking and enlarging the range of targets. With the very act of misclassification itself a ULP, the campaign could set its organizing sights on any port trucking firm without having to find independent acts of employer misconduct to justify a strike.

The second significant impact of the lawyers' NLRB advocacy was that, once the foundation for misclassified driver strikes was firmly laid, the campaign was able to challenge trucking companies at their most vulnerable point: entry to the ports. It was only with NLRB authorization for misclassification strikes that the campaign could legally extend those strikes to port terminals under the doctrine of ambulatory picketing—permitting drivers to exert maximum leverage against trucking firms by delaying or denying truck access to cargo.

Inspiration In the Justice for Port Driver campaign, the strategic counterpart to striking was suing, which involved combining individual wage claims (through the DLSE), public enforcement actions (by the DOL), and private class and mass actions against trucking firm targets (by plaintiff-side lawyers). In each category, movement lawyers played critical roles to *inspire, but not control* litigation in order to build financial liability against trucking firms that, in tandem with economic loss from striking, would promote negotiated labor agreements.

Movement lawyers were key architects of the suing strategy, even though they were not centrally involved in its execution. For example, the Teamsters' Mike Manley and the National Employment Law Center's Rebecca Smith were important voices persuading the DOL that misclassification was an extensive labor problem that warranted a concentrated enforcement effort. Manley played a similar role helping to lay out the legal case for enforcing port driver wage and business expense claims at the DLSE. Manley and LAANE's Sanjukta Paul were also critical sources of support for private misclassification lawsuits.

Manley helped design the campaign's suing model in which private suits would complement organizing goals even while proceeding at arm's length from the campaign. The model was predicated on initial success by individual workers in the DLSE wage claim process to prove that misclassification lawsuits could pay off. That success depended on Paul building a pipeline of DLSE cases through the campaign's legal clinic, creating a track record that, as it became publicized, underscored the potential for economic returns on private lawsuits. And Paul's involvement in the *Talavera v. QTS* lawsuit provided a class action template that could be followed by other private counsel. Because *Talavera* was a suit launched in close concert with the campaign, Manley helped supply its lawyers with strategic research and backup support during its early stage and met with the QTS owners in efforts to facilitate a resolution. In the cases brought by plaintiff-side law firms, like Kabateck Brown Kellner, Manley had more limited conversations, in which he explained the campaign's organizing goals as a way of expanding the lawyers' conception of what constituted a successful settlement. Although the private lawyers have shown varying degrees of interest in the campaign's organizing goals, their lawsuits have consistently, even if unintentionally, bolstered the campaign's central objective: demonstrating to port trucking firms that so long as they continue to misclassify drivers, such suits will continue to be filed, thereby strengthening the campaign's argument that the route to eliminating litigation liability is through bargained-for labor peace.

Progressive Local Regulation in a Fractured Nation

The port struggle is a microcosm of broader national conflict over the transformation of work, the degradation of the environment, and the appropriate regulatory approach to each problem. Operating at the leading edge of globalization—at the exact border where the global economy meets local community—the ports are spaces where the unequal costs and benefits of trade and consumption see their most immediate and palpable impact. To understand that impact—on labor and the environment—and the struggle against its most damaging effects is to peer through a window onto some of the most significant challenges confronting contemporary American society.

In Los Angeles and Long Beach, the ports are emblems of growing class and racial divisions in America. Moving cargo valued at nearly $500 billion per year, the ports' transportation system hinges on a network of roughly twelve thousand short-haul truck drivers, nearly all of whom are Latino and over half of whom are immigrants. As they deliver luxury goods and cheap consumer items to some of the world's most profitable companies, many drivers can barely afford to maintain their trucks, to say nothing of earning a secure living for themselves and their families. It is a paradox that is invisible to those who reap its benefits, going to Target or Walmart to buy low-cost items fresh from factories in East Asia or eagerly awaiting their Amazon delivery from the cavernous warehouses that pockmark the Inland Empire. It is a paradox made all the more unsettling in the context of ongoing investment by the ports and shipping companies in multi-million-dollar facilities that carry forward their slow march into neighboring communities, themselves populated by many of the same people who deliver the goods, but cannot afford to buy them.

One way to understand the turn by the labor and environmental movements to local government as a place to redress these disparities is as an effort to bring the fight to the precise place where the problem exists. The ports are part of cities so cities are the entities that should fix port maladies. This impulse is amplified by the sheer political reality of federal constraint. For labor, there has been complete political gridlock at the national level on issues as simple as raising the minimum wage, which has stagnated at $7.25 per hour. A similar Congressional resistance to environmental reform has stymied efforts to adequately address issues like climate change, forcing progress to be channeled through the executive branch during the Obama era, where the lack of codification through statute has made environmental policy subject to rescission by a Trump administration hostile to its goals.

For these reasons, cities have become the locus of progressive policy change in the areas of labor and environmental regulation. The growth of local workplace regulation—laws passed by subfederal governmental units that directly or indirectly affect employer duties or employee rights—has gained attention among progressive movements as a way to avoid gridlock and change conditions for millions of low-wage workers.[50] As this book has recounted, this regulatory impulse operates in response to the weakness of federal labor and employment law, seeking to augment their rights and protections. Local workplace regulation aims to take advantage of receptive local political decision makers who may use local governmental authority—contracting, land use, and general police powers—to reshape the nature of work. Primarily a big city phenomenon, there are now, for example, roughly 150 municipalities with some version of a minimum wage or living wage law, and 225 with ordinances that prohibit private employer discrimination on the basis of gender identity, which is not a protected category under federal law.[51]

Los Angeles in particular has been a leader in progressive labor reform. It passed the nation's second living-wage ordinance in 1997 and has since then enacted cutting-edge laws limiting big-box development, tying home improvement store permits to the creation of day labor sites, and creating a local office to enforce its newly minted $15 per hour minimum wage law. Why Los Angeles has assumed this leadership role has been the subject of intense interest. Ruth Milkman's influential thesis suggests that Los Angeles's labor movement has benefitted from its historical strength in trade union organizing in industries, such as trucking, in which union density depended on sectoral strategies rather than shop-floor efforts, and by the influx of immigrants whose energy and openness to unionization has powered a renaissance.[52] As the labor movement has amassed victories and built strength, it has reshaped local politics, giving rise to more opportunities to legislate on behalf of workers.

This local political power has had specific implications for the port struggle because of Los Angeles's and Long Beach's status as port owner, which confers significant power to change operational rules while lending political urgency since the ports are such significant economic engines for the cities. The particular governmental structure—the ports are proprietary departments of city government—has permitted Los Angeles and Long

Beach to take innovative regulatory action in ways that have been less possible in places like New York–New Jersey, where joint state control has given more power to state governors who select port authority commissioners, or Seattle, which is governed by an elected body that does not answer to city leaders. In Los Angeles, by contrast, the power to effectuate significant labor reform rests, in part, on the ability of city leaders—deeply influenced by labor movement power—to install members to the board of harbor commissioners, who may enact port rules, subject to approval by a labor-friendly city council. In this sense, local politics and governance authority have created the conditions of possibility for Los Angeles's leadership on local labor regulation generally and port reform in particular.

It is precisely this possibility that has stimulated academic interest in "local labor activism," in which labor-community alliances leverage local government law powers to achieve labor regulation—like minimum wage and big-box ordinances—around the country.[53] Interest in this form of labor localism has intersected with recent legal scholarship focused on "law making from below," in which social movements organize outside of formal legal institutions to create new meanings of justice—for example, around ideals of women's and civil rights[54]—that shape the content of formal law as it becomes codified by legislatures or issued by courts. Arguments in favor of a bottom-up approach to law making in general and labor regulation in particular stress that localism gives affected communities more direct say in and control over regulatory processes, and builds public support by shifting culture before changing law.[55] However, the ports campaign reveals a more complicated and dynamic process in which the fight for control over the content and meaning of law is always influenced by contending local and national legal forces.

In this critical sense, the port struggle teaches a basic but important lesson: localism is never entirely local. To the contrary, a key takeaway from this study is that political control of the federal system—particularly the courts and the administrative state—profoundly affects the power of social movement-driven local regulatory action. Local efforts are dependent on federal law and legal processes, playing out on multiple levels, with proponents of labor reform selecting a hospitable arena for policy change—the progressive city—while opponents try to pull the struggle into venues where they have legal advantages: in the case of the Clean Truck Program, the federal court.

In this battle over work regulation at the ports, one thus hears echoes of Progressive Era legal struggle in which the labor movement sought to mobilize its significant state and local power to pass pro-worker legislation, like minimum wage ordinances and workday limits.[56] The difference now is that the barrier to local labor regulation is not a constitutional vision of substantive due process, as in the famous 1905 Supreme Court case of *Lochner v. New York*, but a legal laissez faire advanced under a deregulatory doctrine of FAAAA preemption. In this struggle over port regulation, courts are placed in the role of policing city policy making, like the Clean Truck Program, which was invalidated not on the basis of a higher affirmative law, but a claimed commitment to the *absence* of law: a prohibition against local trucking law mandated by federal deregulation used as the basis for preemption. This is not preemption because an affirmative regulatory scheme—like labor or environmental or immigration law—occupies the space, but rather because of the *withdrawal* of legal regulation mandated by the FAAAA.

A similar picture of federal-local interaction emerges when examining the Justice for Port Drivers campaign's effort to mobilize misclassified workers under the NLRA. That strategy seeks to avoid the preemption problem, but in doing so forces the campaign to rely on an innovative interpretation of federal labor law to support driver strikes. This work, as already suggested, is dependent on the ideological leanings of personnel in charge of labor law's implementation at the NLRB—now back in the hands of antiunion forces. More drastically, with conservative control over all three branches of federal government, there is a real possibility of national legislation that scales back already-weak federal labor protections and undoes local progress by explicitly preempting local minimum wage and other progressive workplace laws.

This focus on federal-local interaction has important implications. One is that the current version of labor localism, at least as it has been expressed in Los Angeles campaigns challenging low-wage work in the trucking and other service industries, is fundamentally *strategic*—in the sense that localism is a product of political calculation by labor leaders that progressive local reform is the best that can be achieved in an era of national political polarization. From this perspective, localism has advanced less as a first-order political priority of the labor movement than as a second-best strategic necessity. As the earlier discussion of converting legal weakness into strength has already highlighted, a key element of strategic calculation is determining on which institutional level to fight—fundamentally a calculation about institutional advantages and power. The Campaign for Clean Trucks targeted policy change at the port as a city entity precisely because of the campaign's local political leverage. When that was undercut in court, the campaign turned to other favorable venues: first, Congress to change law, and when that failed, to the DLSE and DOL to enforce law. One would presume that organized labor would prefer a national solution to the independent-contractor problem rather than a piecemeal approach. It is the political implausibility of such a solution that has channeled planning and resources to Los Angeles and other local targets.

This is not to suggest that labor and other progressive social movements should abandon progressive city-building as a central goal, nor that such bottom-up investments in organizing strength and regulatory success cannot over time feed into larger national currents of reform. But it is to express caution about risks: one of which is that social movement efforts to build progressive cities may create islands of greater equality amid a larger sea of inequality—causing the gap between the politics of progressive big cities and the politics of everywhere else to grow wider, reinforcing polarization and making it more difficult for progressivism to gain wider audience and political traction. In this sense, progressive city making may deepen the blue-red division across city boundaries even as it reduces the blue-green division within them. Another risk is that the pull of local activism and responsive regulation will obscure the extent of national-level problems in need of national-level solutions. In this regard, although proponents of localism might extol the benefits of local regulatory flexibility, such flexibility is most useful when local conditions are significantly different. Yet with respect to many of the core problems facing American workers, such as the inadequacy and underenforcement of the minimum wage, local variation is not the key issue.

In Richard Schragger's book, *City Power: Urban Governance in a Global Age*, he makes the case for "the possibility of an urban democracy that promotes individuals' participation in economic and political life on terms of equality."[57] The experience of Los Angeles, often held out as a paragon of the progressive city, in many ways validates this possibility. Indeed, one of the important lessons of the port struggle is that bringing together two of the most powerful progressive movements in America today around one goal creates the foundation for an enduring collaboration to achieve reform over time that is impossible when groups work in isolation. In this sense, the port struggle is not just about clean trucks and better conditions for drivers, but about showing political decision makers that a blue-green alliance can deliver on its promise to deepen democratic participation and expand equality over the long term.

To achieve this promise, progressive law making from below asserts greater space for local power, pushing back against the boundaries of federalism. Understanding how to best harness the dynamic relationship between the federal and local spheres is one of the most critical issues arising from the port struggle—a struggle that underscores how local politics is constrained without an energized national counterpart, while also showing how federal law is itself challenged and transformed by robust social movement activism that claims the power to make local law. It is in this sense that the history of economic and environmental activism at the ports illuminates how the social movement pursuit of local regulation necessarily reconstitutes relations with the federal government. By creating new facts on the ground that challenge federal authority, local mobilization is always a legal act that advances a program of national reform: revising the scope of federal law, even as it seeks to escape its reach.

Transcendent Themes

Looking back at the movement to reshape the port trucking industry in Los Angeles and Long Beach, what stands out is the enormous energy, commitment, and ingenuity of the activists, lawyers, policy makers, and community members involved in the fight against the related problems of pollution and poverty at the ports. The decade-long struggle reveals how sophisticated planning and political contingency converge to create moments in which social movements may deploy well-designed plans to challenge the status quo. Although the role of law in such challenges has been the organizing framework of this book, and while law has been—and remains—an essential tool in the ongoing blue-green struggle to redress the problems of port trucking, it is a tool ultimately wielded by people motivated by an elemental commitment to change. This last section reflects on three transcendent themes that emanate from that commitment: the power of expertise, tenacity, and courage to transform the lives of the least privileged.

Expertise

The notion of expertise—or specialized knowledge developed through extensive training and applied experience—has a complicated relation to democratic movements for social change. Elites, by virtue of their privilege and status, have a tendency (even if not intentional) to deploy their expertise to dominate collective decision making, particularly within groups composed of non-elites, who are low-income, less educated, and less socially privileged.

What emerges from the port struggle, however, is a positive vision of expertise, coming from both the bottom up and the top down, to empower drivers and port community residents to engage in collective action. Scholars who have advocated that lawyers pay greater attention to the store of lay knowledge present in the community would appreciate the degree to which community-based expertise was cultivated and valued throughout.[58] For instance, it was a Long Beach community member who proposed putting placards on port trucks with a number to call to report pollution or parking problems—an idea that became an essential part of the Clean Truck Program concession plan. It was the community effort to fight pollution coming from rail yards around Commerce, east of downtown Los Angeles, that sparked the creation of East Yard Communities for Environmental Justice, whose director, Angelo Logan, emerged as a powerful voice in the Campaign for Clean Trucks. And it was the solitary action of individual port drivers, objecting to long hours at low pay and filing DLSE wage claims on their own, before the Justice for Port Drivers campaign was officially underway, who taught campaign leaders that the wage claim process could be an effective avenue of driver redress.

Yet the campaign also relied crucially on the expertise of professionals and elites, without whom it would have been inconceivable to navigate the complex legal rules surrounding local policy making and union organizing, or mount a resource-intensive, multi-prong misclassification campaign against several large trucking firm targets at once. Strategic research by LAANE staffers Jon Zerolnick and later Sheheryar Kaoosji was essential to identifying trucking firms that had unique features making them susceptible to suing and striking: Toll's affiliation with its unionized Australian parent company, TTSI's ownership by the progressive-minded private equity firm Saybrook Capital, and the integration of the XPO firms, IBT, and Shippers within larger asset-rich corporate structures that could absorb the higher costs of employee reclassification.

In relation to professional expertise, it has become fashionable in legal scholarship to diminish the importance of lawyers in social movements in order to raise up the voices of the oppressed in shaping struggle. This critical perspective has provided a valuable corrective to the heroic portrayals of civil rights lawyers valiantly forging new paths to achieve "simple justice."[59] However, there are times when a lawyer's seminal work for social change deserves recognition—and the Justice for Port Drivers campaign provides such a moment.

After the Clean Truck Program failed to effectuate the long-sought-after legislative solution of employee conversion, Teamsters' counsel Mike Manley and his colleagues were forced back to the drawing board to devise new ways to slip the independent-contractor knot. This time, as they once again reviewed the literature on the subject, they started asking different questions. What if port drivers did not wait for employer permission to be treated as employees? Instead, what if they called themselves what they were—employees—and named what the employers were doing for what it was—misclassification? Then, based on their own self-definition as employees, what if drivers asserted their legal rights? What if they walked off the job, claiming the right to strike under labor law as misclassified employees?

These questions led the Justice for Port Drivers campaign to fashion its suing-and-striking strategy, which it mobilized incrementally at first, then more boldly, claiming misclassification as the legal ground for strikes against Pac 9, IBT, and the XPO firms. This has had a profound impact. As NLRB Region 21 has issued complaints on behalf of misclassified drivers and ruled in their favor in the Green Fleet case, port drivers, for decades excluded from labor law, have for the first time been given legal status as misclassified employees with the right to strike in protest of their misclassification—a right they have used to win six union agreements over the past three years. This is, as the book has emphasized, the result of the enormous efforts of drivers and campaign activists. But it is also because of lawyers who asked different questions, developed a vision, and mobilized this vision to change law and thereby enable worker organizing to build power for drivers. That President Trump is now making appointments to the NLRB who will be hostile to this outcome only underscores the central lesson of the story. Movements have to mobilize in the political arena in order to defend law—who gets elected and chooses who staffs administrative agencies matters intensely. But lawyers, like those in the port struggle, must also play a central role in redefining law to empower effective political action.

Tenacity

It is often said that how one handles failure says more about character than how one handles success—that tenacity in the face of adversity is a more essential quality than the capacity to do well in good times. By that standard, the movement for economic and environmental justice at the Los Angeles and Long Beach ports poignantly captures the values of grit and resilience. Social change happens, if at all, because people come together—and stay together—over long periods of time, through success and loss, to keep fighting for a cause they believe is worth it.

At the ports, that fight has spanned nearly forty years. In the 1980s, Wilmington residents began contesting port expansion and advocating for commercial development and community improvement, particularly access to a public beach that was blocked by the port. Important community groups were born out of these early fights, like the Far East Wilmington Improvement Association, which sustained activism against port growth through the 1990s. These groups focused attention on the growing nuisance of port trucks driving through the community and the environmental harm of port ships that spewed exhaust while transporting dangerous bulk items like coal and petroleum coke—adding to the hazards of a community already overpopulated with oil facilities, waste disposal plants, and freeway intersections. Yet residents won few tangible victories as the port growth imperative rolled over community dissent.

At the national level, during this same period, labor leaders endured their own test, watching trucking industry unionization plummet after deregulation in the 1980s. Boxed in by a legal system that fostered the growth of independent-contractor drivers while denying them power to organize for collective benefit, national labor leaders also had to persevere—waiting for a chance to make even a little progress against a port trucking system that was easy to understand but difficult to challenge.

It was precisely because of this long wait in the face of enormous odds that the achievement of the Clean Truck Program was such a high point—and the loss of employee conversion and other elements of the concession plan was experienced as such a devastating blow. What was arguably more impressive than the effort expended to achieve the program in the first instance—motivated by a sense of excitement and possibility—was the effort it took to move forward after the loss—weighed down by disappointment and the knowledge that drivers had been made worse off.

In the immediate aftermath of the Campaign for Clean Trucks, failure to achieve the highest-order goal redirected leaders to backup plans. This required not just grit but also planning and foresight, in which worst-case scenarios were identified and alternatives gamed out. After the Ninth Circuit's unfavorable 2009 preliminary injunction decision, the coalition adapted its strategy to focus on passing federal legislation explicitly exempting the Clean Truck Program from the FAAAA—a move that also failed when the Democrats lost control of Congress in 2010.

The campaign was then forced back to its default position: facilitating misclassification lawsuits to impose costs on companies that wrongfully denied drivers employee status, while simultaneously organizing to unionize trucking firms, like Toll, that already recognized their drivers as employees. Going back to this starting point provoked an existential crisis in the campaign, particularly within organized labor, for which the loss was most difficult to swallow. In the uncertain phase that followed, labor leaders like Change to Win's Nick Weiner and Teamsters Port Division Director Fred Potter had to advocate within the labor movement to maintain political and financial support for a campaign that had just consumed a vast amount of resources without a concrete payoff. This meant devising a credible alternative while dealing with internal critics and asking top labor decision makers for more patience. Yet the campaign's response to failure went beyond individual example, drawing on the power of solidaristic relationships to collectively sustain a struggle and adopt a new strategy deeply informed by lessons of the past.

This learning has been most prominently on display in the current drive for labor peace at the ports. The Justice for Port Drivers' misclassification campaign has thus far targeted "top 10" industry players, imposed economic costs through suing and striking, and promoted the systemic benefits of employee firms as models of a sustainable drayage industry that could bring stability and reliability to the ports. To effectuate this industry shift, the campaign is rebuilding its legal case that the ports should bar companies that fail, under the terms of concession agreements, to guarantee labor peace—which, in the context of misclassification strikes, would be predicated on employee conversion. Such a labor peace policy would undoubtedly reprise the specter of preemption, yet this time around the campaign has laid the groundwork so that any local policy solution would respond to the port's proprietary concern in mitigating labor-based suing and striking, and thus fall more squarely within the judicial interpretation of market participation articulated in the ATA case. In this way, the campaign seeks to refashion the legal doctrine that had thwarted its earlier proposal into a protective shield.

In addition, as the misclassification fight has persisted, elements of the Clean Truck Program have been redesigned to advance a parallel campaign to organize city waste haulers—in which the hard lessons from the port have been used to strengthen the legal grounds for securing victory in a distinct sector of the local trucking industry. Failure to achieve one of the movement's biggest prizes—unionization of port trucking as the supply chain "choke point"—has nonetheless produced important learning that continues to inform ongoing policy cycles in which labor movement actors are repeat political players. In this vein, LAANE has led a campaign to require private sanitation companies servicing Los Angeles businesses and some residential properties to obtain city-issued franchises for waste hauling to and from eleven designated city zones. Titled "Don't Waste L.A.," the campaign has brought together the same environmental and labor alliance that sought to green the ports around the mutual objectives of converting the waste fleet of over one thousand trucks to clean emission technology, while improving conditions for waste drivers and promoting recycling to achieve the city's "zero waste" goal.[60]

However, this time around, the coalition—drawing upon its port experience—has designed the policy to avoid the litigation that undercut the Clean Truck Program. Instead of granting concessions to all companies that agree to employee conversion (among other requirements), the waste hauling plan awards an exclusive franchise for trucking companies selected through a competitive bidding process (one franchise per zone), in which bids are judged based on a range of good business practices that include using clean trucks, following efficient routing, promoting recycling and organic waste diversion, and implementing strong driver work standards. In so doing, the plan operates outside the scope of FAAAA preemption (not regulating motor carriers in the transportation of property) and also fits squarely within the heart of market participation (dealing with the efficient procurement of city services). The waste-hauler plan passed the Los Angeles City Council in April 2014 and, after a rollout process in which the city bureau of sanitation solicited bids,[61] the city approved a final waste collection program in 2016, awarding exclusive franchises to serve distinct city zones to seven garbage hauling companies that have agreed to labor peace. This new achievement is another indication that the labor movement's investment in the Clean Truck Program—though itself disappointing—was not wasted.

Courage

In the final analysis, the essential fuel of the drive for justice at the ports was courage: of the lawyers and activists who devoted their professional lives, often working long hours for little pay, out of a commitment to make their communities better; and most significantly, of the drivers themselves, who broke out of the isolation of their independent-contractor status, who refused to sit idly in their solitary trucks, and who had the audacity to assert that they were employees to a chorus of powerful voices telling them they were not—rising up to demand that they were worth more than industry was offering, risking their own livelihood and often that of their families to do so.

The ports story was thus made possible by the courage of activists like Jon Zerolnick: a Yale graduate, who could have chosen most any path in life, and yet dropped out of graduate school to stand side by side with some of the most vulnerable members of American society, leaving the East Coast to help organize culinary workers in Las Vegas, coming to Los Angeles to help farmworkers, and then agreeing to join LAANE to take the helm of what would be one of the biggest labor fights in Los Angeles history.

It was made possible by Jesse Marquez: someone who grew up in the port community of Wilmington experiencing pain in his lungs when he ran high school track and who, as an adult, believed that his voice should count in port governance; someone who stood up at a port meeting announcing plans to build a wall between the TraPac terminal and his community, shouted "Hell no" and rallied his neighbors to join him in starting a powerful community-based organization, the Coalition for a Safe Environment, that stopped the wall and later helped pass the Clean Truck Program.

It was made possible by driver-turned-organizer Carlos Santamaria: a second-generation port truck driver who lived blocks from the port and whose family breathed its pollution; someone who had experienced the hardship of independent-contractor life, who pursued the benefits of employee status at Swift Transportation after the Clean Truck Program passed, only to see it pulled away when employee conversion was enjoined; someone who had the courage to call out Swift's hypocrisy when it claimed redesignating drivers as contractors would allow them to earn more money, and whose repeated leadership in organizing fellow drivers against mistreatment caught the eye of the Teamsters union, which recruited him as a driver from Toll to play an essential role in the campaign's first organizing victory against that same company.

Most fundamentally, the port struggle would not have occurred without the courage of hundreds of drayage drivers: many immigrant men who lived on the margins, working long hours without rest to provide for families, who believed they deserved more and were willing to stand up to some of the most empowered actors in the global economy to achieve it. Some received recognition, like Dennis Martinez, a driver who was unable to pay for his truck, who nonetheless stood on the April 2014 picket line against Green Fleet, Pac 9, and TTSI, and who, when asked by a *Los Angeles Times* columnist why he was there, responded with steely resolve: "Because I want to finish this fight." Yet most drivers were not named or even seen by most people, but rather walked in silent solidarity with fellow drivers, peering across a chasm of wealth and power at a port complex denying them basic dignity, causing the port, the city, and the rest of the world to take note: a moving symbol of how hollow the bedrock American principle of "equal justice under law" can be—and just what it takes to make it real.

Assessing the Movement

Coda

The movement for economic and environmental justice at the Los Angeles and Long Beach ports is an epic story of the potential, and tradeoffs, of blue-green mobilization and law making at the local level—as well as an example of how such law making is framed, and ultimately constrained, by federal power. This study of social struggle has showed how the legal regime of port governance imposed negative local impacts on low-income communities and low-wage workers, while also creating the possibility for a strategic alliance to advance labor and environmental change. It is a story of success, failure, and redemption. Success came early. The blue-green alliance, formed by deliberate design and unforeseen opportunity, executed sophisticated inside and outside strategies to achieve major policy reform in Los Angeles: the Clean Truck Program converting port trucks to clean vehicles and port drivers to employees. Failure was crushing. The Clean Truck Program was undercut by industry litigation that split the policy interests of the blue-green alliance, preserving the program's environmental provisions (clean trucks), while negating the labor ones (employee drivers). As a result, while the ports are now on track to achieve green growth, it is still on the backs of their most vulnerable workers.

Yet, as the book has emphasized, out of defeat has come redemption—in the form of organized labor's surprisingly successful effort to battle trucking misclassification on the streets and in the courts. This effort has gained considerable momentum at the Los Angeles and Long Beach ports, reopening a dialogue that appears to be leading back toward a policy solution that could once again transform the conditions for drivers who continue to toil long hours for low pay in a model of "ownership" that is largely forced upon them rather than chosen. In this sense, the story of struggle at America's port ends in inspiration: in the face of overwhelming odds and frequent setback, the pursuit of justice for port truck drivers endures.

Looking back, the struggle's endurance is its defining feature. In the drive for justice at America's port, a coalition of labor, environmental, and community groups—held together by a collective commitment to advancing the interests of low-wage workers and protecting the health of low-income communities—took on some of the most powerful economic actors in the global economy. And they won—for a moment, they won big, and even in the end, they have won something substantial. That the benefits of victory have, in some respects, been unevenly distributed between the campaign's environmental and labor constituencies is partly a result of the unequal legal and political power they brought to bear in the first instance. Yet as the labor movement continues to build a new model for organizing independent contractors in the gig economy, one overriding lesson is that the struggle for social change never ends. It only changes form and shifts site.

When the ports campaign emerged from the devastating loss of employee status for drivers at the Port of Los Angeles, labor leaders made a difficult and risky decision to make driver misclassification their new focal point. The first company they chose to organize was Green Fleet Systems, a mid-sized trucking firm with a minority of drivers misclassified as independent contractors. That was 2012. Five years later—after nearly $300,000 in legal liability, four strikes, two reinstated drivers, multiple charges of labor violations, and a corporate bankruptcy—there is now a labor peace agreement in place that paves the way for all of the company's drivers to be unionized. What began as a glimmer of hope—that legal power and organizing grit could build union density in an industry that has not seen it for over thirty years—has now become a model for the port drayage sector to follow. In this sense, Green Fleet may be more than an aptly named coda to a monumental struggle. It may be a glimpse into the future.

Notes

Chapter 1

1. The Port of Los Angeles officially branded itself as "America's Port." In the book, I use "America's port" more broadly to refer to the entire port complex, which includes the port in neighboring Long Beach.

2. By focusing on this movement-countermovement dynamic, the book builds upon scholarship in the burgeoning law and social movement field. For the classics, see, e.g., Joel F. Handler, *Social Movements and the Legal System: A Theory of Law Reform and Social Change* (New York: Academic Press, Inc., 1978); Michael W. McCann, *Rights at Work: Pay Equity Reform and the Politics of Legal Mobilization* (Chicago: University of Chicago Press, 1994); Gerald N. Rosenberg, *The Hollow Hope: Can Courts Bring About Social Change?* (Chicago: University of Chicago Press, 1991). For important recent work, see Tomiko Brown-Nagin, *Courage to Dissent: Atlanta and the Long History of the Civil Rights Movement* (New York: Oxford University Press, 2011); Michael J. Klarman, *From Jim Crow to Civil Rights: The Supreme Court and the Struggle for Racial Equality* (New York: Oxford University Press, 2004); Catherine Albiston, "The Dark Side of Litigation as a Social Movement Strategy," *Iowa Law Review Bulletin* 96 (2011): 61; Anthony V. Alfieri, "Faith in Community: Representing 'Colored Town,'" *California Law Review* 95 (2007): 1829; Sameer M. Ashar, "Public Interest Lawyers and Resistance Movements," *California Law Review* 95 (2007): 1879; Douglas NeJaime, "The Legal Mobilization Dilemma," *Emory Law Journal* 61 (2012): 663; Robert Post and Reva Siegel, "*Roe* Rage: Democratic Constitutionalism and Backlash," *Harvard Civil Rights-Civil Liberties Law Review* 42 (2007): 373. For a comparative view, see *Stones of Hope: How African Activists Reclaim Human Rights to Challenge Global Poverty*, ed. Lucie E. White and Jeremy Perelman (Stanford, CA: Stanford University Press, 2010).

3. When discussing the individual phases of the port struggle, I will refer to them by the specific name given by the coalition (i.e., "Campaign for Clean Trucks" or "Justice for Port Drivers campaign").

4. See Rosenberg, note 2; Stuart Scheingold, *The Politics of Rights: Lawyers, Public Policy, and Political Change* (New Haven, CT: Yale University Press, 1974); Derrick A. Bell, Jr., "Serving Two Masters: Integration Ideals and Client Interests in School Desegregation Litigation," *Yale Law Journal* 85 (1976): 470; Alan David Freeman, "Legitimizing Racial Discrimination Through Antidiscrimination Law: A Critical Review of Supreme Court Doctrine," *Minnesota Law Review* 62 (1978): 1049.

5. See Bell, note 4; see also Gerald P. López, *Rebellious Lawyering: One Chicano's Vision of Progressive Law Practice* (Boulder, CO: Westview Press, 1992); William H. Simon, "Solving Problems vs. Claiming Rights: The Pragmatist Challenge to Legal Liberalism," *William & Mary Law Review* 46 (2004): 127.

6. For an excellent synthesis of these critical views, see Orly Lobel, "The Paradox of Extralegal Activism: Critical Legal Consciousness and Transformative Politics," *Harvard Law Review* 120 (2007): 937.

7. Michael W. McCann, "Legal Mobilization and Social Reform Movements: Notes on Theory and Its Application," *Studies in Law, Politics, and Society* 11 (1991): 225, 232, 234, 237, 241, 244.

8. Steven Boutcher, "Mobilizing in the Shadow of the Law: Lesbian and Gay Rights in the Aftermath of *Bowers v. Hardwick*," *Research in Social Movements, Conflict, and Change* 31 (2011): 175; Lynette Chua, "Pragmatic Resistance, Law, and Social Movements in Authoritarian States: The Case of Gay Collective Action in Singapore," *Law & Society Review* 46 (2012): 713.

9. Michael McCann, "Law and Social Movements: Contemporary Perspectives," *Annual Review of Law and Social Movements* 2 (2006): 17.

10. Michael McCann and Helena Silverstein, "Rethinking Law's 'Allurements': A Relational Analysis of Social Movement Lawyers in the United States," in *Cause Lawyering: Political Commitments and Professional Responsibilities*, ed. Austin Sarat and Stuart Scheingold (New York: Oxford University Press, 1998), 261, 266 (emphasis in original); see also *Cause Lawyers and Social Movements*, ed. Austin Sarat and Stuart Scheingold (Stanford, CA: Stanford University Press, 2006).

11. See Larry D. Kramer, *The People Themselves: Popular Constitutionalism and Judicial Review* (New York: Oxford University Press, 2004); Scott L. Cummings, "Movement Lawyering," *Illinois Law Review* (2017): 1645; Lani Guinier and Gerald Torres, "Changing the Wind: Notes Toward a Demosprudence of Law and Social Movements," *Yale Law Journal* 123 (2014): 2740; Jennifer Gordon, "The Lawyer Is Not the Protagonist: Community Campaigns, Law, and Social Change," *California Law Review* 95 (2007): 2133; see also Luke W. Cole and Sheila R. Foster, *From the Ground Up: Environmental Racism and the Rise of the Environmental Justice Movement* (New York: New York University Press, 2001).

Notes

12. Suzanne Staggenborg, "Coalition Work in the Pro-Choice Movement: Organizational and Environmental Opportunities and Obstacles," *Social Problems* 33 (1986): 374, 388.

13. Suzanne Staggenborg, "The Consequences of Professionalization and Formalization in the Pro-Choice Movement," *American Sociology Review* 53 (1988): 585; see also Suzanne Staggenborg, *The Pro-Choice Movement: Organization and Activism in the Abortion Conflict* (New York: Oxford University Press, 1991).

14. Richard C. Schragger, *City Power: Urban Governance in a Global Age* (New York: Oxford University Press, 2016).

15. Benjamin I. Sachs, "Despite Preemption: Making Labor Law in Cities and States," *Harvard Law Review* 124 (2011): 1153; Katherine Stone and Scott Cummings, "Labor Activism in Local Politics: From CBAs to 'CBAs,'" in *The Idea of Labour Law*, ed. Guy Davidov and Brian Langille (Oxford: Oxford University Press, 2011), 273.

16. See Richard Abel, "Speaking Law to Power: Occasions for Cause Lawyering," in *Cause Lawyering: Political Commitments and Professional Responsibilities*, ed. Austin Sarat and Stuart Scheingold (New York: Oxford University Press, 1998), 69.

17. For a classic account of this movement-countermovement dynamic in the labor context, see William E. Forbath, *Law and the Shaping of the American Labor Movement* (Cambridge, MA: Harvard University Press, 1991). For a more recent discussion in relation to the same-sex marriage movement, see Michael C. Dorf and Sidney Tarrow, "Strange Bedfellows: How an Anticipatory Countermovement Brought Same-Sex Marriage into the Public Arena," *Law & Social Inquiry* 39 (2014): 449.

18. For this book, I conducted interviews with thirty-one key movement actors (pursuant to UCLA Institutional Review Board protocol #G08–06–076–02), reviewed legal materials related to all litigation and administrative proceedings, and reviewed the internal campaign archive made available by the Coalition for Clean and Safe Ports.

19. See Sachs, note 15.

20. See Richard Schragger, "Mobile Capital, Local Economic Regulation, and the Democratic City," *Harvard Law Review* 123 (2010): 482; see also *Working for Justice: The L.A. Model of Organizing and Advocacy*, ed. Ruth Milkman, Joshua Bloom, and Victor Narro (Ithaca, NY: Cornell University Press, 2010).

Chapter 2

1. See Edna Bonacich and Jake B. Wilson, *Getting the Goods: Ports, Labor, and the Logistics Revolution* (Ithaca, NY: Cornell University Press, 2008), 45.

2. Ibid. (noting that, as of 2004, the two ports combined were the third largest in the world). As of 2011, the Port of Los Angeles was the sixteenth largest in the world

by container volume, while the Port of Long Beach was ranked twenty-first; combined, they formed the eighth largest port in the world. "The JOC Top 50 World Container Ports," *Journal of Commerce*, August 20–27, 2012, 24, 26. In 2015, Los Angeles and Long Beach were ranked nineteenth and twenty-first, respectively; combined they were the tenth largest in the world by cargo volume. "The JOC Top 50 World Container Ports," *Journal of Commerce*, August 22, 2016, 24–25.

3. See Steven P. Erie, *Globalizing L.A.: Trade, Infrastructure, and Regional Development* (Stanford, CA: Stanford University Press, 2004), 47. See generally *Board of Harbor Commissioners, The Port of Los Angeles* (1913), 23–28 (providing a history of the Port of Los Angeles); Charles F. Queenan, *The Port of Los Angeles: From Wilderness to World Port* (Los Angeles: Los Angeles Harbor Department, 1983), 27–56 (discussing the growth and evolving role of the Port of Los Angeles between the late nineteenth and early twentieth centuries).

4. See Charles F. Queenan, *Long Beach and Los Angeles: A Tale of Two Ports* (Northridge, CA: Windsor Publications, 1986), 13–16. Cowhide was the major trading commodity. Ibid., 23.

5. Ibid.

6. Ibid.

7. See Remi A. Nadeau, *City-Makers: The Men Who Transformed Los Angeles from Village to Metropolis During the First Great Boom, 1868–76* (Garden City, NY: Doubleday, 1948), 24.

8. Queenan, note 4, 26.

9. See ibid., 29–30. During the war, Banning ceded some of his property to the Union army to build a new base; he was rewarded with a construction contract and he profited from the shipment of military supplies. Ibid., 29.

10. Nadeau, note 7, 26–27.

11. Ibid., 27–29.

12. See Robert M. Fogelson, *The Fragmented Metropolis: Los Angeles, 1850–1930* (Cambridge, MA: Harvard University Press, 1967), 108–109.

13. See Nadeau, note 7, 23–24.

14. See Fogelson, note 12, 52.

15. See Erie, note 3, 49; see also Steven P. Erie, "How the Urban West Was Won: The Local State and Economic Growth in Los Angeles, 1880–1932," *Urban Affairs Review* 27 (1992): 519, 526–527.

16. See Nadeau, note 7, 73.

Notes

17. Ibid., 78; see also "The 'Committee of Thirty,'" *Los Angeles Herald*, February 3, 1883, 3.

18. Nadeau, note 7, 79–86; see also Erie, note 3, 49.

19. Erie, note 3, 49.

20. See ibid., 46; see also Fogelson, note 12, 56 (reporting that, in 1880, the population of the city of Los Angeles was 11,183 and the county population was 33,381).

21. Fogelson, note 12, 63–66. For the classic account of the construction of the California Dream, see Kevin Starr, *Inventing the Dream: California Through the Progressive Era* (New York: Oxford University Press, 1985).

22. Erie, note 3, 49–50; Queenan, note 4, 33.

23. Erie, note 3, 49; Queenan, note 4, 33.

24. Queenan, note 4, 37–44.

25. Ibid., 49–51.

26. Fogelson, note 12, 110.

27. See Queenan, note 4, 51.

28. See Erie, note 3, 50.

29. Ibid., 52–53.

30. Ibid., 53; see also Fogelson, note 12, 112–114.

31. Fogelson, note 12, 112–114.

32. Erie, note 3, 54.

33. Fogelson, note 12, 114–115.

34. Erie, note 3, 60–61.

35. Queenan, note 4, 37, 45–46.

36. Ibid., 61.

37. See Erie, note 3, 65.

38. Fogelson, note 12, 115.

39. Queenan, note 4, 62.

40. Ibid., 62–63.

41. The consolidation required a majority vote of the residents of the annexing city, Los Angeles, and the cities to be annexed. Wilmington was incorporated in 1872. Donna St. George, "Wilmington: Community of Contradictions," *Los Angeles Times*,

Oct. 6, 1985, SB1. San Pedro was incorporated in 1888. Sheryl Stolberg, "No Longer the City It Once Was, San Pedro to Mark 100th Birthday," *Los Angeles Times*, Feb. 26, 1988, M8.

42. See Fogelson, note 12, 115.

43. See Queenan, note 4, 63.

44. See ibid.; see also "Fleming Tells of Campaign to Secure Harbor," *Los Angeles Herald*, Aug. 25, 1909, 1.

45. Erie, note 3, 65.

46. See Queenan, note 4, 74.

47. Erie, note 3, 55.

48. Ibid.

49. L.A., Cal., City Charter art. XVI, § 176 (1911) (amended 1913).

50. Ibid., §§ 168–186. The proprietary nature of the Port of Los Angeles is atypical: only 17 percent of U.S. ports are governed by municipal authorities and of the eight ports in the largest American cities, the Port of Los Angeles is the only one under city control. See Erie, note 3, 31.

51. Erie, note 3, 55, 57–60.

52. Ibid., 55–56.

53. Ibid., 57.

54. Ibid. The development of the port continued to benefit from federal support, with the federal government appropriating another nearly $10 million during this period to "dredge the outer harbor, widen the main channel, and double the length of the breakwater." Ibid., 56.

55. David L. Clark, "Improbable Los Angeles," in *Sunbelt Cities: Politics and Growth Since World War II*, ed. Richard M. Bernard and Bradley R. Rice (Austin: University of Texas Press, 1983), 268, 270.

56. Fogelson, note 12, 78.

57. Ibid., 119. During this time, commercial fishing also became a central industry in the harbor, with the rise of canned tuna drawing new investment and labor, including Japanese fishermen who built an active community on Terminal Island until they were interned during World War II. Queenan, note 4, 66, 117.

58. Erie, note 3, 61; Queenan, note 4, 82–83.

59. Fogelson, note 12, 119. Port trade at this stage was still dominated by lumber imports and oil exports, despite a $15 million bond-financed effort to attract other industries by doubling wharf space. Queenan, note 4, 90.

Notes

60. Fogelson, note 12, 119.

61. Queenan, note 4, 65.

62. Ibid., 67–78.

63. Erie, note 3, 67.

64. Queenan, note 4, 79.

65. Ibid., 82.

66. Erie, note 3, 71.

67. Queenan, note 4, 83.

68. Erie, note 3, 71–72.

69. Ibid., 72.

70. Queenan, note 4, 89. There was some cooperation between the ports, most crucially the joint establishment of the Harbor Belt Line Railroad in 1929, which linked multiple harbor rail lines to permit seamless rail travel around the two ports. Ibid., 92.

71. Erie, note 3, 79.

72. See Gordon Laird, *The Price of a Bargain: The Quest for Cheap and the Death of Globalization* (New York: Palgrave MacMillan, 2009), 130–131. For an analysis of the subsidence problem, which briefly caused Long Beach to earn the title of "The Sinking City," see Queenan, note 4, 119–120 (quoting *Time Magazine*). Earlier oil discoveries in the area had already attracted refineries. See ibid., 83; see also "California Oil Refinery History," *California Energy Commission* (July 13, 2016), http://www.energy.ca.gov/almanac/petroleum_data/refinery_history.html (accessed June 14, 2017) (noting the creation of refineries by Union Oil of California in 1917 and California Petroleum Corporation in 1923).

73. See Queenan, note 4, 91–92.

74. "The Union Oil Co.: The Trend," *Port of Los Angeles: Visual History Tour*, www.laporthistory.org/level4/Berth150/berth150_trend2.html (accessed Apr. 1, 2014).

75. Erie, note 3, 80.

76. Ibid.

77. Erie notes that from 1945 to 1954, the Port of Los Angeles spent $25 million on construction. Ibid. Long Beach financed its improvements with money from its vast oil reserves. Queenan, note 4, 127. In addition to new wharves and other facilities, the harbor breakwater was finally completed in 1949. Ibid., 126.

78. Queenan, note 4, 126, 129 (noting the establishment of Star-Kist Foods Inc. in 1952).

79. Erie, note 3, 81.

80. Queenan, note 4, 123. The amendment occurred after Long Beach voters approved a charter amendment allowing the city to use oil revenues for nonharbor improvements. Erie, note 3, 85. The state sought to block diversion of oil funds under the new law, which the state supreme court upheld. The state then sued to recover back payments and a political compromise was struck in Assembly Bill 77, which let Long Beach keep 50 percent of oil revenues if they were dedicated to harbor improvement. Queenan, note 4, 123. However, over time, the state took more of the oil funds. See Erie, note 3, 85. The Tidelands Trust Act was again amended in 1965 to give the state an even greater percentage of oil revenues, effectively ending Long Beach's reliance on oil for harbor development. Ibid.

81. Queenan, note 4, 128, 137.

82. Interstate Commerce Act, ch. 104, 24 Stat. 379 (1887).

83. Shipping Act, ch. 451, §§ 3, 14–22, 39 Stat. 728, 729, 733–736 (1916) (repealed in part and amended in part 1984).

84. Wayne K. Talley, *Port Economics* (New York: Routledge, 2009), 150.

85. Motor Carrier Act, ch. 498, 49 Stat. 543 (1935).

86. See Michael H. Belzer, *Sweatshops on Wheels: Winners and Losers in Trucking Deregulation* (New York: Oxford University Press, 2000), 60–61.

87. Talley, note 84, 150.

88. See Fred Thompson III, Note, "Challenges to the Legality of Minibridge Transportation Systems," *Duke Law Journal* (1978): 1233, 1236–1237.

89. See Wayne K. Talley, "Wage Differentials of Intermodal Transportation Carriers and Ports: Deregulation Versus Regulation," *Review of Network Economics* 3 (2004): 207, 211–212.

90. See Queenan, note 4, 96.

91. See ibid.

92. Bruce Nelson, *Workers on the Waterfront: Seamen, Longshoremen, and Unionism in the 1930s* (Urbana: University of Illinois Press, 1988), 1.

93. Ibid., 6.

94. David Selvin, *A Terrible Anger: The 1934 Waterfront and General Strikes in San Francisco* (Detroit: Wayne State University Press, 1996), 23.

95. Nelson, note 92, 48–52.

96. Selvin, note 94, 25–29.

Notes

97. Ibid., 31–32.

98. Nelson, note 92, 104.

99. Ibid.

100. Ibid., 112–126; see also Mike Quin, *The Big Strike* (Olema, CA: Olema Publishing Company, 1949).

101. Bonacich and Wilson, note 1, 173.

102. Ibid.

103. Nelson, note 92, 127–129. For a detailed analysis of Bloody Thursday, see Selvin, note 94, 141–153.

104. Selvin, note 94, 238–239.

105. Nelson, note 92, 190.

106. Selvin, note 94, 96.

107. Ibid., 157, 163–164.

108. Ibid., 118.

109. Nelson, note 92, 139.

110. Selvin, note 94, 169.

111. Ibid., 229.

112. Nelson, note 92, 190.

113. Ibid., 224–233.

114. Ibid., 232–233.

115. Bonacich and Wilson, note 1, 173–174; see also Nelson, note 92, 238, 269. The ILA, which became much more conservative, continued to represent East Coast dockworkers. See Howard Kimeldorf, *Reds or Rackets? The Making of Radical and Conservative Unions on the Waterfront* (Berkeley: University of California Press, 1988). In the immediate aftermath of the CIO split, despite the fact that the ILWU had been designated the official bargaining agent of West Coast longshoremen, some workers chose to align themselves with the ILA. Among these workers were fifteen in San Pedro whose refusal to cede control over a local hiring hall produced a nasty spat with Harry Bridges that led to his being fined for criticizing a local court ruling in the ILA workers' favor—a fine that was reversed on First Amendment grounds in a case litigated by the American Civil Liberties Union on Bridges' behalf all the way to the United States Supreme Court. See Charles L. Larrowe, *Harry Bridges: The Rise and Fall of Radical Labor in the United States* (Westport, CT: Lawrence Hill & Company, 1972), 127–130. The CIO split also ruptured relations between the longshoremen

and the Sailor's Union of the Pacific, which remained allied with the AFL. See Nelson, note 92, 239–241.

116. Kimeldorf, note 115, 4. "Combining militancy and radical politics, the ILWU was widely recognized as the strongest bastion of Communist unionism on the West Coast, if not the entire country." Ibid., 5.

117. Ibid., 14.

118. Ibid., 145.

119. Bonacich and Wilson, note 1, at 175–176.

120. For the definitive account of this history, see generally Donald Garnel, *The Rise of Teamster Power in the West* (Berkeley: University of California Press, 1972).

121. Ruth Milkman, *L.A. Story: Immigrant Workers and the Future of the U.S. Labor Movement* (New York: Russell Sage Foundation, 2006), 46.

122. Ibid., 46–49. See generally Louis B. Perry and Richard S. Perry, *A History of the Los Angeles Labor Movement, 1911–1941* (Berkeley: University of California Press, 1963).

123. Milkman, note 121, 47.

124. Ibid., 47–48.

125. Ibid., 49.

126. Ibid., 51. Belzer reports that in the 1970s, the trucking industry was almost completely under union contract. Belzer, note 86, 107.

127. Milkman, note 121, 60 fig.1.2.

128. Ibid., 61.

129. Bonacich and Wilson, note 1, 47.

130. Ibid., 50–54.

131. Ibid., 51.

132. Ibid., 50; see also Queenan, note 4, 133 ("While the average longshore gang of sixteen to eighteen men could handle eight to ten tons of cargo in regular packaging, a five-man team could move 450 tons of containerized goods and expend only a fraction of the effort and energy doing it.").

133. Bonacich and Wilson, note 1, 51.

134. Queenan, note 4, 135.

135. Erie, note 3, 88–89.

136. Queenan, note 4, 143.

Notes

137. Dan Weikel, "Freighters Enter the Age of the Mega-Ship," *Los Angeles Times*, June 15, 1999, A1.

138. Talley, note 84, 150.

139. Erie, note 3, 90.

140. Ibid.; see also Queenan, note 4, 150 (describing the $61 million dredging project).

141. Erie, note 3, 91.

142. Bonacich and Wilson, note 1, 59.

143. Ibid., 59–60.

144. Queenan, note 4, 147, 149; see also Bonacich and Wilson, note 1, 47.

145. Queenan, note 4, 147.

146. See ibid., 149.

147. See Erie, note 3, 22.

148. Bonacich and Wilson, note 1, 47.

149. See Daryl Kelley, "Edgerton on Junket with Port Officials to Asia," *Los Angeles Times*, June 27, 1985, LB8. Politicians also did outreach to other regions, such as Latin America, to promote harbor trade. See "The Region," *Los Angeles Times*, Oct. 2, 1985, http://articles.latimes.com/1985-10-02/news/mn-16051_1_los-angeles-mayor-tom-bradley (accessed June 14, 2017).

150. Bonacich and Wilson, note 1, 61.

151. Ibid., 62–63.

152. Ibid., 63.

153. Ibid., 53.

154. Ibid.

155. Ibid., 54.

156. Belzer, note 86, 64–65.

157. Bonacich and Wilson, note 1, 54, 103.

158. Ibid., 54.

159. Talley, note 84, 150.

160. Ibid., 150.

161. Edward James Taaffe, Howard L. Gauthier, and Morton E. O'Kelly, *Geography of Transportation*, 2nd ed. (Upper Saddle River, NJ: Prentice-Hall, 1996), 161.

162. Talley, note 89, 211–212.

163. See Taaffe, Gauthier, and O'Kelly, note 161, 161–162.

164. Talley, note 89, 212, 214.

165. Erie, note 3, 23 (stating that containerization "placed a premium on the capacity, efficiency, and ground accessibility of local port and airport facilities").

166. See David Jaffee, "Kinks in the Intermodal Supply Chain: Longshore Workers and Drayage Drivers" (unpublished manuscript, June 2010), 16, https://www.unf.edu/uploadedFiles/aa/coas/cci/ports/REPORT_PortPaper10-SASE-KinksintheIntermodalSupplyChain.pdf (accessed June 14, 2017). Bonacich and Wilson cite estimates that 65 percent of Los Angeles-Long Beach containers are bound for U.S. destinations; of these, 25 percent are loaded to rail on dock, while 40 percent are drayed to rail heads. Bonacich and Wilson, note 1, 115.

167. Fogelson, note 12, 118.

168. Whether containers are loaded directly onto rail on dock or go to trucks first depends in part on the condition of the containers. If they come in full, they are typically loaded to rail on dock. If they have to be consolidated with other partially full containers, they are loaded onto trucks and then transported to consolidator warehouses, some of which are located near the port but others are as far away as Riverside. Interview with John Holmes, deputy executive director, Port of Los Angeles (July 19, 2013). Two-thirds of the containers arriving from Asia in 2013 were full. Ibid.

169. Queenan, note 4, 156–157; see also Tim Waters, "Railhead Is Competitive Edge for Port," *Los Angeles Times*, Nov. 23, 1986, SB1 ("The new rail facility was financed primarily through the sale of $53.9 million in bonds that will be paid back with money collected by a $30-per-container gate charge that shippers must pay."). The ICTF is now operated by UP; the other major rail line, Burlington Northern Santa Fe (BNSF), does not use the ICTF, but rather has its Los Angeles Intermodal Facility downtown at Hobart. Bonacich and Wilson, note 1, 108. As discussed more fully in chapter 6, there is a controversial effort by BNSF to create an intermodal facility closer to the port complex. In May 2013, the Los Angeles City Council approved a new rail yard in Wilmington, adjacent to one of the city's major high schools. Dan Weikel, "Rail Yard for Port Complex OKd," *Los Angeles Times*, May 9, 2013, AA3. "The 153-acre project would be capable of handling up to 2.8 million 20-foot shipping containers a year by 2035 and 8,200 trucks a day." Ibid. Public officials and advocates disputed its environmental impact, with proponents claiming that it would reduce the number of truck trips each year by one million, while opponents argued that overall emissions would increase and local community residents would be disproportionately affected. Ibid. As of 2016, the project was on hold after a court ruled that the environmental analysis for the Southern California International Gateway was inadequate. "BNSF to Appeal Ruling in SCIG Case," *BNSF Connects*,

http://www.bnsfconnects.com/blog/entry/bnsf-to-appeal-ruling-in-scig-case (accessed June 14, 2017).

170. Bonacich and Wilson, note 1, 108–109.

171. Ibid., 109.

172. Ibid., 108. State Highway 103 was built in the 1940s to connect the naval base on Terminal Island to the mainland, but eventually became a main conduit for port trucking. See Christine Mai-Duc and Laura J. Nelson, "Turning Freeway to Park?" *Los Angeles Times*, Nov. 20, 2013, A1. The highway was never connected to the interstate freeway system and has become less important with the creation of the Alameda Corridor rail project and the expansion of the 710 freeway. Ibid. In 2013, the Long-Beach owned portion of the highway was set to be decommissioned under a plan that would convert it to greenspace. Greg Yee, "Plans to Decommission Terminal Island Freeway in West Long Beach Unveiled," *Long Beach Press-Telegram*, Oct. 24, 2015, http://www.presstelegram.com/environment-and-nature/20151024/plans-to-decommission-terminal-island-freeway-in-west-long-beach-unveiled (accessed June 14, 2017).

173. Bonacich and Wilson, note 1, 108.

174. Ibid.; Erie, note 3, 93.

175. Bonacich and Wilson, note 1, 102.

176. William Fulton, *The Reluctant Metropolis* (Baltimore, MD: Johns Hopkins University Press, 2001), 133–134.

177. See "Final Harbor Freeway Link to Be Opened," *Los Angeles Times*, Sept. 24, 1962, 25; "Harbor Freeway Link Opens Today," *Los Angeles Times*, July 30, 1952, A1; "South LA: Harbor Freeway Sector Opens with Ceremony," *Los Angeles Times*, Sept. 25, 1958, B1; see also Josh Sides, *L.A. City Limits: African American Los Angeles from the Great Depression to the Present* (Berkeley: University of California Press, 2003), 113–114.

178. See City of South Pasadena v. Slater, 56 F. Supp. 2d 1106 (C.D. Cal. 1999).

179. Erie, note 3, 23.

180. Bonacich and Wilson, note 1, 58–59. In 1985, the Port of Los Angeles led the nation in car imports. James Risen, "Japanese Shipments to GM Soar, L.A. Port Now Leads Nation in Car Imports," *Los Angeles Times*, Sept. 4, 1985, B1.

181. Tim Waters, "Wave of Imported Autos Floods Ports," *Los Angeles Times*, June 5, 1986, SB1.

182. Jane Fritsch, "Trade Missions Prime Pump for L.A. Port," *Los Angeles Times*, May 29, 1990, B1 ("Los Angeles port officials said the presence of Bradley on trade missions gives them a secret weapon not available to port representatives from other

cities. Bradley's stature as mayor of the nation's second-most populous city and now its busiest port opens doors in the Far East that would otherwise be inaccessible, they said.").

183. Douglas Jehl, "Port of L.A. Steams Past New York as Busiest in U.S.," *Los Angeles Times*, Mar. 30, 1990, A1.

184. Bonacich and Wilson, note 1, 107–108. Long Beach proposed additional on-dock rail lines to also address this problem. See Daryl Kelley, "Opposition Rises to Proposal for Dockside Cargo Trains," *Los Angeles Times*, Jan. 16, 1986, SE9. However, the proposal ran up against opposition for creating noise pollution and was ultimately shelved. See Daryl Kelley, "Dockworkers Stage Sickout on Project Delay," *Los Angeles Times*, Dec. 2, 1986, M1; Daryl Kelley, "Plan for Dockside Rail Yard Yanked Amid Opposition," *Los Angeles Times*, Dec. 4, 1986, LB9.

185. Myra L. Frank and Assocs., Inc., *Alameda Corridor: Environmental Impact Report* S-1 (1993); see also Nona Liegeois, Francisca Gonzalez Baxa, and Barbara Corkrey, "Helping Low-Income People Get Decent Jobs: One Legal Services Program's Approach," *Clearinghouse Review* 33 (1999): 279, 289.

186. Liegeois et al., note 185, 289.

187. "Number of Trains Running on the Alameda Corridor," Alameda Corridor Transportation Authority, http://www.acta.org/pdf/CorridorTrainCounts.pdf (accessed Aug. 30, 2013).

188. As of 2013, there were six on-dock rail lines, all of which connected to the Alameda Corridor. Interview with John Holmes, note 168.

189. In addition, the Alameda Corridor also connected to UP's East Los Angeles Yard near downtown. Bonacich and Wilson, note 1, 109. However, it bypassed the BNSF downtown yards, including Hobart, which operate as alternatives to the Alameda Corridor for shippers loading onto BNSF trains off-dock. Ibid.

190. Erie, note 3, 147; see also Daryl Kelley, "L.B. Port Looks to New Growth Spurt in 1985: Harbor's 35-Year Plan Will Expand Size to Match Its Increasing Cargo," *Los Angeles Times*, Jan. 6, 1985, SE1 ("The Port of Long Beach, straining at its seams after two decades of robust growth, is looking to 1985 to complete $150 million in wharf, road and rail construction and to firm up another $400 million in building for 1986.").

191. Erie, note 3, 92; see also Dean Murphy, "Ports Cheer Bill Promising Major Growth of Harbor," *Los Angeles Times*, Oct. 26, 1986, SB1.

192. Dan Weikel, "L.A., Long Beach Ports Will See More Cargo Volume," *Los Angeles Times*, Dec. 18, 1998, C1 (quoting Mark Pisano, executive director of the Southern California Association of Government).

193. Erie, note 3, 119–123.

Notes

194. Ibid., 141 tbl. 5.4.

195. Bonacich and Wilson, note 1, 45.

196. Ibid., 49. China trade increased as part of a concerted effort by port officials. In the early 1990s, Mayor Tom Bradley led a delegation of port officials to expand Los Angeles's share of the booming China import business. George White, "Port of Los Angeles Seeks More China Business," *Los Angeles Times*, Sept. 23, 1991, D3.

197. George White, "Packing in at Port of L.A.," *Los Angeles Times*, May 8, 1989, C1 ("In all, about 99% of the space at the Port of Los Angeles is leased. Of the 3,800 acres of land at the port's locations in Wilmington, San Pedro and Terminal Island, only one berth—a 25-acre site—is unleased.").

198. Tim Waters, "Activity Rises Sharply at Port of Los Angeles," *Los Angeles Times*, Sept. 19, 1985, SD C1 ("The Port of Los Angeles lured six steamship companies from berths at rival Long Beach last year and, as a result, handled a third more general cargo. According to figures released this week, 22.2 million metric tons of general cargo—that is, containerized cargo and all loose cargo except liquid or dry bulk—passed through the Los Angeles port during the 12 months ended June 30, a 34.5% increase from the 16.5 million tons the year before.").

199. Edmund Newton, "The Ship Has Come in: Long Beach Steams Past L.A. to Become Nation's No. 1 Port," *Los Angeles Times*, Apr. 23, 1995, D1.

200. The World Bank and Public-Private Infrastructure Advisory Facility, "Module 2: The Evolution of Ports in a Competitive World," *Port Reform Toolkit*, 2d ed. (2007), 48, http://www.ppiaf.org/sites/ppiaf.org/files/documents/toolkits/Portoolkit/Toolkit/pdf/modules/02_TOOLKIT_Module2.pdf (accessed June 14, 2017).

201. Dan Weikel, "Major Shipping Firm to Leave Long Beach Port for Los Angeles," *Los Angeles Times*, Oct. 28, 1999, B1.

202. Erie, note 3, 142–143.

203. Asian imports continued to lead the way. See "Asian Imports Up 20% Over a Year Ago," *Los Angeles Times*, June 18, 1998, D2; Gregory Stephen, "Imports Climb at L.A., Long Beach," *Los Angeles Times*, Nov. 20, 1998, C2.

204. Bonacich and Wilson, note 1, 120–121.

205. Ibid.

206. See, e.g., Chris Kraul, "Mexican Port Hopes to Be Big in Containers," *Los Angeles Times*, Dec. 17, 1991, D2.

207. Harry Bernstein, "East, Gulf Dock Strike Issues Loom for West," *Los Angeles Times*, Jan. 20, 1969, D1 (noting that more shippers were loading their goods into containers which threatened longshore jobs).

208. Bonacich and Wilson, note 1, 177.

209. Queenan, note 4, 132–133. This reduction in longshoremen was welcomed by the ports, which had seen shippers divert cargo to San Francisco and San Diego as a result of labor unrest in the 1950s. See ibid., 131.

210. Ibid., 143.

211. Ibid., 143–144.

212. See, e.g., James Flanigan, "Striking Costs: Region Has Much to Lose if Shippers Decide to Go Elsewhere," *Los Angeles Times*, July 16, 1997, D1.

213. Harry Bernstein, Merger of Dock Unions Opposed, *Los Angeles Times*, Jan. 5, 1972, A3.

214. Ibid.

215. Harry Bernstein, "West Coast Shipping Off 50% as New Dock Strike Looms," *Los Angeles Times*, Jan. 14, 1972, A3.

216. Harry Bernstein, "Teamsters-Dockers Merger in Works," *Los Angeles Times*, Jan. 15, 1972, A1.

217. Harry Bernstein, "Unions Will Picket Trucks at Border to Tighten Ship Strike," *Los Angeles Times*, Jan. 28, 1972, A1.

218. Harry Bernstein, "Dock Work Stoppages Employed by Bridges," *Los Angeles Times*, Aug. 5, 1972, B10.

219. Talley, note 89, 213–217; see also Tim Waters, "Cargo Moves Again After Accord Ends Port Strike," *Los Angeles Times*, July 8, 1986, B1 (reporting that the strike by longshoremen in support of port office workers over job security was the longest in fifteen years and resulted in a three-year contract, under which "employers agree not to transfer jobs out of the union's jurisdiction").

220. Bonacich and Wilson, note 1, 177–179; see also Talley, note 89, 214–216; Henry Weinstein, "4,000 Dockers Walk Out Over 5 Deaths in a Year," *Los Angeles Times*, June 28, 1985, A1. Because of a similar power to disrupt operations, harbor pilots, a small group that steered cargo ships through the ports, also exerted great leverage to negotiate large salaries and benefits (they were set to make over $140,000 on average in 2001 after a year-long strike in 1997–1998). Dan Weikel, "Port Chief Agrees to Sign Harbor Pilots' Delayed Contract," *Los Angeles Times*, July 10, 1998, B5.

221. Bonacich and Wilson, note 1, 57.

222. Ibid. ("Ports used to invest mainly for the benefit of their region. Now they are being asked to invest for the benefit of the entire country, without the security of knowing that the investment will pay off. Even if a port is successful, the regions that are nearby may have to bear additional costs on top of the financial ones, such as congestions and pollution.").

Notes

223. Erie, note 3, 82–83 tbl. 4.1.

224. Ibid., 86–87 tbl. 4.2. The charter amendments included changes to harbor department rules asserting council approval over budgets and department salaries, and shifting control over the department from the city manager to the mayor, who was given the power to appoint commissioners subject to term limits. Ibid.

225. Ibid., 127.

226. Ibid., 109; see also Mark Gladstone and Ralph Frammolino, "Funds of San Diego, Other Port Districts Under Siege," *Los Angeles Times*, July 8, 1992, A1.

227. Erie, note 3, 124; see also Mark Gladstone and Greg Krikorian, "Port Cities Might Dodge a Fiscal Bullet," *Los Angeles Times*, Sept. 3, 1992, J9 (noting that state legislation would permit four charter cities with ports to use port revenues for two years to fund services such as fire and police, allowing them to balance budgets in light of lost state funding; the city of Los Angeles was expected to receive $44 million from the Port of Los Angeles). A class action was filed arguing that the state legislation was unconstitutional and seeking recovery of $69 million to the Port of Los Angeles and $21 million to the Port of Long Beach. Susan Woodward, "Harbor to Help L.A., Long Beach Fight Suit," *Los Angeles Times*, Sept. 29, 1994, J15.

228. Jeff Leeds and John Cox, "L.A. Harbor Panel Votes to Pay City $80 Million in Fees," *Los Angeles Times*, Aug. 25, 1995, B3 ("The Los Angeles Harbor Commission, hoping to end a two-year dispute, has decided to pay the city about $80 million in fees after a private study found that the port has underpaid for municipal services since 1977.").

229. Dan Weikel, "City to Repay $62 Million to Port of L.A.," *Los Angeles Times*, Jan. 20, 2001, B1 (stating that the State Lands Commission and some shipping companies sued Los Angeles in 1996, arguing that the payments violated the state Tidelands Trust Act requiring that port revenue be used only on harbor projects; that suit was settled in 2001, with the city agreeing to repay $62 million to the port).

230. Erie, note 3, 126–129.

231. Yuko Aoyama, Samuel Ratick, and Guido Schwarz, "Organizational Dynamics of the U.S. Logistics Industry: An Economic Geography Perspective," *Professional Geographer* 58 (2006): 327, 335.

232. Jean-Paul Rodrigue, Claude Comtois, and Brian Slack, *The Geography of Transport Systems*, 2nd ed. (New York: Routledge, 2009), 64.

233. For these concepts, see Jaffee, note 166, 6.

234. Ibid.

235. Ibid.

236. Ibid., 12.

237. Palos Verdes, within the unincorporated area of Los Angeles County, was largely undeveloped in the early twentieth century—used for cattle and farming. Its purchase by development interests in the early 1900s and eventual sale to a mining company led to the creation of a master plan for real estate development in the 1950s; as a building boom commenced, developers and homeowners pushed for separate incorporation to both facilitate and control the expansion. The last incorporation—the city of Rancho Palos Verdes—was completed in 1973, after dramatic litigation that went to the California Supreme Court, which determined that municipal incorporation had to occur based on a vote of a majority of individual residents and not a majority of the ownership of assessed land value. Curtis v. Bd. of Supervisors, 501 P.2d 537 (Cal. 1972).

238. The port, in an effort to mitigate the negative impact of its facilities in San Pedro, created the Cabrillo Beach Recreational Complex on the peninsula's southern tip, which includes a park, picnic area, bird sanctuary, and marina. Queenan, note 4, 157. Developers built up the surrounding area, which included an $18 million, 216-room hotel two blocks from the port. "$18-Million Hotel Planned in San Pedro," *Los Angeles Times*, Nov. 3, 1985, H26.

239. In 2013, Palos Verdes Peninsula High School was ranked as the 286th best public or private high school in the United States. See "America's Most Challenging High Schools: Palos Verdes," *Washington Post*, May 5, 2017, http://apps.washingtonpost.com/local/highschoolchallenge/schools/2011/list/national/palos-verdes-palos-verdes-estates-ca/ (accessed Oct. 2, 2013).

240. See "Mapping L.A., Palos Verdes Estates," *Los Angeles Times*, http://maps.latimes.com/neighborhoods/neighborhood/palos-verdes-estates/ (accessed Oct. 24, 2013); "Mapping L.A., San Pedro," *Los Angeles Times*, http://projects.latimes.com/mapping-la/neighborhoods/neighborhood/sanpedro/?q=San+Pedro%2C+Los+Angeles%2C+CA%2C+USA&lat=33.7360619&lng=-118.2922461&g=Geocodify (accessed Oct. 2, 2013).

241. Bridges was known for his visionary leadership and commitment to promoting a broad-based movement for economic justice. He was also a Communist who spent most of his adult life fighting against the government's effort to deport him.

242. "Mapping L.A., Wilmington," *Los Angeles Times*, http://projects.latimes.com/mapping-la/neighborhoods/neighborhood/wilmington/?q=Wilmington,+CA,+USA&lat=33.7857948&lng=118.2643567&g=Geocodify (accessed Oct. 2, 2013).

243. See generally Bonacich and Wilson, note 1, 159–240; Arin Dube et al., "On the Waterfront and Beyond: Technology and the Changing Nature of Cargo-Related Employment in the West Coast" (unpublished manuscript, 2004).

244. See The World Bank and Public-Private Infrastructure Advisory Facility, "Module 1: Framework for Port Reform," *Port Reform Toolkit*, 2d ed. (2007), http://www.ppiaf.org/sites/ppiaf.org/files/documents/toolkits/Portoolkit/Toolkit/pdf/

Notes

modules/01_TOOLKIT_Module1.pdf (accessed June 14, 2017); see also Bonacich and Wilson, note 1, 199–224.

245. See Jaffee, note 166, 15. Power resides in the "upstream/downstream impact of failure to deliver goods." Beverly J. Silver, *Forces of Labor: Workers' Movements and Globalization Since 1870* (New York: Cambridge University Press, 2003), 100; see also Edna Bonacich, "Pulling the Plug: Labor and the Global Supply Chain," *New Labor Forum* 12 (2003): 41.

246. Jaffee, note 166, 15.

247. Talley, note 89, 215.

248. Nancy Cleeland, "Dispute Shows a Union Firmly Anchored at West Coast Ports," *Los Angeles Times*, Oct. 13, 2002, C1.

249. Nancy Cleeland, "Long Lines at the Ports Are Gone," *Los Angeles Times*, Nov. 11, 2002, C2.

250. Talley, note 89, 216.

251. Jaffee, note 166, 14.

252. David Bensman, *Port Trucking Down the Low Road: A Sad Story of Deregulation* (Dēmos, 2009), 3, http://www.demos.org/sites/default/files/publications/PortTrucking DowntheLowRoad.pdf (accessed June 14, 2017).

253. Bonacich and Wilson, note 1, 209–210.

254. Ibid., 211–212.

255. This was part of an overall employer strategy to promote labor flexibility. Jaffee, note 166, 11.

256. Despite the "independent" designation, researchers have suggested that contract relationships between trucking companies and drivers are not that distinct from the pre-deregulatory employment system. See Rebecca Smith, David Bensman, and Paul Alexander Marvy, *The Big Rig: Poverty, Pollution, and the Misclassification of Truck Drivers at America's Ports* (2010), 26, report from the National Employment Law Project and Change to Win (concluding that "[t]rucking companies exert a high degree of control over the work activities of the truck drivers"); see also Belzer, note 86, 37; Bensman, note 252, 11–12. Jaffee notes that independent-contractor drivers are typically prevented from working for more than one company. Jaffee, note 166, 17. Milkman and Wong state that drivers rely on companies to finance the acquisition of trucks and insurance. Ruth Milkman and Kent Wong, "Organizing Immigrant Workers: Case Studies from Southern California," in *Rekindling the Movement: Labor's Quest for Relevance in the 21st Century*, ed. Lowell Turner, Harry C. Katz, and Richard W. Hurd (Ithaca, NY: Cornell University Press, 2001), 99.

257. Jaffee, note 166, 17.

258. Bonacich and Wilson, note 1, 104.

259. See, e.g., Columbia River Packers Ass'n v. Hinton, 315 U.S. 143 (1942); see generally Note, "Employee Bargaining Power Under the Norris-LaGuardia Act: The Independent Contractor Problem," *Yale Law Journal* 67 (1957): 98 (arguing that there should be exceptions to the ban on organizing for independent contractors). On the challenges to unionization after deregulation, see Michael H. Belzer, "Collective Bargaining After Deregulation: Do the Teamsters Still Count?" *Industrial and Labor Relations Review* 48 (1995): 636.

260. See, e.g., Ronald D. White, "The Ports' Short-Haul Truckers Endure Long Hours, High Costs," *Los Angeles Times*, Nov. 21, 2005, C1. In a study of the overall trucking industry, Belman and Monaco reported that truckers' wages fell by 21 percent from 1973 to 1995, and that one-third of that decrease was attributable to deregulation. Dale L. Belman and Kristen A. Monaco, "The Effects of Deregulation, De-Unionization, Technology, and Human Capital on the Work and Work Lives of Truck Drivers," *Industrial and Labor Relations Review* 54 (2001): 502.

261. Jaffee, note 166, 18.

262. Bensman, note 252, 1.

263. Some studies of the overall industry, not just Los Angeles-Long Beach, put the wage rate much lower. See Ted Prince, "Endangered Species," *Journal of Commerce*, May 9, 2005, 12–16; see also Bonacich, note 245, 46. Studies of drivers at other ports have found similarly low pay. See, e.g., East Bay Alliance for a Sustainable Economy, *Taking the Low Road: How Independent Contracting at the Port of Oakland Endangers Public Health, Truck Drivers, & Economic Growth* (2007), http://www.forworkingfamilies.org/sites/pwf/files/publications/archive/ebase/TakingTheLowRoad.pdf (accessed June 14, 2017) (finding that Oakland drivers made $10.69 per hour on average and that one-quarter made less than $7.64 per hour).

264. Kristen Monaco, *Incentivizing Truck Retrofitting in Port Drayage: A Study of Drivers at the Ports of Los Angeles and Long Beach* (2008), 18–19, final report for Metrans Project 06-02 (finding that 91.24 percent of port truck drivers were Hispanic and 44 percent were noncitizens). Bonacich and Wilson cite sources estimating that 90 percent of port truckers are from Central America, while the remainder are Mexican. Bonacich and Wilson, note 1, 218.

265. Bonacich and Wilson, note 1, 212 (quoting Ernesto Nevarez, a port trucking activist).

266. Ibid.

267. Ibid., 212–213.

268. Ibid.

Notes

269. Ibid., 223–224 (noting also that some longshoremen viewed immigrants as responsible for the labor movement's decline).

270. See Bensman, note 252, 8–10.

271. See ibid., 10; see also Antoine Frémont, "Empirical Evidence for Integration and Disintegration of Maritime Shipping, Port and Logistics Activities" (2009) (OECD/ITF Joint Transport Research Centre Discussion Paper); David Bensman, "Barriers to Innovation in Global Logistics: The Deregulated Port Trucking Sector" (unpublished presentation to the Industry Studies Association, 2009).

272. A study by Monaco and Grobar found that, for each trip, port truck drivers spent more time waiting than driving. Kristen Monaco and Lisa Grobar, *A Study of Drayage at the Ports of Los Angeles and Long Beach* (2005), 11.

273. Jaffee, note 166, 24.

274. Ibid., 9.

275. See Deborah Schoch, "Study Details Port Pollution Threat," *Los Angeles Times*, Mar. 22, 2004, C1.

276. Smith, Bensman, and Marvy, note 256, 10 (quoting Art Marroquin, "Judge Rules Port of L.A. Can Fully Implement Clean Trucks Program," *Daily Breeze*, Aug. 26, 2010). In 2008, Monaco reported that the median model year of trucks driven by independent contractors in Los Angeles was 1995–1996. Monaco, note 264, 12. Port Director of Operations John Holmes described the market this way: "Generally, big companies like Swift would own trucks for five years, then sell to regional carriers which would own them for a few years, then the trucks would be sold into the drayage market." Interview with John Holmes, note 168. A survey commissioned by the ports in 2007 found that the "vast majority of drivers engaged in Port drayage" were independent owner operators and that the average truck was from 1994. CGR Management Consultants, *A Survey of Drayage Drivers Serving the San Pedro Bay Ports* (2007), 1.

277. See, e.g., Diane Bailey et al., "Driving on Fumes: Truck Drivers Face Elevated Health Risks from Diesel Pollution" NRDC Issue Paper, Dec. 2007, 8 tbl. 1, http://www.nrdc.org/health/effects/driving/driving.pdf (accessed June 14, 2017).

278. Ibid., 4.

279. Pingkuan Di, *Diesel Particulate Matter Exposure Assessment Study for the Ports of Los Angeles and Long Beach* (2006), 2, final report for the California Environmental Protection Agency, Air Resources Board, https://www.arb.ca.gov/ports/marinevess/documents/portstudy0406.pdf (accessed June 14, 2017).

280. Ibid.

Chapter 3

1. Guido G. Weigend, "Some Elements in the Study of Port Geography," *Geographical Review* 48 (1958): 185.

2. Jean-Paul Rodrigue, Claude Comtois, and Brian Slack, *The Geography of Transport Systems*, 2nd ed. (New York: Routledge, 2009), 343.

3. David Jaffee, "Kinks in the Intermodal Supply Chain: Longshore Workers and Drayage Drivers" (unpublished manuscript, June 2010), 12–13, https://www.unf.edu/uploadedFiles/aa/coas/cci/ports/REPORT_PortPaper10-SASE-KinksintheIntermodalSupplyChain.pdf (accessed June 14, 2017); see also Mary R. Brooks, "The Governance Structure of Ports," *Review of Network Economics* 3 (2004): 168.

4. See Jaffee, note 3, 13.

5. Ibid.

6. *Port of Los Angeles Handbook & Business Directory* (2011–2012), 34–42.

7. See City of Long Beach v. Morse, 188 P.2d 17, 21–22 (Cal. 1947); see also Marks v. Whitney, 491 P.2d 374, 380 (Cal. 1971) (holding that the trust also encompasses using the tidelands for open space, ecological preservation, scientific study, and recreation).

8. See *Morse*, note 7, at 19.

9. Under the current Los Angeles city charter, the Board of Harbor Commissioners is granted "possession, management and control of all navigable waters and all tidelands and submerged lands." L.A., Cal., City Charter art. VI, § 651(a) (2014).

10. Ibid., § 502(a), (d).

11. Ibid., § 655.

12. Ibid.

13. Ibid., § 604(a).

14. Ibid., § 508(e). The manager possesses the right to appeal termination to the City Council.

15. Ibid., § 652(e).

16. Ibid., § 653(a).

17. Ibid., §§ 605(a), 606.

18. Ibid., § 654(a)(1).

19. See City of Los Angeles Harbor Department, "Adopted Budget: Fiscal Year 2013–2014" (2013), 4 (stating that 87.2 percent of port revenues come from shipping ser-

Notes

vices and 11.0 percent from rentals; another 1.8 percent come from royalties and fees, and other operating revenues).

20. See Rodrigue, Comtois, and Slack, note 2, 181; see also Nicholas Miranda, "Concession Agreements: From Private Contract to Public Policy," *Yale Law Journal* 117 (2007): 510. The Port of Los Angeles also leases property to "various shipyards, fish markets, boat repair yards, railroads, restaurants and other similar operations." City of Los Angeles Harbor Department, "Financial Policies," (2013), 7.

21. See Nelson Lichtenstein, *State of the Union: A Century of American Labor* (Princeton, NJ: Princeton University Press, 2002), 260–264.

22. See Norris LaGuardia Act of 1932, Pub. L. No. 72–65, 47 Stat. 70 (codified as amended at 29 U.S.C. §§ 101–115 (2012)); see also National Labor Relations (Wagner) Act of 1935, ch. 372, 49 Stat. 449 (codified as amended at 29 U.S.C. §§ 151–69 (2012)).

23. See generally NLRB v. Jones & Laughlin Steel Corp., 301 U.S. 1 (1937).

24. See Cynthia L. Estlund, "The Ossification of American Labor Law," *Columbia Law Review* 102 (2002): 1527, 1571.

25. See Catherine Fisk, "Law and the Evolving Shape of Labor: Narratives of Expansion and Retrenchment," *Law, Culture and the Humanities* 8 (2012): 1, 2, 9.

26. See Taft-Hartley Act of 1947, Pub. L. No. 80–101, 61 Stat. 136 (codified as amended in scattered sections of 29 U.S.C. (2012)); see also Landrum-Griffin Act of 1959, Pub. L. No. 86–257, § 101, 73 Stat. 519 (codified as amended in scattered sections of 29 U.S.C. (2012)).

27. See NLRB v. Mackay Radio & Tel. Co., 304 U.S. 333 (1938) (validating the use of permanent replacements for striking workers).

28. See Benjamin I. Sachs, "Employment Law as Labor Law," *Cardozo Law Review* 29 (2008): 2685, 2694–2700.

29. Ibid.

30. Sherman Antitrust Act, 15 U.S.C. § 1 (West 2004).

31. See Sanjukta M. Paul, "The Enduring Ambiguities of Antitrust Liability for Worker Collective Action," *Loyola University Chicago Law Journal* (2016): 969 (questioning the historical underpinnings of the labor exemption).

32. Steven P. Erie, *Globalizing L.A.: Trade, Infrastructure, and Regional Development* (Stanford, CA: Stanford University Press, 2004), 106–109.

33. Ibid.

34. Ibid.

35. The California Coastal Commission, established in 1972, also has the power to approve all development in the coastal zone. Ibid., 108.

36. See 42 U.S.C. § 7604(a) (2012).

37. See generally California Clean Air Act, California Health and Safety Code §§ 40910–40930 (West 2006).

38. South Coast Air Quality Management District, Authority, http://www.aqmd.gov/home/about/authority (accessed July 3, 2017).

39. Ibid.

40. California Health and Safety Code §§ 40000, 43103(b), 43018 (West 2006). CARB exercises its authority over motor vehicles under a Clean Air Act waiver permitting it to set its own on-road vehicle emission standards. See 42 U.S.C. § 7543(b) (2012).

41. California Health and Safety Code § 40000 (West 2006).

42. Shipping Act, 46 U.S.C. § 41102(c) (2012); see also Marva Jo Wyatt, "Ports, Politicians and the Public Trust: The Los Angeles Port Funds Controversy Comes Face to Face with Federal Law," *University of San Francisco Maritime Law Journal* 9 (1997): 357, 363.

43. 46 U.S.C. § 41307(b)(1) (2012).

44. 49 U.S.C. § 14501 (2012).

45. Ibid., § 14501(a)(1).

46. Californians for Safe & Competitive Dump Truck Transp. v. Mendonca, 152 F.3d 1184, 1187 (9th Cir. 1998).

47. 49 U.S.C. § 14501(c)(1).

48. See Ann E. Carlson, "Federalism, Preemption, and Greenhouse Gas Emissions," *U.C. Davis Law Review* 37 (2003): 281, 283–285; Catherine L. Fisk, "The Anti-Subordination Principle of Labor and Employment Law Preemption," *Harvard Law and Policy Review* 5 (2011): 17, 18.

49. See U.S. Constitution art. 6, clause 2.

50. See, e.g., Cipollone v. Liggett Grp., Inc., 505 U.S. 504, 516 (1992).

51. Ibid.

52. Pacific Gas & Elec. Co. v. State Energy Res. Conservation & Dev. Comm'n, 461 U.S. 190, 204 (1983).

53. San Diego Bldg. Trades Council v. Garmon, 359 U.S. 236, 247 (1959).

Notes

54. See Lodge 76, Int'l Ass'n of Machinists & Aerospace Workers v. Wis. Emp't Relations Comm'n, 427 U.S. 132 (1976).

55. See Jim Rossi and Thomas Hutton, "Federal Preemption and Clean Energy Floors," *North Carolina Law Review* 91 (2013): 1283, 1294–1303.

56. Rowe v. N.H. Motor Transp. Ass'n, 552 U.S. 364, 367–368 (2008).

57. See Michael Burger, "'It's Not Easy Being Green': Local Initiatives, Preemption Problems, and the Market Participant Exception," *University of Cincinnati Law Review* 78 (2010): 835, 843–844.

58. Wis. Dep't of Indus., Labor & Human Relations v. Gould Inc., 475 U.S. 282, 289 (1986).

59. Bldg. & Constr. Trades Council v. Associated Builders & Contractors of Mass./ R.I., Inc., 507 U.S. 218, 227 (1993). The market participation doctrine had its origins in dormant commerce clause cases, where it was used to permit some local action impacting interstate commerce if done for proprietary reasons. See generally Hughes v. Alexandria Scrap Corp., 426 U.S. 794 (1976). Although the market participation doctrine developed to permit state law making, some jurisdictions, like the Ninth Circuit, applied the exception to municipal government entities as well. See Big Cnty. Foods, Inc. v. Bd. of Educ. 952 F.2d 1173, 1178–1179 (9th Cir. 1992).

60. See Engine Mfrs. Ass'n v. S. Coast Air Quality Mgmt. Dist., 498 F.3d 1031, 1041 (9th Cir. 2007).

61. *Gould*, 475 U.S. at 287 (finding that a Wisconsin law prohibiting the state from doing business with companies that had violated the NLRA served "plainly as a means of enforcing the NLRA" and thus was preempted by it).

62. *Bldg. & Constr. Trades Council*, 507 U.S. at 229, 232; see also Chamber of Commerce of U.S. v. Brown, 554 U.S. 60, 70 (2008). The Ninth Circuit adopted a similar market participation test. See Johnson v. Rancho Santiago Cmty. Coll. Dist., 623 F.3d 1011, 1023–24 (9th Cir. 2010) (citing Cardinal Towing & Auto Repair, Inc. v. City of Bedford, 180 F.3d 686, 693 (5th Cir. 1999)).

63. See Associated Gen. Contractors of Am. v. Metro. Water Dist. of S. Cal., 159 F.3d 1178, 1183–84 (9th Cir. 1998); see also *Brown*, 554 U.S. at 70.

64. Engine Mfrs. Ass'n v. S. Coast Air Quality Mgmt. Dist., 541 U.S. 246 (2004). The SCAQMD's jurisdiction is over the South Coast Air Basin, which includes most of Los Angeles County, all of Orange County, and parts of San Bernardino and Riverside Counties.

65. Engine Mfrs. Ass'n v. S. Coast Air Quality Mgmt. Dist., No. Cv00-09065FMC(BQRX), 2005 WL 1163437, at *12 (C.D. Cal. May 5, 2005), *aff'd* Engine Mfrs. Ass'n v. S. Coast Air Quality Mgmt. Dist., 498 F.3d 1031, 1041 (9th Cir. 2007).

66. Tocher v. City of Santa Ana, 219 F.3d 1040 (9th Cir. 2000), *abrogated on other grounds by* City of Columbus v. Ours Garage & Wrecker Serv., Inc., 536 U.S. 424 (2002) (reversing that part of *Tocher* which held that a state may not delegate its regulatory authority to a municipality under the FAAAA's safety exception).

67. See, e.g., Estlund, note 24, 1571 (noting that the market participant exception is so difficult to ascertain that even incremental state and local reforms are subject to challenge as preempted).

Chapter 4

1. See Charles F. Queenan, *Long Beach and Los Angeles: A Tale of Two Ports* (Northridge, CA: Windsor Publications, 1986), 63.

2. See Donna St. George, "Wilmington: Community of Contradictions," *Los Angeles Times*, Oct. 6, 1985, SB1 ("In that era—as near as the community ever came to a heyday—Wilmington attracted tourists who traveled on cruise lines from the port to Santa Catalina Island, Hawaii and the South Pacific. What is now a pawn shop was a J.C. Penney store. The Don Hotel, where rooms now rent for $15 a night, catered to affluent steamship passengers.").

3. See Deborah Belgum, "A School's Not-So-Golden Anniversary," *Los Angeles Times*, July 5, 1996, B10.

4. Kenneth J. Garcia and Janet Rae-Dupree, "Aerospace Slump Casts Its Shadow," *Los Angeles Times*, Jan. 5, 1992, B3; see also Janet Rae-Dupree, "These Trying Times: Job Security Evaporates as Economic Ripple Effect Hits Home," *Los Angeles Times*, May 19, 1991, B3.

5. Dean Murphy, "Port in a Storm: Harbor Profits Are Rising, but So Are Neighbors' Complaints," *Los Angeles Times*, Nov. 27, 1987, SB12; Nancy Yoshihara, "2020: A Southland Port Plan for the Long Haul," *Los Angeles Times*, Oct. 19, 1987, E3.

6. See Murphy, note 5.

7. See St. George, note 2.

8. Ibid.

9. Ibid.

10. Sheryl Stolberg, "Wilmington, Harbor City Plan Will Be Revised Again: L.A. Planners Delay Action After Hearing," *Los Angeles Times*, June 30, 1989, B10 South Bay ed.

11. St. George, note 2.

12. Ibid.

13. See Murphy, note 5.

14. St. George, note 2.

15. Donna St. George, "Wilmington—Battered but Not Broken: Pride and Community Spirit Persevere Despite Area's Problems," *Los Angeles Times*, Oct. 10, 1985, SB1. The *Los Angeles Times* reported that "[a]bout 45 percent of Wilmington's 11,518 dwelling units are owner-occupied homes"; "the area remains largely blue-collar and union-oriented, with 63 percent of its population employed as laborers and 12 union halls located in the community"; "Latinos now make up at least 67 percent of Wilmington's population, compared to 27.5 percent citywide." Ibid. The *Los Angeles Times* also reported that the remaining population was 22 percent Anglo, 8 percent Asian and American Indian, and 4 percent black; undocumented immigrants were estimated to be 10–20 percent of the population and unemployment stood at 8 percent. Ibid.

16. Murphy, note 5.

17. Donna St. George, "Moratorium on Building Sought in Wilmington," *Los Angeles Times*, Nov. 3, 1985, SB1.

18. Donna St. George, "Wilmington Activists Chalk Up Another Win," *Los Angeles Times*, Oct. 24, 1985, SB1.

19. Donna St. George, "Wilmington Woes to Get Closer Look in City Plan," *Los Angeles Times*, Jan. 19, 1986, SB1.

20. "Wilmington: Hamilton to Conduct Study," *Los Angeles Times*, Feb. 12, 1987, SB2.

21. Murphy, note 5.

22. Sheryl Stolberg, "Cinderella-by-the-Sea," *Los Angeles Times*, Dec. 13, 1987, U24.

23. Sheryl Stolberg, "Homeowners Offer Plan to Upgrade Community," *Los Angeles Times*, Dec. 13, 1987, M24 South Bay ed.

24. Sheryl Stolberg, "Flores Unveils Plans for Port Traffic Study, Other Projects," *Los Angeles Times*, Jan. 22, 1988, SB8 (indicating Flores' plan offered $750,000 in government funding to develop a port transportation plan, waterfront promenade, and athletic fields).

25. Dean Murphy, "Flores Has an Anchor: She's Coasting Toward Reelection in Diverse District," *Los Angeles Times*, Apr. 2, 1989, B1.

26. Sheryl Stolberg, "Catch-22 for Wilmington Hazardous Facilities Bar Recreation Uses," *Los Angeles Times*, Jan. 28, 1988, 8.

27. Sheryl Stolberg, "Rezoning Debate: Harbor Activists See Opportunity," *Los Angeles Times*, Mar. 13, 1988, M6 South Bay ed.

28. Dean Murphy, "L.A. Port Urged to Relocate Waterfront Petroleum Plant," *Los Angeles Times*, Apr. 21, 1988, 8.

29. Sheryl Stolberg, "Wilmington Residents Win in Bid for Waterfront Access," *Los Angeles Times*, July 28, 1988, SB8.

30. Sheryl Stolberg, "Long-Stagnant Industrial Park Rushing to Life in Wilmington," *Los Angeles Times*, Oct. 15, 1988, H22; see also Sheryl Stolberg, "Harbor Plan Brings New Wilmington a Little Closer," *Los Angeles Times*, Dec. 25, 1988, SB6 (noting that the city also initiated plans for the further redevelopment of Avalon Boulevard as a commercial corridor).

31. Sheryl Stolberg, "Revival of Wilmington Depends on 2 Key Projects: Officials Say Hotel, Waterfront Plans Will Affect Growth," *Los Angeles Times*, May 28, 1989, N8.

32. Dean Murphy, "Area Plan Revision Inches Slowly Ahead: Wilmington, Harbor City Waiting 6 Years," *Los Angeles Times*, June 23, 1989, J8A.

33. Sheryl Stolberg, "Final Polish Put on Harbor Area Plan Before Vote," *Los Angeles Times*, Dec. 1, 1989, B3.

34. Clay Evans, "Groups in Wilmington Oppose Junkyard Pact," *Los Angeles Times*, Apr. 5, 1990, B6.

35. Clay Evans, "Proposal for Former Cannery Rejected by Harbor Director," *Los Angeles Times*, Apr. 26, 1990, B3.

36. Greg Krikorian, "Cannery Monument Proposal Is Canned," *Los Angeles Times*, Aug. 4, 1990, B13.

37. See Clay Evans, "They Call This Part of Wilmington 'Third World,'" *Los Angeles Times*, May 6, 1990, B1.

38. Greg Krikorian, "East Wilmington Business Owner Sues City over Area's Blight, Crime," *Los Angeles Times*, Sept. 6, 1991, B3.

39. Susan Woodward, "This 'Third World' May Be the Most Run-Down Section of Los Angeles," *Los Angeles Times*, Nov. 10, 1994, 9.

40. Dan Weikel, "Wilmington Still Waits for Its Ship to Come In," *Los Angeles Times*, May 28, 2000, B1.

41. Ibid.

42. Ibid.

43. See Dean Murphy, "As Wilmington Suspected, Study Finds Port Gets More Containers than It Ships," *Los Angeles Times*, Nov. 16, 1989.

44. Sheryl Stolberg, "End to Traffic Sought: Wilmington Residents Hungry for Truck Stop," *Los Angeles Times*, Oct. 18, 1987, SB1.

Notes

45. Ibid.

46. Sheryl Stolberg, "Cement Plant OKd Despite Traffic Fear," *Los Angeles Times*, Nov. 19, 1987, SE14; see also Murphy, note 5.

47. Tim Waters, "Police Crack Down on Faulty Big Rigs in Port Area Traffic," *Los Angeles Times*, Aug. 30, 1990, B3 South Bay ed. ("For the fourth time in eight months, law-enforcement officers fanned out in the Harbor area Tuesday to crack down on errant truck drivers traveling into and out of the ports of Los Angeles and Long Beach.").

48. Greg Krikorian, "Wilmington Bans Trucks on 3 Streets," *Los Angeles Times*, July 11, 1991, B3 South Bay ed. ("Under the restrictions adopted Tuesday, the trucks will be banned from Avalon Boulevard, between B Street and the Carson border; Wilmington Boulevard, from C Street, to the Carson border, and on Anaheim Street, from Eubank Avenue to Figueroa Street, unless they have local deliveries or pickups in Wilmington.").

49. "Port Traffic Study Ordered," *Los Angeles Times*, Apr. 21, 1988, M10 South Bay ed.

50. Susan Woodward, "Harbor Area: Report on Improving Truck Traffic Approved," *Los Angeles Times*, Aug. 4, 1994, J7 South Bay ed.

51. Murphy, note 43 (reporting that over a recent twenty-month period, the Los Angeles port received 117,000 more containers than it shipped out, raising concerns about storage in Wilmington).

52. See "Curb Sought on Containers," *Los Angeles Times*, July 14, 1988, M9 South Bay ed.

53. Ibid.

54. Sheryl Stolberg, "Thefts of Container Cargo Soaring in Harbor Area," *Los Angeles Times*, Dec. 22, 1989, B3A.

55. Greg Krikorian, "Plan to Expand Ports Meets with Public Criticism," *Los Angeles Times*, Oct. 11, 1990, B3 South Bay ed.

56. Ibid.

57. Greg Krikorian, "Hiuka Drops Its Plans for Scrap Yard in Wilmington," *Los Angeles Times*, Sept. 4, 1992, B3 South Bay ed. The Hiuka America Corporation Plant, which shipped 500,000 tons of scrap through the Port of Los Angeles each year, was forced to leave San Pedro after residents mounted a successful five-year campaign to have the scrap yard declared a nonconforming use. Dean Murphy, "San Pedro Zone Change May Push Scrap Yard Out," *Los Angeles Times*, Aug. 2, 1990, B3 South Bay ed.

58. Susan Woodward, "Scrap Metal Firm's Lease Creates a Ruckus," *Los Angeles Times*, Aug. 25, 1994, 3.

59. Sheryl Stolberg, "L.A. Planners Won't Downzone Gaffey St. for Light Industry," *Los Angeles Times*, Feb. 2, 1990, B3 South Bay ed.

60. Tim Waters, "Dust Pollution Problem Remains in Doubt: Firms May Evade Petroleum Coke Rule," *Los Angeles Times*, Jan. 10, 1985, SE4; Tim Waters, "Kaiser Ordered to Stop Storing Coal, Petroleum Coke at Port," *Los Angeles Times*, Mar. 31, 1985, SB2.

61. "Petroleum Coke OKd," *Los Angeles Times*, Oct. 24, 1987, M2.

62. Murphy, note 5. The struggle over coal and petroleum coke lasted a decade, with residents challenging the approval of a new coal export facility in the Los Angeles Export Terminal. Kevin O'Leary, "Air Board Seeks More Surveys on Port Coal," *Los Angeles Times*, Nov. 16, 1997, B3. Terminal operators agreed to build domes to cover stored coke, Dan Weikel, "Coal Terminal Has New Plan to Build Dust Control Domes," *Los Angeles Times*, Sept. 12, 1998, B3, yet disputes remained, with local investigations and a lawsuit filed by the San Pedro-Wilmington Coalition for Environmental Justice and Santa Monica BayKeeper over the environmental impact of the coal terminal. See "Environmentalists Sue Port over Coal Terminal," *Los Angeles Times*, June 24, 2000, B4; "Feuer Urges Study of Coke Dust Hazard at Port," *Los Angeles Times*, Jan. 8, 2000, B4.

63. Murphy, note 5.

64. Ibid.

65. Ibid.

66. Ibid. ("When completed in the next few years, the new development will provide slips for more than 3,000 private boats and will include a hotel, a youth aquatics camp, restaurants, shops, offices, parks, a salt marsh and bicycle paths.").

67. St. George, note 2; see also Sheryl Stolberg, "L.A. Port's New Terminal Ready for Cruise Ships, Era of Growth," *Los Angeles Times*, Apr. 29, 1988, E1.

68. Tim Waters, "San Pedro Hotel Seen as 'Jewel' for Business District," *Los Angeles Times*, July 18, 1985, A1 South Bay ed.

69. David Ferrell, "Redevelopment Projects Brighten San Pedro's Once Dreary Horizon," *Los Angeles Times*, Aug. 1, 1985, SB1; Bob Williams, "The Changing Face of South Bay: Downtowns," *Los Angeles Times*, Jan. 4, 1987, SB2.

70. Sheryl Stolberg, "On the Waterfront: Wave of Development in San Pedro Threatens Entrenched Ethnic Community," *Los Angeles Times*, June 19, 1988, B1 (noting that some feared redevelopment would threaten the community's ethnic distinctiveness, with immigrants from Italy, Yugoslavia, and Latin America).

Notes

71. Sheryl Stolberg, "Drop In, Stay Awhile: With Its Traditional Industries in Decline, San Pedro Chases the Tourist Dollar," *Los Angeles Times*, Aug. 6, 1989, M4.

72. Louis Sahagun, "Face Lift to Raze Shops at Harbor," *Los Angeles Times*, Aug. 15, 2001, B3; Susan Woodward, "Some Have Reservations about Ambitious Port Plan," *Los Angeles Times*, May 11, 1995, 13.

73. "Judge Clears Way for Ports of Call Redevelopment," *Los Angeles Times*, May 26, 1999, B4 ("According to port statistics, revenue has declined from $25.5 million a year in 1989 to $8.2 million in 1998. Vacancy rates soared from 8% to 50% over the same period.").

74. See Jerry Hirsch, "Ships Trying to Get Everybody on Board," *Los Angeles Times*, June 23, 2003, C1.

75. See Eric Malnic, "Harbor Area Secession Drive Nears Key Number of Backers," *Los Angeles Times*, July 3, 1999, B5. Wilmington had joined an earlier effort by westside communities to secede in 1990. Greg Krikorian, "Grass-Roots Unrest Wilmington, Westside Areas to Join in Bid to Secede from L.A.," *Los Angeles Times*, Oct. 3, 1990, B3 South Bay ed.

76. See Sharon Bernstein, "Report Bolsters Bid to Secede," *Los Angeles Times*, Mar. 7, 2002, B1.

77. Dan Weikel, "Port Officials Extend Olive Branch to Communities," *Los Angeles Times*, Jan. 11, 2001, B1.

78. Ibid., B9 (quoting Commissioner Jonathan Y. Thomas).

79. Matea Gold and Patrick McGreevy, "Hahn, Council Woo Harbor Area with Visit," *Los Angeles Times*, Dec. 13, 2001, B3. In August 2001, it was revealed that the harbor commission had scaled back its contract with the community relations firm hired to promote a better community partnership because the firm was too supportive of community concerns. Louis Sahagun, "Local Storm Clouds Brew over Port," *Los Angeles Times*, Aug. 10, 2001, B4.

80. Patrick McGreevy, "New Slant on Harbor Area Bottom Line," *Los Angeles Times*, June 10, 2001, B1.

81. Kristina Sauerwein, "Fever for Secession in Harbor Area Cools," *Los Angeles Times*, Sept. 15, 2002, B3; see also Patrick McGreevy, "Harbor Cityhood Found Unfeasible," *Los Angeles Times*, Apr. 26, 2001, B1; McGreevy, note 80.

82. Sharon Bernstein, "Panel Urges Port Remain with L.A.," *Los Angeles Times*, Apr. 10, 2002, B5.

83. Dan Weikel, "Barrier Rift," *Los Angeles Times*, Jan. 30, 2001, B1.

84. See telephone interview with Jesse Marquez, executive director, Coalition for a Safe Environment (May 20, 2013).

85. Weikel, note 83, B4.

86. See telephone interview with Jesse Marquez, note 84.

87. Ibid.

88. Ibid.

89. Ibid.

90. Ibid.

91. See Louis Sahagun, "Evictions Loom for Many in Wilmington's 'Third World,'" *Los Angeles Times*, May 13, 2001, B1.

92. Telephone interview with Angelo Logan, co-director, East Yard Communities for Environmental Justice (May 1, 2013).

93. Ibid.

94. Ibid.

95. Ibid.

96. Ibid.

97. Ibid.

98. Ibid.

99. Ibid.

100. See Edna Bonacich and Jake B. Wilson, *Getting the Goods: Ports, Labor, and the Logistics Revolution* (Ithaca, NY: Cornell University Press, 2008), 210–211.

101. James Peoples, "Deregulation and the Labor Market," *Journal of Economic Perspectives* 12 (1998): 111, 111–112.

102. Bonacich and Wilson, note 100, 211.

103. See Chris Woodyard, "Truckers Told to Avoid Violence in Port Strike," *Los Angeles Times*, July 22, 1988, D1 (stating that the union's goal was to get companies to agree to talks with container terminals to decrease wait times of up to eight hours; the union asked the Federal Maritime Commission to make terminals pay $68 per hour for wait time).

104. Bonacich and Wilson, note 100, 219.

105. See Anthony Millican, "IRS Probing Trucking Companies: Possible Reclassification of Drivers as Employees Could Cost Area Firms Millions," *Los Angeles Times*, Aug. 29, 1991, B3.

106. Ibid.

Notes

107. Ibid.

108. Ibid.

109. Ibid.

110. See Jesus Sanchez, "Wildcat Strike Idles Cargo at L.A.-Area Ports," *Los Angeles Times*, Nov. 13, 1993, D1.

111. Ruth Milkman, *L.A. Story: Immigrant Workers and the Future of the U.S. Labor Movement* (New York: Russell Sage Foundation, 2006), 179.

112. Ibid.

113. Ibid.

114. Ibid.

115. Bonacich and Wilson, note 100, 220.

116. Milkman, note 111, 179–180.

117. "Court Order Limits Picketing of Truckers at Port Gates," *Los Angeles Times*, May 9, 1996, B5.

118. Milkman, note 111, 180–181.

119. Ibid., 181.

120. Jeff Leeds, "Long Beach, L.A. Ports Face Crisis in Labor Dispute," *Los Angeles Times*, May 8, 1996, A1.

121. Milkman, note 111, 182.

122. Ibid., 183–184.

123. Ruth Milkman and Kent Wong, "Organizing Immigrant Workers: Case Studies from Southern California," in *Rekindling the Movement: Labor's Quest for Relevance in the 21st Century*, ed. Lowell Turner, Harry C. Katz, and Richard W. Hurd (Ithaca, NY: Cornell University Press, 2001), 99, 101–102. Milkman and Wong note that as a result of the campaign, the ILWU was able to win the right to represent intra-harbor truckers. Ibid., 129.

124. Bonacich and Wilson, note 100, 222.

125. Milkman, note 111, 181.

126. Jeff Leeds, "Transport Firms Sued on Behalf of Port Truck Drivers," *Los Angeles Times*, July 12, 1996, B3.

127. Ibid. In addition to claims under the Labor Code, the drivers also brought claims for unfair business practices under section 17200 of the California Business and Professions Code. Albillo v. Intermodal Container Servs., Inc., 8 Cal. Rptr. 3d 350, 354 (Cal. Ct. App. 2003).

128. Leeds, note 126.

129. Ibid.

130. Ibid. ("Cardoza, 34, said he grossed $42,000 last year, but that deductions and the costs of maintaining the truck left him with only $12,000. With his daughters Samantha, 3, and Jennifer, 11, at his feet, he told reporters that he has been borrowing money from family members to make ends meet.")

131. Nancy Cleeland, "Harbor Drivers Independent, Panel Says," *Los Angeles Times*, Jan. 19, 2000, C2. Littler Mendelson represented defendants Intermodal Container Service Inc., Interstate Consolidation Inc., and Cartage Service.

132. *Albillo*, 8 Cal. Rptr. 3d at 353.

133. Ibid.

134. Ibid., 354.

135. Ibid.

136. Ibid., 362.

137. Milkman, note 111, 184.

138. Cleeland, note 131.

139. Telephone interview with Michael Manley, staff attorney, International Brotherhood of Teamsters (Feb. 20, 2013).

140. Ibid.

141. Ibid.

142. Ibid.

143. Doreen Hemlock, "Strike Ends at Port of Miami," *Sun-Sentinel*, Feb. 23, 2000, http://articles.sun-sentinel.com/2000-02-23/business/0002220813_1_independent-truckers-miami-dade-county-strike (accessed July 5, 2017).

144. Ibid.

145. Nancy Cleeland and Dan Weikel, "Truckers at Area Ports Demanding Wage, Health and Pension Benefits," *Los Angeles Times*, Feb. 18, 2000, C1.

146. Ibid.; see also Bonacich and Wilson, note 100, 221–222.

147. Bonacich and Wilson, note 100, 211, 222.

148. Telephone interview with Michael Manley, note 139.

149. Ibid.

150. Ibid.

Notes

151. Ibid.

152. Ibid.

153. Ibid.

154. Ibid.

155. Ibid.

156. Ibid.

157. Ibid.

158. Telephone interview with John Canham-Clyne, campaign director, Change to Win (May 20, 2013).

159. Telephone interview with Nick Weiner, national campaigns organizer, Change to Win (Apr. 17, 2013).

160. Ibid.

161. Ibid.

162. Canham-Clyne and Weiner also worked with Rich Yeselson. Ibid.; see also telephone interview with Michael Manley, note 139.

163. Telephone interview with John Canham-Clyne, note 158.

164. Ibid.

165. Ibid.

166. Weiner had started in the AFL-CIO's Food and Allied Service Trade department. Telephone interview with Nick Weiner, note 159. In 2004, HERE joined with the Union of Needletrades, Industrial, and Textile Employees (UNITE) to form UNITE HERE, which withdrew from the AFL-CIO to join Change to Win a year later.

167. Ibid.

168. Telephone interview with John Canham-Clyne, note 158.

169. Ibid.

170. Telephone interview with Nick Weiner, note 159.

171. See Harold Meyerson, "Hard Labor," *American Prospect*, June 18, 2006, http://www.prospect.org/article/hard-labor (accessed July 5, 2017).

172. Telephone interview with Michael Manley, note 139.

173. Telephone interview with John Canham-Clyne, note 158.

174. Ibid.

175. Ibid.

176. Telephone interview with Michael Manley, note 139.

177. Telephone interview with John Canham-Clyne, note 158.

178. Telephone interview with Michael Manley, note 139.

179. Telephone interview with Nick Weiner, note 159.

180. Telephone interview with Michael Manley, note 139.

181. Ibid.

182. Ibid.

183. Ibid.

184. Telephone interview with Nick Weiner, note 159. A similar concept had been used in connection with city development subsidies to attach card-check neutrality to new development agreements. Scott L. Cummings and Steven A. Boutcher, "Mobilizing Local Government Law for Low-Wage Workers," *University of Chicago Legal Forum* (2009): 187.

185. Attorney Richard McCracken had advised UNITE HERE on the market participant exception in connection with living wage and other local laws. Telephone interview with John Canham-Clyne, note 158.

186. Engine Mfrs. Ass'n v. S. Coast Air Quality Mgmt. Dist., No. CV00–09065FMC(BQRX), 2005 WL 1163437, at *12 (C.D. Cal. May 5, 2005).

187. For example, infrastructure development had eroded the tidal ecosystem. See "A Proposal Worth Pursuing," *Los Angeles Times*, July 28, 1985, A2 (noting that the port and Pacific Texas Pipeline Company had offered to spend $10 million restoring the Batiquitos Lagoon in San Diego County in compliance with a federal regulation requiring mitigation for destroying tidal lands).

188. Sheryl Stolberg, "Stiff Rules Focus on Port's Many Risks," *Los Angeles Times*, Feb. 24, 1990, A30. In 1998, the port and one of its contractors paid a $1 million fine to settle a federal action against them for illegally dumping polluted sediments into the ocean. Dan Weikel, "L.A. Port, Firm Fined in Dredging Case," *Los Angeles Times*, July 22, 1998, B3.

189. A 1985 spill of nearly 19,000 gallons of crude oil occurred after a Mobil employee accidentally left a valve open at a storage facility. Tim Waters, "Mobil Corp. Faces Fine in Spill: Mop-Up Continues After Oil Fouls San Pedro Beach," *Los Angeles Times*, Jan. 12, 1985, B5. A 1991 spill of 12,000 gallons was the result of a tanker's fuel tank overflow. Judy Pasternak, "Oil Spill Damage Estimate Expands," *Los Angeles Times*, Jan. 16, 1991, B1. Port congestion produced other near-misses. George Hatch, "Near-Misses at Sea Spur Call for 'Traffic Cop,'" *Los Angeles Times*,

Notes 391

Feb. 3, 1991, B1 (describing proposal passed by legislature to create private vessel tracking system to guide port traffic).

190. "Sparks Start L.A. Port Blast, Fire," *Los Angeles Times*, May 9, 1985, A1.

191. Lisa Richardson, "Neighbors Want Scrap Yard to Clean Up Its Act," *Los Angeles Times*, Aug. 6, 1993, B3 ("Harbor area boat and homeowner groups have asked the Port of Los Angeles to demand sharp reductions in noise, air and water pollution at the Hugo Neu-Proler scrap yard, which wants the port to renew its 27-year Terminal Island lease. Labor leaders, meanwhile, have warned harbor officials not to take steps that would endanger the 165 jobs at the yard, where discarded cars, refrigerators and other metal refuse are shredded and the scrap is loaded aboard ships for export."). The residents ultimately failed to prevent the lease renewal, while union leaders later charged Hugo with labor violations. See Dan Weikel, "Firm at Port in Hot Water over Lawsuit," *Los Angeles Times*, Jan. 6, 1999, C2.

192. Greg Krikorian, "City Council Approves Port of L.A. Coal Facility Despite Long Beach Suit," *Los Angeles Times*, July 29, 1993, B3.

193. George Stein, "Tomorrow's Menu: Today's Traffic Jam," *Los Angeles Times*, Dec. 26, 1985, B7.

194. Federal environmental laws requiring cleanup before new development projects could proceed placed the ports in the position of forcing leaseholders to remediate contamination—or face the prospect of having to pay for remediation themselves. George Hatch, "Port Bedeviled by Pollution Sins of the Past," *Los Angeles Times*, Aug. 17, 1991, B1 (stating that the Port of Los Angeles had to pay $12 million to clean up a scrap metal site vacated by National Metal and Steel before the port knew of the extent of environmental contamination).

195. California Coastal Act of 1976, California Public Resources Code §§ 30000–30900 (West 2007).

196. Greg Krikorian, "Coastal Commission Delays Action on Plan for $2-Billion Port Project," *Los Angeles Times*, Aug. 13, 1992, B4 (noting that the waterway was "home to a variety of sea life, including the endangered California least tern and brown pelican").

197. Greg Krikorian, "Coastal Panel Staff Urges Approval of Dredging Plan," *Los Angeles Times*, Oct. 14, 1992, B7; see also Lisa Richardson, "State OKs Port of L.A. Plan for Expansion, Phase by Phase," *Los Angeles Times*, Oct. 16, 1992, B3 ("The Harbor Department must now seek final federal approval for the project and then ask Congress for $100 million—the federal share of the $580-million plan. The balance will be paid with port funds.").

198. Deborah Schoch, "Port Air Cleanup Plan May Become a Model," *Los Angeles Times*, Apr. 1, 2003, B1 ("On an average day, 16 ships arrive at Los Angeles and Long Beach, releasing more pollution than a million cars.").

199. See, e.g., Bigelow v. Virginia, 421 U.S. 809 (1975). The United States failed to ratify a 1978 treaty setting global environmental standards for ships. "Pioneering Cleanup at Ports," *Los Angeles Times*, Mar. 17, 2003, B10.

200. Control of Emissions from New Marine Compression-Ignition Engines at or Above 30 Liters Per Cylinder, 68 Fed. Reg. 9746, 9759 (Feb. 28, 2003) (codified at 40 C.F.R. pts. 9 & 94).

201. Engine Mfrs. Ass'n v. E.P.A., 88 F.3d 1075 (D.C. Cir. 1996).

202. Daryl Kelley, "Port Employers Attack Plan to Curb Ship Smog," *Los Angeles Times*, Mar. 8, 1987, SE1.

203. Ibid.

204. 42 U.S.C. § 7543(d) (2012).

205. Coal. for Clean Air v. S. Cal. Edison Co., 971 F.2d 219, 222 (9th Cir. 1992).

206. Ibid.

207. Abramowitz v. EPA, 832 F.2d 1071, 1073 (9th Cir. 1987). A number of environmental groups intervened on the side of the petitioner in the case, including the Sierra Club, Citizens for a Better Environment, and the Coalition for Clean Air.

208. *Coal. for Clean Air*, 971 F.2d at 222–223.

209. The EPA tried to wiggle out of the settlement agreement after the Clean Air Act Amendments of 1990 were passed by Congress, arguing that the amendments effectively reset the clock for compliance; however, the Ninth Circuit ultimately ruled that the EPA remained bound. Ibid., 230.

210. Marla Cone, "EPA Smog Plan Is Just the Start of Negotiations," *Los Angeles Times*, Feb. 21, 1994, A3.

211. James Flanigan, "Deals in Smoke-Filled Rooms Could Help Clear the Region's Air," *Los Angeles Times*, Nov. 9, 1994, D1.

212. Ibid.

213. Ibid.

214. Ibid.

215. South Coast Air Quality Management District, Final 1999 Amendment to the 1997 Ozone State Implementation Plan for the South Coast Air Basin ES-2 (1999), http://www.aqmd.gov/docs/default-source/clean-air-plans/ozone-plans/ozone-plan-final-1999-amendment.pdf?sfvrsn=2 (accessed July 5, 2017). The amended state implementation plan was accompanied by an agreement settling the environmental lawsuit. A revised plan, based on new scientific modeling data that increased emission projections, was submitted in 2003, though portions related to emission attain-

Notes

ment goals were later withdrawn. The EPA disapproved this part of the plan in 2008, but did not order California to submit a revised attainment plan, prompting yet another lawsuit by environmental groups, which resulted in still another court order mandating that EPA request a new state implementation plan. Ass'n of Irritated Residents v. EPA, 686 F.3d 668 (9th Cir. 2012).

216. In 2000, the EPA issued regulations requiring diesel trucks to dramatically reduce emissions beginning in the 2007 model year. EPA, "Regulatory Announcement: Final Emission Standards for 2004 and Later Model Year Highway Heavy-Duty Vehicles and Engines" (2000).

217. Marla Cone, "Diesel—the Dark Side of Industry," *Los Angeles Times*, May 30, 1999, A30.

218. Ibid.

219. South Coast Air Quality Management District, *Multiple Air Toxics Exposure Study Final Report (Mates II)* (2000) [hereinafter MATES II], http://www.aqmd.gov/docs/default-source/air-quality/air-toxic-studies/mates-ii/mates-ii-contents-and-executive-summary.pdf?sfvrsn=4 (accessed July 5, 2017).

220. Ibid., ES-5.

221. Ibid., ES-5–ES-12. Some efforts to reduce pollution were spurred by the MATES II findings. For example, Marine Terminals Corp. purchased five low-emission trucks under a state incentive program. "Cargo Terminal Operator Using Low-Emission Trucks," *Los Angeles Times*, Sept. 16, 2000, B4. In 2001, the EPA settled a suit by Bluewater Network under which it agreed to "begin developing rules to cut smog-forming exhaust from the largest, diesel powered ships, including cargo vessels, tankers and cruise liners." Gary Polakovic, "EPA Settlement Seeks to Curb Air Pollution from Big Ships," *Los Angeles Times*, Jan. 17, 2001, A15.

222. See Schoch, note 198.

223. See Martha M. Matsuoka and Robert Gottlieb, "Environmental and Social Justice Movements and Policy Change in Los Angeles: Is an Inside-Outside Game Possible?" in *New York and Los Angeles: The Uncertain Future*, ed. David Halle and Andrew A. Beveridge (New York: Oxford University Press, 2013), 445, 452.

224. Jeff Leeds, "Long Beach Port Faces Rising Tide of Criticism," *Los Angeles Times*, Mar. 24, 1997, B1.

225. To secure COSCO's business, each port sent delegations to lobby Chinese officials. Ibid.

226. Jeff Leeds, "Ruling May Block Long Beach Port Project," *Los Angeles Times*, May 21, 1997, B1.

227. Leeds, note 224.

228. Leeds, note 226.

229. Ibid. The naval station had been closed in 1991 and turned over to the port in 1995.

230. Leeds, note 224.

231. Leeds, note 226; Jeff Leeds, "Coastal Commission OKs Permit for Long Beach Port Terminal," *Los Angeles Times*, Jan. 9, 1997, B3.

232. Leeds, note 226.

233. City of Vernon v. Bd. of Harbor Comm'rs of Long Beach, 74 Cal. Rptr. 2d 497, 500 (Cal. Ct. App. 1998).

234. Ibid.

235. Ibid., 501.

236. Ibid.

237. Jeff Leeds, "Harbor Panel OKs Terminal for Chinese Line," *Los Angeles Times*, Mar. 25, 1997, A1.

238. *City of Vernon*, 74 Cal. Rptr. 2d at 501; Jeff Leeds, "Port Cancels Lease with Chinese Firm," *Los Angeles Times*, Apr. 22, 1997, B1.

239. Leeds, note 226.

240. *City of Vernon*, 74 Cal. Rptr. 2d at 502.

241. Douglas P. Shuit, "Cities, Counties Join Long Beach in Port Fight," *Los Angeles Times*, Jan. 15, 1998, B11.

242. Dan Weikel, "Port Shifts Focus After Cosco Deal for Base Unravels," *Los Angeles Times*, Sept. 19, 1998, B1.

243. Ibid.

244. NRDC v. City of Los Angeles, 126 Cal. Rptr. 2d 615, 622 (Cal. Ct. App. 2002).

245. Louis Sahagun, "Lawsuit Seeks to Block Shipping Terminal Plan," *Los Angeles Times*, June 15, 2001, B5.

246. Louis Sahagun, "Judge Halts Work on New Port of L.A. Terminal," *Los Angeles Times*, July 24, 2002, B3; Sahagun, note 245.

247. *NRDC*, 126 Cal. Rptr. 2d at 622.

248. Ibid.; Sahagun, note 246.

249. Schoch, note 198.

250. Ibid.

Notes

251. Petition for Writ of Mandate, NRDC v. City of Los Angeles, No. BS070017, 2002 WL 34340562 (Cal. Super. Ct. May 30, 2002) (filed June 14, 2001).

252. Ibid., 13–19. Petitioners also argued that the port's approval of the project violated the city's General Plan. See ibid., 20. Petitioners filed an amended complaint with two additional causes of action, one for abuse of discretion for approving a project inconsistent with the port's master plan and the second for violating the coastal act. Amended Petition for Writ of Mandate at 26–29, NRDC, No. BS070017, 2002 WL 34340562 (filed Oct. 19, 2001).

253. For CEQA rules, see California Public Resources Code §§ 21000 et seq. (West 2007).

254. Petition for Writ of Mandate, note 251.

255. Sahagun, note 245.

256. Amended Petition for Writ of Mandate, note 252, at 9.

257. NRDC v. City of Los Angeles, 126 Cal. Rptr. 2d 615, 617–20 (Cal. Ct. App. 2002).

258. Ibid., 620. In April 2001, the Army Corps issued China Shipping a permit to build the first wharf, which the coalition also challenged under NEPA. Louis Sahagun, "Work to Resume on Port of L.A.," *Los Angeles Times*, July 27, 2002, B4.

259. *NRDC*, 126 Cal. Rptr. 2d at 622 (citation omitted).

260. Ibid.

261. Ibid.

262. Ibid., 623.

263. Ibid.

264. Ibid., 622.

265. Answer to Amended Petition for Writ of Mandate, NRDC v. City of Los Angeles, No. BS070017, 2002 WL 34340562 (Cal. Super. Ct. May 30, 2002) (filed Dec. 31, 2001).

266. Louis Sahagun, "Anger at Hahn Brings Unusual Allies Together," *Los Angeles Times*, Feb. 21, 2002, B3.

267. Ibid. Hahn's father was longtime County Board of Supervisor Kenneth Hahn, who served the predominately African-American Baldwin Hills community. Ibid.

268. Ibid.

269. Ibid.

270. Ibid.

271. Judgment Denying Petition for Writ of Mandate, NRDC v. City of Los Angeles, No. BS070017, 2002 WL 34340562 (Cal. Super. Ct. May 30, 2002).

272. Ibid., 9–11.

273. Appellants' Opening Brief at 3, NRDC v. City of Los Angeles, 126 Cal. Rptr. 2d 615 (Cal. Ct. App. 2002) (No. B159157) (filed Aug. 23, 2002).

274. Ibid., 1.

275. Respondents' Brief at 39–40, NRDC, 126 Cal. Rptr. 2d 615 (No. B159157) (filed Sept. 18, 2002). Defendants argued that there were three approvals issued: the first a "use" approval to redesign the West Basin that was clearly within the 1997 EIR; the second an "occupancy" approval encompassed in the China Shipping lease; and the third a "construction" approval for phase I, evidenced in the port's issuance of a coastal development permit for that phase only in October 2001. The defendants contended that the prior EIRs addressed potential impacts from these three decisions. Ibid., 2–3.

276. Ibid., 4.

277. Ibid., 5.

278. Ibid., 9.

279. Amicus Brief of the State of California, Ex Rel. Attorney General Bill Lockyer in Support of Appellants at 4, NRDC, 126 Cal. Rptr. 2d 615 (No. B159157) (filed Oct. 2, 2002).

280. Petition for Writ of Supercedeas or Other Appropriate Stay Order, and for an Immediate Stay; Memorandum of Points and Authorities; Supporting Declaration, NRDC, 126 Cal. Rptr. 2d 615 (No. B159157) (filed June 7, 2002).

281. Order Denying Petition for Writ of Supersedeas and Setting Expedited Briefing Schedule and Oral Argument for Appeal, NRDC, 126 Cal. Rptr. 2d 615 (No. B159157) (filed Aug. 5, 2002).

282. "Court Expected to Rule on Port Injunction," *Los Angeles Times*, July 23, 2002, B4.

283. Sahagun, note 246.

284. Ibid.

285. Sahagun, note 258.

286. Eric Malnic, "Environmentalists Win a Battle over L.A. Port Terminal," *Los Angeles Times*, Oct. 24, 2002, B4.

287. NRDC, 126 Cal. Rptr. 2d at 625.

288. Ibid., 627.

Notes

289. Ibid., 625.

290. Ibid., 617.

291. Louis Sahagan, "Court Halts Work at L.A. Port," *Los Angeles Times*, Oct. 31, 2002, B1.

292. Order Denying Petition for Rehearing, NRDC v. City of Los Angeles, No. B159157, 2002 Cal. App. Lexis 5076 (Cal. Ct. App. Nov. 18, 2002). In its petition, the city argued that the court had failed to adopt an appropriately deferential standard of review and misstated crucial facts. Petition for Rehearing, Modification of Stay and/or Depublication at 10, *NRDC*, No. B159157, 2002 Cal. App. Lexis 5076 (Cal. Ct. App. Nov. 18, 2002) (filed Nov. 14, 2002).

293. Order Denying Application for Stay and Petition for Review, NRDC v. City of Los Angeles, No. S111953, 2002 Cal. Lexis 8631 (Cal. Super. Ct. Dec. 18, 2002).

294. *NRDC*, 126 Cal. Rptr. 2d at 628.

295. Louis Sahagun, "Shipping Cranes Cleared to Land at Port," *Los Angeles Times*, Nov. 6, 2002, B3.

296. Schoch, note 198.

297. Stipulated Judgment, Modification of Stay, and Order Thereon, NRDC v. City of Los Angeles, No. BS070017, 2002 WL 34340562 (Cal. Super. Ct. May 30, 2002) (filed Mar. 6, 2003).

298. Ibid., 6.

299. Ibid., 16.

300. Ibid., 17.

301. Ibid., 18.

302. Ibid., 18–19.

303. Ibid., 19–23.

304. Ibid., 20–22.

305. Deborah Schoch, "Port Project Suit Settled," *Los Angeles Times*, Mar. 6, 2003, B1. The environmental groups also settled the federal action, contingent on the approval of the state court settlement, which required the U.S. Army Corps of Engineers to conduct "a full environmental impact review of the project and reconsider its issuance of the permits in light of the review." NRDC, press release, "City of Los Angeles and Community and Environmental Groups Reach Record Settlement of Challenge to China Shipping Terminal Project at Port" (Mar. 3, 2005), http://www.nrdc.org/media/pressreleases/030305.asp (accessed July 5, 2017).

306. Deborah Schoch and Peter Nicholas, "Plans for 1st 'Green' Ship Terminal in U.S. Stall," *Los Angeles Times*, June 14, 2003, B3.

307. Ibid.

308. Ibid.

309. Ibid.

310. Ibid.

311. Deborah Schoch, "Port's Settlement of Environmental Suit Gets Costlier," *Los Angeles Times*, July 10, 2003, B3.

312. Deborah Schoch, "Port Officials Dispute Chick on Settlement Cost," *Los Angeles Times*, July 11, 2003, B4.

313. Deborah Schoch, "Residents Feel Ignored in L.A. Harbor Deal," *Los Angeles Times*, Mar. 11, 2004, B11.

314. Ibid.

315. Deborah Schoch, "Accord Clears Way for '03 Plan to Clean Up Air at Port," *Los Angeles Times*, Mar. 12, 2004, B3. The final amended agreement was filed with the court on June 21, 2004. Amended Stipulated Judgment, Modification of Stay, and Order Thereon, NRDC v. City of Los Angeles, No. BS070017, 2002 WL 34340562 (Cal. Super. Ct. May 30, 2002) (filed June 21, 2004).

316. See Deborah Schoch, "Port and Shipper End Fight," *Los Angeles Times*, May 26, 2005, B6. Although the terminal opened in May, the legal wrangling continued, with the port settling a lawsuit by China Shipping for over $20 million to compensate the company for delays. Ibid.

317. See "Port of Los Angeles Hosts First Plugged In Container Ship," Environmental News Service (June 21, 2004), http://www.ens-newswire.com/ens/jun2004/2004-06-21-04.asp (accessed July 5, 2017). The port circulated draft EIRs in July 2006 and then again in April 2008. Environmental Management Division, Port of Los Angeles Regulatory Branch, U.S. Army Corps of Engineers, LA Division, "Berth 97–109 [China Shipping] Container Terminal Re-Circulated Draft EIS/EIR" (2008), http://www.portoflosangeles.org/EIR/ChinaShipping/DEIR/_Public_Meeting_Presentation.pdf (accessed July 5, 2017). The city approved them at the end of 2008. Port of Los Angeles, press release, "China Shipping Container Terminal Expansion Is Approved by Port of Los Angeles" (Dec. 19, 2008), http://www.portoflosangeles.org/newsroom/2008_releases/news_121908cs.asp (accessed July 5, 2017). China Shipping completed phase II in 2011. See "China Shipping Celebrates Major Terminal Expansion at Port of Los Angeles," *Longshore & Shipping News*, Apr. 28, 2011, http://www.longshoreshippingnews.com/2011/04/china-shipping-celebrates-major-terminal-expansion-at-port-of-los-angeles/ (accessed July 5, 2017).

Notes

318. "Port of Los Angeles Hosts First Plugged In Container Ship," note 317.

319. Control of Air Regulation from New Motor Vehicles, 66 Fed. Reg. 5002, 5002 (Jan. 18, 2001) (codified at 40 C.F.R. pts. 69, 80, 96) (claiming that new regulation "will reduce particulate matter and oxides of nitrogen emissions from heavy duty engines by 90 percent and 95 percent below current standard levels, respectively").

320. Control of Emissions of Air Pollution from Nonroad Diesel Engines and Fuel, 69 Fed. Reg. 38,958, 38,960 (June 29, 2004) (codified at 40 C.F.R. pts. 9, 69, 80, 86, 89, 94, 1039, 1048, 1051, 1065, 1068) (regulating nonroad diesel fuel by (among other controls) limiting sulfur levels to fifteen parts per million); see also Jerilyn Lopez Mendoza, testimony on EPA's proposed rulemaking for "Control of Emissions of Air Pollution from Nonroad Diesel Engines and Fuel," (June 17, 2003) (stressing the importance of the fifteen part per million limit for locomotives and commercial marine engines).

321. Deborah Schoch, "City Downplays Port Pollution, Critics Say," *Los Angeles Times*, July 9, 2004, B4.

322. Ibid.

323. Deborah Schoch, "Plan to Cut Port Smog to Be Unveiled," *Los Angeles Times*, Dec. 27, 2004, B1.

324. Deborah Schoch, "Hahn Shift on Port Cleanup Is Criticized," *Los Angeles Times*, Nov. 16, 2004, B3.

325. Louis Sahagun, "Hahn Outlines His Vision for Port," *Los Angeles Times*, Dec. 7, 2002, B4.

326. Ibid.

327. See Diane Bailey, Thomas Plenys, Gina M. Solomon, Todd R. Campbell, Gail Ruderman Feuer, Julie Masters, and Bella Tonkonogy for NRDC, *Harboring Pollution: The Dirty Truth about U.S. Ports* (2004), 11, http://www.nrdc.org/air/pollution/ports1/ports.pdf (accessed July 5, 2017).

328. Natural Resource Defense Council, press release, "New Study Says U.S. Seaports Are Largest Urban Polluters (Mar. 22, 2004), http://www.nrdc.org/media/press releases/040322.asp (accessed July 5, 2017).

329. Diane Bailey, Thomas Plenys, Gina M. Solomon, Todd R. Campbell, Gail Ruderman Feuer, Julie Masters, and Bella Tonkonogy for NRDC, *Harboring Pollution: Strategies to Clean Up U.S. Ports* (2004), ix–xii, http://www.nrdc.org/air/pollution/ports/ports2.pdf (accessed July 5, 2017).

330. Ibid., 43.

331. Ibid., 21.

332. Ibid., 49.

333. Ibid., 73. The idling bill was supported by the Teamsters. Bill Mongelluzzo, "Big Win for Truckers," *Teamsters Nation Blog*, Sept. 9, 2002, http://www.teamster.org/content/joc-big-win-truckers (accessed July 5, 2017).

334. California Air Resources Board, "Final Statement of Reasons for Rulemaking, Public Hearing to Consider Proposed Regulation Order: Airborne Toxic Control Measure to Limit Diesel-Fueled Commercial Motor Vehicle Idling" (2004), http://www.arb.ca.gov/regact/idling/fsor.pdf (accessed July 5, 2017) (discussing rationales for and evolution of proposed regulation entitled "Airborne Toxic Control Measure to Limit Diesel-fueled Commercial Motor Vehicle Idling"). The rule was codified at California Code of Regulations title 13, § 2485 (2012). CARB later approved a rule requiring 2008 and newer trucks to be equipped with a sleeper switch to automatically shut down idling trucks after five minutes. Cal. Code Regs. tit. 13, § 1956.8 (2012).

335. "News: Assemblyman Lowenthal Introduces Bill to Require Zero Net Increase in Air Pollution from Future Growth at Ports of LB and LA," *LBReport.com: News*, Feb. 21, 2004, http://www.lbreport.com/news/feb04/lowprtai.htm (accessed July 5, 2017). The bill was part of a trio of bills designed to reduce port emissions, including AB 2041, which imposed a fee on containers shipped by trucks during working hours, and AB 2043, which established a task force to deal with port growth and environmental issues. Ibid.

336. Port of Long Beach and Port of Los Angeles, Air Pollution: Hearing on A.B. 2042 Before the Assemb. Comm. on Transp., 2003–2004 Leg. Reg. Sess. (Cal. April 12, 2004) (comments of the Pacific Merchant Shipping Association).

337. Ibid.

338. Ibid.

339. "News in Depth: LB Port & LB Council on Collision Course—Again—This Time Over Assemblyman Lowenthal's AB 2042 For Zero Net Increase in Port Air Pollution," *LBReport.com: News*, May 1, 2004, http://www.lbreport.com/news/may04/lowbilz.htm (accessed July 5, 2017).

340. Ibid.

341. Assemb. B. 2042, 2003–2004 Leg., Reg. Sess. (Cal. 2004) (as amended by Leg. Assembly, May 5, 2004).

342. *LBReport.com: News*, May 3, 2004, http://www.lbreport.com/news/may04/lowbilz2.htm (accessed July 5, 2017).

343. "Council Backs AB 2042 (Lowenthal Zero Net Increase in Port Air Pollution) Bill & And Two Related Port-Pollution Bills," *LBReport.com: News*, May 5, 2004, http://www.lbreport.com/news/may04/lowbilz3.htm (accessed July 5, 2017).

Notes

344. Ibid.

345. Rick Holguin, "New Terminal Opens, Along with Scores of Jobs," *Los Angeles Times*, Apr. 15, 1993, J1.

346. Ibid.

347. Deborah Schoch, "Residents Fight Port Expansion," *Los Angeles Times*, Sept. 12, 2004, B1.

348. Notice of Availability for the Revised Draft Environmental Impact Statement/Environmental Impact Report for the Pier J South Marine Terminal Expansion Project, Los Angeles County, CA, 68 Fed. Reg. 48344 (Aug. 13, 2003).

349. Letter from Steve Smith, program supervisor, California Environmental Quality Act Section, South Coast Air Quality Management District, to Dr. Robert Kanter, director of planning, Port of Long Beach (Feb. 7, 2003).

350. Letter from Susan Nakamura, planning and rules manager, South Coast Air Quality Management District, to Dr. Robert Kanter, director of planning, Port of Long Beach (Oct. 8, 2003).

351. Ibid.

352. The port had assumed in its emissions analysis that as of 2007, the first year the EPA rule went into effect, all trucks entering the port would have 2007-model-year-compliant engines, even though the rule only applied to the production of new trucks—not their purchase. Ibid.

353. Letter from Susan Nakamura, planning and rules manager, South Coast Air Quality Management District, to Dr. Robert Kanter, director of planning, Port of Long Beach (July 30, 2004) ("The AQMD staff remains concerned that operational emissions are underestimated for on-road vehicles.").

354. Deborah Schoch, "Port's Effort to Cut Smog Is Criticized," *Los Angeles Times*, Sept. 15, 2004, B1.

355. Schoch, note 347.

356. W. James Gauderman et al., "The Effect of Air Pollution on Lung Development from 10 to 18 Years of Age," *New England Journal of Medicine* 351 (2004): 1057.

357. Deborah Schoch, "County Cancer Pockets Are a Puzzle," *Los Angeles Times*, Sept. 3, 2004, B1.

358. "Port Pummeled in Hearing on Pier J EIR; Council Gives Both Sides until Nov. 16 to Discuss Issues Before a Council Vote," *LBReport.com: News*, Sept. 14, 2004, http://www.lbreport.com/news/sep04/pierj.htm (accessed July 5, 2017).

359. Long Beach had recently entered into a voluntary agreement with BP to convert two of its vessels to cold ironing by 2006. Deborah Schoch, "Long Beach Port Goes 'Green,'" *Los Angeles Times*, Aug. 31, 2004, B3.

360. "Port Pummeled in Hearing on Pier J EIR," note 358.

361. Ibid.

362. Ibid.

363. Letter from Barry Wallerstein, executive officer, South Coast Air Quality Management District, to Dr. Robert Kanter, director of planning, Port of Long Beach (Sept. 22, 2004).

364. Ibid. In addition, the SCAQMD flatly rejected the port's claim that the SCAQMD had previously accepted the port's emission calculation. Specifically, the SCAQMD denied that it had received a letter from the port addressing its emission calculation, which the port distributed at the September 14 council meeting to suggest that the SCAQMD had approved of the port's methodology. Deborah Schoch, "Expansion of Port Faces Vote," *Los Angeles Times*, Sept. 29, 2004, B1.

365. Schoch, note 364.

366. "In Depth: LB Bd of Harbor Rescinds Pier J EIR; Port Staff Will Revise It; Enviro Groups Urge Broad Revision; Several Harbor Commissioners Pledge Revision Will Address All Issues," *LBReport.com: News*, Sept. 29, 2004, http://www.lbreport.com/news/sep04/pierjei2.htm (accessed July 5, 2017).

367. Ibid.

368. Ibid.

369. Telephone interview with Adrian Martinez, staff attorney, Natural Resources Defense Council (Apr. 1, 2010).

370. Emissions from ocean vessels, cargo handling equipment, railroad cars, and trucks were to be included. Assemb. B. 2042, 2003–2004 Leg., Reg. Sess. § 40459.1(a)(2) (Cal. 2004) (the bill was passed by the Legislative Assembly on August 25, 2004).

371. Under the revised version, the SCAQMD was required to develop a Memorandum of Agreement (MOA) with CARB and the ports of Los Angeles and Long Beach that would include a "requirement that, on or before January 1, 2006, and on or before January 1 of each year thereafter, the level of air pollution at the Port of Los Angeles and the Port of Long Beach not exceed the baseline." However, the amended bill did not impose sanctions for failure to enter into an MOA; instead, in the event an MOA could not be negotiated with regulators, the law permitted the ports to develop their own emission baselines and to "operate ... in a manner that prevents the level of air pollution at the port from exceeding the baseline." Ibid. §§ 40459.1(c)(1), 40459.2(b), 40459.3(b).

372. AB 2042 Veto Statement by Governor Arnold Schwarzenegger (Sept. 29, 2004), http://www.leginfo.ca.gov/pub/03-04/bill/asm/ab_2001-2050/ab_2042_vt_20040929.html (accessed July 5, 2017).

Notes

373. Ibid. The following year, Schwarzenegger appointed NRDC's Feuer to be a judge on the Los Angeles County Superior Court. "Gov. Schwarzenegger Appoints NRDC's Gail Ruderman Feuer to Judgeship on L.A. County Superior Court," *LBReport.com: News*, July 27, 2005, http://www.lbreport.com/news/jul05/feuer.htm (accessed July 5, 2017).

374. Schoch, note 321.

375. See Port of Los Angeles Community Advisory Committee, "Joint Subcommittee Meeting with Wilmington Waterfront Development Subcommittee Traffic Committee: Minutes" (Jan. 13, 2005).

376. Ibid.

377. Ibid. Truck pollution remained a significant concern despite the Alameda Corridor rail project, which was proving disappointing. By August 2004, only 40 trains per day ran on the corridor, which was built for 150; in contrast, there were 47,285 trucks per weekday traveling on the 710 freeway, a number expected to increase to 99,300 in 2020. This was the result of changes in the shipping industry in which shippers, instead of loading cargo directly to trains, hauled "most of their imports by truck to hubs in Riverside and San Bernardino counties," where the cargo was repackaged before being sent to recipients, such as large retail chains. Sharon Bernstein and Deborah Schoch, "New Routes Just for Trucks Urged," *Los Angeles Times*, Aug. 22, 2004, B1.

378. Deborah Schoch, "Mayor Tells Port to Create New Air Plan," *Los Angeles Times*, Aug. 13, 2004, B3.

379. Patrick McGreevy and Deborah Schoch, "L.A. Port Director Resigns," *Los Angeles Times*, Sept. 18, 2004, B1.

380. Schoch, note 324.

381. Schoch, note 323.

382. Jack Leonard and Deborah Schoch, "Plans for L.A. Port Focus on Pollution," *Los Angeles Times*, Dec. 30, 2004, B3.

383. Deborah Schoch, "Meeting Delay for Hahn Task Force Stirs Concern," *Los Angeles Times*, Feb. 10, 2005, B6.

384. Deborah Schoch, "Port Clean-Air Plan Nearly Set," *Los Angeles Times*, Mar. 3, 2005, B3.

385. Deborah Schoch, "Panel Backs Plan to Curb Pollution at Port," *Los Angeles Times*, Mar. 4, 2005, B3.

386. Ibid.

387. Schoch, note 384.

388. Deborah Schoch, "2 Ports Split on How to Clear the Air," *Los Angeles Times*, Mar. 13, 2005, B1.

389. Deborah Schoch, "Hahn's Harbor Pollution Plan Faces an Uncertain Future," *Los Angeles Times*, June 22, 2005, B10.

390. Ibid.

391. *No Net Increase Task Force, Report to Mayor Hahn and Councilwoman Hahn* (2005), http://www.portoflosangeles.org/DOC/REPORT_NNI_Final.pdf (accessed July 5, 2017).

392. Ibid.

393. Ibid., ES-2. The control measures included proposals to move ocean vessel engines to low-sulfur fuel, mandate low-emission rail engines, electrify the Alameda Corridor, expand the low-emission truck conversion program, retrofit diesel trucks with filters, and impose truck-idling-reduction measures. Ibid., 3–4 to 3–9.

394. Ibid., 5–1 to 5–100.

395. Ibid., 5–44.

396. Ibid., 5–50.

397. Deborah Schoch, "Hahn Supports Port Task Force's Plan," *Los Angeles Times*, June 30, 2005, B10.

398. Ibid.

399. Deborah Schoch and Richard Fausset, "Villaraigosa's Port Panel Choices Suggest New Direction," *Los Angeles Times*, July 27, 2005, B3.

400. Telephone interview with Jerilyn López Mendoza, commissioner, Los Angeles Board of Public Works, former commissioner, Los Angeles Board of Harbor Commissioners (Apr. 26, 2013).

401. Ibid.

402. Ibid.

403. Ibid. Mendoza had previously been involved in helping negotiate the first-ever community benefits agreement in connection with the development of the L.A. Live complex in the downtown area.

404. Ibid.

405. Ibid.

406. Ibid.

407. Ibid.

408. Ibid.

Notes

409. Ibid.

410. Schoch and Fausset, note 399.

411. Ibid.

412. Deborah Schoch, "A Radical Shift in Tone for L.A. Harbor Panel," *Los Angeles Times*, Sept. 16, 2005, B4.

413. Deborah Schoch, "Wilmington Looks to Step Out from under Port's Shadow," *Los Angeles Times*, Sept. 14, 2005, B4.

414. Schoch, note 412.

415. Deborah Schoch, "New Harbor Panel Aims to Cut Pollution While Expanding Port," *Los Angeles Times*, Sept. 29, 2005, B6.

416. Deborah Schoch, "Panel to Target Air Pollution at Southland Ports," *Los Angeles Times*, Nov. 5, 2005, B3.

417. Ibid.

418. Air Resources Board, California Environmental Protection Agency, *Diesel Particulate Matter Exposure Assessment Study for the Ports of Los Angeles and Long Beach"* (2005), 2, http://www.arb.ca.gov/ports/marinevess/documents/100305draftexposrep.pdf (draft version) (accessed July 5, 2017).

419. Deborah Schoch, "Study Links Diesel Fumes to Illnesses," *Los Angeles Times*, Dec. 3, 2005, B3.

420. Air Resources Board, note 418, 4.

421. Janet Wilson, "Trade Boom's Unintended Costs," *Los Angeles Times*, Apr. 23, 2006, B1.

422. Janet Wilson, "Pollution Plan on Haulers Nears OK," *Los Angeles Times*, Apr. 20, 2006, B3.

423. Ibid.

424. Wilson, note 421.

425. Ibid.

426. Ibid.

427. Ronald D. White, "Growing Problems Give Ports a Bad Reputation," *Los Angeles Times*, May 4, 2005, C1 (referencing the L.A. County Economic Development Corporation study indicating that "the ports and their related industries continue to be a reliable job generator, adding 42,600 jobs in the five-county area last year to a total of 404,600 workers").

428. Telephone interview with S. David Freeman, interim general manager, Los Angeles Department of Water and Power Water Systems and former commissioner, Los Angeles Board of Harbor Commissioners (Apr. 29, 2013).

429. Telephone interview with Jerilyn López Mendoza, note 400.

430. Ibid.

431. Telephone interview with David Libatique, senior director, government affairs, Port of Los Angeles (June 2, 2013).

432. Ibid.

433. Ibid.

434. Ibid. Libatique also vetted the candidates for the harbor commission.

435. Ibid.

436. Jim Newton, "Once Rivals, Local Ports Clear Air in Partnership," *Los Angeles Times*, July 4, 2006, A1.

437. Ibid.

438. Telephone interview with Adrian Martinez, note 369.

439. Telephone interview with David Libatique, note 431.

440. Ibid.

441. Newton, note 436.

442. Telephone interview with S. David Freeman, note 428.

443. At the Port of Los Angeles, approximately 50 percent of containers were routed locally to Los Angeles, Riverside, San Bernardino, and Ventura Counties. Interview with John Holmes, deputy executive director, Port of Los Angeles (July 19, 2013).

444. Newton, note 436.

445. Ibid.

446. Ibid.

447. Other CAAP proposals were also important, including the recommendation to require ships to burn sulfur fuel within twenty miles of the port and dock with electrical power. Port of Long Beach and Port of Los Angeles, *San Pedro Bay Ports Clean Air Action Plan: Technical Report* (2006), 6, 87, http://www.portoflosangeles.org/CAAP/CAAP_Tech_Report_Final.pdf (accessed July 5, 2017).

448. Janet Wilson, "Diesel Trucks Target of Port Plan," *Los Angeles Times*, Nov. 7, 2006, B3.

449. Port of Long Beach and Port of Los Angeles, note 447, 4.

Notes

450. Ibid., 57.

451. Ibid., 59. The report also assumed that 500 trucks would be replaced through the China Shipping-created Gateway Cities program.

452. Ibid., 58.

453. Ibid., 62–63.

454. Ibid., 71.

455. Ibid., 59. The report estimated that LNG trucks would cost $188,500, while clean diesel trucks would cost $129,500. Retrofitting was estimated at $19,500 per truck. Ibid., 60.

456. Ibid., 67–70.

457. Ibid., 68.

458. Ibid., 68–70.

459. The deadlines were "so that the community would know that we were taking their concerns seriously but also so that our business contacts, our tenants and our customers would know and have certainty about what was going to be expected of them in terms of delivering cleaner air to the public." Telephone interview with Jerilyn López Mendoza, note 400.

460. Port of Long Beach and Port of Los Angeles, note 447, 73.

461. See "San Pedro Bay Ports Clean Air Action Plan Joint Board Meeting" (Nov. 20, 2006), http://portofla.granicus.com/MediaPlayer.php?view_id=9&clip_id=406 (accessed Dec. 1, 2017).

462. The mayor urged the port to "grow green, but grow indeed," noting that port growth would support 1.9 million regional jobs. Janet Wilson, "Port Panels OK Plan to Cut Pollution," *Los Angeles Times*, Nov. 21, 2006, B3.

463. "Statements of the Presidents of the Los Angeles Board of Harbor Commissioners and the Long Beach Board of Harbor Commissioners," in Port of Los Angeles and Port of Long Beach, note 447.

464. Ibid.

465. Ibid.

466. Ibid.

467. Louis Sahagun, "Port OKs 'Green' Cargo Fee," *Los Angeles Times*, Dec. 18, 2007, B1. As part of the emission-reduction effort, the Los Angeles port and SCAQMD funded the production of electric drayage trucks in conjunction with Balqon Corporation. See Port of Los Angeles, press release, "Mayor Villaraigosa Drives First Heavy-Duty, Electric Port Drayage Truck off the Assembly Line at New Harbor City

Factory" (Feb. 24, 2009), http://www.portoflosangeles.org/newsroom/2009_releases/news_022409_etruck.asp (accessed July 5, 2017).

468. Janet Wilson, "Trucks Targeted in Clean-Air Drive," *Los Angeles Times*, Nov. 12, 2006, B1.

469. Ibid.

Chapter 5

1. Carl Pope, "A New Blue-Green Alliance Is Born," *Huffington Post*, June 8, 2006, http://www.huffingtonpost.com/carl-pope/a-new-bluegreen-alliance-_b_22558.html (accessed July 10, 2017).

2. The video was narrated by Diane Keaton. See Natural Resources Defense Council, "Terminal Impact," *YouTube*, Sept. 1 2006, http://www.youtube.com/watch?v=qOUbj1ssjKs (accessed July 10, 2017).

3. Telephone interview with Nick Weiner, national campaigns organizer, Change to Win (Apr. 17, 2013).

4. Ibid.

5. Ibid.

6. Ibid.

7. Telephone interview with John Canham-Clyne, campaign director, Change to Win (May 20, 2013).

8. Jon Zerolnick, "The Clean and Safe Ports Campaign: False Dichotomies and the Underground Economy Versus Coalition-Building and the Power of Local Government" (unpublished manuscript), 9.

9. Telephone interview with Nick Weiner, note 3.

10. Ibid.

11. Telephone interview with John Canham-Clyne, note 7.

12. Ibid.

13. Ibid.

14. Interview with Patricia Castellanos, director of Clean and Safe Ports Project (Feb. 23, 2012).

15. Interview with Jon Zerolnick, senior research and policy analyst, L.A. Alliance for a New Economy (Feb. 23, 2012).

16. Ibid.

Notes

17. Ibid.

18. Ibid.

19. Ibid.

20. Interview with Patricia Castellanos, note 14.

21. Telephone interview with John Canham-Clyne, note 7.

22. Interview with Patricia Castellanos, note 14.

23. Ibid.

24. Ibid.

25. Ibid.

26. Ibid.

27. Ibid.

28. Telephone interview with Nick Weiner, national campaigns organizer, Change to Win (Apr. 25, 2013).

29. Telephone interview with Nick Weiner, note 3.

30. Interview with Jon Zerolnick, note 15.

31. Telephone interview with Michael Manley, staff attorney, International Brotherhood of Teamsters (Feb. 20, 2013).

32. Ibid.

33. Ibid.

34. Telephone interview with Nick Weiner, note 3.

35. See, e.g., Colleen Callahan, "Clean Trucks Program Case Study" (unpublished manuscript).

36. See International Brotherhood of Teamsters, press release, "Teamsters and ILWU Announce Port Legislation Strategy" (Feb. 4, 2002), http://teamster.org/content/teamsters-and-ilwu-announce-port-legislation-strategy (accessed July 10, 2017) (noting that the unions were joining to support state bills to force terminal operators to pay fines for making truckers idle while waiting for cargo and to require trucks to be safety certified).

37. See "Teamsters Deny Any Role in Planning Work Shutdown for L.A. Port Drivers," *American Shipper*, Apr. 28, 2004, http://www.labornet.org/news/0504/lateamst.htm (accessed July 10, 2017).

38. See Scott Martindale, "Port Truckers Rally Against Off-Hours Plan," *Daily Breeze*, July 25, 2005, http://teamster.org/content/daily-breeze-port-truckers-rally-against-hours-plan (accessed July 10, 2017).

39. Telephone interview with Adrian Martinez, staff attorney, Natural Resources Defense Council (Apr. 1, 2010).

40. Telephone interview with Melissa Lin Perrella, staff attorney, Natural Resources Defense Council (Apr. 2, 2010).

41. Interview with Patricia Castellanos, note 14.

42. See Dep't of Transp. v. Public Citizen, 541 U.S. 752 (2004) (holding that neither NEPA nor Clean Air Act requires the Federal Motor Carrier Safety Administration to evaluate environmental impact of cross-border trucking).

43. Telephone interview with Adrian Martinez, note 39.

44. Engine Mfrs. Ass'n v. S. Coast Air Quality Mgmt. Dist., 498 F.3d 1031 (9th Cir. 2007).

45. Telephone interview with Adrian Martinez, staff attorney, Natural Resources Defense Council (Apr. 2, 2010).

46. Ibid.

47. Ibid.

48. Telephone interview with Adrian Martinez, note 39.

49. Telephone interview with David Pettit, senior attorney, Urban Program, Natural Resources Defense Council (Apr. 5, 2010).

50. Ibid.

51. Telephone interview with Melissa Lin Perrella, note 40.

52. Ibid.

53. Telephone interview with Adrian Martinez, note 45.

54. Telephone interview with Melissa Lin Perrella, note 40.

55. Ibid.

56. Telephone interview with Adrian Martinez, note 39.

57. Ibid.

58. Telephone interview with Melissa Lin Perrella, note 40.

59. Telephone interview with Elina Green-Nasser, administrator, UCLA School of Public Health, and former project manager, Long Beach Alliance for Children with Asthma (Apr. 23, 2013) (Green changed her name to Green-Nasser after the campaign); telephone interview with Angelo Logan, co-director, East Yard Communities for Environmental Justice (May 1, 2013).

60. Interview with Jon Zerolnick, note 15.

Notes

61. Telephone interview with Nick Weiner, note 3.

62. Telephone interview with John Canham-Clyne, note 7.

63. Ibid.

64. Telephone interview with Tom Politeo, volunteer, Sierra Club Harbor Vision Task Force (Mar. 26, 2013).

65. Ibid.

66. Ibid.

67. Ibid.

68. Ibid.

69. Ibid.

70. Ibid.

71. Ibid.

72. Ibid.

73. Ibid.

74. Ibid.

75. Telephone interview with Candice Kim, senior campaign associate, Coalition for Clean Air (Apr. 26, 2013).

76. Telephone interview with Tom Politeo, note 64. The group was coordinated by Martha Matsuoka, a Ph.D. candidate in urban policy at UCLA and now professor at Occidental College. Telephone interview with Candice Kim, note 75.

77. Telephone interview with Tom Politeo, note 64.

78. Ibid.; telephone interview with Jesse Marquez, executive director, Coalition for a Safe Environment (May 20, 2013).

79. Telephone interview with Jesse Marquez, note 78.

80. Ibid.

81. Ibid. Security concerns were fueled by news of foreign entrants into the terminal operations market. In the winter of 2006, the company Dubai Ports World publicized its plan to purchase twenty-two U.S. port terminals. Although its plan did not include Los Angeles and Long Beach, it raised security concerns about the regulation of immigrant drivers—concerns that organizers tried to use to promote employee conversion. See Judith Lewis, "A Heavy Load," *LA Weekly*, July 25, 2007, http://www.laweekly.com/2007-07-26/news/a-heavy-load/ (accessed July 10, 2017).

82. Telephone interview with Colleen Callahan, deputy director, UCLA Luskin Center for Innovation and former manager of air quality policy, American Lung Association (Apr. 15, 2013).

83. Ibid.

84. Ibid.

85. Ibid.

86. Telephone interview with Elina Green-Nasser, note 59.

87. Ibid.

88. Ibid.

89. Telephone interview with Tom Politeo, note 64.

90. Ibid.

91. Ibid.

92. Ibid.

93. Telephone interview with Jonathan Klein, executive director, Clergy and Laity United for Economic Justice (May 8, 2013).

94. Port truckers had heavily participated in the 2006 May Day immigration demonstrations, angered by recent immigration raids.

95. Interview with Patricia Castellanos, note 14.

96. Ibid.

97. Ibid.

98. Ibid.

99. Interview with Jon Zerolnick, note 15.

100. Ibid.

101. Interview with Patricia Castellanos, note 14.

102. Ibid.

103. Ibid.

104. Ibid.

105. Ibid.

106. Ibid.

107. Telephone interview with Colleen Callahan, note 82. Callahan stated that the American Lung Association senior managers wanted to make sure "the campaign was truly about ... clean air and not just about ... labor issues." Ibid.

Notes

108. Telephone interview with Elina Green-Nasser, note 59.

109. Telephone interview with Angelo Logan, note 59.

110. Ibid.

111. Coalition for Clean & Safe Ports, Draft Mission Statement (unpublished document).

112. Telephone interview with Adrian Martinez, note 39.

113. Interview with Jon Zerolnick, note 15.

114. Ibid.

115. Ibid.

116. Telephone interview with Adrian Martinez, note 39.

117. Interview with Jon Zerolnick, note 15.

118. Letter from LAANE et al. to Geraldine Knatz, executive director, Port of Los Angeles, and Richard D. Steinke, executive director, Port of Long Beach (Aug. 28, 2006). The partners included at that point were Change to Win, CLUE, Coalition for Clean Air, Coalition for Humane Immigrant Rights of Los Angeles, Communities for a Better Environment, Harbor-Watts Economic Development Corporation, the Teamsters, the Los Angeles County Federation of Labor, and NRDC.

119. Ibid., 2.

120. Ibid., 3–4.

121. Ibid., 5–6.

122. Ibid., 6.

123. Ibid., 7.

124. Interview with Jon Zerolnick, note 15.

125. Interview with Patricia Castellanos, note 14; interview with Jon Zerolnick, note 15.

126. Interview with Jon Zerolnick, note 15.

127. Interview with Patricia Castellanos, note 14 (noting that End Oil joined the coalition after the initial launch).

128. There were approximately thirty initial members in the coalition, which grew to around forty members. See ibid.

129. Telephone interview with Adrian Martinez, note 45.

130. Coalition for Clean & Safe Ports, Structure and Decision-Making (unpublished document).

131. Ibid. The first steering committee, not yet at full strength, included Adrian Martinez of NRDC, Elina Green of LBACA, Louis Diaz from Teamsters Local 848, Nativo Lopez from Hermandad Mexicana, Rafael Pizarro of Coalition for Clean Air, a representative from Teamsters Local 63, and Tom Politeo from the Sierra Club.

132. Interview with Patricia Castellanos, note 14.

133. Steering committee decisions were by consensus; if no consensus could be achieved, decisions went to the full coalition. See Coalition for Clean & Safe Ports, Structure and Decision-Making, note 130.

134. Interview with Patricia Castellanos, note 14.

135. Port of Long Beach and Port of Los Angeles, "San Pedro Bay Ports Clean Air Action Plan Implementation Stakeholder Meeting" (PowerPoint presentation).

136. See Port of Los Angeles and Port of Long Beach, San Pedro Bay Ports Clean Air Action Plan, "CAAP Stakeholder Group Members."

137. See "San Pedro Bay Ports Clean Air Action Plan Implementation Stakeholder Meeting," note 135.

138. Telephone interview with Melissa Lin Perrella, note 40.

139. Interview with Jon Zerolnick, note 15.

140. Ibid.

141. Ibid.

142. Ibid.

143. Telephone interview with Angelo Logan, note 59.

144. Interview with Jon Zerolnick, note 15.

145. Interview with Patricia Castellanos, note 14.

146. Coalition for Clean & Safe Ports, Request for Proposals: Port Drayage Service Contract, Executive Summary (unpublished document, Apr. 2007), 5.

147. Ibid., 5–6.

148. Interview with Jon Zerolnick, note 15.

149. Ibid.

150. Coalition for Clean & Safe Ports, note 146, 2.

151. Ibid., 3.

152. Ibid.

153. Interview with Jon Zerolnick, note 15.

Notes

154. Coalition for Clean & Safe Ports, note 146, 10–11.

155. Ibid., 4.

156. Ibid., 3. The RFP also stated that 25 percent of the converted fleet must be natural gas trucks.

157. Ibid., 12–13.

158. Ibid., 4.

159. Telephone interview with Jon Zerolnick, director of Clean & Safe Ports Project, L.A. Alliance for a New Economy (Feb. 20, 2013).

160. Janet Wilson and Ronald D. White, "2 Ports Aim to Slash Diesel Exhaust," *Los Angeles Times*, Apr. 14, 2007, B1.

161. Ibid.

162. Port of Long Beach and Port of Los Angeles, *Ports of Long Beach and Los Angeles Proposed Clean Trucks Program* (2007), 1.

163. Ibid., 1–2.

164. Ibid., 2.

165. Port of Long Beach and Port of Los Angeles, "Proposed Clean Trucks Program Fact Sheet" (2007).

166. Port of Long Beach and Port of Los Angeles, note 162, 8.

167. Ports of Los Angeles and Long Beach, "Clean Truck Program: Program Elements for Stakeholder Discussion, CAAP HDV1" (PowerPoint presentation, Apr. 11, 2007).

168. Interview with Jon Zerolnick, note 15.

169. Wilson and White, note 160.

170. Interview with Jon Zerolnick, note 15.

171. Ibid.

172. Letter from Coalition for Clean & Safe Ports to Dr. Geraldine Knatz, executive director, Port of Los Angeles, and Richard D. Steinke, executive director, Port of Long Beach (May 10, 2007), 1.

173. Ibid., 2.

174. Ibid. The letter also reiterated the coalition's argument for imposing minimum business standards on LMCs, requiring labor peace agreements, mandating some alternative fuel trucks, and developing an off-street parking and community impact plan. Ibid., 5–9.

175. Interview with Patricia Castellanos, note 14.

176. Interview with Jon Zerolnick, note 15.

177. Telephone interview with David Pettit, note 49 (recalling statement by port general counsel Tom Russell).

178. Interview with Jon Zerolnick, note 15.

179. Louis Sahagun and Ronald D. White, "Port Drivers Steer Toward Clean-Truck Program," *Los Angeles Times*, June 6, 2007, B2.

180. Interview with Jon Zerolnick, note 15.

181. See "Agenda of the Regular Meeting of the Los Angeles Board of Harbor Commissioners" (Oct. 12, 2007), http://lacity.granicus.com/MediaPlayer.php?view_id=14&clip_id=2191&meta_id=27885 (accessed July 10, 2017).

182. See telephone interview with Elina Green-Nasser, note 59.

183. Tiffany Hsu and Rong-Gong Lin II, "Taking a Message to the Streets," *Los Angeles Times*, June 28, 2007, B1.

184. Telephone interview with Candice Kim, note 75.

185. Telephone interview with Nick Weiner, note 28.

186. Telephone interview with Candice Kim, note 75.

187. Interview with Patricia Castellanos, note 14.

188. Telephone interview with Jerilyn López Mendoza, commissioner, Los Angeles Board of Public Works, former commissioner, Los Angeles Board of Harbor Commissioners (Apr. 26, 2013).

189. Telephone interview with Colleen Callahan, note 82.

190. Telephone interview with Elina Green-Nasser, note 59.

191. Ibid.

192. Interview with Patricia Castellanos, note 14.

193. Ibid.

194. Telephone interview with Angelo Logan, note 59; telephone interview with Jesse Marquez, note 78.

195. Telephone interview with Colleen Callahan, note 82.

196. Ibid.; telephone interview with Elina Green-Nasser, note 59.

197. Telephone interview with Larry Frank, president, Los Angeles Trade Technical College (Nov. 4, 2016).

198. See Larry Frank and Kent Wong, "Dynamic Political Mobilization: The Los Angeles County Federation of Labor," *WorkingUSA: Journal of Labor & Society* 8 (2004): 159, 171 (noting that the occasional voter strategy was pioneered by Marshall Ganz in San Diego and utilized by Anthony Thigpenn, who went on to lead an important South Los Angeles organizing group, AGENDA, which has supported labor-community organizing efforts in the African American community).

199. Telephone interview with Larry Frank, note 197.

200. Ibid.

201. Ibid.

202. Ibid.

203. Ibid.

204. Ibid.

205. Ibid.

206. Ibid.

207. Ibid.

208. Ibid.

209. Ibid.

210. David Zahniser and Louis Sahagun, "Truckers' Status Is a Hitch in Port Plan," *Los Angeles Times*, Mar. 6, 2008, B1.

211. Telephone interview with Larry Frank, note 197.

212. Ibid.

213. Ibid.

214. Ibid.

215. Ibid.

216. Telephone interview with Sean Arian, president, Eos Consulting (Apr. 26, 2013).

217. Telephone interview with Larry Frank, note 197.

218. Telephone interview with Nick Weiner, note 28.

219. Telephone interview with David Libatique, senior director, government affairs, Port of Los Angeles (June 2, 2013).

220. Ibid.

221. Ibid.

222. Telephone interview with Larry Frank, note 197.

223. Telephone interview with Sean Arian, note 216.

224. Ibid.

225. Telephone interview with Michael Manley, note 31.

226. Ibid.

227. Ibid.

228. Ibid.

229. Telephone interview with Nick Weiner, note 28.

230. Telephone interview with Adrian Martinez, note 39.

231. Telephone interview with David Pettit, note 49.

232. Email from Joy Crose, assistant general counsel, Office of the City Attorney, to Scott Cummings, professor, UCLA School of Law (May 7, 2013).

233. Telephone interview with Steven S. Rosenthal, partner, Complex Commercial Litigation Department, Kaye Scholer (Dec. 16, 2013).

234. Ibid.

235. Telephone interview with Nick Weiner, note 3.

236. Telephone interview with Adrian Martinez, note 39.

237. Ibid.

238. Telephone interview with David Libatique, note 219; see also telephone interview with Sean Arian, note 216 ("[W]e were very confident that we had strong legal justification for [the program] to pass.").

239. Howard Blume and Joel Rubin, "Judge Tosses Out Mayor's Takeover of L.A. Schools," *Los Angeles Times*, Dec. 22, 2006, A1.

240. Telephone interview with Larry Frank, note 197.

241. Ibid.

242. Louis Sahagun, "Ports Complex Plans to Grow Bigger, Cleaner," *Los Angeles Times*, May 28, 2007, B1.

243. Ibid.

244. Ibid.

245. Ibid.

246. Ibid.

Notes

247. Ibid.

248. Interview with Patricia Castellanos, note 14.

249. Ronald D. White, "Record Southland Imports Predicted," *Los Angeles Times*, May 1, 2007, C6.

250. Ronald D. White, "Exporters Making Waves over Ports' Clean-Air Plan," *Los Angeles Times*, June 1, 2007, C3.

251. John E. Husing, Thomas E. Brightbill, and Peter A. Crosby, *San Pedro Bay Ports Clean Air Action Plan, Economic Analysis: Proposed Clean Truck Program* (2007), http://www.polb.com/civica/filebank/blobdload.asp?BlobID=4397 (accessed July 10, 2017).

252. Ibid., iv–v.

253. Ibid., 6.

254. Ibid., iii.

255. Ibid., 39–41.

256. Ibid., 66–69.

257. Ibid., v.

258. Ibid., iv.

259. Interview with Jon Zerolnick, note 15.

260. Jon Zerolnick, *The Road to Shared Prosperity: The Regional Economic Benefits of the San Pedro Bay Ports' Clean Trucks Program*, 6–7 (2007), https://perma.cc/34NM-PEQJ (accessed July 10, 2017).

261. Husing, Brightbill, and Crosby, note 251, i.

262. Ibid., 74.

263. Ibid., 75.

264. Ibid., vi, 17 (predicting loss of 376 "mostly smaller LMCs" out of a total of 800 to 1,200 LMCs overall).

265. Ronald D. White, "Plan to Cut Port Air Pollution Assailed," *Los Angeles Times*, Sept. 28, 2007, C3.

266. Ronald D. White and Janet Wilson, "Opposition Grows to Ports' Clean-Air Plan," *Los Angeles Times*, Sept. 29, 2007, C1.

267. Ronald D. White, "Changes Urged in Proposal for Ports," *Los Angeles Times*, Oct. 10, 2007, C2.

268. White and Wilson, note 266.

269. Ibid.

270. Telephone interview with David Libatique, note 219.

271. Telephone interview with Sean Arian, note 216.

272. Ibid.

273. Telephone interview with Larry Frank, note 197.

274. Telephone interview with Sean Arian, note 216.

275. Ibid.

276. Ibid.

277. Ibid.

278. Telephone interview with David Libatique, note 219.

279. Telephone interview with Nick Weiner, note 28.

280. Telephone interview with Sean Arian, note 216.

281. Telephone interview with John Holmes, deputy executive director, Port of Los Angeles (Apr. 29, 2013).

282. Ibid.

283. Ibid. The placard and parking programs were also in the version that Holmes received.

284. Ibid.

285. Ibid.

286. Telephone interview with Nick Weiner, note 28.

287. Ibid.

288. Ibid.

289. Ibid.

290. Ibid.

291. Ibid.

292. Ibid.

293. Ibid.

294. Ibid.

295. Telephone interview with John Holmes, note 281.

296. Telephone interview with Sean Arian, note 216.

Notes

297. Telephone interview with Nick Weiner, note 28.

298. Ibid.

299. Telephone interview with Sean Arian, note 216. During the same time frame, the port also commissioned a study from Beacon Economics, which was funded by the Hewlett Foundation and released in February 2008, finding that the overall benefits of the program outweighed the costs. Jon Haveman and Christopher Thornberg, Beacon Economics, *Clean Trucks Program: An Economic Policy Analysis* (2008), https://perma.cc/8AZ6-TJ2G (accessed July 10, 2017).

300. Telephone interview with Sean Arian, note 216.

301. Ibid.

302. Telephone interview with John Holmes, note 281.

303. Interview with John Holmes, deputy executive director, Port of Los Angeles (July 19, 2013).

304. Ibid.

305. Ibid.

306. Ibid.

307. Ibid.

308. Ibid.

309. Telephone interview with John Holmes, note 281.

310. Ibid.

311. Interview with John Holmes, note 303.

312. Telephone interview with John Holmes, note 281.

313. Ibid.

314. Telephone interview with Jerilyn López Mendoza, note 188.

315. Telephone interview with Sean Arian, note 216.

316. Telephone interview with Jerilyn López Mendoza, note 188.

317. Telephone interview with S. David Freeman, interim general manager, Los Angeles Department of Water and Power and former commissioner, Los Angeles Board of Harbor Commissioners (Apr. 29, 2013).

318. Ibid.

319. Ibid.

320. Telephone interview with Jerilyn López Mendoza, note 188.

321. Ibid.

322. Telephone interview with S. David Freeman, note 317.

323. Louis Sahagun, "L.A. Panel OKs Plan to Cut Port Truck Soot," *Los Angeles Times*, Nov. 2, 2007, B1.

324. Memorandum from Ralph G. Appy and Michael R. Christensen, Environmental Management Division, Permanent Order Amending Port of L.A. Tariff No. 4 (Oct. 29, 2007).

325. L.A. Bd. of Harbor Comm'rs., Order 6935, Items 2010, 2015, 2020 (Nov. 1, 2007). The progressive ban operated by barring drayage trucks built before 1989 by October 1, 2008; barring unretrofitted trucks built before 2004 (and retrofitted trucks built before 1994) by January 1, 2010; and completely barring any trucks that did not meet 2007 model year standards by January 1, 2012. Ibid.

326. L.A., Cal., Ordinance 165789 (Apr. 10, 1990) (adopting L.A. Bd. of Harbor Comm'rs, Order No. 5837 (July 12, 1989) (adopting Port of L.A. Tariff No. 4).

327. L.A. Bd. of Harbor Comm'rs, Order 6935, Items 2010, 2015, 2020.

328. Ibid., ¶ 12.

329. Ibid., ¶¶ 13, 14.

330. Ibid., Items 2000, 2005, 2025.

331. Long Beach Board of Harbor Commissioners, "Minutes of a Special Meeting," *Port of Long Beach* (Nov. 5, 2007), 20. At this meeting, during which the board approved the ordinance's first reading, the coalition turned out several drivers, who spoke, as did NRDC's Martinez, the Clean Air Coalition's Kim, and the American Lung Association's Callahan. Ibid., 3–10. The Long Beach harbor commissioners approved the ordinance's "second and final reading" on November 12. Long Beach Board of Harbor Commissioners, "Minutes," *Port of Long Beach* (Nov. 12, 2007), 9. The ordinance, as adopted, was No. HD-1997, which amended Long Beach Tariff No. 4 to include findings and policy language that were identical to Los Angeles Order 6935. Long Beach, Cal., Ordinance HD-1997 (Nov. 12, 2007) (amending Ordinance HD-1357, Tariff No. 4 (Dec. 27, 1983)). The Long Beach Clean Truck Program was drafted by city attorney Robert Shannon.

332. Louis Sahagun, "Long Beach Joins Port Ban on Old Trucks," *Los Angeles Times*, Nov. 6, 2007, B4.

333. Sahagun, note 323.

334. Ibid.

335. Ibid.

336. Sahagun, note 332.

Notes

337. Louis Sahagun, "Port OKs 'Green' Cargo Fee," *Los Angeles Times*, Dec. 18, 2007, B1.

338. Ibid. Lowenthal reintroduced the measure, which Schwarzenegger vetoed again in September 2008. Louis Sahagun and Ronald D. White, "Local Ports Initiate Antipollution Program," *Los Angeles Times*, Oct. 2, 2008, B2; Sahagun, note 337.

339. Editorial, "Long Beach and L.A. Port Officials Should Vote for Container Fees That Will Lead to Cleaner Air," *Los Angeles Times*, Dec. 17, 2007, A18 (noting that a disproportionately small amount of Proposition 1B funds had been allocated to Southern California).

340. Sahagun, note 337. The fee ordinance, No. HD-2005, was unanimously approved in its second and final reading (with one commissioner absent) on January 7, 2008. See Long Beach, Cal., Ordinance No. HD-2005 (Jan. 9, 2008); Long Beach Board of Harbor Commissioners, "Minutes," *Port of Long Beach* (Jan. 7, 2008), 9.

341. Long Beach Bd. of Harbor Comm'rs, Clean Truck Tariff Amendment and Fee, Item 1030 (Dec. 11, 2007), http://www.polb.com/civica/filebank/blobdload.asp?BlobID=4708 (accessed July 10, 2017).

342. Sahagun, note 337.

343. Long Beach Bd. of Harbor Comm'rs, Clean Truck Tariff Amendment and Fee, note 341, Item 1035.

344. L.A. Bd. of Harbor Comm'rs Order No. 6943 (Dec. 20, 2007). This order also followed a strong staff endorsement. See memorandum from Ralph G. Appy and Michael R. Christensen to L.A., Cal. Board of Harbor Commissioners (Dec. 20, 2007).

345. Telephone interview with Nick Weiner, note 28.

346. Telephone interview with S. David Freeman, note 317.

347. Ibid.

348. Sahagun, note 337.

349. Louis Sahagun, "Ports Turn Over a New, Green Leaf," *Los Angeles Times*, Dec. 25, 2007, A1.

350. Louis Sahagun, "Unsafe Trucks Stream Out of L.A.'s Ports," *Los Angeles Times*, Jan. 21, 2008, A1.

351. Sahagun, note 349.

352. Ibid.

353. Ibid.

354. Ibid.

355. See Environmental Management Division, Port of Los Angeles, *Draft Findings of Fact and Statement of Overriding Considerations, Berths 136–147 [TraPac] Container Terminal Project* (2007).

356. Los Angeles Board of Harbor Commissioners, "Minutes of the Special Meeting of the Los Angeles Board of Harbor Commissioners," *Port of Los Angeles* (Dec. 6, 2007).

357. Letter from David Pettit et al. to Members of the Los Angeles City Council, Re: Appeal from Board of Harbor Commissioners Decision to Approve the Final EIR for TraPac Container Terminal (Dec. 14, 2007).

358. Telephone interview with Kathleen Woodfield, member, Sierra Club (May 14, 2013). Woodfield was also a member of the Port Community Advisory Committee. Ibid.

359. Ibid.

360. See Am. Trucking Ass'ns, Inc. v. City of Los Angeles, No. CV 08–4920 CAS (RZx), 2010 WL 3386436, at *8 (C.D. Cal. Aug. 26, 2010).

361. David Zahniser and Louis Sahagun, "Harbor Reaches Pollution Accord," *Los Angeles Times*, Apr. 3, 2008, B1.

362. Louis Sahagun, "Long Beach Port Faces Suit Threat," *Los Angeles Times*, Feb. 7, 2008, B3.

363. Ibid.

364. Ibid.

365. Louis Sahagun, "Officials of Area Ports Split Over Truck Issue," *Los Angeles Times*, Feb. 19, 2008, B1.

366. Ibid.

367. Telephone interview with Larry Frank, note 197.

368. Long Beach, Cal., Ordinance No. HD-2011, ¶ 14 (Mar. 17, 2008).

369. Ibid., Item 1030.

370. Long Beach Board of Harbor Commissioners, "Minutes," *Port of Long Beach* (Feb. 19, 2008), 5–6.

371. A second and final reading of the ordinance was approved on March 17, 2008, with two commissioners (Topsy-Elvord and Walter) absent. Long Beach Board of Harbor Commissioners, "Minutes," *Port of Long Beach* (Mar. 17, 2008), 6–7. The board authorized the port director to execute the concession agreements on June 2. Long Beach Board of Harbor Commissioners, "Minutes," *Port of Long Beach* (June 2, 2008), 10.

Notes

372. Louis Sahagun, "Public Health, Labor Groups Decry Harbor Panel's Air Plan," *Los Angeles Times*, Feb. 20, 2008, B4.

373. The Long Beach port's concession agreement stated that a concessionaire "shall give a hiring preference to drivers with a history of providing drayage services to the port," but did not require conversion. Long Beach Bd. of Harbor Comm'rs, Port of Long Beach Concession Agreement § III(e). Like the Los Angeles program that would follow, the Long Beach agreement mandated compliance with truck routes and parking restrictions, a maintenance plan, and a placard requirement, though it reduced the concession fee to $250 per concessionaire. Ibid. §§ III(f), (g), (m) and 2.1.1.

374. Sahagun, note 365.

375. Sahagun, note 372. The coalition appealed the board's decision to the city council to no avail. NRDC then filed a Freedom of Information Act suit to get all documents relevant to the Clean Truck Program. "We just wanted to find out what was going on because they just kept on making these bizarre decisions behind closed doors, with no public process." Telephone interview with Adrian Martinez, note 45.

376. Telephone interview with Adrian Martinez, note 45.

377. Sahagun, note 365.

378. Ibid.

379. Editorial, "A Storm in Every Port," *Los Angeles Times*, Feb. 24, 2008, M2.

380. Ibid.

381. Ibid.

382. Ibid.

383. Martin Schlageter, Letter to the Editor, "Delays in Cleaning Up the Ports," *Los Angeles Times*, Mar. 1, 2008, A20.

384. Telephone interview with Nick Weiner, note 28.

385. Telephone interview with David Pettit, note 49.

386. L.A., Cal., Ordinance 179707 (Feb. 27, 2008) (adopting L.A. Bd. of Harbor Comm'rs, Order No. 6935 (Nov. 1, 2007) (progressive truck ban)); L.A., Cal., Ordinance 179708 (Feb. 27, 2008) (adopting L.A. Bd. of Harbor Comm'rs Order No. 6943 (Dec. 20, 2007) (clean truck fee)).

387. Zahniser and Sahagun, note 210.

388. Ibid.

389. Ibid.

390. Ibid.

391. Ibid.

392. The Boston Consulting Group, *San Pedro Bay Ports Clean Truck Program: CTP Options Analysis* (2008), 6.

393. Ibid.

394. Ibid., 10.

395. Ibid., 9.

396. Ibid.

397. Ibid., 68.

398. Ibid., 70, 79.

399. Ibid., 74.

400. Ibid., 9.

401. Ibid., 74.

402. Ronald D. White and Louis Sahagun, "Risk Seen in Port Plan," *Los Angeles Times*, Mar. 8, 2008, C1.

403. Telephone interview with Larry Frank, note 197.

404. Ibid.

405. Ibid.

406. Memorandum from John Holmes, deputy executive director of operations, and Molly Campbell, deputy executive director of finance and administration, to the Los Angeles Board of Harbor Commissioners (Mar. 12, 2008).

407. Telephone interview with Sean Arian, note 216.

408. Editorial, "Harbor No Illusions," *Los Angeles Times*, Mar. 20, 2008, A20.

409. Telephone interview with David Libatique, note 219.

410. The presentation of the concession plan provided at the meeting drew heavily on the BCG findings. See *The Port of Los Angeles Clean Truck Program: Program Overview & Benefits*, http://www.portoflosangeles.org/ctp/CTP_O&B.pdf (accessed Apr. 2, 2014).

411. L.A. Bd. of Harbor Comm'rs, Order No. 6956, ¶ 3 (Mar. 20, 2008).

412. L.A. Bd. of Harbor Comm'rs, Port of L.A., Tariff No. 4, Item No. 2040 (Mar. 20, 2008).

413. Ibid., Item No. 220(b).

Notes

414. L.A. Bd. of Harbor Comm'rs, Order No. 6956, ¶¶ 16, 19.

415. Ibid., ¶¶ 21, 23.

416. Ibid., ¶ 24.

417. Transmittal 1, Port of Los Angeles Drayage Truck Concession Requirements, ¶ (b).

418. Ibid., ¶¶ (f), (m), (o).

419. L.A. Bd. of Harbor Comm'rs, Order No. 6956, ¶ 26.

420. Ibid., ¶ 25. The order also made a technical change to the definition of "cargo owner" that affected application of the fee. Ibid., ¶ 28.

421. L.A. Bd. of Harbor Comm'rs, Port of L.A., Tariff No. 4, Item No. 2095 (Mar. 20, 2008).

422. Telephone interview with S. David Freeman, note 317.

423. L.A. Bd. of Harbor Comm'rs, Resolution 6522 (Mar. 20, 2008), 1.

424. Ibid., ¶¶ a–o.

425. Ibid., 7.

426. Ibid.

427. Louis Sahagun, "Port Shifts Plan's Cost to Shippers," *Los Angeles Times*, Mar. 21, 2008, B5.

428. Telephone interview with Tom Politeo, note 64.

429. Telephone interview with Kathleen Woodfield, note 358.

430. Ibid.

431. Telephone interview with Tom Politeo, note 64.

432. See Port of Los Angeles, press release, "Mayor Villaraigosa, Councilwoman Hahn Announce Historic Agreement That Will Allow TraPac Terminal Renovations to Go Forward at Port of L.A." (Apr. 3, 2008); see also Zahniser and Sahagun, note 361.

433. Telephone interview with Kathleen Woodfield, note 358. Woodfield also noted the contributions of Serena Lin, a lawyer at Public Counsel Law Center, to the TraPac settlement. Ibid.

434. L.A. Harbor Dep't Agreement 09-2764, Memorandum of Understanding, 4 (Apr. 2, 2008); see also Zahniser and Sahagun, note 361.

435. See L.A. Harbor Dep't Agreement 09-2764, note 434, 4.

436. Ibid.

437. Ibid., 3; see also Los Angeles Board of Harbor Commissioners, "Minutes of a Special Meeting," *Port of Los Angeles* (Oct. 26, 2010), 15–16 (approving creation of nonprofit organization to administer the Mitigation Trust Fund).

438. See L.A. Harbor Dep't Agreement 09–2764, note 434, 2–3.

439. Zahniser and Sahagun, note 361.

440. Ibid.

441. L.A. Bd. of Harbor Comm'rs, Drayage Services Concession Agreement for Access to the Port of Los Angeles, ¶ III(d).

442. Ibid.

443. Ibid., ¶¶ III(f), (g), (l), (n) and § 2.1.1.

444. Ibid., §§ 4.3 and 4.4.

445. See Los Angeles Board of Harbor Commissioners, "Minutes of the Regular Meeting," *Port of Los Angeles* (May 15, 2008), 19–20.

446. See L.A., Cal., Ordinance No. 179981 (June 17, 2008).

447. Phil Willon, "Mayor Signs Law to Clean Port Air," *Los Angeles Times*, June 27, 2008, B4. There were a number of subsequent technical amendments that clarified exemptions to the Clean Truck Fee, clarified the basis for charging the fee, delayed its implementation, and made other adjustments. See L.A., Cal., Ordinance No. 180681 (Aug. 21, 2008); L.A., Cal., Ordinance No. 180679 (May 5, 2009); L.A., Cal., Ordinance No. 180923 (Oct. 14, 2009); L.A., Cal., Ordinance No. 1809253 (Oct. 14, 2009); L.A., Cal., Ordinance No. 180942 (Oct. 27, 2009); L.A., Cal., Ordinance No. 181125 (Mar. 12, 2010); L.A., Cal., Ordinance No. 181126 (Mar. 12, 2010); L.A., Cal., Ordinance No. 181255 (June 27, 2010).

448. Telephone interview with Nick Weiner, note 28.

449. Telephone interview with Jon Zerolnick, note 159.

450. Ronald D. White and Louis Sahagun, "National Trucking Group to Sue Ports Over Cleanup Plan," *Los Angeles Times*, July 26, 2008, B9.

451. Complaint for Declaratory Judgment and Injunctive Relief, Am. Trucking Ass'ns, Inc. v. City of L.A., 577 F. Supp. 2d 1110 (C.D. Cal. 2008) (No. CV 08–04920 CAS (CTx)) (filed July 28, 2008). The ATA was represented by its in-house counsel, Robert Digges; outside counsel law firm Constantine Cannon LLP, appearing pro hac vice; and local counsel Christopher C. McNatt, Jr. of Scopelitis, Garvin, Light, Hanson & Feary, LLP, in Pasadena. Ibid.

452. 49 U.S.C. § 14501(c)(1) (2012).

Notes

453. Complaint for Declaratory Judgment and Injunctive Relief, note 451, ¶ 2.

454. Ibid., ¶ 3.

455. Ibid., ¶ 4.

456. Ibid.

457. Louis Sahagun, "Truck Group Sues Ports," *Los Angeles Times*, July 29, 2008, B4.

458. Complaint for Declaratory Judgment and Injunctive Relief, note 451, ¶¶ 37–47 (Count I), ¶¶ 48–54 (Count II). Count III argued that both plans were preempted by the Commerce Clause. Ibid., ¶¶ 55–66.

459. Notice of Motion and Motion for Preliminary Injunction on Counts I and II of Plaintiff's Complaint, *Am. Trucking Ass'ns, Inc.*, 577 F. Supp. 2d 1110 (No. CV 08–04920 CAS (CTx)) (filed July 30, 2008).

460. Louis Sahagun and Ronald D. White, "Truckers and Ports Head to Court," *Los Angeles Times*, Sept. 8, 2008, B3.

461. Telephone interview with Adrian Martinez, note 45.

462. Notice of Motion and Motion to Intervene of Natural Resources Defense Council, Sierra Club and Coalition for Clean Air at 1–2, *Am. Trucking Ass'ns, Inc.*, 577 F. Supp. 2d 1110 (No. CV 08–04920 CAS (CTx)) (filed July 31, 2008).

463. Opposition of Proposed Defendant-Intervenors Natural Resources Defense Council, Sierra Club and Coalition for Clean Air to Plaintiff's Motion for Preliminary Injunction at 3, *Am. Trucking Ass'ns, Inc.*, 577 F. Supp. 2d 1110 (No. CV 08–04920 CAS (CTx)) (filed Aug. 20, 2008).

464. Ibid., 5.

465. Telephone interview with Melissa Lin Perrella, note 40.

466. Telephone interview with Adrian Martinez, note 45.

467. Ibid.

468. Ibid.

469. Telephone interview with David Pettit, note 49.

470. Telephone interview with Melissa Lin Perrella, note 40.

471. Telephone interview with David Pettit, note 49.

472. Telephone interview with Melissa Lin Perrella, staff attorney, Natural Resources Defense Council (Apr. 23, 2013).

473. Telephone interview with Melissa Lin Perrella, note 40.

474. Ibid. Rosenthal suggested that although the city was "aligned" with NRDC in the litigation, they had "differing interests," noting that "NRDC certainly supported what we were doing, but NRDC sues the port as well." Telephone interview with Steven S. Rosenthal, note 233.

475. Telephone interview with Melissa Lin Perrella, note 472.

476. Telephone interview with Michael Manley, note 31.

477. Telephone interview with Michael Manley, staff attorney, International Brotherhood of Teamsters (Mar. 29, 2013).

478. The outside counsel was Paul L. Gale from Ross, Dixon & Bell, LLP, and C. Jonathan Benner and Mark E. Nagle, appearing pro hac vice, from Troutman Sanders, LLP. See Defendants' Opposition to Plaintiff's Motion for Preliminary Injunction, *Am. Trucking Ass'ns, Inc.*, 577 F. Supp. 2d 1110 (No. CV 08–04920 CAS (CTx)) (filed Aug. 20, 2008).

479. Ibid., 11.

480. Ibid., 17.

481. Ibid., 22.

482. Engine Mfrs. Ass'n v. S. Coast Air Quality Mgmt. Dist., 498 F.3d 1031 (9th Cir. 2007).

483. Defendants' Opposition to Plaintiff's Motion for Preliminary Injunction, note 478, 26–29.

484. 49 U.S.C. § 14501(c)(2)(A) (2012).

485. Defendants' Opposition to Plaintiff's Motion for Preliminary Injunction, note 478, 38.

486. Declaration of John M. Holmes in Support of Defendants' Opposition to Plaintiff's Motion for Preliminary Injunction at 4, *Am. Trucking Ass'ns, Inc.*, 577 F. Supp. 2d 1110 (No. CV 08–04920 CAS (CTx)) (filed Aug. 20, 2008). Affidavits of other port officials, including Los Angeles Executive Director Geraldine Knatz and Long Beach Executive Director Richard Steinke, were also submitted to support the defendants' broader market participant arguments. Declaration of Dr. Geraldine Knatz in Support of Defendants' Opposition to Plaintiff's Motion for Preliminary Injunction, *Am. Trucking Ass'ns, Inc.*, 577 F. Supp. 2d 1110 (No. CV 08–04920 CAS (CTx)) (filed Aug. 20, 2008); Declaration of Richard Steinke in Support of Defendants' Opposition, *Am. Trucking Ass'ns, Inc.*, 577 F. Supp. 2d 1110 (No. CV 08–04920 CAS (CTx)) (filed Aug. 20, 2008).

487. Susan Gordon, "Beverly Hills Bar Association Honors the Honorable Christina A. Snyder and Top Trial Attorney Marshall B. Goldman," *WestsideToday*, Feb.

Notes

6, 2013, http://westsidetoday.com/2013/02/06/beverly-hills-bar-association-honors-the-honorable-christina-a-snyder-and-top-trial-attorney-marshall-b-goldman/ (accessed July 11, 2017).

488. Ibid.

489. Order Denying Plaintiff's Motion for Preliminary Injunction at 26, *Am. Trucking Ass'ns, Inc.*, 577 F. Supp. 2d 1110 (No. CV 08–04920 CAS (CTx)) (filed Sept. 9, 2008).

490. *Am. Trucking Ass'ns, Inc.*, 577 F. Supp. 2d at 1117.

491. Ibid., 1125.

492. Ibid., 1124–1125.

493. Ibid., 1118.

494. Ibid., 1121–1123.

495. Preliminary Injunction Appeal, *Am. Trucking Ass'ns, Inc.*, 577 F. Supp. 2d 1110 (No. CV 08–04920 CAS (CTx)) (filed Sept. 10, 2008).

496. Consumer Federation of California, League of United Latin American Citizens, Los Angeles Alliance for a New Economy, and National Association for the Advancement of Colored People, *Foreclosure on Wheels: Long Beach's Truck Program Puts Drivers at High Risk for Default* (2008), http://laane.org/downloads/B568P314C.pdf (accessed July 11, 2017).

497. Ibid., 2.

498. Ibid.

499. Louis Sahagun, "Long Beach Port's Truck Loans Criticized," *Los Angeles Times*, Aug. 20, 2008, B3.

500. Louis Sahagun, "Truckers Blast Long Beach Loan Program," *Los Angeles Times*, Aug. 23, 2008, B4.

501. Ronald D. White and Louis Sahagun, "2 Big Haulers Accept L.A. Port's Clean-Truck Criteria," *Los Angeles Times*, Aug. 22, 2008, C2.

502. Ronald D. White, "Truckers on Board with Clean-Air Plans," *Los Angeles Times*, Sept. 6, 2008, C1.

503. Sahagun and White, note 338 (noting a 9.9 percent decrease in container traffic at the Long Beach port and 4.6 percent decrease at the Los Angeles port).

504. Francisco Vara-Orta, "Clean Air Program Revs Up Truck Sales," *Los Angeles Business Journal*, May 25, 2009.

505. Ronald D. White, "Truckers Unload Fuel Costs," *Los Angeles Times*, Apr. 10, 2012, A1.

506. Sahagun and White, note 338.

507. Ibid.

508. Brief of Appellant Am. Trucking Ass'ns, Inc. at 15, Am. Trucking Ass'ns, Inc. v. City of Los Angeles, 559 F.3d 1046 (9th Cir. 2009) (No. 08–56503) (filed Oct. 8, 2008).

509. Opening Brief for Appellees, *Am. Trucking Ass'ns, Inc.*, 559 F.3d 1046 (No. 08–56503) (filed Nov. 26, 2008).

510. Intervenor-Appellees' Brief at 3–4, *Am. Trucking Ass'ns, Inc.*, 559 F.3d 1046 (No. 08–56503) (filed Nov. 5, 2008).

511. Brief for the U.S. as Amicus Curiae in Support of Reversal, *Am. Trucking Ass'ns, Inc.*, 559 F.3d 1046 (No. 08–56503) (filed Oct. 21, 2008).

512. Brief of Amicus Curiae for Nat'l Indus. Transp. League in Support of Appellant Am. Trucking Ass'ns, Inc., *Am. Trucking Ass'ns, Inc.*, 559 F.3d 1046 (No. 08–56503) (filed Oct. 20, 2008).

513. Brief of Amicus Curiae for Nat'l Ass'n of Waterfront Employers in Support of Appellant Am. Trucking Ass'ns, *Am. Trucking Ass'ns, Inc.*, 559 F.3d 1046 (No. 08–56503) (filed June 30, 2009).

514. Brief of the Attorney Gen. of the State of Cal. as Amicus Curiae in Support of Appellees the City of L.A., *Am. Trucking Ass'ns, Inc.*, 559 F.3d 1046 (No. 08–56503) (filed Dec. 18, 2008). At the same time, CARB was finalizing its own emission rules to phase out old, dirty diesel trucks by requiring all diesel trucks to meet 2010 standards by the year 2023. Margot Roosevelt, "Community Groups, State Battle Pollution," *Los Angeles Times*, Dec. 11, 2008, B1. This was viewed by the coalition as a backstop, requiring trucks to move away from diesel irrespective of the outcome of the ATA litigation. See telephone interview with Melissa Lin Perrella, note 472. The *Los Angeles Times* editorial board praised the CARB plan, but also used it to take another jab at the Los Angeles port's employee conversion program: "The port needs a separate truck plan because it has a separate mechanism for funding cleaner vehicles, but it would be better off imitating state regulators and focusing on cleaning the air, not trying to reinvent the steering wheel." Editorial, "A New Day for Diesel," *Los Angeles Times*, Dec. 12, 2008, A32. Industry groups challenged the CARB standard and NRDC intervened to support it. Telephone interview with Melissa Lin Perrella, note 472. The standard was approved in 2008, though by 2014 it still had not been fully implemented. Tony Barboza, "Date for Cleaner Trucks Delayed," *Los Angeles Times*, Apr. 26, 2014, AA1 (stating that CARB agreed to postpone the compliance deadline for "small fleets, lightly used trucks and those operating in rural areas"). The Port of Los Angeles subsequently amended Tariff No. 4 to be consistent with the CARB standards. L.A. Bd. of Harbor Comm'rs, Order No. 09–7031 (Dec. 8, 2009).

Notes

515. *Am. Trucking Ass'ns, Inc.*, 559 F.3d at 1060.

516. Ibid., 1057.

517. Ibid., 1053.

518. Ibid., 1055.

519. Ibid.

520. Ibid., 1056.

521. Ibid.

522. Ibid., 1056–1057.

523. Ibid., 1057–1058.

524. Ibid., 1059–1060.

525. Ibid., 1060.

526. Notice of Plaintiff's Motion on Remand for Entry of a Preliminary Injunction on Counts I and II of Complaint, Am. Trucking Ass'ns, Inc. v. City of L.A., No. CV 08–4920 CAS (CTx), 2009 WL 1160212 (C.D. Cal. Apr. 28, 2009) (filed Apr. 3, 2009).

527. Los Angeles Defendants' Opposition to ATA's Motion on Remand for a Preliminary Injunction at 6–11, *Am. Trucking Ass'ns, Inc.*, No. CV 08–4920 CAS (CTx), 2009 WL 1160212 (filed Apr. 13, 2009). The provisions of Los Angeles's concession plan not addressed by the Ninth Circuit order included those requiring concessionaires to prepare maintenance plans, ensure driver enrollment in TWIC, guarantee compliance with federal, state, and local laws (including other provisions of the Clean Truck Program), and post placards.

528. Long Beach Defendants' Memorandum in Opposition to Plaintiff's Motion on Remand for Preliminary Injunction at 2–3, *Am. Trucking Ass'ns, Inc.*, No. CV 08–4920 CAS (CTx), 2009 WL 1160212 (filed Apr. 13, 2009). Long Beach also argued that its plan should be examined in the first instance by the Secretary of Transportation, who had statutory authority to determine whether its plan could be enforced. Ibid., 7.

529. Reply of ATA in Support of Motion on Remand for Preliminary Injunction at 22–23, *Am. Trucking Ass'ns, Inc.*, No. CV 08–4920 CAS (CTx), 2009 WL 1160212 (filed Apr. 17, 2009).

530. Editorial, "Let's Get Truckin'," *Los Angeles Times*, Apr. 24, 2009, A32. As he had in the past, the Coalition for Clean Air's Martin Schlageter responded to the *Times*: "Without a systemic fix, today's new trucks will be tomorrow's broken-down trucks." Martin Schlageter, "The Road We're On," *Los Angeles Times*, Apr. 28, 2009, A22 (letter to editor).

531. *Am. Trucking Ass'ns, Inc.*, No. CV 08–4920 CAS (CTx), 2009 WL 1160212.

532. Ibid., *20–21. The court also enjoined Long Beach's driver health insurance provision, and both ports' truck tariffs and concession fees. Ibid. However, in a subsequent ruling, responding to a motion by the ATA to modify the April 28 order so that it would not have to post bond, the court reinstituted the Port of Los Angeles's concession fee. Am. Trucking Ass'ns, Inc. v. City of L.A., No. CV 08–4920 CAS (CTx), 2009 WL 2412578, at *2–3 (C.D. Cal. Aug. 4, 2009).

533. *Am. Trucking Ass'ns, Inc.*, No. CV 08–4920 CAS (CTx), 2009 WL 1160212, at *11–18. In total, the court let stand nine provisions requiring concessionaires to: (1) be LMCs; (2) use permitted trucks; (3) ensure driver compliance; (4) prepare a maintenance plan; (5) ensure driver enrollment in TWIC; (6) ensure that trucks have compliance tags; (7) ensure compliance with security laws; (8) post placards; and (9) keep records not related to safety. Ibid., *21.

534. Ibid., *19–20.

535. Ronald D. White, "Judge Restricts Ports' Truck Plan," *Los Angeles Times*, Apr. 30, 2009, B2.

536. Brief of Appellant Am. Trucking Ass'ns at 13–16, Am. Trucking Ass'ns, Inc. v. City of L.A., 596 F.3d 602 (9th Cir. 2010) (No. 09–55749) (filed June 11, 2009).

537. Stipulation of Settlement and Joint Motion for Voluntary Dismissal with Prejudice between Plaintiff ATA and Long Beach Defendants, Exhibit A, Motor Carrier Registration and Agreement at 1–2, Am. Trucking Ass'ns, Inc. v. City of L.A., No. CV 08–4920 CAS (CTx) (filed Oct. 19, 2009). The settlement was approved by the board after a closed session meeting, with Commissioner Cordero voting against. Long Beach Board of Harbor Commissioners, "Minutes of a Special Meeting," *Port of Long Beach* (Oct. 19, 2009), 1.

538. Ronald D. White, "Port Settles Truckers Lawsuit," *Los Angeles Times*, Oct. 21, 2009, B4.

539. Exhibit A, Motor Carrier Registration and Agreement, note 537, 2–4. The settlement also required the port to amend provisions of its Clean Truck Program to be consistent with the settlement terms, which it did after a full hearing at which Pettit, Zerolnick, Schlageter, and other coalition members spoke out against the changes. Long Beach Board of Harbor Commissioners, "Minutes," *Port of Long Beach* (Nov. 16, 2009), 9–11.

540. Order of Voluntary Dismissal with Prejudice of Long Beach Defendants, *Am. Trucking Ass'ns, Inc.*, No. CV 08–4920 CAS (CTx) (filed Oct. 20, 2009).

541. Ronald D. White, "Ports Split on Union Stance," *Los Angeles Times*, Nov. 28, 2009, B1.

Notes

542. NRDC, Inc. v. City of Long Beach, No. CV 10–826 CAS (PJWx), 2011 WL 2790261, at *1–2 (C.D. Cal. July 14, 2011).

543. Ibid., *5.

544. Thomas Johnson Environmental Consultant LLC, *Port of Long Beach ATA Litigation Settlement Agreement: Initial Study* (2011), http://www.polb.com/civica/filebank/blobdload.asp?BlobID=9236 (accessed July 11, 2017).

545. Joint Stipulation Regarding Plaintiff's Motion to Compel Production of Documents Withheld Due to the Deliberate Process Privilege at 1–3, *Am. Trucking Ass'ns, Inc.*, No. 08–4920 CAS (RZx), 2010 WL 3386436 (filed Oct. 19, 2009).

546. Order Re: Plaintiff's Motion for Review of the Order of the Magistrate Judge Declining to Require Production of Documents Withheld Due to the Deliberative Process Privilege at 13, *Am. Trucking Ass'ns, Inc.*, No. 08–4920 CAS (RZx), 2010 WL 3386436 (filed Dec. 21, 2009).

547. Seema Mehta, "Truckers Protest New Rules," *Los Angeles Times*, Nov. 14, 2009, A3.

548. Bill Mongelluzzo, "ILWU Backs Long Beach Clean Trucks Program," *Journal of Commerce*, Jan. 7, 2010, http://www.joc.com/maritime-news/ilwu-backs-long-beach-clean-trucks-program_20100107.html (accessed July 11, 2017).

549. Am. Trucking Ass'ns, Inc. v. City of Los Angeles, 596 F.3d 602, 606 (9th Cir. 2010). This time, the panel consisted of a staunch liberal, Harry Pregerson, appointed by President Jimmy Carter; Ronald M. Gould, a conservative appointed by President Bill Clinton in a deal to break a nominations impasse; and Myron H. Bright, a senior Eighth Circuit Judge (appointed by President Lyndon Johnson) sitting by designation.

550. Ibid.

551. Ibid., 606–607.

552. On February 25, 2010, the district court denied both the ATA and the Port of Los Angeles's cross motions for summary judgment, clearing the way for trial. Civil Minutes—General, *Am. Trucking Ass'ns, Inc.*, No. 08–4920 CAS (RZx), 2010 WL 3386436 (proceedings in chambers on Feb. 25, 2010).

553. Telephone interview with Jon Zerolnick, note 159.

554. Ibid.

555. Ibid.

556. Ibid.

557. Ibid.

558. Patrick J. McDonnell, "Truckers Caught in a Tight Spot," *Los Angeles Times*, Nov. 27, 2009, A18. Critics charged that most port incentives went to fund big trucking companies (Swift Transportation received $12 million for 591 clean trucks, while Knight Transportation got $4.4 million for 172 trucks), although the port stated that $200 million went to small firms. Ibid.

559. Ibid.

560. Ibid.

561. Patrick J. McDonnell, "Truckers Assail 'Green' Cost," *Los Angeles Times*, Dec. 7, 2010, A1.

562. Ibid.

563. Ibid.

564. Ibid. Similar reports came out of Long Beach. See Kristopher Hanson, "Straining Under the Load," *Long Beach Press-Telegram*, Aug. 23, 2009, 3A. The article quotes a driver stating: "Between payments for the new truck, insurance, fuel, taxes and the lack of work, I'm barely making it." Ibid.

565. See 46 U.S.C. §§ 40301(b), 40302, 40304, 40307 (2012).

566. Fed. Mar. Comm'n, *Los Angeles and Long Beach Port Infrastructure and Environmental Programs Cooperative Working Agreement*, Federal Maritime Commission, Agreement No. 2011170 (original effective date Aug. 10, 2006), 1.

567. Fed. Mar. Comm'n, *Los Angeles and Long Beach Port Infrastructure and Environmental Programs Cooperative Working Agreement*, Federal Maritime Commission, Amended and Restated Agreement No. 2011170–001 (filed Aug. 1, 2008), 3.

568. Ronald D. White and Louis Sahagun, "Ports' Truck Plan May Be Delayed," *Los Angeles Times*, Sept. 13, 2008, B3.

569. Telephone interview with Adrian Martinez, note 45.

570. Ronald D. White, "Agency Objects to Clean Truck Program," *Los Angeles Times*, Oct. 30, 2008, at C1.

571. Complaint for an Injunction Pursuant to Section 6(h) of the Shipping Act of 1984 46 U.S.C. § 41307, Fed. Mar. Comm'n v. City of L.A., 607 F. Supp. 2d 192 (D.D.C. 2009) (No. 08–1895 (RJL)) (filed Oct. 31, 2008).

572. 46 U.S.C. § 41307(b)(1) (2012).

573. Complaint for an Injunction Pursuant to Section 6(h) of the Shipping Act of 1984 46 U.S.C. § 41307, note 571, 27–28.

574. Ibid., 10.

Notes

575. Kristopher Hanson, "Obama and Clinton Push for Truckers' Rights at Port," *Long Beach Press-Telegram*, Jan. 12, 2008, A7.

576. See Coalition for Clean & Safe Ports, press release, "Federal Maritime Commission's Move to Drop Lawsuit Against Clean Trucks Program Signals President Obama's Support to Protect Port Drivers, Public Health" (June 16, 2009), https://perma.cc/9DVF-W5QG (accessed July 11, 2017).

577. Plaintiff's Motion for Preliminary Injunction at 1, *Fed. Mar. Comm'n*, 607 F. Supp. 2d 192 (No. 08–1895 (RJL)) (filed Nov. 17, 2008).

578. Memorandum in Support of Plaintiff's Motion for Preliminary Injunction—Expedited Hearing Requested—Pursuant to Local Rule LCvR 65.1(d) at 19–20, *Fed. Mar. Comm'n*, 607 F. Supp. 2d 192 (No. 08–1895 (RJL)) (filed Nov. 17, 2008).

579. Ibid., 29.

580. On December 22, 2008, both ports moved to dismiss the complaint. Memorandum in Support of the Los Angeles Defendants' Motion to Dismiss, *Fed. Mar. Comm'n*, 607 F. Supp. 2d 192 (No. 08–1895 (RJL)) (filed Dec. 22, 2008); Memorandum in Support of Long Beach Defendants' Motion to Dismiss, *Fed. Mar. Comm'n*, 607 F. Supp. 2d 192 (No. 08–1895 (RJL)) (filed Dec. 22, 2008). In January 2009, the FMC filed an opposition to the motion to dismiss, Plaintiff's Reply to Motions to Dismiss of Defendants Los Angeles and Long Beach, *Fed. Mar. Comm'n*, 607 F. Supp. 2d 192 (No. 08–1895 (RJL)) (filed Jan. 15, 2009), and the ports filed reply briefs, Los Angeles Defendants' Reply Memorandum in Support of Their Motion to Dismiss, *Fed. Mar. Comm'n*, 607 F. Supp. 2d 192 (No. 08–1895 (RJL)) (filed Jan. 29, 2009); Long Beach Defendants' Reply to Plaintiff's Opposition to Motions to Dismiss, *Fed. Mar. Comm'n*, 607 F. Supp. 2d 192 (No. 08–1895 (RJL)) (filed Jan. 29, 2009). The FMC filed an amended complaint adding further allegations of coordinated activity by port staff. First Amended Complaint for an Injunction Pursuant to Section 6(h) of the Shipping Act of 1984 46 U.S.C. §41307, *Fed. Mar. Comm'n*, 607 F. Supp. 2d 192 (No. 08–1895 (RJL)) (filed Feb. 11, 2009).

581. Carol J. Williams, "Court Refuses to Halt Clean-Truck Program," *Los Angeles Times*, Apr. 16, 2009, A10.

582. *Fed. Mar. Comm'n*, 607 F. Supp. 2d at 197–204.

583. Ibid., 200–201. The court also held that the FMC had not shown irreparable harm and that the balance of equities and the public interest weighed in favor of upholding the programs. Ibid., 202–204.

584. Plaintiff's Motion to Dismiss Proceeding and for Vacatur of the Court's April 15, 2009 Order and Memorandum Opinion at 2, *Fed. Mar. Comm'n*, 607 F. Supp. 2d 192 (No. 08–1895 (RJL)) (filed June 16, 2009).

585. Ibid., 1–2.

586. Ibid.

587. Ibid., 7.

588. See Ronald D. White, "Port's Clean-Rig Program Is Running on Empty," *Los Angeles Times*, Jan. 27, 2009, C1. Despite the FMC's interference with the ports' collection of fees, the Clean Truck Programs succeeded in adding 3,000 new clean diesel trucks to the port fleet within the first five months of its implementation. Ronald D. White, "Cleanup at Ports Starts to Pay Off," *Los Angeles Times*, Feb. 23, 2009, C1. One year after the programs' launch, port officials "said they expect to reduce truck emissions at both ports by 80 percent by the end of 2010—a year ahead of schedule." Phil Willon, "Diesel Emissions Are Down Dramatically at Port Complex," *Los Angeles Times*, Oct. 2, 2009, A11. The trucks were touted as improving driver comfort and safety, though some independent contractors complained about the prospect of being barred from the Port of Los Angeles. See Ronald D. White, "Reaping Benefits of Clean Trucking," *Los Angeles Times*, Apr. 6, 2009, B1.

589. Plaintiff's Motion to Dismiss Proceeding and for Vacatur of the Court's April 15, 2009 Order and Memorandum Opinion, note 584, 7.

590. Ibid.

591. Stipulation of Voluntary Dismissal, *Fed. Mar. Comm'n*, 607 F. Supp. 2d 192 (No. 08–1895 (RJL)) (filed July 24, 2009). In a startling turnaround, less than one year later, the FMC issued its inaugural "Chairman Earth Day Award" to the Port of Los Angeles for its Clean Truck Program. Coalition for Clean & Safe Ports, press release, "National 'Blue-Green' Coalition Applauds Key Obama Appointee's Inaugural Earth Day Award to LA Clean Truck Program" (Apr. 21, 2010), https://perma.cc/2YSE-B4QR (accessed July 11, 2017). In another twist, President Obama appointed then Long Beach Harbor Commissioner Mario Cordero to chair the FMC in 2013. "Chairman Mario Cordero," Federal Maritime Commission, http://www.fmc.gov/bureaus_offices/commissioner_mario_cordero.aspx (accessed April 1, 2014).

592. Sejal Patel, *From Clean to Clunker: The Economics of Emissions Control* (2010), 5, https://perma.cc/M8QN-WBQR (accessed July 11, 2017).

593. Plaintiff's L.R. 16–10 Trial Brief at i, Am. Trucking Ass'ns, Inc. v. City of L.A., No. 08–4920 CAS (RZx), 2010 WL 3386436 (C.D. Cal. Aug. 26, 2010) (filed Apr. 13, 2010). That the commerce argument was relegated to a subsidiary position did not surprise NRDC's Martinez, who reflected that the freight industry "lawyers will scream commerce clause violations to the top of their lungs during the advocacy or administrative stage, ... [b]ut ultimately I haven't found them to want to litigate it because ... there's fear of bad precedent ... [since] I don't think their case is that strong." Telephone interview with Adrian Martinez, note 45.

594. *Am. Trucking Ass'ns, Inc.*, No. 08–4920 CAS (RZx), 2010 WL 3386436, at *2–3.

595. Castle v. Hayes Freight Lines, Inc., 348 U.S. 61, 63–64 (1954).

Notes

596. Am. Trucking Ass'ns, Inc. v. City of Los Angeles, 596 F.3d 602, 607 (9th Cir. 2010).

597. Plaintiff's L.R. 16–10 Trial Brief, note 593, 25.

598. Defendants' Trial Brief (L.R. 16–10) at 1, *Am. Trucking Ass'ns, Inc.*, No. 08–4920 CAS (RZx), 2010 WL 3386436 (filed Apr. 13, 2010).

599. Telephone interview with David Pettit, note 49.

600. Defendants' Trial Brief (L.R. 16–10), note 598, 22–23 n.10, 24–25.

601. Telephone interview with David Pettit, note 49.

602. Transcript of Defendant The City of Los Angeles' Opening Statement at 35, *Am. Trucking Ass'ns, Inc.*, No. 08–4920 CAS (RZx), 2010 WL 3386436.

603. Ibid., 28, 33.

604. Ibid., 30, 41.

605. Transcript of Intervenor Natural Resources Defense Council's Opening Statement at 49, *Am. Trucking Ass'ns, Inc.*, No. 08–4920 CAS (RZx), 2010 WL 3386436.

606. Ibid., 52–53.

607. Transcript of Testimony of Defendant's Witness, Dr. Geraldine Knatz, *Am. Trucking Ass'ns, Inc.*, No. 08–4920 CAS (RZx), 2010 WL 3386436.

608. Transcript of Testimony of Defendant's Witness, Simon David Freeman, *Am. Trucking Ass'ns, Inc.*, No. 08–4920 CAS (RZx), 2010 WL 3386436.

609. Transcript of Testimony of Defendant's Witness, John Merrill Holmes at 84, *Am. Trucking Ass'ns, Inc.*, No. 08–4920 CAS (RZx), 2010 WL 3386436.

610. Transcript of Testimony of Defendant's Expert Witness, Jeffrey Walter Brown, *Am. Trucking Ass'ns, Inc.*, No. 08–4920 CAS (RZx), 2010 WL 3386436; Transcript of Testimony of Defendant's Expert Witness, James Evan Hall, *Am. Trucking Ass'ns, Inc.*, No. 08–4920 CAS (RZx), 2010 WL 3386436; Transcript of Testimony of Defendant's Witness, Bruce Charles Wargo, *Am. Trucking Ass'ns, Inc.*, No. 08–4920 CAS (RZx), 2010 WL 3386436.

611. Transcript of Testimony of Defendant's Expert Witness, Dr. Elaine Chang, *Am. Trucking Ass'ns, Inc.*, No. 08–4920 CAS (RZx), 2010 WL 3386436.

612. Transcript of Testimony of Intervenor's Witness, Bernice Banares, *Am. Trucking Ass'ns, Inc.*, No. 08–4920 CAS (RZx), 2010 WL 3386436.

613. Telephone interview with Melissa Lin Perrella, note 472.

614. *Am. Trucking Ass'ns, Inc.*, No. 08–4920 CAS (RZx), 2010 WL 3386436, at *20.

615. Ibid., *22–23.

616. Ibid., *19–22.

617. Ibid., *26.

618. Ibid.

619. Ibid., *27.

620. Ibid., *28.

621. Ibid., *29.

622. Ibid., *29–32. In reaching this conclusion, the court dismissed the port's tideland trust power claim. Ibid., *23.

623. Memorandum of Points & Authorities on Motion to Stay Final Judgment Pending Appeal, *Am. Trucking Ass'ns, Inc.*, No. 08–4920 CAS (RZx), 2010 WL 3386436 (filed Sept. 24, 2010).

624. Civil Minutes—General on Plaintiff's Motion for Injunction Pending Appeal at 8, *Am. Trucking Ass'ns, Inc.*, No. 08–4920 CAS (RZx), 2010 WL 3386436 (proceedings in chambers Oct. 25, 2010).

625. Compare Brief of Appellant Am. Trucking Ass'ns, Inc., Am. Trucking Ass'ns, Inc. v. City of L.A., 660 F.3d 384 (9th Cir. 2011) (No. 10–56465) (filed Dec. 28, 2010), with Brief for Appellees, *Am. Trucking Ass'ns, Inc.*, 660 F.3d 384 (No. 10–56465) (filed Jan. 31, 2011) and Brief for Intervenor-Appellees, *Am. Trucking Ass'ns, Inc.*, 660 F.3d 384 (No. 10–56465) (filed Jan. 31, 2011).

626. Intermodal Ass'n of North America, Inc. to Participate as Amicus Curiae, *Am. Trucking Ass'ns, Inc.*, 660 F.3d 384 (No. 10–56465) (filed Jan. 4, 2011).

627. Motion for Leave to File Amicus Brief in Support of Appellant by Raymond Porras, et al., *Am. Trucking Ass'ns, Inc.*, 660 F.3d 384 (No. 10–56465) (filed Jan. 4, 2011).

628. Brief of the Owner-Operator Indep. Drivers Ass'n, Inc., as Amicus Curiae in Support of Appellant Am. Trucking Ass'ns, Inc., for Reversal of District Court's Findings of Fact and Conclusions of Law, *Am. Trucking Ass'ns, Inc.*, 660 F.3d 384 (No. 10–56465) (filed Jan. 4, 2011).

629. Brief Amicus Curiae of Ctr. for Constitutional Jurisprudence and Harbor Trucking Ass'n in Support of Plaintiff-Appellant at 9, *Am. Trucking Ass'ns, Inc.*, 660 F.3d 384 (No. 10–56465) (filed Jan. 4, 2011). The California Attorney General again supported the port. See Amicus Curiae Brief of the State of California, ex rel. Attorney General Kamala Harris in Support of Appellees and Affirmance, *Am. Trucking Ass'ns, Inc.*, 660 F.3d 384 (No. 10–56465) (filed Feb. 17, 2011).

630. *Am. Trucking Ass'ns, Inc.*, 660 F.3d 384 (filed Sept. 26, 2011). Dissenting, Judge Smith would have held that the entire plan was preempted, that market participant

Notes

status did not apply because the relevant market was trucking services (in which the port did not participate), and that the safety exception was precluded by *Castle*. Ibid., 410–415 (Smith, J., dissenting).

631. Ibid., 401.

632. Ibid., 401–402.

633. Ibid., 402.

634. Ibid., 402–403.

635. Ibid., 403–409.

636. Ibid., 408.

637. Ibid.

638. Ibid.

639. Ibid.

640. Ibid.

641. Louis Sahagun, "Panel Throws Out Part of Port's Clean Truck Program," *Los Angeles Times*, Sept. 27, 2011, AA3.

642. Editorial, "Truckin' Toward a Cleaner Port," *Los Angeles Times*, Sept. 28, 2011, A12.

643. Telephone interview with David Pettit, note 49.

644. Telephone interview with Adrian Martinez, note 45.

645. Telephone interview with Adrian Martinez, staff attorney, Natural Resources Defense Council (March 29, 2013).

646. Petition for a Writ of Certiorari, Am. Trucking Ass'ns, Inc. v. City of L.A., 133 S. Ct. 2096 (2013) (No. 11-798) (filed Dec. 22, 2011).

647. Telephone interview with Melissa Lin Perrella, note 472.

648. Ibid.

649. Petition for a Writ of Certiorari, note 646, 2. In their amicus briefs, the Chamber of Commerce and National Industrial Transportation League, Center for Constitutional Jurisprudence and Harbor Trucking Association, and Airlines for America (the airline trade group) all sided with the ATA to limit the market participant exception. See Brief for the Chamber of Commerce of the United States of America and Nat'l Indus. Transp. League as Amici Curiae in Support of Petitioner, *Am. Trucking Ass'ns, Inc.*, 133 S. Ct. 2096 (No. 11-798) (filed Jan. 23, 2012); Motion for Leave to File and Brief Amicus Curiae of Ctr. of Constitutional Jurisprudence and Harbor Trucking Ass'n in Support of Petitioner, *Am. Trucking Ass'ns, Inc.*, 133 S. Ct. 2096

(No. 11–798) (filed Jan. 23, 2012); Brief of Amicus Curiae Airlines for America in Support of Petitioner, *Am. Trucking Ass'ns, Inc.*, 133 S. Ct. 2096 (No. 11–798) (filed Jan. 23, 2012).

650. Petition for a Writ of Certiorari, note 646, 25–26.

651. City of L.A. Brief in Opposition at 9–38, *Am. Trucking Ass'ns, Inc.*, 133 S. Ct. 2096 (No. 11–798) (filed Feb. 21, 2012).

652. NRDC Brief in Opposition at 5–14, *Am. Trucking Ass'ns, Inc.*, 133 S. Ct. 2096 (No. 11–798) (filed Feb. 21, 2012).

653. Telephone interview with Adrian Martinez, note 645.

654. Brief for the United States as Amicus Curiae Supporting Reversal, *Am. Trucking Ass'ns, Inc.*, 133 S. Ct. 2096 (No. 11–798) (filed Feb. 22, 2013).

655. David G. Savage, "High Court to Hear Case on Port's Clean Truck Program," *Los Angeles Times*, Jan. 12, 2013, at B2.

656. Brief for the United States as Amicus Curiae Supporting Reversal, note 654, 6.

657. The ATA technically did not seek review of the Ninth Circuit holding that the truck maintenance provision fell under the safety exception; it did ask for review of the financial capacity provision, but the court refused to grant it. *Am. Trucking Ass'ns, Inc.*, 133 S. Ct. at 2102 n.3.

658. Telephone interview with Melissa Lin Perrella, note 472.

659. Brief for the City of L.A. Respondents at 15, *Am. Trucking Ass'ns, Inc.*, 133 S. Ct. 2096 (No. 11–798) (filed Mar. 18, 2013).

660. Ibid., 29–30.

661. Brief for Respondents Natural Res. Def. Council, et al. at 10, *Am. Trucking Ass'ns, Inc.*, 133 S. Ct. 2096 (No. 11–798) (filed Mar. 18, 2013).

662. Telephone interview with Michael Manley, note 477.

663. Telephone interview with Melissa Lin Perrella, note 472. Perrella also recalled circulating information on the Supreme Court case on a ports listserv, responding to listserv questions, and explaining the case to coalition members. Ibid.

664. Transcript of Oral Argument at 30, *Am. Trucking Ass'ns, Inc.*, 133 S. Ct. 2096.

665. Ibid., 31.

666. Telephone interview with Melissa Lin Perrella, note 472.

667. Transcript of Oral Argument, note 664, 39.

668. Ibid., 44–51.

Notes

669. *Am. Trucking Ass'ns, Inc.*, 133 S. Ct. at 2102.

670. Ibid.

671. Ibid., 2103.

672. Ibid.

673. Ibid.

674. Ibid., 2104.

675. Ibid.

676. Ibid.

677. Ibid., 2105.

678. Ibid.

679. Ibid. In a concurrence, Justice Thomas, while agreeing entirely with the opinion, noted an issue that the port failed to raise but, in his view, should have: that the FAAAA's application to intrastate trucking was quite likely unconstitutional under the Commerce Clause and thus the FAAAA itself would lack preemptive force. Ibid., 2106 (Thomas, J., concurring). "Although respondents waived any argument that Congress lacks authority to regulate the placards and parking arrangements of drayage trucks using the port, I doubt that Congress has such authority." Ibid.

680. Ibid., 2102 n.4 (majority opinion).

681. Ibid., 2102.

682. Interview with John Holmes, note 303.

683. Ronald D. White, "Cleaner Port Air, But How?" *Los Angeles Times*, Jan. 9, 2010, B1 (noting that the Los Angeles port had given out $44 million in incentives); David Zanhiser, "Trucking Group to Appeal Port Ruling," *Los Angeles Times*, Aug. 28, 2010, AA3 (stating that the harbor commission had given out $57 million to subsidize vehicles).

684. The Clean Truck Program did not mandate that truck drivers purchase and maintain their own trucks. To the contrary, it directed terminal operators to bar noncompliant dirty trucks, while stating that trucking companies "shall be responsible for vehicle condition and safety and shall ensure that the maintenance of all Permitted Trucks ... is conducted in accordance with manufacturer's instructions." L.A. Bd. of Harbor Comm'rs, Drayage Services Concession Agreement for Access to the Port of Los Angeles, note 441, ¶ III(g). The program did not preclude trucking companies from passing on the purchase and maintenance costs to drivers and, as a matter of practice, that is what the companies did. See McDonnell, note 561.

685. Telephone interview with Melissa Lin Perrella, note 40.

686. Am. Trucking Ass'ns, Inc. v. City of L.A., 559 F.3d 1046, 1056 (9th Cir. 2009).

687. Louis Sahagun and Ronald D. White, "Groups to Ask Congress for Help on Port," *Los Angeles Times*, Apr. 15, 2009, A12.

688. Telephone interview with Jon Zerolnick, note 159.

689. Telephone interview with Melissa Lin Perrella, note 40.

690. Coalition for Clean & Safe Ports, Background on the Port of Los Angeles Clean Trucks Program (July 15, 2009).

691. Art Marroquin, "Port Panel Extends Lobbying Contract," *Pasadena Star-News*, Jan. 20, 2010, http://www.pasadenastarnews.com/20100121/port-panel-extends-lobbying-contract (accessed July 11, 2017).

692. Telephone interview with Jon Zerolnick, note 159.

693. LAANE, Briefing Book, Clearing the Roadblocks: A Map to Green and Grow a Key American Industry to Create 85,000 Middle-Class Jobs at Our Nation's Ports.

694. Coalition for Clean & Safe Ports, Background on the Port of Los Angeles Clean Tucks Program, note 690.

695. Letter from Zoe Lofgren et al., Representatives from the State of California, United States House of Representatives, to James L. Oberstar, Chairman of the Transportation and Infrastructure Committee, United States House of Representatives (Nov. 4, 2009).

696. Antonio Villaraigosa, "Clean Trucks: One Year Later," *Huffington Post Green*, Dec. 1, 2009, http://www.huffingtonpost.com/antonio-villaraigosa/clean-trucks-one-year-lat_b_307158.html (accessed July 11, 2017).

697. Geraldine Knatz, "Port Clean-Truck Program Is Goal Worth Fighting For," *Los Angeles Daily News*, Dec. 8, 2009, A15.

698. White, note 538.

699. Ibid.

700. Ronald D. White, "L.A. Port Urged to Stop Lobbying Over Clean Truck Program," *Los Angeles Times*, Aug. 26, 2009, B2.

701. Telephone interview with Nick Weiner, note 28.

702. Telephone interview with Jonathan Klein, note 93.

703. Telephone interview with Nick Weiner, note 28.

704. Ibid.

705. Letter from LAANE et al. to James B. Oberstar, Chairman of the Transportation and Infrastructure Committee, United States House of Representatives, and John

Notes

Mica, Ranking Member of the Transportation and Infrastructure Committee, United States House of Representatives (Apr. 22, 2010).

706. Ibid.

707. Assessing the Implementation and Impacts of the Clean Truck Programs at the Port of Los Angeles and the Port of Long Beach: Hearing Before the Comm. on Transp. and Infrastructure, 111th Cong. 2 (2010).

708. Telephone interview with Jon Zerolnick, note 159.

709. Telephone interview with Nick Weiner, note 28.

710. Clean Ports Act of 2010, H.R. 5967, 111th Cong. (2010); see also Darren Goode, "E2 Morning Round-up: Green Groups Highlight Oil Accidents, Spill Response Debate Heats Up, Nadler Floats 'Clean Ports' Bill and Oil Spill Threatens Lake Michigan," *The Hill*, July 29, 2010, http://thehill.com/blogs/e2-wire/111569-e2-morning-round-up-green-groups-highlight-oil-accidents-spill-response-debate-heats-up-nadler-floats-clean-ports-bill-and-oil-spill-threatens-lake-michigan (accessed July 11, 2017).

711. H.R. 5967.

712. Coalition for Clean & Safe Ports, *Longer Hours, Lower Wages & Little Hope*, 1.

713. Ibid.

714. Telephone interview with Nick Weiner, note 28.

715. Clean Ports Act of 2011, H.R. 572, 112th Cong. (2011). The new legislation amended the last clause of proposed section 14501(c)(2)(A) so that it now exempted local or state laws related to environmental, traffic, or operational concerns so long as "adoption or enforcement of such requirements" did not conflict with other federal laws. Ibid.

716. Kristopher Hanson, "Congress Considers Law Allowing Other Ports to Mimic Clean Trucks Program," *Long Beach Press-Telegram*, Dec. 19, 2011, http://www.presstelegram.com/technology/20111220/congress-considers-law-allowing-other-ports-to-mimic-clean-trucks-program (accessed July 11, 2017).

717. Coalition for Clean & Safe Ports, "Support the Clean Ports Act of 2011" (Dec. 7, 2011).

718. Coalition for Clean & Safe Ports, "The Clean Ports Act of 2011 (H.R. 572)" (May 2011).

719. Ibid.

720. Clean Ports Act of 2011, S. 2011, 112th Cong. (2011). Gillibrand had five cosponsors: Senators Barbara Boxer (CA), Sherrod Brown (OH), Al Franken (MN), Robert Menendez (NJ), and Charles Schumer (NY).

721. Assemb. B. 950, 2011–2012 Leg., Reg. Sess. (Cal. 2011).

722. Assembly Third Reading, Assemb. B. 950 (Feb. 18, 2011); Assemb. Comm. on Lab. and Employment, Hearing on Assemb. B. 950 (May 4, 2011).

723. "California Bill Seeking to Ban Independent Truck Operators Shelved," *Pacific Maritime Online*, June 7, 2011, http://www.pmmonlinenews.com/2011/06/california-bill-seeking-to-ban.html (accessed July 11, 2017).

724. Ibid.

725. Ibid.

Chapter 6

1. A 2007 report found that only 9 percent of Los Angeles and Long Beach port truck drivers worked as employees. See Greenberg Quinlan Rosner Research, *Demographic Overview of Truck Drivers at the Ports of Los Angeles and Long Beach* (2007), 3, https://perma.cc/B6CD-2FAT (accessed July 20, 2017).

2. These are factors outlined in the IRS's right-of-control test for determining employee status for tax purposes. Rev. Rul. 87–41, 1987–1 C.B. 296.

3. Telephone interview with Jon Zerolnick, director of Clean & Safe Ports Project, L.A. Alliance for a New Economy (Feb. 20, 2013).

4. See Lambert v. Ackerley, 180 F.3d 997, 1012 (9th Cir. 1999) (FLSA); S.G. Borello & Sons, Inc. v. Dept. of Indus. Relations, 769 P.2d 399, 403–05 (Cal. 1989) (California Labor Code). Under federal labor law, the NLRB and the courts "employ a fact-specific approach to determining whether truck driver owner-operators are employees or independent contractors. Although 'right to control' is still a factor, the courts also focus on the extent of the owner-operators' entrepreneurial opportunity for gain or loss." Brent Garren, John E. Higgins, Jr., and David A. Kadela, eds., *How to Take a Case Before the NLRB*, 9th ed. (Arlington, VA: BNA Books, 2016), 1-9.

5. California Department of Justice, press release, "Attorney General Brown Sues Three Trucking Companies in Ongoing Worker Abuse Crackdown at Los Angeles and Long Beach Ports" (Oct. 27, 2008), https://oag.ca.gov/news/press-releases/attorney-general-brown-suesw-three-trucking-companies-ongoing-worker-abuse (accessed July 17, 2017).

6. The companies sued by Attorney General Brown included Pacifica Trucks, Guasimal Trucking, Jose Maria Lira Trucking, Esdmundo Lira Trucking, and Noel and Emma Moreno Trucking. California Department of Justice, press release, "Brown Wins Fifth Suit Against Port Trucking Companies that Violated Workers' Rights" (Feb. 4, 2010), http://oag.ca.gov/news/press-releases/brown-wins-fifth-suit-against-port-trucking-companies-violated-workers-rights (accessed July 17, 2017).

Notes

7. California Business & Professions Code §§ 17200–17210 authorize injunctive relief and civil restitution against anyone engaged in "unfair competition," defined broadly as "any unlawful, unfair or fraudulent business act or practice."

8. California Department of Justice, press release, note 6. The California Attorney General also reached a settlement with Pacifica Trucks, pursuant to which the company was "enjoined permanently from misclassifying as independent contractors truck drivers who operate trucks that are provided, owned, or leased by Pacifica Trucking." Final Judgment at 2, Brown v. Pacifica Trucks, L.L.C., No. BC428934 (Cal. Super. Ct. Jan. 5, 2010).

9. Telephone interview with Michael Manley, staff attorney, International Brotherhood of Teamsters (Mar. 29, 2013).

10. See Complaint for Restitution, Penalties and Injunctive Relief, State v. Pac Anchor Transp., Inc., No. BC397600 (Cal. Super. Ct. Sept. 4, 2008), http://oag.ca.gov/system/files/attachments/press_releases/n1606_complaint_pac_anchor.pdf (accessed July 17, 2017).

11. "Truckers Claim Brown's Frivolous Lawsuit Designed to Curry Political Favor," *Legal Newsline*, Nov. 25, 2008, http://legalnewsline.com/news/217658-truckers-claim-browns-frivolous-lawsuit-designed-to-curry-political-favor (accessed July 17, 2017).

12. See People *ex rel.* Harris v. Pac Anchor Transp., Inc., 125 Cal. Rptr. 3d 709, 716 (Cal. Ct. App. 2011), *review granted by* 329 P.3d 154 (Cal. 2011). The case was set for argument at the California Supreme Court on May 28, 2014.

13. PAC Anchor Transportation, Inc. v. California, *SCOTUSblog*, http://www.scotusblog.com/case-files/cases/pac-anchor-transportation-inc-v-california/ (accessed July 23, 2017).

14. Telephone interview with Nick Weiner, national campaigns organizer, Change to Win (Nov. 2, 2016).

15. Ibid.

16. Ibid.

17. Rebecca Smith, David Bensman, and Paul Alexander Marvy, *The Big Rig: Poverty, Pollution, and the Misclassification of Truck Drivers at America's Ports* (2010), 13–14, 21–22.

18. Ibid., 6. Other research linked misclassification to the concept of "wage theft." UCLA researchers found that nearly one-third of Los Angeles workers in a typical week were deprived of the minimum wage. Ruth Milkman, Ana Luz González, and Victor Narro, *Wage Theft and Workplace Violations in Los Angeles* (2010), 2 (finding that 30 percent of L.A. workers were paid less than minimum wage in the week preceding the survey, while 21.3 percent were not paid overtime). The UCLA report

urged a "move toward proactive, 'investigation-driven' enforcement in low-wage industries, rather than simply reacting to complaints." Ibid., 56.

19. Isaías Alvarado, "We Are at the Mercy of God," *La Opinión*, Dec. 8, 2010, https://web.archive.org/web/20120106200332/http://cleanandsafeports.org/fileadmin/pdf/La_Opinion_december_article_Eng.pdf (accessed July 17, 2017).

20. James Rufus Koren, "Truckers Want Break on Audits," *Los Angeles Business Journal*, Oct. 1, 2012, 1, 35.

21. Marc Lifsher, "Her Job: Putting a Stop to Wage Theft," *Los Angeles Times*, May 22, 2013, B1.

22. See Bureerong v. Uvawas, 922 F. Supp. 1450, 1459–61 (C.D. Cal. 1996).

23. See interview with Nick Weiner, note 14.

24. Ibid.

25. Telephone interview with Sanjukta Paul, legal coordinator, L.A. Alliance for a New Economy (May 9, 2013).

26. Ibid. Initial funding for the LAANE legal coordinator position came from the Public Welfare Foundation.

27. Ibid.

28. Ibid.

29. Ibid.

30. Telephone interview with Michael Manley, note 9.

31. Telephone interview with Jonathan Klein, executive director, Clergy and Laity United for Economic Justice (May 8, 2013).

32. See telephone interview with Nick Weiner, note 14.

33. California Labor Code § 2802 (Deering 2013). Although truckers also claimed minimum wage and overtime violations under state law, those amounts were dwarfed by the quantity of illegal deductions.

34. Telephone interview with Nick Weiner, national campaigns organizer, Change to Win (Apr. 25, 2013).

35. Telephone interview with Michael Manley, note 9.

36. James Rufus Koren, "Port Access Still Drives Teamsters," *Los Angeles Business Journal*, Dec. 12, 2011, 1.

37. Ibid.

38. Ibid.

Notes

39. Ibid., 44.

40. Howard Fine, "Leasing Could Be Roadblock for Trucking Companies," *Los Angeles Business Journal*, Oct. 24, 2011, 51.

41. Telephone interview with Sanjukta Paul, note 25.

42. California Labor Code § 2802 (Deering 2013) ("An employer shall indemnify his or her employee for all necessary expenditures or losses incurred by the employee in direct consequence of the discharge of his or her duties.").

43. Telephone interview with Sanjukta Paul, note 25.

44. California Labor Code §§ 226.8 and 2753 (Deering 2013).

45. Howard Fine, "Contracting May Squeeze Employers," *Los Angeles Business Journal*, Oct. 24, 2011, 1, 51.

46. The Willful Misclassification Law imposed penalties of up to $15,000 for individual violations and $25,000 for violations that showed a "pattern or practice" of misclassification. California Labor Code § 226.8 (b) and (c) (Deering 2013).

47. California Labor Code § 266.8(a)(2) (Deering 2013). This law was in addition to the existing law making it illegal to deduct business expenses from employee compensation. California Labor Code § 2802 (Deering 2013).

48. Sanjukta Paul, Memorandum, Synopsis of the California Willful Misclassification Law, 3.

49. See Fiona Smith, "Truckers Sue over Alleged 'Wage Theft,'" *Los Angeles Daily Journal*, Nov. 6, 2009, 3; see also Class Action Complaint, Montoya v. Total Transp. Servs., Inc., No. BC 425121 (Cal. Super. Ct. Nov. 3, 2009).

50. Coalition for Clean & Safe Ports, press release, "Class Action Wage-and-Hour Suit Filed Against Sun Pacific Trucking, Inc., and Pacific Green Trucking, Inc. on Behalf of Southern CA Port Drivers" (June 30, 2010), https://perma.cc/YYH9-DPSC (accessed July 17, 2017). Attorney Scot D. Bernstein was co-counsel on the Sun Pacific class action suit.

51. "Another Case Against Port Trucking Firm Underscores Widespread Industry Abuse, Disregard of Labor Laws, Teamsters Charge," *PR Newswire*, June 30, 2010, http://www.prnewswire.com/news-releases/another-case-against-port-trucking-firm-underscores-widespread-industry-abuse-disregard-of-labor-laws-teamsters-charge-97502984.html (accessed July 17, 2017).

52. Ibid.

53. The case against Sun Pacific and Pacific Green Trucking settled in 2011. The case against Total Transportation Services, Inc. was settled in October 2013 and approved

by the court in spring 2014. Notice of Settlement of Entire Case, Montoya v. Total Transportation Services, Inc., No. BC425121 (Cal. Super. Ct., filed Oct. 28, 2013).

54. Order of the Labor Commissioner, Garcia v. Seacon Logix, Inc., No. 05–52821-LT (Cal. Labor Comm'r Jan. 10, 2012); Order of the Labor Commissioner, Urbina v. Seacon Logix, Inc., No. 05–53002-LT (Cal. Labor Comm'r Jan. 10, 2012).

55. Telephone interview with Sanjukta Paul, clinical fellow, UCLA School of Law (Jan. 23, 2014).

56. Ibid.

57. Tom Gilroy, Daily Labor Report, "Shipping Company Drivers Are Employees, Not Contractors, California Court Decides," *Bloomberg BNA* 42, Mar. 4, 2013, A-5.

58. California Division of Labor Standards Enforcement, press release, "California Labor Commissioner Prevails in Misclassification Case Against Port Trucking Company" (Mar. 1, 2013), http://www.dir.ca.gov/DIRNews/2013/IR2013-11.html (accessed July 17, 2017).

59. U.S. Department of Labor, news release, "US Labor Department, California Sign Agreement to Reduce Misclassification of Employees as Independent Contractors" (Feb. 9, 2012), http://www.dol.gov/opa/media/press/whd/WHD20120257.htm (accessed July 17, 2017).

60. Koren, note 20. The DOL investigations resulted in some misclassification findings and orders of back pay. See, e.g., Wage & Hour Div. Investigation Findings Letter, Container Connection of S. Cal., Inc., Case No. 1634525 (Jan. 22, 2013) (finding that two workers were owed one day of back pay for $211.52 and $26.17 respectively).

61. Koren, note 20.

62. Solis v. Shippers Transport Express, Inc., No. 13-cv-04255 (C.D. Cal., filed Aug. 13, 2012). In August 2012, the DOL also brought suit against Shippers Transport Express on behalf of Oakland port truckers. See Complaint for Violations of the Fair Labor Standards Act, Solis v. Shippers Transp. Express, Inc., No. 12–4249 (N.D. Cal. Aug. 13, 2012). That case was subsequently transferred to the Central District of California and merged with the case related to the Los Angeles and Long Beach ports.

63. Koren, note 20.

64. Cameron W. Roberts and Sean Brew, Roberts & Kehagiaras LLP, "Understanding the Clean Truck Litigation Part VI" (Powerpoint presentation, Nov. 13, 2013), 1, 30–31, https://perma.cc/E7YZ-Z2NR (accessed July 18, 2017).

Notes

65. Bill Mongelluzzo, "Feds Take on Drayage Misclassification," *Journal of Commerce*, July 16, 2012, http://www.teamsters492.org/docs/FedsTakeonDrayageMisclassification.pdf (accessed July 18, 2017).

66. Complaint for Declaratory and Injunctive Judgment at 12–13, Clean Truck Coal., LLC v. Su, No. 12–08949 (C.D. Cal. Oct. 17, 2012).

67. Joseph Lapin, "Notes from the Underground Economy: Are Companies at the Port of Long Beach Cheating Truckers Out of Their Rightful Wages?," *OC Weekly*, Jan. 10, 2013, http://www.ocweekly.com/news/notes-from-the-underground-economy-6425266 (accessed July 18, 2017).

68. Telephone interview with Nick Weiner, note 14.

69. Telephone interview with Sanjukta Paul, note 25.

70. Telephone interview with Sanjukta Paul, note 55.

71. Telephone interview with Sanjukta Paul, note 25; see also Joseph Lapin, "Local Coalition Tries to Organize Misclassified Port Truckers," *Long Beach Post*, Aug. 24, 2012, https://lbpost.com/news/2000000890-local-coalition-tries-to-organize-misclassified-port-truckers (accessed August 21, 2017).

72. Telephone interview with Sanjukta Paul, note 25. At the initial legal clinic meeting, Paul was joined by Carlos Bowker, deputy labor commissioner at the DLSE; Abel Gervacio, an investigator with the DOL; Victor Narro, project director at the UCLA Labor Center; Peter Riley, regional manager of the California Division of Occupational Safety and Health Region 3; Rebecca Smith, attorney at the National Employment Law Project; and Mike Manley from the Teamsters.

73. See Teamsters Local 848, Know Your Rights Workshop, Sept. 14, 2012 (asking: "What is employee misclassification and what can you do to end it at the ports?").

74. Telephone interview with Sanjukta Paul, note 25.

75. Telephone interview with Sanjukta Paul, note 55.

76. Ibid.

77. Ibid.

78. Class Action Complaint at 17, Talavera, Jr. v. QTS, Inc., No. BC501571 (Cal. Super. Ct. Feb. 22, 2013). Defendants sought to prevent class members from participating in the suit by seeking to enforce releases drivers were required to sign as a condition of continuing to work for defendants. Plaintiffs' Notice for Motion and Motion for Declaratory Relief and Curative Order at 1, *Talavera, Jr.*, No. BC501571 (filed Jan. 21, 2014).

79. Taylor v. Shippers Transport Express, No. BC477047 (Cal. Super. Ct., filed Jan. 31, 2012). In the wake of the class action, the lead plaintiff, Grayling Taylor, was

fired by Shippers, prompting a retaliation lawsuit filed by the law firm of Alexander Krakow + Glick LLP in March 2013. Taylor v. Shippers Transport Express, Inc., No. BC502231 (Cal. Super. Ct., filed Mar. 2013). The settlement for that case was filed in December 2014 and finalized in May 2015.

80. Class Action Complaint, Hernandez v. Gold Point Transportation, Inc., No. BC477445 (Cal. Super. Ct., filed Mar. 9, 2012).

81. First Amended Complaint for Damages, Restitution, Etc., Class Action, Arellano v. Container Connection of Southern California, Inc., No. BC500675 (Cal. Super. Ct., filed Apr. 8, 2013).

82. Ricardo Lopez, "Truck Drivers Sue for Overtime, Meal Breaks," *Los Angeles Times*, May 15, 2013, B2.

83. Telephone interview with Sanjukta Paul, note 55.

84. Telephone interview with Michael Manley, note 9.

85. The campaign has kept a tally of trucking companies converting to an employee driver model. See Sheheryar Kaoosji, "Port Trucking Companies Big and Small Changing Their Business Strategy, Convert to Company Driver Model," Nov. 23, 2015, Los Angeles Alliance for a New Economy, *Port Innovations*, https://perma.cc/7F8S-3JP6 (accessed July 18, 2017).

86. Evelyn Larrubia, "Union Protests Truckers' Firings," *Los Angeles Times*, Mar. 3, 2009, B1.

87. Ricardo Lopez, "Truck Drivers Set for 2-Day Strike at Ports," *Los Angeles Times*, Apr. 27, 2014, A25; see also Bill Mongelluzzo, "Teamster-backed Drayage Company with Employee Drivers Launches," May 4, 2015, http://www.joc.com/trucking-logistics/drayage/teamster-backed-drayage-company-employee-drivers-launches_20150504.html (accessed July 18, 2017).

88. Telephone interview with Jonathan Klein, note 31.

89. Ibid.

90. Telephone interview with Nick Weiner, note 34.

91. See "America's Port Truckers Deliver a Resounding Yes Winning Union Recognition as Teamsters in Historic Vote," press release, *The Grim Truth at Toll Group* (Apr. 11, 2012), https://perma.cc/4A33-TV9Q (accessed July 18, 2017); "Jobs Can't Be Good or Green If They're Not Union!," *LAANE*, Mar. 20, 2012, https://perma.cc/5ETU-7JC7 (accessed July 18, 2017).

92. Telephone interview with Nick Weiner, note 34; see also telephone interview with Michael Manley, note 9.

Notes

93. James Rufus Koren, "Toll Group Drivers File with NLRB," *Los Angeles Business Journal*, Jan. 27, 2012, http://labusinessjournal.com/news/2012/jan/27/toll-group-drivers-file-nlrb (accessed July 18, 2017).

94. "Toll Group in Dispute with Los Angeles Truck Drivers," *The Australian Business Review*, Mar. 10, 2012, http://www.theaustralian.com.au/business/toll-group-in-dispute-with-los-angeles-truck-drivers/story-e6frg8zx-1226295773360 (accessed July 18, 2017).

95. "Jobs Can't Be Good or Green If They're Not Union!," note 91. Teamsters complaints also triggered a federal investigation into Toll driver misclassification at the New York and New Jersey ports. See U.S. Department of Labor, Wage and Hour Division, *Compliance Action Report, Toll Global Forwarding*, Case ID 1647833 (May 1, 2012).

96. Telephone Interview with Jon Zerolnick, note 3.

97. Telephone interview with Jonathan Klein, note 31.

98. Telephone interview with Michael Manley, note 9.

99. "America's Port Truckers Deliver a Resounding Yes Winning Union Recognition as Teamsters in Historic Vote," press release, note 91; Laura Clawson, "Truck Drivers Win Union Representation at Toll Group," *Daily Kos*, Apr. 12, 2012, https://perma.cc/AF7H-P3GQ (accessed July 18, 2017).

100. Telephone interview with Nick Weiner, note 34.

101. "America's Port Truckers Deliver a Resounding Yes Winning Union Recognition as Teamsters in Historic Vote," press release, note 91.

102. "Toll Drivers Unanimously Ratify First Time Contract," *Teamsters Local 848* (Dec. 30, 2012), https://perma.cc/NP9V-GKGL (accessed Feb. 16, 2014).

103. Coalition for Clean & Safe Ports, "The Road to the Middle Class: Teamster Contract with Toll Group Fuels Port Driver Hope" (2012).

104. Ibid.

105. Telephone interview with Michael Manley, note 9.

106. Steve Lopez, "Truckers Want a Fair Shake; Drivers at the Port of L.A. Are Willing to Fight What They Call Wage Theft," *Los Angeles Times*, May 18, 2014, A2.

107. Telephone interview with Sheheryar Kaoosji, director, Project for Clean & Safe Ports, L.A. Alliance for a New Economy (Oct. 15, 2016).

108. Justice for Port Drivers, http://justiceforportdrivers.com/ (accessed July 18, 2017).

109. Garren, Higgins, Jr., and Kadela, note 4, 1-2.

110. Ibid., 1-3.

111. National Labor Relations Act (NLRA), 29 U.S.C. § 157 (Section 7) (2012).

112. Garren, Higgins, Jr., and Kadela, note 4, 1-13.

113. Northeast Beverage Corp. v. NLRB, 554 F.3d 133 (D.C. Cir. 2009).

114. Garren, Higgins, Jr., and Kadela, note 4, 1-14.

115. NLRA, 29 U.S.C. § 158(a)(1) (Section 8) (2012).

116. Garren, Higgins, Jr., and Kadela, note 4, 1-23–1-24.

117. Cynthia L. Estlund, "The Ossification of American Labor Law," *Columbia Law Review* 102 (2002): 1527.

118. Lydia DePillis, "Labor Law Has Been Frozen for 60 Years. Democrats Are Trying to Crack It Open," *Washington Post*, Sept. 16, 2015, https://perma.cc/AJT4-CVW3 (accessed July 18, 2017); Lydia DePillis, "Seven Themes for the Working World in 2015," *Washington Post*, Dec. 28, 2015, https://perma.cc/T7K8-ETB4 (accessed July 18, 2017).

119. NLRA, 29 U.S.C. § 158(b)(7).

120. Garren, Higgins, Jr., and Kadela, note 4, 20-3–20-7.

121. Eileen B. Goldsmith, "The Role of Regional Directors in the National Labor Relations Board," https://perma.cc/34VG-HUNA (accessed July 18, 2017).

122. Sherman Antitrust Act, 15 U.S.C. § 1 (2012).

123. Telephone interview with Nick Weiner, note 14. For his part, Weiner advocated to maintain the campaign in the face of the devastating Clean Truck Program loss and the internal squabbling that provoked: "I don't like losing. And my tolerance for dealing with bullshit is high." Ibid.

124. Choi's position was funded as a three-year campaign fellowship by the Rosenberg Foundation to be supervised by Julie Gutman Dickinson. Email from Julie Gutman Dickinson, partner, Bush Gottlieb, to Scott Cummings, professor, UCLA School of Law (Jan. 16, 2017).

125. Telephone interview with Sheheryar Kaoosji, note 107.

126. Telephone interview with Nick Weiner, note 14.

127. Ibid.

128. Ibid.

129. Telephone interview with Sheheryar Kaoosji, note 107.

130. Ibid.

Notes

131. Telephone interview with Nick Weiner, note 14.

132. Telephone interview with Michael Manley, staff attorney, International Brotherhood of Teamsters (Nov. 23, 2016).

133. Between August 2003 and 2013, the NLRB was not fully staffed. National Labor Relations Board, "Board Members Since 1935," https://perma.cc/8DJ3-WQU4 (accessed July 18, 2017); see also Josh Hicks, "Senate Committee Approves Obama's NLRB Nominees Despite GOP Opposition," *Washington Post*, May 23, 2013, https://perma.cc/JJ8B-R7K9 (accessed July 18, 2017).

134. Partisanship left the NLRB quite literally gridlocked in 2013 as litigation progressed through the D.C. Circuit up to the U.S. Supreme Court to decide the fate of members appointed during congressional recesses. See Mark Landler and Steven Greenhouse, "Vacancies and Partisan Fighting Put Labor Relations Agency in Legal Limbo," *New York Times*, July 15, 2013, http://www.nytimes.com/2013/07/16/us/politics/vacancies-and-partisan-fighting-put-labor-relations-agency-in-legal-limbo.html (accessed July 18, 2017); see also Hicks, note 133.

135. The recess appointment language is found at Article II, Section 2, Clause 3 of the United States Constitution.

136. Noel Canning v. NLRB, 705 F.3d 490 (D.C. Cir. 2013).

137. The Supreme Court ultimately affirmed the appellate ruling on different grounds. NLRB v. Noel Canning, 134 S. Ct. 2550, 2574 (2014) (holding that, for purposes of the Recess Appointments Clause, the Senate is in session "when it says that it is, provided that, under its own rules, it retains the capacity to transact Senate business").

138. President Obama waged a similar battle to appoint an NLRB General Counsel. His first choice, Lafe Solomon, was appointed Acting General Counsel in 2010 under the Federal Vacancies Reform Act of 1998 after the previous general counsel resigned. When the president formally nominated Solomon, the Senate refused to act. Obama subsequently withdrew Solomon's nomination and, as part of the deal to break the logjam at the NLRB, appointed Richard Griffin, who was confirmed by the Senate on October 29, 2013.

139. Telephone interview with Michael Manley, note 132.

140. Ibid.

141. Ibid.

142. Ibid.

143. Ibid.

144. Ibid.

145. Ibid.

146. Ibid.

147. Ibid.

148. Ibid.

149. Ibid.

150. Telephone interview with Sheheryar Kaoosji, note 107.

151. Telephone interview with Nick Weiner, note 14.

152. Telephone interview with Sheheryar Kaoosji, note 107.

153. Ibid.

154. Telephone interview with Nick Weiner, note 14.

155. Ibid.

156. Telephone interview with Fred Potter, director, Teamsters Port Division (Nov. 22, 2016).

157. Ibid.

158. Ibid.

159. Ibid.

160. Ibid.

161. Telephone interview with Carlos Santamaria, lead organizer, Teamsters Port Campaign/Justice for Port Drivers (Nov. 21, 2016).

162. Ibid.

163. Ibid.

164. Ibid.

165. Ibid.

166. Ibid.

167. Ibid.

168. Ibid.

169. Ibid.

170. Santamaria recalls identifying Toll as a "good target" to Nick Weiner, who gave him the "green light." Because a lot of the Toll drivers had come from Swift, Santamaria believed the Teamsters had an organizing advantage because the drivers "wanted a union" and "they were pissed" because the Toll system, which gave

incentives for loads, rewarded drivers who were "kiss asses" with the dispatcher. In organizing the Toll drivers, Santamaria's "rap" was to tell them that the only way to fix the problem was to get a union contract "in black and white." Ibid.

171. Ibid.

172. Ibid.

173. Telephone interview with Nick Weiner, note 14.

174. Telephone interview with Sheheryar Kaoosji, note 107.

175. Ibid. During his master's program, Kaoosji did research projects with the UCLA Downtown Labor Center, where he was acquainted with Larry Frank, UCLA sociologist Ruth Milkman, and lawyer-activist Victor Narro; he also began working with LAANE on projects related to land use and economic development. Ibid.

176. Ibid.

177. Ibid.

178. Ibid.

179. Kaoosji worked on the first phase of Green Fleet from Change to Win, and then officially joined the campaign as LAANE ports director in 2014. Ibid.

180. Stephanie West, "NLRB Issues Complaint Against Green Fleet," *Laborpress*, June 24, 2014, https://perma.cc/T5FK-QAYY (accessed July 18, 2017). Six additional Green Fleet drivers subsequently brought DLSE claims the union said were potentially worth over $900,000. "Teamsters: NLRB Complaint Issued Against Green Fleet Systems," *Teamsters*, June 20, 2014, https://perma.cc/Y5FZ-72ZP (accessed July 18, 2017).

181. Telephone interview with Nick Weiner, note 14.

182. Ibid.

183. Ibid.

184. Telephone interview with Julie Gutman Dickinson, partner, Bush Gottlieb (Dec. 15, 2016).

185. Ibid.

186. Ibid.

187. Ibid.

188. Ibid.

189. Ibid.

190. Ibid.

191. Ibid.

192. Ibid. In addition to assisting on the NLRB work, Gutman Dickinson also represented six Green Fleet drivers at the DLSE. Ibid.

193. Order Revoking Settlement Agreement, Order Consolidating Cases, Consolidated Complaint and Notice of Hearing at 4–5, Green Fleet Systems, LLC and Int'l Brotherhood of Teamsters, Cases 21-CA-100003, 115910, 119154, 121368 (NLRB Region 21, June 18, 2014), https://perma.cc/FD2H-VTQJ (accessed July 20, 2017).

194. Ricardo Lopez, "Truckers Strike, Push to Unionize," *Los Angeles Times*, Aug. 27, 2013, B1.

195. Telephone interview with Nick Weiner, note 14.

196. Paula Winicki, "Port Truck Drivers Owed Millions in Wages," *City Watch*, Sept. 13, 2013, http://www.citywatchla.com/index.php/archive/5708-port-truck-drivers-owed-millions-in-wages (accessed July 18, 2017).

197. Telephone interview with Sheheryar Kaoosji, note 107.

198. See NLRB Office of the General Counsel, Advice Memorandum from Barry J. Kearney, associate general counsel, Division of Advice to Olivia Garcia, director, Region 21, Re: Pacific 9 Transportation, Inc., Case No. 21-CA-150875 (Dec. 18, 2015), 3.

199. "Port of L.A. Truck Drivers Stage 36-Hour Strike," *Los Angeles Times*, Nov. 18, 2013, http://articles.latimes.com/2013/nov/18/local/la-me-ln-port-truck-drivers-strike-20131118 (accessed July 18, 2017); see also Greg Yee, "Port Truck Drivers at 3 Companies on Strike," *Long Beach Press-Telegram*, Nov. 19, 2013, 3.

200. See Teamsters Local 287 (*Buck's Butane-Propane Service Inc.*), 186 N.L.R.B. 187 (1987).

201. See Sailors Union of the Pacific (*Moore Dry Dock Co.*), 92 N.L.R.B. 547 (1950) (holding that a union may picket a struck company at premises of another company where struck company employees are working).

202. Telephone interview with Nick Weiner, note 14.

203. Ibid.

204. Ibid.

205. Yee, note 199.

206. See Janice Hahn, Op-Ed., "To Keep on Trucking," *Los Angeles Times*, Nov. 25, 2013, A11.

207. "Teamsters: NLRB Complaint Issued Against Green Fleet Systems," note 180.

Notes

208. Confidential Communication Via E-Mail to Olivia Garcia, regional director, Bruce Hill, assistant regional director, William Pate, regional attorney, and Sylvia Meza, board agent, NLRB Region 21, from Bush Gottlieb, Re: Request for 10(j) Relief in Cases 21-CA-100003, -119154 and -121368 (Apr. 9, 2014).

209. Pursuant to typical practice, the Teamsters filed an amicus brief in support of the NLRB petition. Amicus Curiae Memorandum of International Brotherhood of Teamsters, Port Division, Garcia v. Green Fleet Systems, LLC, No. CV 14-622 PSG JEM (C.D. Cal., filed Aug. 22, 2014).

210. Telephone interview with Carlos Santamaria, note 161.

211. Ibid.

212. Ibid. Santamaria described questioning the drivers to help them arrive at their own conclusions about the value of filing claims and organizing. He would ask drivers: "If I give you $100,000 right now, how long is it going to last you? ... But if you work to get an employee job, with benefits, with insurance, with vacations, how long does that last you? ... So in the end what gives you more? ... What's best for you?" Ibid.

213. Ibid.

214. Karen Robes Meeks, "Report Slams Port Truck Labor Practices," *Long Beach Press-Telegram*, Feb. 20, 2014, 1.

215. Rebecca Smith, Paul Alexander Marvy, and Jon Zerolnick, *The Big Rig Overhaul: Restoring Middle-Class Jobs at America's Ports Through Labor Law Enforcement* (2014), 14; see also Dan Weikel, "Campaign Questions Status of Port Truck Drivers; Labor-Backed Study Says Independent Contractors Are Actually Employees," *Los Angeles Times*, Feb. 19, 2014, AA3. According to campaign records, in 2013, the DLSE awarded $550,714 to five Seacon Logix drivers, $299,152 to three TTSI drivers, and $18,057 to one Western Freight Carrier Driver. "Justice for Port Drivers Campaign, Ports Campaign—Misclassification Work—Southern California, California Labor Commissioner" (DLSE) (2016), 1.

216. Smith, Marvy, and Zerolnick, note 215, 20–22; see also Ricardo Lopez, "Port Truckers Load Up on Labor Suits," *Los Angeles Times*, July 5, 2014, B1.

217. Class Action Complaint at 10-19, Mendoza v. Pacer Cartage, Inc., No. 37-2013-00063453-CU-OE-CTL (Super. Ct. Cal., filed Aug. 19, 2013).

218. Class Action Complaint at 2, Castro v. Pacific 9 Transportation Inc., No. BC537252 (Cal. Super. Ct., filed Feb. 24, 2014).

219. Ibid., 10.

220. The second class action suit, Aguilar v. Pacific 9 Transportation Inc., No. BC 566080 (Cal. Super. Ct., filed Dec. 8, 2014), was brought by Farrah Mirabel, a solo plaintiff-side lawyer in Orange County.

221. NLRB Office of the General Counsel, Advice Memorandum, note 198, 3.

222. Pat Maio, "Port Trucking Company, Drivers Agree to Labor Deal," *Orange County Register*, Mar. 23, 2014, D.

223. Karen Robes Meeks, "NLRB Finds for Truck Drivers—Federal Agency Says Pacific 9 Transportation Threatened Workers over Union Organizing Bids," *Long Beach Press-Telegram*, Mar. 22, 2014, 3.

224. James Rainey, "A Strong Shift for Truckers; In a 'Historic' Action, Drivers Are Recognized as Employees Subject to Labor Protections," *Los Angeles Times*, Mar. 22, 2014, AA1; see also Meeks, note 223.

225. Meeks, note 223.

226. Justice for Port Drivers, press release, "Federal Agency Concludes Truck Drivers at Nation's Busiest Port Are Illegally Misclassified as 'Independent Contractors,'" (Mar. 21, 2014), https://perma.cc/P5UP-QCU5 (accessed July 18, 2017).

227. Maio, note 222.

228. Rainey, note 224.

229. NLRB Office of the General Counsel, Advice Memorandum, note 198, 3.

230. Ibid., 4.

231. Telephone interview with Sheheryar Kaoosji, note 107.

232. Order, Decision or Award of the Labor Commissioner, Cristobal Cardona Barrera v. Total Transportation Services, Inc., No. 05–55410 LT (Cal. Labor Comm'r Feb. 28, 2013); Order, Decision or Award of the Labor Commissioner, Jose Montero v. Total Transportation Services, Inc., No. 05–54135 LT (Cal. Labor Comm'r Feb. 28, 2013).

233. Order, Decision or Award of the Labor Commissioner, Saul Alvarado v. Total Transportation Services, Inc., No. 05–56129 (Cal. Labor Comm'r Oct. 14, 2013).

234. Signed Charge Against Employer, Total Transportation Services, Inc., Case No. 21-CA-127303 (filed Apr. 24, 2014), *NLRB*, https://www.nlrb.gov/case/21-CA-127303 (accessed July 19, 2017). These charges would be followed by several more over the succeeding year. TTSI was represented by counsel from labor defense powerhouse Littler Mendelson.

235. Telephone interview with Nick Weiner, note 14.

236. Telephone interview with Michael Manley, note 132.

237. Telephone interview with Nick Weiner, note 14.

238. Ibid.

Notes

239. Karen Robes Meeks, "Port Truckers Decry 'Hazards'—Drivers Urge Officials to Address Conditions," *Long Beach Press-Telegram*, Apr. 17, 2014, 3.

240. "Truckers Speak Out before L.A. Harbor Commission—Drivers Urge Panel to Look into Working Conditions at One Company," *Long Beach Press-Telegram*, Apr. 18, 2014, 4.

241. Ibid.

242. Paul Rosenberg, "Port Truckers: A Year of Epic Labor Struggle," *Random Lengths*, Jan. 8, 2015, 3, 6–7.

243. Brian Watt, "CA Labor Office Sides with Port Truck Drivers," *89.3 KPCC*, Apr. 2, 2014, https://perma.cc/M7UP-P6DM (accessed July 18, 2017).

244. Ibid.

245. Ibid.

246. Ibid. Dines added: "If [drivers] can get five or six turns in a day, then this would present itself as a wonderful small business opportunity." Ibid.

247. Lopez, note 87; Pat Maio, "Strike at L.A., Long Beach Ports Costs Millions, Officials Say," *Orange County Register*, Apr. 28, 2014, http://www.ocregister.com/2014/04/29/strike-at-la-long-beach-ports-costs-millions-officials-say/ (accessed July 18, 2017). Drivers in Savannah planned to strike in coordination with the Southern California drivers. Ibid.; see also Mary Carr Mayle, "Savannah Port Truckers: 'We Are Employees,'" *Savannah Morning News*, Apr. 28, 2014, https://perma.cc/JW3Z-CXMQ (accessed July 18, 2017).

248. Maio, note 247.

249. Ibid. The longshoremen's contract was set to expire in July 2014. There had been rumblings of discontent in late 2012, when 800 clerical workers represented by the ILWU walked off the job to protest outsourcing that threatened their positions. Walter Hamilton and Ronald D. White, "Port Strike Part of Bigger Fight," *Los Angeles Times*, Dec. 6, 2012, A1. Contract talks between the ILWU and Pacific Maritime Association began in May 2014. Karen Robes Meeks, "Contract Talks between Longshore Workers, Employers Begin," *Long Beach Press-Telegram*, May 12, 2014, http://www.presstelegram.com/business/20140512/contract-talks-between-longshore-workers-employers-begin (accessed July 19, 2017).

250. Maio, note 247.

251. Ibid.; telephone interview with Nick Weiner, note 14.

252. Maio, note 247.

253. Ibid.

254. Telephone interview with Sheheryar Kaoosji, note 107.

255. Teamsters: NLRB Complaint Issued Against Green Fleet Systems, note 180; see also "Out of Step: How Skechers Hurts Its California Supply Chain Workers," May 2014, http://www.laane.org/wp-content/uploads/2014/06/Skechers-Report.pdf (accessed July 18, 2017).

256. Ricardo Lopez, "Truck Drivers Picket Port Haulers," *Los Angeles Times*, Apr. 29, 2014, B1. After the strike ended, the *Long Beach Press-Telegram* published an op-ed by a driver making the case for "why many truckers like me choose to be in business for themselves." Robert Astorga, Commentary, "Why Truckers Choose Independence," *Long Beach Press-Telegram*, June 5, 2014, 12.

257. Pat Maio, "L.A. Port Spokesman Said It Was 'Close to Business as Usual' on Night Before Strike Was Scheduled to End," *Orange County Register*, Apr. 30, 2014, F; see also Ricardo Lopez, "Truck Drivers End 2-Day Port Protest," *Los Angeles Times*, May 1, 2014, B2 ("The protest was the third in the last year, and more are planned in the coming months, organizers said. ... The demonstration ... was the largest yet by the group, escalating tensions between drivers and trucking companies.").

258. Telephone interview with Sheheryar Kaoosji, note 107.

259. Ibid.

260. Steve Lopez, "Truckers Want a Fair Shake; Drivers at the Port of L.A. Are Willing to Fight What They Call Wage Theft," *Los Angeles Times*, May 18, 2014, A2.

261. Order Revoking Settlement Agreement, Order Consolidating Cases, Consolidated Complaint and Notice of Hearing, Green Fleet Systems, LLC and Int'l Brotherhood of Teamsters, Port Division, note 193. The complaint alleged numerous violations of the drivers' rights, including claims of retaliation, discharge, and even death threats. Green Fleet Systems, LLC, Case No. 21-CA-131831, *NLRB*, https://www.nlrb.gov/case/21-CA-131831 (accessed July 19, 2017); see also Dan Weikel, "Carson Firm Faces Labor Claims; U.S. Says Green Fleet Truckers Fired Drivers for Their Pro-Union Activities and Planted an Operative," *Los Angeles Times*, June 23, 2014, AA3.

262. Karen Robes Meeks, "Judge Will Hear Union's Claim: Complaint Alleges Rancho Dominguez Firm Harassed Drivers," *Long Beach Press-Telegram*, June 20, 2014, 3.

263. Telephone interview with Michael Manley, note 132.

264. Karen Robes Meeks, "Teamsters Violated Law in Pickets of Carson Firm," *Long Beach Press-Telegram*, June 28, 2014, 1.

265. Ibid.

266. Ibid.

Notes

267. Karen Robes Meeks, "Protests Cited as Reason Trucking Firm Loses Lease," *Long Beach Press-Telegram*, July 1, 2014, 4.

268. Ibid.

269. Telephone interview with Nick Weiner, note 14.

270. Ibid.

271. Brian Watt, "Los Angeles/Long Beach Port Truck Drivers Launch Strike with No Planned End," *89.3 KPCC*, July 7, 2014, https://perma.cc/LQF4-2H3S (accessed July 18, 2017); see also Andrew Khouri, "Some Port Truckers Walk Off Job," *Los Angeles Times*, July 8, 2014, B2.

272. Karen Robes Meeks, "Port Truckers Launch Strike—Grievances: Workers Say Three Companies Skirt Laws by Classifying Them as Contractors," *Long Beach Press-Telegram*, July 8, 2014, 1.

273. Ibid.

274. Meeks, note 272; Karen Robes Meeks, "Strike—Truckers, Port Workers Briefly Shut Terminals," *Long Beach Press-Telegram*, July 9, 2014, 1.

275. Andrew Khouri, "Port Terminals Briefly Close," *Los Angeles Times*, July 9, 2014, B2; Meeks, "Strike—Truckers, Port Workers Briefly Shut Terminals," note 274.

276. Ned Resnikoff, "Port Truckers' Strike Sends Ripples Through Labor World," *MSNBC*, July 9, 2014, https://perma.cc/8MWU-SZYQ (accessed July 18, 2017).

277. Karen Robes Meeks, "Raging Against the Trucking Companies," *Long Beach Press-Telegram*, July 10, 2014, 3.

278. Karen Robes Meeks, "Some Port Drivers Defend Firms—Amid Picketing, about 50 from Two of Three Firms Say They're Treated Fairly," *Long Beach Press-Telegram*, July 11, 2014, 3.

279. Karen Robes Meeks, "Truck Drivers Strike for Fifth Day—Los Angeles Mayor Has Asked Harbor Board to Investigate Allegations Made by Protestors," *Long Beach Press-Telegram*, July 12, 2014, 3.

280. Ibid.

281. Karen Robes Meeks, "Drivers, Firms Agree to Truce—L.A. Mayor Brokers Deal for Cooling Off Period; Trucks to Roll," *Long Beach Press-Telegram*, July 13, 2014, 1; see also Andrew Khouri, "Truckers Put Breaks on Protest at Twin Ports," *Los Angeles Times*, July 12, 2014, 30.

282. Verified Complaint for Injunctive Relief Against Retaliation for Pursuit of DLSE Claims at 3, Menendez v. Total Transportation Services Inc., No. BC553064 (Cal. Super. Ct., filed July 29, 2014).

283. NLRB Office of the General Counsel, Advice Memorandum, note 198, 4.

284. Karen Robes Meeks, "Some Port Truckers Say Firms Continue Reprisals—Workers Seeking to Unionize Complain that Employers Aren't Following Terms of a Truce," *Long Beach Press-Telegram*, Aug. 30, 2014, 3.

285. "Justice for Port Drivers Campaign, Ports Campaign—Misclassification Work—Southern California, California Labor Commissioner" (DLSE), note 215, 1.

286. Meeks, note 284.

287. Ibid.

288. Paul Rosenberg, "TTSI Illegally Fires 33 Truckers, Breaks 'Cooling Off' Agreement," *Random Lengths*, Sept. 18, 2014, 7.

289. Rosenberg, note 242.

290. Karen Robes Meeks, "Port Truckers Pray, Fast During Protest—Workers Continue Demonstrations Against 3 Firms Over Employment Status," *Long Beach Press-Telegram*, Sept. 30, 2014, 3.

291. Ibid.

292. Ibid.

293. Taylor v. Shippers Transport Express, Inc., 2014 WL 7499046, No. CV 13–02092 BRO (PLAx) (C.D. Cal. Sept. 30, 2014).

294. Order Granting in Part and Denying in Part Petitioner's Petition for a Preliminary Injunction, Olivia Garcia v. Green Fleet Systems, LLC, No. CV 14-6220 PSG (JEMx) (C.D. Cal., filed Oct. 10, 2014); see also Karen Robes Meeks, "Judge Orders Trucking Firm to Reinstate Two Fired Port Drivers," *Long Beach Press-Telegram*, Oct. 15, 2014, http://www.presstelegram.com/business/20141015/judge-orders-trucking-firm-to-reinstate-two-fired-port-drivers (accessed July 18, 2017).

295. Order Granting in Part and Denying in Part Petitioner's Petition for a Preliminary Injunction, Olivia Garcia v. Green Fleet Systems, LLC, note 294, 23.

296. Telephone interview with Julie Gutman Dickinson, note 184.

297. Amicus Curiae Memorandum of International Brotherhood of Teamsters, Port Division, note 209, 23–24.

298. Order Granting in Part and Denying in Part Petitioner's Petition for a Preliminary Injunction, Olivia Garcia v. Green Fleet Systems, LLC, note 294, 30.

299. Jordan England-Nelson, "2 Truckers Return to Work—Following Federal Injunction, Rancho Dominguez Company Rehires the Drivers as Employees," *Long Beach Press-Telegram*, Nov. 8, 2014, 3.

300. Ibid.

Notes

301. Telephone interview with Nick Weiner, note 14.

302. Karen Robes Meeks, "Ports' Congestion Is Forcing Action—Exemption: Trucking Group Is Seeking a Waiver of Federal Rules on When Drivers Can Work," *Long Beach Press-Telegram*, Nov. 7, 2014, 1.

303. Ibid.

304. Ibid.

305. Karen Robes Meeks, "Truckers Say Port Strikes May Resume—Harbor Leader Says Finding Solutions to Drivers' Issues Remains a Priority," *Long Beach Press-Telegram*, Nov. 11, 2014, 3.

306. Ibid.

307. Andrew Khouri, "L.A. Truck Drivers Strike as Tension Mounts at Ports," *Los Angeles Times*, Nov. 13, 2014, http://www.latimes.com/business/la-fi-trucker-strike-20141113-story.html (accessed July 18, 2017). The TTSI and Pac 9 strike happened against the backdrop of ongoing negotiations between the PMA and ILWU.

308. Karen Robes Meeks, "L.A. Mayor Brokers Truce in Driver Strike: Teamsters and One of Three Trucking Firms in Dispute Agree to Cooling-Off," *Long Beach Press-Telegram*, Nov. 14, 2014, 3.

309. Khouri, note 307.

310. Steve Lopez, "Port Truckers Drive Home Difference between Contractors and Employees," *Los Angeles Times*, Nov. 15, 2014, http://www.latimes.com/local/california/la-me-1116-lopez-trucking-20141116-column.html (accessed July 18, 2017).

311. Karen Robes Meeks, "Truckers Protest Expands at Ports—Drivers Seek Classification as Employees, Possible Teamsters Representation," *Long Beach Press-Telegram*, Nov. 18, 2014, 1 ("Meanwhile, truckers and their supporters were stationed at LBCT, ITS and TTI terminals at the Port of Long Beach and Maersk, Evergreen and APL terminals and the ICTF transfer facility at the Port of Los Angeles. Officials from both ports reported minimal disruption.").

312. The Harbor Rail Transport case was filed in March 2014, alleging willful misclassification, violations of wage-and-hour laws, and unfair competition on behalf of roughly eighty plaintiffs. Complaint, Disus v. Intermodal Container Service Inc. dba Harbor Rail Transport, No. BC540538 (Cal. Super. Ct., filed Mar. 26, 2014). The same day it filed against Harbor, the Gomez Law Group also filed a mass action suit against PDS Trucking, Inc., also owned by XPO but now defunct, which remains pending. Complaint, Lopez v. PDS Trucking, Inc., BC540537 (Cal. Super. Ct., filed Mar. 26, 2014). In December 2014, Gomez brought a virtually identical case against Pacer Cartage on behalf of more than fifty drivers in order to protect state law claims

imperiled by the Pacer class action's removal to federal court. Complaint, Contreras v. Pacer Cartage, Inc., No. BC567807 (Cal. Super. Ct., filed Dec. 23, 2014).

313. Telephone interview with Sheheryar Kaoosji, note 107. While Harbor Rail Transport specialized in moving goods to and from the ports, Pacer Cartage primarily transported cargo to intermodal rail yards.

314. Meeks, note 311.

315. Ibid.

316. Karen Robes Meeks, "Drivers Take Strikes to Ports' Rail Yards—Truckers Protest Classification of Drivers as Independent Contractors," *Long Beach Press-Telegram*, Nov. 19, 2014, 3 ("For more than a year, drivers have been joined by the Teamsters, the union that wants to represent them as employees, in a movement against area trucking companies employing independent contractors. Until recently, they had set their sights on three companies: Total Transportation Services Inc., Green Fleet Systems and Pacific 9 Transportation. … Last week, the three companies either agreed to a cooling-off period or to meet with the Teamsters and Garcetti's office.").

317. Ibid.

318. Rex Richardson, Commentary, "Wage Theft Shouldn't Be Tolerated," *Long Beach Press-Telegram*, Nov. 19, 2014, 13.

319. Karen Robes Meeks, "Port Drivers Urge Harbor Leaders to Support Fight—Commissioners Say They Sympathize but Can't Do Much to Help," *Long Beach Press-Telegram*, Nov. 20, 2014, 3.

320. Karen Robes Meeks, "Truckers Call End to 8-Day Strike—Decision Will Have Little Effect on Easing Supply Chain Congestion," *Long Beach Press-Telegram*, Nov. 22, 2014, 1.

321. Ibid.

322. Ibid.

323. Ibid.

324. Ibid.

325. Ibid.

326. Karen Robes Meeks, "The Missing Link in Port Tie-Ups—A New Way of Handling an Old Technology Contributes to the Slowdown," *Long Beach Press-Telegram*, Mar. 9, 2015, 1, 5.

327. Karen Robes Meeks, "Leaders to Tackle Congestion—Harbor Board Will Look at Three Ways to Deal with Bottlenecks to Help Truck Drivers and Shippers," *Long Beach Press-Telegram*, Dec. 21, 2014, 3.

328. Karen Robes Meeks, "Truck Chassis Plan Advances—Commissioners Hope Availability of Trailers Will Relieve Shipping Tie-Ups," *Long Beach Press-Telegram*, Dec. 23, 2014, 3.

329. Karen Robes Meeks, "Depot Will Ease Bottlenecks at the Port," *Long Beach Press-Telegram*, Dec. 31, 2014, 3.

330. Telephone interview with Fred Potter, note 156.

331. Justice for Port Drivers, "Regulatory Action and Litigation Addressing Truck Driver Misclassification at the Ports of Los Angeles and Long Beach," Jan. 2015, 1.

332. Ibid. These included awards in favor of two Container Freight drivers (over $92,965.62), one LACA Express driver, three Sterling Express drivers ($284,824.91), seven Pacer Cartage drivers ($2.2 million), four Pac 9 drivers (over $254,627.12), four QTS drivers (over $167,050.59), fourteen TTSI drivers ($954,953.62), and one WinWin Logistics driver ($204,323.48). Justice for Port Drivers Campaign, "Ports Campaign—Misclassification Work—Southern California, California Labor Commissioner" (DLSE), note 215, 1.

333. Justice for Port Drivers, note 331, 1.

334. Ibid., 2

335. Ibid.

336. Lambey v. Shippers Transport Express, Inc., BC549530 (Cal. Super. Ct., filed June 20, 2014).

337. Gaspard v. Shippers Transport Express, Inc., BC564329 (Cal. Super. Ct., filed Nov. 18, 2014).

338. Consent Judgment & Order at 9, Perez v. Shippers Transport Express, Inc., No. 2:13-CV-04255-BRO-PLA (C.D. Cal., filed Nov. 17, 2014). The judgment also ordered Shippers to pay $188,587.29 in compensatory and liquidated damages. Ibid., 11.

339. "Port Truck Drivers with Shippers Transport Express Join Teamsters Local 848," *Teamsters*, Jan. 9, 2015, https://perma.cc/EW7R-FTBA (accessed July 18, 2017).

340. Karen Robes Meeks, "Truckers to Decide on Unionizing—Agreement Opens the Door for Teamsters to Negotiate on Behalf of Port Drivers at Carson Firm," *Long Beach Press-Telegram*, Jan. 7, 2015, 3.

341. Editorial, "Port Truckers Stuck in Middle of Labor Dispute," *Long Beach Press-Telegram*, Jan. 9, 2015, 10.

342. Karen Robes Meeks, "Carson Truckers Opt to Join Teamsters—About 80 Percent of Shippers Transport Express Drivers Support Union Organizing," *Long Beach Press-Telegram*, Jan. 10, 2015, 3; see also Brian Watt, "LA and Long Beach Ports: Truckers Win Employee Status, Quickly Vote to Join Teamsters," *89.3 KPCC*, Jan. 9, 2015, https://perma.cc/X66B-PWC5 (accessed July 18, 2017).

343. Meeks, note 342.

344. Ibid.

345. Ibid. (Leonardo Mejia was a driver at the ports for approximately fifteen years and drove for Shippers for four years).

346. Ibid.

347. Ibid.

348. Karen Robes Meeks, "First Labor Contract gets OK'd—Wage Increases and Full Medical Coverage Are Coming for Shippers Transport Express Drivers," *Long Beach Press-Telegram*, Feb. 10, 2015, 3. The Shippers collective bargaining agreement was effective on February 8, 2015.

349. A notice of settlement in the willful misclassification and wrongful death suit brought by Hadsell & Stormer was filed in February 2016.

350. Letter from Fredrick Potter, director, Teamsters Port Division, to Jon Slangerup, executive director, Port of Long Beach (Jan. 13, 2015), 1.

351. Ibid.

352. Ibid., 2. Potter later wrote a similar letter to the Los Angeles port director, Eugene Seroka, arguing that the same companies were in violation of the Clean Truck Concession Agreement provision requiring trucks to comply with "all applicable federal, state and municipal laws, statutes, ordinances, rules and regulations that govern Motor Carrier's operations. …" Letter from Fredrick Potter, director, Teamsters Port Division, to Eugene D. Seroka, executive director, Port of Los Angeles (Jan. 14, 2016), 1.

353. Ronald D. White, "Short-Haul Trucker Julio Cervantes, Fighting for Employee Status, Needs to Start Attending to Needs of Wife and Daughters," *Los Angeles Times*, Jan. 25, 2015, C1.

354. Ibid.

355. Ibid.

356. Rosenberg, note 242, 3.

357. Telephone interview with Nick Weiner, note 14.

358. Ibid.

359. Ibid.

360. "Judge: Truckers Are Owed $2 Million—Decision Says Port Drivers Are Employees Not Independent Contractors," *Long Beach-Press Telegram*, Jan. 30, 2015, 1.

361. Ibid., 6.

Notes

362. Ibid.

363. First Amended Class Action Complaint at 3-5, Mendoza v. Pacer Cartage, No. 13cv2344-LAB (JMA) (S.D. Cal., filed Jan. 30, 2015).

364. PAC Anchor Transportation, Inc. v. California, note 13. Shortly after the California Supreme Court denied cert, Julie Gutman Dickinson gave a presentation to the *Journal of Commerce* Trans Pacific Maritime (TPM) conference arguing against the sustainability of the independent-contractor model for port drayage in which she asserted that the "federal pre-emption argument has been lost." Julie Gutman Dickinson, "Is the 'Independent Contractor' Model in Port Drayage Sustainable?" JOC TPM Conference, Mar. 3, 2015 (PowerPoint Presentation).

365. Andrew Edwards, "Teamsters Union Promises to Aid Truckers—Thousands of Nonunion Drivers Would Be Without Work, and Potentially Benefits, If Shutdown Occurs," *Long Beach Press-Telegram*, Feb. 19, 2015, 1, 6.

366. Ibid., 1.

367. Ibid.

368. Peter Jamison, "Clearing Cargo Backlog at Southland Ports May Take Three Months," *Los Angeles Times*, Feb. 23, 2015, http://www.latimes.com/business/la-fi-ports-deal-20150224-story.html (accessed July 18, 2017); see also Karen Robes Meeks, "Port Union Delegates Pore over New Deal," *Long Beach Press-Telegram*, Mar. 31, 2015, 1 (describing ILWU meeting to review proposed agreement with PMA).

369. Jamison, note 368.

370. Editorial, "Port Dispute Is Over, but Damage Lingers," *Long Beach Press-Telegram*, Feb. 27, 2015, 13.

371. Meeks, note 326.

372. Karen Robes Meeks, "Official: U.S. Must Invest in Long-Term Port Solution—Labor Dispute Cited as Just One Factor in Congestion," *Long Beach Press-Telegram*, Mar. 20, 2015, 1.

373. Ibid.

374. Karen Robes Meeks, "Port Cargo Flow Focus of Talks," *Long Beach Press-Telegram*, Mar. 26, 2015, 3, 9.

375. Karen Robes Meeks, "LB and L.A. Ports Taming Congestion; Only 6 Parked at Sea," *Long Beach Press-Telegram*, Apr. 10, 2015, 1.

376. Karen Robes Meeks, "Feds Target Port Tie-Ups—Agency Tackles Charges that Penalize Importers, Exporters for Congestion," *Long Beach Press-Telegram*, Apr. 15, 2015, 1, 7.

377. Karen Robes Meeks, "Forum Focuses on Slow Ports—More than 70 Stakeholders Attend Event to Discuss Challenges," *Long Beach Press-Telegram*, Apr. 23, 2015, 3.

378. Ben Bergman, "Congestion Still Plagues Ports of LA and Long Beach," *89.3 KPCC*, Apr. 23, 2015, https://perma.cc/62X6-CESC (accessed July 18, 2017). The 11th annual Pulse of the Ports: Peak Season Forecast in Long Beach a week later also focused on congestion and the coming impact of megaships. Karen Robes Meeks, "Twin Ports Facing 'New Normal'—Backlog from Dispute Being Cleared Out, but Bigger Ships to Become More Frequent," *Long Beach Press-Telegram*, Apr. 30, 2015, 3.

379. Decision at 62-64, Green Fleet Systems, LLC and International Brotherhood of Teamsters, Port Division, Cases 21-CA-100003, 21-CA-115910, 21-CA-119154, 21-CA-121368 (NLRB Apr. 9, 2015).

380. Ibid., 52.

381. Telephone interview with Fred Potter, note 156.

382. Ibid.

383. Ibid.

384. Ibid.

385. Ibid.

386. Karen Robes Meeks, "Truckers to Vote on Strike Authorization—Dispute Lingers over Independent Contractor Status of Some Drivers," *Long Beach Press-Telegram*, Apr. 25, 2015, 5.

387. At the time of the April 27, 2015 strike, IBT was subject to three suits, one a state law class action by the law firms of Sayas and Hayes (also counsel in the Shippers class action), which did not include a willful misclassification claim but did include a Private Attorney General Act claim. Correa v. Intermodal Bridge Transport, Inc., No. BC566279 (Cal. Super. Ct., filed Dec. 10, 2014). The *Correa* case remains pending. In addition, the Gomez Law Group had filed two cases: one a class action (Flores v. Intermodal Bridge Transport, Inc., No. BC573630 (Cal. Super. Ct., filed Feb. 24, 2015) and the other a mass action (Flores v. Intermodal Bridge Transport, Inc., No. BC574719 (Cal. Super. Ct., filed Mar. 6, 2015). Both of the Gomez law firm cases asserted the same claims on behalf of the same clients. The class action was dismissed in April 2016, while the mass action remains pending.

388. Andrew Edwards, "Port Drivers Strike against 4 Firms—Classifications of Truckers as Contractors Are at Root of Latest Walkout," *Long Beach Press-Telegram*, Apr. 28, 2015, 3; Andrew Khouri, "Port Truck Drivers Strike at 4 Firms," *Los Angeles Times*, Apr. 28, 2015, C2. The four targeted companies—Pac 9, Pacer Cartage, Harbor Rail Transport, and IBT—had 500 drivers total, constituting roughly 3.5 percent of the port drayage market. Andrew Khouri, "After Four-Day Stoppage, Most Port Truck Drivers Return to Work," *Los Angeles Times*, May 2, 2015, C3; see also Cole Stangler,

"Labor Troubles Return to Los Angeles, Long Beach Ports as Truck Drivers Strike," *International Business Times*, Apr. 27, 2015, http://www.ibtimes.com/labor-troubles-return-los-angeles-long-beach-ports-truck-drivers-strike-1898273 (accessed July 19, 2017).

389. Bill Mongelluzzo, "Drayage Drivers Picket LA, Long Beach Marine Terminals," *JOC.com*, Apr. 27, 2015, http://www.joc.com/port-news/us-ports/port-los-angeles/drayage-drivers-picket-la-long-beach-marine-terminals_20150427.html (accessed July 19, 2017); see also Brian Watt, "Truck Drivers at Ports of LA and Long Beach Strike Again," *89.3 KPCC*, Apr. 27, 2015, https://perma.cc/L8NL-3G4N (accessed July 19, 2017).

390. Khouri, "Port Truck Drivers Strike at 4 Firms," note 388.

391. Telephone interview with Carlos Santamaria, note 161.

392. Ibid.

393. Ibid.

394. Khouri, "After Four-Day Stoppage, Most Port Truck Drivers Resume Work," note 388.

395. Karen Robes Meeks, "L.A. Mayor Touts Upstart Hauler as a Labor Model," *Long Beach Press-Telegram*, May 5, 2015, 1, 11. In April 2015, before the announced TTSI agreement, drivers in the *Menendez* class action settled their claims in relation to the company's planned retaliatory discharge the previous year. After the deal was struck, another group of drivers filed suit against TTSI alleging a variety of employment violations. Rojas v. Total Transportation Services Inc., No. NC060391 (Cal. Super. Ct., filed Dec. 9, 2015). Resolution of that case is still pending as TTSI's bankruptcy plan is finalized.

396. Mongelluzzo, note 87.

397. Ibid.

398. Meeks, note 395.

399. Mongelluzzo, note 87.

400. Stas Margaronis, "Will Rosenthal and Eco Flow Re-engineer Harbor Trucking?" *American Journal of Transportation*, June 8–21, 2015, 1.

401. Meeks, note 395, 11.

402. Erica E. Philips, "Port Truckers Revive Drive for Employee Status," *Wall Street Journal*, May 28, 2015, http://www.wsj.com/articles/port-truckers-revive-drive-for-employee-status-1432848972 (accessed July 19, 2017).

403. Ibid.

404. Mongelluzzo, note 87.

405. Karen Robes Meeks, "Port Truckers Suing Carson Firm," *Long Beach Press-Telegram*, May 7, 2015, 6.

406. Motion for Substantive Consolidation, In re QTS, Inc., Petition No. 2:14-bk-26361-NB (C.D. Cal., filed May 5, 2015).

407. Karen Robes Meeks, "'Stepchild of the Industry'—Trucking: Short-Haulers Who Get Goods from Port to Warehouse Are Often Finding It Harder Just to Get By," *Long Beach Press-Telegram*, May 17, 2015, 1, 5; see also Brian Watt, "For Truck Drivers at the Ports of Los Angeles and Long Beach, It's a Waiting Game," *89.3 KPCC*, June 2, 2015, https://perma.cc/6RTH-2TAY (accessed July 19, 2017).

408. Karen Robes Meeks, "Truckers Hit Hard by Port Tie-Ups—Megaships and Other Changes Causing Bottlenecks Leave Haulers Struggling to Stay Afloat," *Long Beach Press-Telegram*, May 18, 2015, 1, 5. The Harbor Trucking Association announced its own program to distribute 200 chassis to companies at the Los Angeles and Long Beach ports. Andrew Edwards, "Truckers Group Forms Chassis Partnership," *Long Beach Press-Telegram*, May 21, 2015, 1.

409. Sarah Favot, "Supes Mull How to Aid Truckers—County Will Explore Program to Investigate Complaints of Workers Classed as Contractors," *Long Beach Press-Telegram*, May 20, 2015, 3, 7.

410. Chris Kirkham and Andrew Khouri, "Cargo Crush; With Docks Clogged, Local Ports Lose to Rivals," *Los Angeles Times*, June 2, 2015, C1.

411. Ibid.

412. Ibid.

413. Ibid.

414. Brian Watt, "As Cargo Traffic Grows, the Ports Face a Shortage of Drivers," *89.3 KPCC*, June 3, 2015, https://perma.cc/9H3A-9CUZ (accessed July 19, 2017).

415. Editorial, "Truckers Need Voice in Port Controversy," *Long Beach Press-Telegram*, June 11, 2015, 12.

416. Ibid.

417. Roger Hernández, Commentary, "Port Drivers Misclassified as Independent Contractors," *Long Beach Press-Telegram*, June 7, 2015, 16.

418. Erica E. Phillips, "Trucking Companies Try New Approach at Congested California Ports," *Wall Street Journal*, July 1, 2015, http://www.wsj.com/articles/trucking-companies-try-new-approach-at-congested-california-ports-1435760577 (accessed July 19, 2017).

419. Ibid.

Notes

420. Ibid.

421. Telephone interview with Michael Manley, note 132.

422. Ibid.

423. Ibid.

424. Ibid.

425. Letter from Michael T. Manley, staff attorney, International Brotherhood of Teamsters, to Olivia Garcia, regional director, National Labor Relations Board (June 12, 2015), 1.

426. Ibid., 7.

427. Ibid., 7–10.

428. Daina Beth Solomon, "Port Truck Drivers Plan Sixth Strike Against Company," *Los Angeles Times*, July 20, 2015, http://www.latimes.com/business/la-fi-port-trucker-protest-20150720-story.html (accessed July 19, 2017).

429. Ibid.

430. Ibid.

431. Ibid.

432. "Drivers for Ports to Picket Again—Union Organizers Say Latest Work Stoppage Could Last Indefinitely," *Long Beach Press-Telegram*, July 21, 2015, 3; see also Cole Stangler, "Truckers at Ports of Los Angeles, Long Beach Strike Again, Oppose Classification as Independent Contractors," *International Business Times*, July 21, 2015, http://www.ibtimes.com/truckers-ports-los-angeles-long-beach-strike-again-oppose-classification-independent-2018189 (accessed July 19, 2017).

433. Andrew Edwards, "Panel to Examine Trucking Issues," *Long Beach Press-Telegram*, July 27, 2015, 3.

434. Andrew Edwards, "Port Truckers Get Key Backing," *Long Beach Press-Telegram*, July 30, 2015, 8.

435. Ibid.

436. Brian Watt, "Wage Theft Enforcement Bill Approved by California Legislature," *89.3 KPCC*, Sept. 10, 2015, https://perma.cc/C3LX-W22U (accessed July 19, 2017).

437. "Truckers Protest Is in Week 10," *Long Beach Press-Telegram*, Sept. 15, 2015, 4.

438. Stephanie Rivera, "Teamster Officials Announce Support of Port Drivers Amid More Strikes," *Long Beach Post*, Oct. 27, 2015, https://perma.cc/VKG5-2BSK (accessed July 19, 2017).

439. "Trucker Strike Begins at Port," *Long Beach Press-Telegram*, Oct. 26, 2016, http://www.presstelegram.com/business/20151026/trucker-strike-begins-at-ports-of-long-beach-and-los-angeles (accessed July 20, 2017).

440. Rivera, note 438.

441. Intermodal Bridge Transport, Case No. 21-CA-157647, *NLRB*, https://www.nlrb.gov/case/21-CA-157647 (accessed July 20, 2017). The original charge was filed on August 10, 2015.

442. Brian Watt, "Port Truck Drivers and Warehouse Workers Strike Together for Better Wages and 'Employee' Status," *89.3 KPCC*, Oct. 27, 2015, https://perma.cc/C576-VDT8 (accessed July 19, 2017).

443. Sheheryar Kaoosji, "Special Report on California Cartage: While Leading Drayage Companies Modernize, Litigation and Government Investigations Expose Risk and Hamper Operations Nationwide," Dec. 2, 2015, Los Angeles Alliance for a New Economy, *Port Innovations*, https://perma.cc/JY93-NAKG (accessed July 19, 2017). Cal Cartage's subsidiaries included: California Cartage Express, LLC, K & R Transport, LLC, California Multimodal Inc., ContainerFreightEIT, LLC, and F&S Distributing. Cal Cartage was also under investigation by the DOL for labor violations at the port in Savannah, Georgia. Sheheryar Kaoosji, "Update: Special Report on California Cartage: Safety Violations Expose New Risks to Port Logistics Operations," Dec. 14, 2015, Los Angeles Alliance for a New Economy, *Port Innovations*, https://perma.cc/7DSZ-FEB8 (accessed July 19, 2017).

444. Telephone interview with Kaoosji, note 107.

445. Karen Robes Meeks, "Judge to Rule in LB's Case over Rail Yard," *Long Beach Press-Telegram*, Nov. 18, 2015, 6.

446. Ibid.

447. Kaoosji, "Update: Special Report on California Cartage," note 443.

448. Greg Yee, "Long Beach Police Release Details in Port Truck Driver's Death," *Long Beach Press-Telegram*, Sept. 23, 2015, http://www.presstelegram.com/general-news/20150923/long-beach-police-release-details-in-port-truck-drivers-death (accessed July 20, 2017).

449. California Department of Industrial Relations, Division of Occupational Safety and Health, Citation and Notification of Penalty to California Cartage Company, LLC, Inspection # 1079165 (July 20, 2015–Nov. 17, 2015), https://perma.cc/3TZW-DFES (accessed July 20, 2017).

450. Kaoosji, "Update: Special Report on California Cartage," note 443.

451. Watt, note 442.

452. "Hoffa Joins Port Truck Drivers Strike at Long Beach/Los Angeles Ports," Oct. 27, 2015, *Teamsters*, https://perma.cc/KUD4-QN8P (accessed July 20, 2017).

453. James Rufus Koren, "Port Drivers Win Back Wages; State Orders Carson Trucking Firm to Pay Out $7 Million to 38 Misclassified Workers," *Los Angeles Times*, Dec. 22, 2015, C1; Donna Littlejohn, "Truckers to Collect $7M in Back Pay," *Long Beach Press-Telegram*, Dec. 24, 2015, 1; see also Justice for Port Drivers, press advisory, "CA Labor Commissioner Rules on Landmark Misclassification & Wage Theft Case," Dec. 21, 2015; "Wage Theft at the Ports: $6.9 Million Awarded to 38 Striking Truck Drivers," *American Journal of Transportation*, https://perma.cc/EF8E-8VW6 (accessed July 20, 2017).

454. The parties' motion to approve a preliminary settlement agreement was granted in February 2016.

455. NLRB Office of the General Counsel, Advice Memorandum, note 198, 1. In the advice memorandum, Kearney cited to *FedEx Home Delivery*, 361 N.L.R.B. No. 55 (2014), an NLRB decision that determined the delivery company's drivers were statutory employees and not independent contractors. Ibid., 5. Kearney applied the same nine-factor test laid out under the Restatement (Second) of Agency § 220 to determine that the Pac 9 drivers were effectively employees. Ibid. The campaign hailed the memo as proof that the federal government was getting more involved in policing misclassification. Sheheryar Kaoosji, "As Government Enforcement Increases—Both at State and Federal Level—The Legal and Financial Liability Stemming from Port Driver Misclassification Continues to Spread," May 20, 2016, Los Angeles Alliance for a New Economy, *Port Innovations*, https://perma.cc/958E-C94K (accessed July 20, 2017). The memo was released to the public in August 2016, after the Pac 9 case had closed. Lawrence E. Dubé, Daily Labor Report, "NLRB Memo Lays Out Misclassification Theory," *Bloomberg BNA*, Aug. 29, 2016, https://perma.cc/3DVN-Q7TW (accessed July 20, 2017). The *FedEx Home Delivery* case relied on in the Advice Memorandum was later vacated by the D.C. Circuit in FedEx Home Delivery v. NLRB, 849 F.3d 1123 (D.C. Cir. 2017), although the NLRB subsequently petitioned the full D.C. Circuit for a rehearing en banc.

456. NLRB Office of the General Counsel, Advice Memorandum, note 198, 11.

457. Ibid.

458. Sheheryar Kaoosji, "Anti-Innovation: 2 Strikes in 2 Weeks by Port Drivers … and It Is Only February," Feb. 29, 2016, Los Angeles Alliance for a New Economy, *Port Innovations*, https://perma.cc/8F3C-3C4X (accessed July 20, 2017).

459. Sheheryar Kaoosji, "Will Latest Round of DLSE Decisions Spur Port Efficiencies?," Jan. 27, 2016, Los Angeles Alliance for a New Economy, *Port Innovations*, https://perma.cc/EQ3X-RE6V (accessed July 20, 2017).

460. On December 10, 2015, with assistance from LAANE and the law firm Rosen Saba, driver Johel Climaco Valencia won a SLAPP (Strategic Lawsuit Against Public Participation) motion to dismiss a suit by his company, Sterling Express, which had sued him for breach of contract after Valencia had filed a claim with the DLSE asserting his employee status. Sheheryar Kaoosji, "Sterling Express Port Driver Successfully Strikes Former Employer's Effort to Chill His Workplace Rights," Dec. 17, 2015, Los Angeles Alliance for a New Economy, *Port Innovations*, https://perma.cc/M64B-BUWJ (accessed July 20, 2017).

461. Assemb. B. 970, 2015–2016 Leg., Reg. Sess. (Cal. 2015).

462. Assemb. B. 1513, 2015–2016 Leg., Reg. Sess. (Cal. 2015).

463. Assemb. B. 1509, 2015–2016 Leg., Reg. Sess. (Cal. 2015). See generally Sheheryar Kaoosji, "2015 California Legislative Roundup," Oct. 23, 2015, Los Angeles Alliance for a New Economy, *Port Innovations*, https://perma.cc/7VLT-X6SC (accessed July 20, 2017).

464. California Labor Code § 2750.8 (Deering 2013).

465. Julio Cervantes, Commentary, "Long Beach Should Enforce Wage Laws for All," *Long Beach Press-Telegram*, Jan. 17, 2016, 14.

466. Andrew Edwards, "Election 2016—LB Council Hopefuls Keep on Trucking," *Long Beach Press-Telegram*, Jan. 30, 2016, 6.

467. Justice for Port Drivers, "XPO Logistics Misclassification-Related Liability," Feb. 25, 2016.

468. Kaoosji, note 458.

469. Katy Stech, "Port Trucker Pacific 9 Seeks Bankruptcy Protection," *Wall Street Journal*, Apr. 27, 2016, https://www.wsj.com/articles/port-trucker-pacific-9-seeks-bankruptcy-protection-1461790641 (accessed July 20, 2017).

470. Erica A. Phillips, "Hub Group Drops Port Trucking Operation, Cites Driver Costs," *Wall Street Journal*, Feb. 4, 2016, https://www.wsj.com/articles/hub-group-drops-port-trucking-operation-cites-driver-costs-1454627531 (accessed July 20, 2017).

471. Erica E. Phillips, "Port Trucking Company Files for Bankruptcy Protection," *Wall Street Journal*, Mar. 16, 2016, https://www.wsj.com/articles/california-port-freight-hauler-files-for-chapter-11-1458169628 (accessed July 20, 2017).

472. Ibid.

473. Telephone interview with Fred Potter, note 156.

474. Ibid.

475. Ibid.

Notes

476. Ibid.

477. Pac 9 Transportation Inc., Docket No. 2:16-bk-15447 (Bankr. C.D. Cal. Apr. 26, 2016).

478. Rhonda Smith, Daily Labor Report, "Pac 9, Teamsters Agree to End Dispute over Drivers," *Bloomberg BNA*, May 6, 2016, https://perma.cc/5CML-V7SB (accessed July 20, 2017) ("'This agreement effectively ends the drivers' strike that began in July 2015 and allows the company and its valued customers to move forward in confidence that there will be no further interruption of the movement of customer cargo,' the statement said. 'Pac 9 management and the drivers will work collaboratively to rebuild and stabilize the workforce to provide a high level of service to company customers.'").

479. Stech, note 469.

480. Smith, note 478.

481. Ibid.

482. Ibid.

483. The Pac 9 bankruptcy plan had not yet been finalized as of the middle of 2017.

484. Smith, note 478.

485. California Department of Industrial Relations, news release, "Labor Commissioner Accepting Applications from Port Trucking Companies for Motor Carrier Employer Amnesty Program" (May 5, 2016), https://perma.cc/329X-SFAF (accessed July 20, 2017).

486. Erica E. Phillips, "Port-Trucking Firms Run into Labor Dispute," *Wall Street Journal*, May 11, 2016, https://www.wsj.com/articles/port-trucking-firms-run-into-labor-dispute-1462959003 (accessed July 20, 2017); see also Sheheryar Kaoosji, "Bankruptcies and Amnesty Fuel More Changes to the LA/LB Port Trucking Industry," May 26, 2016, Los Angeles Alliance for a New Economy, *Port Innovations*, https://perma.cc/6XJC-3VM8 (accessed July 20, 2017).

487. Phillips, note 486 (quoting Noel Perry, economist with FTR Transportation Intelligence).

488. Marron Lawyers, "Misclassification 3.0" (PowerPoint Presentation, Mar. 16, 2016).

489. Ibid.

490. Ibid.

491. Ibid.

492. Ibid.

493. Ibid.

494. Ibid.

495. Intermodal Bridge Transport, Case No. 21-CA-157647, note 441.

496. Memorandum GC 16-01 from Richard F. Griffin, Jr., general counsel, NLRB to All Regional Directors, Officers-in-Charge, and Resident Officers, Mar. 22, 2016, 1.

497. Ibid., 2.

498. Kaoosji, "As Government Enforcement Increases—Both at State and Federal Level—The Legal and Financial Liability Stemming from Port Driver Misclassification Continues to Spread," note 455.

499. Justice for Port Drivers, press release, "NLRB Issues Historic Complaint against Company Alleging that Misclassification Violates Workers' Right to Form a Union" (Apr. 20, 2016).

500. Complaint and Notice of Hearing at 3, Intermodal Bridge Transport and International Brotherhood of Teamsters, Case No. 21-CA-157647 (NLRB Region 21, Apr. 18, 2016); see also Daniel Wiessner, "Misclassifying Workers Violates Bargaining Rights—NLRB Official," *Reuters Legal*, Apr. 22, 2016, http://www.reuters.com/article/labor-misclassification-idUSL2N17P0I9 (accessed August 25, 2017).

501. Justice for Port Truck Drivers, note 499; see also Daily Labor Report, "Intermodal Bridge Transport Drivers Strike in California, Alleging Unfair Labor Practices," *Bloomberg BNA 76*, Apr. 20, 2016, 1.

502. Kaoosji, "As Government Enforcement Increases—Both at State and Federal Level—The Legal and Financial Liability Stemming from Port Driver Misclassification Continues to Spread," note 455.

503. The day after the NLRB General Counsel's mandatory submissions memo, Region 21 issued a complaint against the California Cartage warehouse and its staffing agency, Orient Tally Company, treating them as a "single-integrated business enterprise" under the NLRA, thus subjecting California Cartage to potential labor law violations. Order Consolidating Cases, Consolidated Complaint and Notice of Hearing at 3, Orient Tally Company, Inc. and California Cartage Company, LLC, A Single Employer and Warehouse Workers Resource Center, Cases 21-CA-160242, 21-CA-162991 (NLRB Region 21, Mar. 23, 2016), https://perma.cc/8L8R-NBZ7 (accessed July 20, 2017).

504. City News Service, "Port Truck Drivers to Return to Work after Week-Long Strike," *Long Beach Press-Telegram*, June 6, 2016, http://www.presstelegram.com/article/LB/20160606/NEWS/160609719 (accessed July 20, 2017); Clarissa Hawes, "Striking Port Truckers Protest Ends with Return to Work," *Trucks.com*, June 6, 2016, https://www.trucks.com/2016/06/06/striking-port-truckers-protest-march-ends-return-work/ (accessed July 20, 2017).

505. Hawes, note 504.

506. Order Consolidating Cases, Consolidated Complaint and Notice of Hearing at 5, Intermodal Bridge Transport and International Brotherhood of Teamsters, Cases 21-CA-157647, 21-CA-177303 (NLRB Region 21, July 5, 2016).

507. Order Consolidating Cases, Consolidated Complaint and Notice of Hearing, XPO Cartage, Inc. and International Brotherhood of Teamsters at 3-5, Cases 21-CA-150873, 21-CA-164483 (NLRB Region 21, July 14, 2016).

508. Order Consolidating Cases, Consolidated Complaint and Notice of Hearing at 3–4, XPO Port Services, Inc. and International Brotherhood of Teamsters, Cases 21-CA-150878, 21-CA-163614, 21-CA-169753 (NLRB Region 21, July 26, 2016). This complaint also contained a misclassification as ULP allegation. Ibid., 3. In a sign of the changed politics of labor law enforcement, after the 2016 presidential election, XPO moved to hold the cases against it in abeyance in order to permit President Trump to appoint new, less labor-friendly, NLRB members; the NLRB judge in the case rejected the motion.

509. Motion for Final Approval of Class Action Settlement, Molina v. Pacer Cartage, Inc., No. 13-cv-2344-LAB (JMA) (S.D. Cal., filed June 13, 2016); Order Granting Final Approval of Class Action Settlement, Dismissing Action and Judgment, Molina v. Pacer Cartage, Inc., No. 13-cv-2344-LAB (JMA) (S.D. Cal., filed Oct. 12, 2016). In early 2016, the Kabateck firm had also filed a state class action against Pacer, Harbor Rail Transport, and (the now-defunct) PDS Transportation (also owned by XPO), which was dismissed in September 2016 after the *Molina* settlement had been presented to the court. Another suit was brought under the Private Attorney General Act against XPO, Pacer, and Harbor Rail Transport in May 2016 by the law firm of Harris & Ruble. Law Enforcement Action Complaint, Cortez v. XPO Logistics, Inc., No. BC621798 (Cal. Super. Ct., filed May 26, 2016). That case remains outstanding.

510. It was because of the lead plaintiff's opt out that the case was renamed, from *Mendoza* to *Molina v. Pacer Cartage, Inc.* See Order Amending Case Caption and Setting Status Conference, Molina v. Pacer Cartage, No. 13-cv-2344-LAB (JMA) (S.D. Cal., filed June 28, 2016).

511. Rachel Uranga, "Port Drivers Win $5 Million Class-Action Lawsuit Against Trucking Companies," *Long Beach Press-Telegram*, July 15, 2016, http://www.presstelegram.com/general-news/20160715/port-drivers-win-5-million-class-action-lawsuit-against-trucking-companies (accessed July 20, 2017).

512. Natalie Kitroeff, "Port Truck Drivers Get a Victory; Trucking Firms Will Pay $5 Million to Settle a Lawsuit," *Los Angeles Times*, July 15, 2016, C1. Under the terms of the settlement, which received final court approval in July 2017, approximately 240 drivers received an average award of $13,052.

513. Ibid.

514. Ibid.

515. Complaint for Declaratory and Injunctive Relief, California Trucking Ass'n v. Julie Su, No. 16-CV-1866 CAB MDD (S.D. Cal., filed July 22, 2016), https://perma.cc/WV7X-8BY8 (accessed July 27, 2017).

516. Order Granting Motion to Dismiss, California Trucking Ass'n v. Julie Su, No. 16-CV-1866 CAB MDD (S.D. Cal., filed July 22, 2016), https://perma.cc/CL5H-UD86 (accessed July 27, 2017).

517. "Truckers Sue Labor Commissioner for Public Records Act and Due Process Violations," *Highland Community News*, Jan. 3, 2017, http://www.highlandnews.net/news/political/truckers-sue-labor-commissioner-for-public-records-act-and-due/article_3f0bcd54-d1f7-11e6-a310-e30955a47600.html (accessed July 27, 2017).

518. Kitroeff, note 512. In September, a misclassification class action was filed against California Multimodal. Class Action Complaint, Melendez v. California Multimodal, LLC, No. BC633972 (Cal. Super. Ct., filed Sept. 14, 2016). In the class action against Container Connection, the court certified the class in October 2016. Sheheryar Kaoosji, "Driver Classification Updates, Chassis Shortages Plague Both Coasts and Hanjin Empties," Oct. 25, 2016, Los Angeles Alliance for A New Economy, *Port Innovations*, https://perma.cc/3L5K-HLFL (accessed July 20, 2017).

519. The Teamsters are currently working with Gold Point to effectuate driver reclassification. In the meantime, the parties in the state law class action suit have moved for final settlement approval. Motion for Final Approval of Settlement, Hernandez v. Gold Point Transportation, Inc., No. BC477445 (Cal. Super. Ct., filed Aug. 17, 2017). However, two more suits—one class action by the Gomez Law Group and one Private Attorney General Act suit—have been filed in 2017 and are pending.

520. Daily Labor Report, "California Ports Strike Coincides with Federal Contractor Ruling," *Bloomberg BNA*, Oct. 25, 2016, https://perma.cc/69AZ-57GE (accessed July 20, 2017).

521. Memorandum from Mike Manley, International Brotherhood of Teamsters Legal Department to Law Enforcement Personnel and Management of Companies and Port Terminals, Oct. 25, 2016.

522. Complaint, Gutierrez et al. v. Everport Terminal Services et al., Case No. NC060891 (Cal. Super. Ct., filed Nov. 14, 2016).

523. Charging Party's Post-Hearing Brief at 1, In re: Intermodal Bridge Transport and International Brotherhood of Teamsters, Cases 21-CA-157647, 21-CA-177303 (NLRB Region 21, April 2017).

524. Sheheryar Kaoosji, "Bankruptcies and Amnesty Fuel More Changes to the LA/LB Port Trucking Industry," May 26, 2016, Los Angeles Alliance for a New Economy, *Port Innovations*, https://perma.cc/G32H-287R (accessed July 20, 2017). In addition

Notes

to DLSE claims, there were other enforcement actions against port trucking firms. On April 28, 2016, Metro Worldwide, Inc. owners were arrested for insurance fraud and charged in the Los Angeles County Superior Court. Sheheryar Kaoosji, "Port Trucking Company Owners Arrested in Criminal Case for Insurance," June 2, 2016, Los Angeles Alliance for a New Economy, *Port Innovations*, https://perma.cc/G7W7-FMXR (accessed July 20, 2017) (noting that Metro Worldwide, Inc. had a Clean Truck Program concession). The Metro owners were charged with underreporting payroll from 2009 to 2013 by $4.6 million. Andrew Edwards, "Prosecutors Accuse Owners of Long Beach Trucking Firm with Fraud," *Long Beach Press-Telegram*, Apr. 28, 2016, http://www.presstelegram.com/general-news/20160428/prosecutors-accuse-owners-of-long-beach-trucking-firm-with-fraud (accessed Aug. 17, 2017).

525. "Justice for Port Drivers Campaign, Ports Campaign—Misclassification Work—Southern California, California Labor Commissioner" (DLSE), note 215, 2. Through the end of 2015, the campaign reported over $30 million in awards to drivers, with an average award amount of $142,000. Ibid., 1.

526. Sea-Logix, owned by the Pasha Group, is also under union contract.

527. As of December 2016, Pac 9 was in bankruptcy with its survival "still in question," though if the company continues there is an agreement drivers can access to organize. Telephone interview with Sheheryar Kaoosji, note 107.

528. With both Green Fleet and Pac 9, the Teamsters union is limited in its ability to coordinate with the companies' management in supporting business reorganization without violating NLRA section 8(a)(2), since, without a union contract, the union does not have bargaining representative status.

529. Telephone interview with Nick Weiner, note 14.

530. Telephone interview with Michael Manley, note 132.

531. Ibid.

532. Telephone interview with Sheheryar Kaoosji, note 107.

533. Ibid.

534. Telephone interview with Nick Weiner, note 14.

535. Telephone interview with Sheheryar Kaoosji, note 107.

536. Ibid.

537. Ibid.

538. Ibid.

539. Ibid.

540. Ibid.

541. Telephone interview with Carlos Santamaria, note 161.

542. Ibid.

543. Most recently, the campaign has asked that the ports bar any trucking company that does not recognize its drivers as employees. Rachel Uranga, "Striking Port Truckers Take Their Protest to LA City Hall," *Long Beach Press-Telegram*, June 23, 2017, http://www.presstelegram.com/social-affairs/20170623/striking-port-truckers-take-their-protest-to-la-city-hall (accessed July 28, 2017).

544. In June 2017, 100 drivers conducted a five-day strike targeting three companies: "Connecticut-based XPO Logistics, Chinese-owned Intermodal Transport and California Cartage Company, a trucking and distribution company that leases land from the Los Angeles port." Rachel Uranga, "Striking LA, Long Beach Port Truckers Picket for 2nd Day," *Long Beach Press-Telegram*, June 20, 2017, http://www.presstelegram.com/general-news/20170620/striking-la-long-beach-port-truckers-picket-for-2nd-day (accessed July 28, 2017).

545. Noam Scheiber, "Trump Takes Steps to Undo Obama Legacy on Labor," *New York Times*, June 20, 2017, https://www.nytimes.com/2017/06/20/business/nlrb-trump-labor.html?_r=0 (accessed July 17, 2017).

546. Fielding Buck, "Port Looks at Rail Instead of Trucks to Move Cargo," *Long Beach Press-Telegram*, Apr. 20, 2016, 1.

547. See Port of Long Beach, Clean Trucks, http://www.polb.com/environment/cleantrucks/default.asp (accessed July 20, 2017); The Port of Los Angeles, Air Quality Report Card, 2005–2011, http://www.portoflosangeles.org/pdf/2011_Air_Quality_Report_Card.pdf (accessed July 20, 2017).

548. Neela Banerjee, "New Air Pollution Standards Aim to Reduce Soot Particles," *Los Angeles Times*, Dec. 15, 2012, A15 (describing the Obama administration's new air emission standards seeking reduction of soot from coal-fired power plants and diesel vehicles).

549. Rong-Gong Lin, "Wilmington Celebrates Portside Park's Opening," *Los Angeles Times*, June 5, 2011, A33 (noting creation of $55 million Wilmington Waterfront Park after ten years of planning); Christine Mai-Duc and Laura J. Nelson, "Turning Freeway to Park?," *Los Angeles Times*, Nov. 20, 2013, A1 (describing plan to rip out part of Terminal Island Freeway and convert to green space); Matt Stevens, "Work Is Flashing Forward," *Los Angeles Times*, Oct. 8, 2011, AA3 (noting that the Angels Gate lighthouse was repainted with funding from the China Shipping settlement).

550. Ryan Faughnder, "L.A. Port Numbers Fall, Long Beach's Rise," *Los Angeles Business Journal*, May 15, 2013, http://labusinessjournal.com/news/2013/may/15/l-port-numbers-fall-long-beachs-rise/ (accessed July 20, 2017) (stating that the number of containers through Los Angeles was down 9.5 percent, and up 13 percent in Long Beach, although Los Angeles still had more units overall (640,330 to 519,464)).

Notes

551. Phillips, note 486 (stating that, in 2015, container volume at the Port of Long Beach was 7.2 million TEUs and 8.2 million at Los Angeles).

552. Holmes and Knatz resigned in the fall of 2013. David Zahniser and Dan Weikel, "Executive Who Runs L.A. Port Will Retire," *Los Angeles Times*, Oct. 4, 2013, AA1. Before he did, Holmes indicated that the port was working on an iPhone app to replace the placard struck down by the Supreme Court and was considering eliminating criminal sanctions for tariff violations—the enforcement mechanism stressed by the Supreme Court in justifying its decision that the concession agreement created as part of the Clean Truck Program was a product of government regulation not market participation. Interview with John Holmes, deputy executive director, Port of Los Angeles (July 19, 2013).

553. James Preston Allen, "What's Wrong with Ending PCAC?," *City Watch*, May 20, 2013, http://www.citywatchla.com/index.php/archive/5110-what-s-wrong-with-ending-pcac-why-the-port-has-taken-a-wrong-turn (accessed July 20, 2017) (noting that the Los Angeles harbor commission voted to dissolve the Port Community Advisory Committee on May 2); see also telephone interview with Kathleen Woodfield, member, Sierra Club (May 14, 2013) (stating that the committee, on which Woodfield sat, was dissolved after the port kept adding industry representatives who failed to come, causing the committee to consistently not have a quorum).

554. "Port of Los Angeles Announces Plans to Spend $400 Million on Projects," *Daily Breeze*, June 5, 2013, http://www.dailybreeze.com/general-news/20130606/port-of-los-angeles-announces-plans-to-spend-400-million-on-projects (listing expenditures to electrify terminals, create a more fuel-efficient rail yard, and create a new waterfront promenade).

555. Editorial, "Megaships, When Big Isn't Always Better," *Long Beach Press-Telegram*, Dec. 30, 2015, 13; see also Editorial, "Megaships Are Big Gamble for Ports," *Long Beach Press-Telegram*, Aug. 12, 2016, 10.

556. See, e.g., Karen Robes Meeks, "Ports of Long Beach, Los Angeles Invest Millions to Accommodate Ships," *Long Beach Press-Telegram*, June 7, 2014, http://www.presstelegram.com/business/20140607/ports-of-long-beach-los-angeles-invest-millions-to-accommodate-ships (accessed July 20, 2017) (stating that the ports were "spending $1 million to $2.5 million a day on infrastructure projects to accommodate … megaships").

557. Pier S Marine Terminal and Back Channel Improvements Draft EIS/EIR (Sept. 2011), http://www.polb.com/civica/filebank/blobdload.asp?BlobID=8731 (accessed July 20, 2017).

558. See Louis Sahagun, "Railroads Sued Over Soot," *Los Angeles Times*, Oct. 21, 2011, AA5; Dan Weikel, "Rail Yard Near L.A. Harbor Won't Cut Pollution, Foes Say," *Los Angeles Times*, Oct. 22, 2012, AA1; Dan Weikel, "Rail Yard Near Port Is Approved," *Los Angeles Times*, Mar. 8, 2013, AA1.

559. The Port of Los Angeles, *Southern California International Gateway Project, Final Environmental Impact Report* (Feb. 2013), 1-3, https://www.portoflosangeles.org/EIR/SCIG/FEIR/01_Introduction_SCIG_FEIR.pdf (accessed July 23, 2017).

560. Telephone interview with Morgan Wyenn, staff attorney, Climate & Air Program, Natural Resources Defense Council (Dec. 13, 2016).

561. Ibid.

562. Ibid.

563. Green L.A. Port Working Group, "Stop the SCIG!," https://perma.cc/FDQ2-QXKG (accessed July 20, 2017).

564. Letter from Port Working Group, Green LA Coalition, to Chris Cannon, director of Environmental Management, Port of Los Angeles, Re: Southern California International Gateway Project Draft Environmental Impact Report (Feb. 1, 2011), 1, https://perma.cc/MF5D-C62Z (accessed July 20, 2017).

565. Telephone interview with Morgan Wyenn, note 560.

566. Karen Robes Meeks, "LB Port Rail Project Completed," *Long Beach Press-Telegram*, Sept. 17, 2015, 1.

567. Morgan Wyenn, "Long Beach Port Investing in On-Dock Rail and Zero Emissions Technology," NRDC, *Expert Blog*, Oct. 17, 2012, https://perma.cc/AN67-Y5FX (accessed July 20, 2017).

568. Morgan Wyenn, "Los Angeles Community Asks Warren Buffett for Help to Fight Mega Railyard," NRDC, *Expert Blog*, May 18, 2012, https://perma.cc/25AL-FG6J (accessed July 20, 2017).

569. Morgan Wyenn, "Long Beach Councilmember Johnson Holds a Packed-House Hearing for Community Members to Discuss BNSF's Proposed Los Angeles-area SCIG Rail Yard," NRDC, *Expert Blog*, Nov. 8, 2012, https://perma.cc/G5WT-QF4X (accessed July 20, 2017).

570. Fastline Transportation, Inc. v. City of Los Angeles et al., No. CIVMSN 140300, 2016 WL 4417206, at *2 (Cal. Super. Ct. July 26, 2016).

571. Ibid.

572. Dan Weikel, "Ruling on Rail Yard Is Appealed; Port of L.A., Railroad Seek to Overturn a Judge's Decision that Has Stalled the Project," *Los Angeles Times*, Aug. 3, 2016, B5 ("If built, the 153 acre rail yard will be in Wilmington next to California 103 and east of Alameda Street between Sepulveda Boulevard and Pacific Coast Highway.").

573. Ibid.

574. Editorial, "Cutting Corners at the Port?," *Los Angeles Times*, Apr. 6, 2016, A14.

Notes

575. Weikel, note 572.

576. Telephone interview with Morgan Wyenn, note 560.

577. Karen Robes Meeks, "Clean Shipping Methods Proposed," *Long Beach Press-Telegram*, Jan. 22, 2014, 3; see also California Cleaner Freight Coalition, *Moving California Forward: Zero and Low-Emissions Freight Pathways* (2014), https://perma.cc/PMF7-LHUK (accessed July 20, 2017).

578. Morgan Wyenn, "Southern California's Air Agency to Decide Whether to Regulate Ports, or Just Let the Ports 'Agree' to Clean Up," NRDC, *Expert Blog*, Feb. 1, 2013, https://perma.cc/H6U6-2NVQ (accessed July 20, 2017).

579. During the Obama administration, advocates and state regulators also lobbied the EPA to promote near-zero emissions for trucks, many purchased out of state, but received predictable pushback from the diesel industry. The diesel industry has remained a holdout in California, which has the fewest new trucks on the road meeting the 2010 EPA standard for emissions. Steve Scauzillo, "What's at Stake in Southern California's New War on Diesel Truck Pollution," *San Gabriel Valley Tribune*, June 12, 2016, https://perma.cc/82LX-VP6K (accessed July 20, 2017).

580. Telephone interview with Melissa Lin Perrella, co-director, Environmental Justice Program, Natural Resources Defense Council (Dec. 15, 2016).

581. Morgan Wyenn, "Enforcement of Key Maintenance Provision Lacking under Port of LA's Clean Truck Program," NRDC, *Expert Blog*, July 23, 2014, https://perma.cc/U84T-U7GY (accessed July 20, 2017). The truck maintenance provision of the Los Angeles Clean Truck Program was upheld by the Ninth Circuit and not challenged by the American Trucking Associations in its petition for review by the Supreme Court.

582. Tony Barboza, "Hahn Calls for Oversight of Port," *Los Angeles Times*, Dec. 17, 2015, B5; see also Tony Barboza, "Better Oversight of Port Operations Urged," *Los Angeles Times*, Dec. 16, 2015, B2 (stating that China Shipping violated eleven out of fifty-two environmental standards in the course of its 2011 expansion). The environmental violations began in 2009 and persisted into 2014, despite the fact that Los Angeles was hailed globally as a model green container terminal.

583. Tony Barboza, "Pollution at the Port of L.A.; A Broken Pledge," *Los Angeles Times*, Dec. 15, 2015, B1.

584. Ibid.

585. Barboza, "Hahn Calls for Oversight of Port," note 582.

586. Tony Barboza and Jack Dolan, "Port Terminal Misses Mark on Air Pollution," *Los Angeles Times*, Feb. 2, 2016, B1.

587. Jack Dolan and Tony Barboza, "Port of L.A.'s Bad Bet," *Los Angeles Times*, March 24, 2016, B1.

588. Ibid., B6.

589. Ibid.

590. "Cutting Corners at the Port?," note 574 ("The Port of L.A. is already under scrutiny after officials disclosed that they'd let two major shipping companies skirt rules to cut dirty diesel pollution from their operations. The court ruling raises concerns that city and harbor officials are again cutting corners on environmental issues and public health.").

591. Rachel Uranga, "Brown Pushes 'Green' Shipping—Governor Aims to Make Freight-Moving Industry Run Cleaner," *Long Beach Press-Telegram*, July 30, 2016, 1.

592. Ibid.

593. Ibid.

594. "Comments Released on Clean Air Action Plan Update," *The Port of Los Angeles* (Feb. 7, 2017), https://perma.cc/Y9FJ-HNF4 (accessed July 23, 2017).

595. Telephone interview with Morgan Wyenn, note 560.

596. Morgan Wyenn, "LA and Long Beach Ports Adopt Clean Ship Programs in Hopes to Reduce Air Pollution," NRDC, *Expert Blog*, May 10, 2012, https://perma.cc/Q94Q-MM9W (accessed July 23, 2017).

597. Morgan Wyenn, "Port of Long Beach to Test Promising Technology to Reduce Ship Emissions," NRDC, *Expert Blog*, Mar. 20, 2014, https://perma.cc/L3L7-5X7J (accessed July 23, 2017).

598. Rachel Uranga, "Garcetti Unveils New Push to Cut Pollution—Mayor Announces Freight Advisory Panel as Port Project to Create a Near-Zero Emissions Terminal Gets Underway," *Long Beach Press-Telegram*, July 13, 2016, 1.

599. Ibid.

600. Ibid.

601. Ibid.

602. According to *Port Innovations*, the status of the Los Angeles and Long Beach port complex as the cleanest in the nation is being challenged by legislative efforts to ban old, dirty trucks from the Port of New York and New Jersey. Listserv Email from PortInnovations@LAANE.org to Scott Cummings, professor, UCLA School of Law, Re: NJ Bill Will Require Cleaner trucks than at LA/Long Beach Ports (Sept. 16, 2016); see also note 642 and corresponding text describing the legislation.

603. San Pedro Bay Ports, Clean Air Action Plan 2017, DRAFT Discussion Document (Nov. 2016), 9, http://www.cleanairactionplan.org/wp-content/uploads/2016/11/CAAP-2017-Draft-Discussion-Document-FINAL.pdf (accessed July 23, 2017).

604. Senate B. 1, § 18, 2017–2018 Leg., Reg. Sess. (Cal. 2017).

605. San Pedro Bay Ports, *Clean Air Action Plan 2017, Draft Final Clean Air Action Plan Update* (July 2017), 10.

606. Ibid., 32.

607. Sheheryar Kaoosji, "CA Labor Commissioner Ruling and China Shipping Settlement Could Help Ports Get Ready for Mega-Ships," Los Angeles Alliance for a New Economy, *Port Innovations*, Jan. 4, 2016, https://perma.cc/7JCL-T3HG (accessed July 23, 2017).

608. Donna Littlejohn, "Pollution Level at L.A. Port Sees Big Drop over Decade," *Long Beach Press-Telegram*, Oct. 2, 2015, 1.

609. Andrew Khouri and James F. Peltz, "Hanjin Bankruptcy Disrupts Shipping at L.A. and Long Beach Ports," *Los Angeles Times*, Sept. 1, 2016, http://www.latimes.com/business/la-fi-hanjin-bankruptcy-20160901-snap-story.html (accessed July 23, 2017).

610. Leslie Shaffer, "Shipshape? What You Need to Know About Hanjin's Troubles," *CNBC.com*, Sept. 2, 2016, http://www.cnbc.com/2016/09/02/shipshape-what-you-need-to-know-about-hanjins-troubles.html?utm_source=Port+Innovations+9.30.16&utm_campaign=Port+Innovations+%231&utm_medium=email (accessed July 23, 2017).

611. Ibid.

612. Jim Puzzanghera and Andrew Khouri, "Funds Sought to Ease Cargo Crisis; Hanjin's Parent Plans to Raise $90 Million to Pay for the Unloading of Goods from Ships," *Los Angeles Times*, Sept. 7, 2016, C1.

613. Matt Stevens, "End to Region's Port Crisis is Sought; Lawmakers Call for Flow of Cargo to the Southland to Resume after Shipping Giant's Bankruptcy Filing," *Los Angeles Times*, Sept. 5, 2016, B1; see also Puzzanghera and Khouri, note 612 ("Southern California officials are worried that Hanjin Shipping's bankruptcy could threaten local union workers' jobs and retailers' access to important goods as the holiday season approaches.").

614. Stevens, note 613; see also Costas Paris and In-soo Nam, "Hanjin Shipping Bankruptcy Unlikely to Ease Glut of Vessels; Experts Say Boost in Shipping Rates Will Be Short-Lived," *Wall Street Journal*, Sept. 3, 2016, https://www.wsj.com/articles/hanjin-shipping-bankruptcy-further-consolidation-unlikely-to-help-industrys-capacity-glut-1472841698 (accessed July 23, 2017) ("Not knowing whether they

would get paid, ports and handlers from South Korea to China, the U.S., Canada, Spain and elsewhere have refused to handle [Hanjin Shipping's] cargo.").

615. Stevens, note 613; see also Rachel Urgana, "7 Questions Key Players in the Hanjin Mess Hope Judge Will Answer," *Long Beach Press-Telegram*, Sept. 8. 2016, http://www.presstelegram.com/business/20160908/7-questions-key-players-in-hanjin-mess-hope-judge-will-answer (accessed July 23, 2017).

616. Tom Corrigan, "Hanjin Shipping to Pay Handlers to Unload U.S.-Bound Ships; Cargo Had Been Stranded at Sea Since Hanjin Bankruptcy," *Wall Street Journal*, Sept. 9, 2016, http://www.wsj.com/articles/hanjin-shipping-to-pay-handlers-to-unload-u-s-bound-ships-1473436745 (accessed July 23, 2017) (stating the U.S. Bankruptcy Court approved bankruptcy protection on an interim basis under chapter 15, which gave the South Korean company the legal protections necessary to unload ships at U.S. ports).

617. Rachel Urgana, "Long Beach and L.A. Ports—Marshals Seize Stranded Hanjin Container Ship," *Long Beach Press-Telegram*, Sept. 16, 2016, 1; see also Bill Mongelluzzo, "Hanjin Ship Call at Long Beach Lifts U.S. Importer Hopes," *JOC.com*, Sept. 12, 2016, http://www.joc.com/maritime-news/container-lines/ckyhe-alliance/hanjin-ship-call-long-beach-lifts-us-importer-hopes_20160910.html?destination=node/3302721 (accessed July 23, 2017).

618. Rachel Urgana, "Hanjin Ship Begins Unloading Cargo at Port of Long Beach," *Long Beach Press-Telegram*, Sept. 10, 2016, http://www.presstelegram.com/business/20160910/hanjin-ship-begins-unloading-cargo-at-port-of-long-beach (accessed July 23, 2017).

619. Natalie Kitroeff, "Hanjin Bankruptcy Is Tip of the Iceberg; Supply of Ships and Their Capacity Is Out of Whack with Demand," *Los Angeles Times*, September 18, 2016, C1.

620. Ibid.

621. Natalie Kitroeff, Q&A, "The Port of L.A. Is Having a Dramatic Year. Here's What Its Chief Sees in Its Future," *Los Angeles Times*, Oct. 7, 2016, http://www.latimes.com/business/la-fi-port-trade-shipping-20160928-snap-story.html (accessed July 23, 2017); see also In-Soo Nam, "Hanjin Shipping's Asia-U.S. Route Assets to Be Put on Sale," *Wall Street Journal*, Oct. 13, 2016, http://www.wsj.com/articles/hanjin-shippings-asia-u-s-route-assets-to-be-put-on-sale-1476329135?mod=djemlogistics&utm_source=Copy+of+Port+Innovations+9.30.16&utm_campaign=Port+Innovations+%231&utm_medium=email (accessed July 23, 2017).

622. Costas Paris, "Hanjin in Talks to Sell Stake in Long Beach Terminal," *Wall Street Journal*, Oct. 21, 2016, http://www.wsj.com/articles/hanjin-in-talks-to-sell-stake-in-long-beach-terminal-1477058872 (accessed July 23, 2017) ("Hanjin [owns a] 54% stake in Total Terminals International LLC, which runs Long Beach Terminal."); see

also Listserv Email from PortInnovations@LAANE.org to Scott Cummings, professor, UCLA School of Law, Re: Hanjin Bankruptcy Throws Ports into Chaos; Is Supply Chain Consolidating? (Sept. 30, 2016).

623. Rachel Uranga, "Long Beach Port Cargo Surges; Hanjin Crisis Finally Fading?," *Long Beach Press-Telegram*, Feb. 9, 2017, http://www.presstelegram.com/business/20170209/long-beach-port-cargo-surges-hanjin-crisis-finally-fading (accessed July 17, 2017).

624. Kirkham and Khouri, note 410.

625. Editorial, "Megaships Are a Big Gamble for Ports," *Long Beach Press-Telegram*, Aug. 12, 2016, 10.

626. Kirkham and Khouri, note 410.

627. Kitroeff, note 619 ("The goal was to shore up profits by doing business on a larger scale as global trade bounced back after the recession. But the new business never came. Freight rates dropped and shippers' revenues plunged."); see also Mark Szakonyi, "Shippers Look Deeper than Carrier Losses to Avoid Next Hanjin," *JOC.com*, Oct. 22, 2016, http://www.joc.com/maritime-news/container-lines/hanjin-shipping/shippers-look-deeper-carrier-losses-avoid-next-hanjin_20161022.html (accessed July 23, 2017).

628. Patrick J. McDonnell and Natalie Kitroeff, "Panama Canal Overhaul Could Reshape Global Trade Routes," *Los Angeles Times*, June 24, 2016, A1.

629. Hugh R. Morley, "NY-NJ Terminal Doubling Capacity Ahead of Bridge Raising," *JOC.com*, Feb. 7, 207, http://www.joc.com/port-news/terminal-operators/port-newark-container-terminal/ny-nj-terminal-doubling-capacity-ahead-bridge-raising_20170207.html (accessed July 23, 2017); see also Natalie Kitroeff, "Sea of Rivals; Even as L.A. and Long Beach Post Sharp Gains in Cargo Traffic, Competitors Across U.S. Work to Lure Shipping Lines," *Los Angeles Times*, Apr. 27, 2016, C1.

630. Watt, note 414.

631. Karen Robes Meeks, "Life on the Road—Truckers Hit Hard by Port Tie-Ups—Megaships and Other Changes Causing Bottlenecks Leave Haulers Struggling to Stay Afloat," *Long Beach Press-Telegram*, May 18, 2015, 1, 5 ("This particular problem is so bad that Jon Slangerup, the head of the Long Beach port who once ran Canada's FedEx logistics, compared finding cargo to picking different colored eggs from an Easter basket.").

632. Bill Mongelluzzo, "LA-LB Hanjin Box Congestion Worsens," *JOC.com*, Oct. 10, 2016, http://www.joc.com/maritime-news/container-lines/hanjin-shipping/la-lb-hanjin-box-congestion-worsens_20161010.html (accessed July 23, 2017); see also Sheheryar Kaoosji, "Driver Classification Updates, Chassis Shortages Plague Both Coasts and Hanjin Empties," Oct. 25, 2016, Los Angeles Alliance for a New

Economy, *Port Innovations*, https://perma.cc/VR4E-UBS5 (accessed August 25, 2017) ("The aftermath of Hanjin's bankruptcy continues to plague Southern California as congestion at the Ports of Los Angeles and Long Beach worsens because a solution to clearing the estimated 20,000 empty Hanjin Containers remains elusive.").

633. Rachel Uranga, "Hanjin Shipping Collapse Leaves Up to 15,000 Cargo Containers Piled Up," *Long Beach Press-Telegram*, Sept. 30, 2016, http://www.presstelegram.com/business/20160930/hanjin-shipping-collapse-leaves-up-to-15000-cargo-containers-piled-up (accessed July 23, 2017).

634. Rachel Uranga, "Trucker Turn Times Ease Up—PierPass Appointment System Seeing Success," *Long Beach Press-Telegram*, July 14, 2016, 3.

635. Ibid.; see also Rachel Uranga, "Trucking Program Oversight Bill Pulled," *Long Beach Press-Telegram*, July 7, 2016, 3. According to *Port Innovations*: "Critics of appointment systems point to the fact that if drayage companies and drivers are the only ones with 'skin in the game' to comply with these systems, there will be little motivation for other stakeholders to change their behavior, leaving the drayage community to yet again bear the increased costs in the supply chain." Sheheryar Kaoosji, "Do Truck Appointment Systems Decrease Turn Times? Preliminary Data Shows No Correlation," Mar. 11, 2016, Los Angeles Alliance for a New Economy, *Port Innovations*, https://perma.cc/3NVH-W5EU (accessed July 23, 2017).

636. Uranga, note 634.

637. Buck, note 546, 1. A single double-stacked train is roughly equivalent to 750 trucks hauling the same volume. Ibid.

638. Board of Port Commissioners, City of Oakland, Port Ordinance 4112 (adopted Oct. 20, 2009).

639. See Rexford B. Sherman, *Seaport Governance in the United States and Canada*, http://www.aapa-ports.org/files/PDFs/governance_uscan.pdf (accessed July 23, 2017).

640. Joseph Bonney, "Port of NY-NJ Plans Expansion of Clean-Trucks Program," *JOC.com*, Feb. 10, 2015, http://www.joc.com/port-news/us-ports/port-new-york-and-new-jersey/port-ny-nj-plans-expansion-clean-trucks-program_20150210.html (accessed July 23, 2017).

641. Joseph Bonney, "Revised NY-NJ Clean-Trucks Plan Stirs Up Controversy," *JOC.com*, Jan. 14, 2016, http://www.joc.com/port-news/us-ports/port-new-york-and-new-jersey/revised-ny-nj-clean-trucks-plan-stirs-controversy_20160114.html (accessed July 23, 2017).

642. Senate B. 2507, 2016-2017 Leg., Reg. Sess. (N.J. 2016).

643. Telephone interview with Melissa Lin Perrella, note 580.

Notes

644. Coral Davenport and Eric Lipton, "Scott Pruitt Is Carrying Out His E.P.A. Agenda in Secret, Critics Say," *New York Times*, Aug. 11, 2017, A1.

645. International Brotherhood of Teamsters, press release, "Drivers at XPO Logistics in Aurora, Illinois Choose Teamster Representation" (Oct. 13, 2016), https://perma.cc/5PDB-X9UC (accessed July 23, 2017); International Brotherhood of Teamsters, press release, "XPO Logistics Drivers in Philadelphia Seek Teamster Representation" (Oct. 11, 2016), https://perma.cc/8SN6-5UHA (accessed July 23, 2017).

646. "New York State Takes Aim at Worker Classification: The Commercial Goods Transportation Industry Fair Play Act," *Proskauer.com*, http://www.proskauer.com/publications/client-alert/new-york-state-takes-aim-at-worker-misclassification/ (accessed July 23, 2017); see also Thomas R. Revnew, "The Push to Correct the Misclassification of Commercial Truck Drivers," *Federal Lawyer* 62 (2015): 16 (discussing the New York State Commercial Goods Transportation Industry Fair Play Act, which presumes drivers are employees unless the employer can meet a test establishing that drivers qualify as a separate business entity).

647. Robbie Whelan, "Trucking Industry, Unions Clash Over Driver Classification in New Jersey," *Wall Street Journal*, June 8, 2016, http://www.wsj.com/articles/trucking-industry-unions-clash-over-driver-classification-in-new-jersey-1465397048 (accessed July 23, 2017).

648. Rachel Uranga, "Port Truck Drivers, Gig-Economy Workers Have Much in Common," *Los Angeles Daily News*, Oct. 29, 2016, http://www.dailynews.com/business/20161029/port-truck-drivers-gig-economy-workers-have-much-in-common (accessed July 23, 2017).

649. National Employment Law Project, Independent Contractor Misclassification Imposes Huge Costs on Workers and Federal and State Treasuries, Fact Sheet, July 2015, 1.

650. Mike Isaac, "Judge Overturns Uber's Settlement with Drivers," *New York Times*, Aug. 19, 2016, B6; David Streitfield, "Uber Drivers Inch Forward in Seeking Court Status," *New York Times*, July 13, 2017, B8.

651. Noam Scheiber, "Uber Drivers Ruled Eligible for Jobless Payments in New York State," *New York Times*, Oct. 12, 2016, http://www.nytimes.com/2016/10/13/business/state-rules-2-former-uber-drivers-eligible-for-jobless-payments.html (accessed July 23, 2017).

652. Ibid.; see also Chris Roberts, "Another Uber Driver Awarded Unemployment Benefits," *SF Weekly*, Mar. 4, 2016, http://archives.sfweekly.com/thesnitch/2016/03/04/uber-driver-awarded-unemployment-benefits-first-known-case-in-state (accessed July 23, 2017).

653. Erica E. Phillips, "Delivery Drivers Sue Amazon, Alleging Violation of Labor Laws," *Wall Street Journal*, Oct. 6, 2016, http://www.wsj.com/articles/delivery-drivers

-sue-amazon-alleging-violation-of-labor-laws-1475714956 (accessed July 23, 2017) ("The lawsuit comes as Amazon is laying the groundwork for its own shipping business, which its executives have said will add delivery capacity.").

654. Sheheryar Kaoosji, "'Gig' Economy Leader Signaling It Wants More Control Over Supply Chain," Feb. 1, 2016, Los Angeles Alliance for a New Economy, *Port Innovations*, https://perma.cc/N62C-HN9W (accessed July 23, 2017).

655. Mike Esterl, "FedEx Agrees to $240 Million Settlement with Drivers in 20 States," *Wall Street Journal*, June 16, 2016, https://www.wsj.com/articles/fedex-agrees-to-240-million-settlement-with-drivers-in-20-states-1466123381 (accessed July 23, 2017).

656. See Jennifer Pinsof, "A New Take on an Old Problem: Employee Misclassification in the Modern Gig-Economy," *Michigan Telecommunications & Technology Law Review* 22 (2016): 341 (arguing that courts should adopt the "ABC Test" used in state unemployment insurance cases to promote consistency in employee determinations and create a presumption in favor of employee status). For additional scholarly treatments of misclassification, see David Bauer, "The Misclassification of Independent Contractors: The Fifty-Four Billion Dollar Problem," *Rutgers Journal of Law & Public Policy* 12 (2015): 138; Peter Tran, "The Misclassification of Employees and California's Latest Confusion Regarding Who Is an Employee or an Independent Contractor," *Santa Clara Law Review* 56 (2016): 677.

Chapter 7

1. Brian K. Obach, *Labor and the Environmental Movement: The Quest for Common Ground* (Cambridge, MA: The MIT Press, 2004).

2. See Marco G. Giugni, "Was It Worth the Effort? The Outcomes and Consequences of Social Movements," *Annual Review of Sociology* 24 (1998): 371; see also *The Consequences of Social Movements*, ed. Lorenzo Bosi, Marco Giugni, and Katrin Uba (New York: Cambridge University Press, 2016); Donatella della Porta and Mario Diani, *Social Movements: An Introduction* (Malden, MA: Blackwell Publishing, 2d ed. 2006), 226–239. For classic definitions of movement outcomes, see William A. Gamson, *The Strategy of Social Protest* (Homewood, IL: The Dorsey Press, 1975), 28–29; J. Craig Jenkins, "Resource Mobilization Theory and the Study of Social Movements," *Annual Review of Sociology* 9 (1983): 527, 544.

3. See Robert Gottlieb, *Reinventing Los Angeles: Nature and Community in the Global City* (Cambridge, MA: The MIT Press, 2007); Robert Gottlieb, Mark Vallianatos, Regina M. Freer, and Peter Dreier, *The Next Los Angeles: The Struggle for a Livable City* (Berkeley: University of California Press, 2005); Edward W. Soja, *Seeking Spatial Justice* (Minneapolis: University of Minnesota Press, 2010); *Working for Justice: The L.A. Model of Organizing and Advocacy*, ed. Ruth Milkman, Joshua Bloom, and Victor Narro (Ithaca, NY: Cornell University Press, 2010).

Notes

4. See Catherine Albiston, "The Dark Side of Litigation as a Social Movement Strategy," *Iowa Law Review Bulletin* 96 (2011): 61.

5. Richard C. Schragger, "Is a Progressive City Possible? Reviving Urban Liberalism for the Twenty-First Century," *Harvard Law & Policy Review* 7 (2013): 231.

6. Katherine V. W. Stone and Scott L. Cummings, "Labor Activism in Local Politics: From CBAs to 'CBAs'," in *The Idea of Labour Law*, ed. Guy Davidov and Brian Langille (Oxford: Oxford University Press, 2011), 273.

7. On law making from below, see David Cole, *Engines of Liberty: The Power of Citizen Activists to Make Constitutional Law* (New York: Basic Books, 2016).

8. See David S. Meyer, *The Politics of Protest: Social Movements in America* (New York: Oxford University Press, 2007), 82; Lee A. Smithey, "Social Movement Strategy, Tactics, and Collective Identity," *Sociology Compass* 3 (2009): 658.

9. Jeff A. Larson, "Social Movements and Tactical Choice," *Sociology Compass* 7 (2013): 866.

10. David A. Snow, E. Burke Rochford, Jr., Steven K. Worden, and Robert D. Benford, "Frame Alignment Processes, Micromobilization, and Movement Participation," *American Sociological Review* 51 (1986): 464.

11. David A. Snow and Robert D. Benford, "Master Frames and Cycles of Protest," in *Frontiers in Social Movement Theory*, ed. Aldon D. Morris and Carol McClurg Mueller (New Haven, CT: Yale University Press, 1992), 133.

12. Martha Davis, "Law, Issue Frames and Social Movements: Three Case Studies," *University of Pennsylvania Journal of Law and Social Change* 14 (2011): 363; see also Amy Kapczynski, "The Access to Knowledge Mobilization and the New Politics of Intellectual Property," *Yale Law Journal* 117 (2008): 804.

13. Della Porta and Diani, note 2, 193–222.

14. Chris Hilson, "New Social Movements: The Role of Legal Opportunity," *Journal of European Public Policy* 9 (2002): 238.

15. Lisa Vanhala, "Legal Opportunity Structures and the Paradox of Legal Mobilization by the Environmental Movement in the UK," *Law & Society Review* 46 (2012): 523.

16. 49 U.S.C. § 14501(c)(1).

17. See Orly Lobel, "The Paradox of Extralegal Activism: Critical Legal Consciousness and Transformative Politics," *Harvard Law Review* 120 (2007): 937.

18. The two-decade trend of recess appointments and long-standing vacancies has become the new normal at the NLRB. William B. Gould IV, "Politics and the Effect on the National Labor Relations Board's Adjudicative and Rulemaking Processes,"

Emory Law Journal 64 (2015): 1501, 1520–1527 (arguing that the degree of political interference is partially attributable to Congress's desire to influence the NLRB).

19. The composition of the NLRB and its prioritization of employee misclassification set the stage in the last phase of the Obama administration for robust labor enforcement that underwrote campaign progress on the misclassification issue. See Noam Scheiber, "As His Term Wanes, Obama Champions Workers' Rights," *New York Times*, Aug. 31, 2015, http://www.nytimes.com/2015/09/01/business/economy/as-his-term-wanes-obama-restores-workers-rights.html (accessed Dec. 18, 2017) ("With little fanfare, the Obama administration has been pursuing an aggressive campaign to restore protections for workers that have been eroded by business activism, conservative governance and the evolution of the economy in recent decades."); see also James Rainey, "A Strong Shift for Truckers; In a 'Historic' Action, Drivers Are Recognized as Employees Subject to Labor Protections," *Los Angeles Times*, Mar. 22, 2014, A1.

20. Marc Galanter, "Why the 'Haves' Come Out Ahead: Speculations on the Limits of Legal Change," *Law & Society Review* 9 (1974): 95, 100.

21. Scott L. Cummings, "Hemmed In: Legal Mobilization in the Los Angeles Anti-Sweatshop Movement," *Berkeley Journal of Employment and Labor Law* 30 (2009): 1, 61.

22. See Joel F. Handler, *Social Movements and the Legal System: A Theory of Law Reform and Social Change* (New York: Academic Press, 1978), 191–222.

23. Jon B. Gould and Scott Barclay, "Mind the Gap: The Place of Gap Studies in Sociolegal Scholarship," *Annual Review of Law and Social Science* 8 (2012): 323.

24. Scott L. Cummings, "Empirical Studies of Law and Social Change: What Is the Field? What Are the Questions?," *Wisconsin Law Review* (2013): 171, 187–193.

25. Am. Trucking Ass'ns, Inc. v. City of L.A., 660 F.3d 384, 408 (9th Cir. 2011).

26. See Benjamin I. Sachs, "Despite Preemption: Making Labor Law in Cities and States," *Harvard Law Review* 124 (2011): 1153.

27. See Doug McAdam, Sidney Tarrow, and Charles Tilley, *Dynamics of Contention* (New York: Cambridge University Press, 2001); Sidney Tarrow, *Power in Movement: Social Movements, Collective Action and Politics* (New York: Cambridge University Press, 1994).

28. See Michael C. Dorf and Sidney Tarrow, "Strange Bedfellows: How an Anticipatory Countermovement Brought Same-Sex Marriage into the Public Arena," *Law & Social Inquiry* 39 (2014): 449.

29. Armendariz v. Foundation Health Psychcare Services, Inc., 24 Cal. 4th 83 (2000). The validity of *Armendariz* is in question after the Supreme Court's decision in AT&T Mobility LLC v. Concepcion, 563 U.S. 333 (2011), holding that the Federal Arbitra-

Notes

tion Act preempted California's contract law rule that class arbitration waivers in consumer contracts were unconscionable.

30. See Meyer, note 8, 73–79.

31. See Scott L. Cummings, "Law in the Labor Movement's Challenge to Wal-Mart: A Case Study of the Inglewood Site Fight," *California Law Review* 95 (2007): 1927, 1991–1997.

32. Bldg. & Constr. Trades Council v. Associated Builders & Contractors of Mass./ R.I., Inc., 507 U.S. 218 (1993).

33. Engine Mfrs. Ass'n v. S. Coast Air Quality Mgmt. Dist., No. Cv00-09065FMC(BQRX), 2005 WL 1163437, at *12 (C.D. Cal. May 5, 2005), *aff'd* Engine Mfrs. Ass'n v. S. Coast Air Quality Mgmt. Dist., 498 F.3d 1031, 1041 (9th Cir. 2007).

34. See Katherine Stovel and Lynette Shaw, "Brokerage," *Annual Review of Sociology* 38 (2012): 141.

35. David Luban, *Lawyers and Justice: An Ethical Study* (Princeton, NJ: Princeton University Press, 1988), 319.

36. Derrick A. Bell, Jr., "Serving Two Masters: Integration Ideals and Client Interests in School Desegregation Litigation," *Yale Law Journal* 85 (1976): 470.

37. Ibid., 504.

38. See Jennifer Gordon, *Suburban Sweatshops: The Fight for Immigrant Rights* (Cambridge, MA: Harvard University Press, 2007), 185–236.

39. Luke W. Cole and Sheila R. Foster, *From the Ground Up: Environmental Racism and the Rise of the Environmental Justice Movement* (New York: NYU Press, 2001).

40. See Tomiko Brown-Nagin, *Courage to Dissent: Atlanta and the Long History of the Civil Rights Movement* (New York: Oxford University Press, 2011); Susan D. Carle, *Defining the Struggle: National Organizing for Racial Justice, 1880–1915* (New York: Oxford University Press, 2013); Anthony V. Alfieri, "Faith in Community: Representing 'Colored Town,'" *California Law Review* 95 (2007): 1829; Sameer M. Ashar, "Public Interest Lawyers and Resistance Movements," *California Law Review* 95 (2007): 1879; Lynette J. Chua, "Pragmatic Resistance, Law, and Social Movements in Authoritarian States: The Case of Gay Collective Action in Singapore," *Law & Society Review* 46 (2012): 713; Lani Guinier and Gerald Torres, "Changing the Wind: Notes Toward a Demosprudence of Law and Social Movements," *Yale Law Journal* 123 (2014): 2740.

41. Michael W. McCann, *Rights at Work: Pay Equity Reform and the Politics of Legal Mobilization* (Chicago: University of Chicago Press, 1994), 5–12.

42. Stuart A. Scheingold, *The Politics of Rights: Lawyers, Public Policy, and Political Change* (New Haven, CT: Yale University Press, 1974), 214.

43. See Sally Engle Merry, Peggy Levitt, Mihaela Serban Rosen, and Diana H. Yoon, "Law from Below: Women's Human Rights and Social Movements in New York City," *Law & Society Review* 44 (2010): 101.

44. See Michael J. Klarman, *From Jim Crow to Civil Rights: The Supreme Court and the Struggle for Racial Equality* (New York: Oxford University Press, 2004); Gerald N. Rosenberg, *The Hollow Hope: Can Courts Bring About Social Change?* (Chicago: University of Chicago Press, 1991); Scheingold, note 42.

45. See generally Scott L. Cummings, "Movement Lawyering," *University of Illinois Law Review* (2017): 1645.

46. Gordon, note 38; Scott L. Cummings, "Litigation at Work: Defending Day Labor in Los Angeles," *UCLA Law Review* 58 (2011): 1617; Douglas NeJaime, "The Legal Mobilization Dilemma," *Emory Law Journal* 61 (2012): 663.

47. Austin Sarat and Stuart Scheingold, "What Cause Lawyers Do *For*, and *To*, Social Movements: An Introduction," in *Cause Lawyers and Social Movements*, ed. Austin Sarat and Stuart A. Scheingold (Stanford, CA: Stanford University Press, 2006), 1.

48. Charles F. Sabel and William H. Simon, "Destabilization Rights: How Public Law Litigation Succeeds," *Harvard Law Review* 117 (2004): 1015.

49. Ibid., 1077.

50. Scott L. Cummings and Steven A. Boutcher, "Mobilizing Local Government Law for Low-Wage Workers," *The University of Chicago Legal Forum* (2009): 187.

51. See Human Rights Campaign, Cities and Counties with Non-Discrimination Ordinances that Include Gender Identity, https://www.hrc.org/resources/cities-and-counties-with-non-discrimination-ordinances-that-include-gender (accessed Oct. 9, 2017); National Employment Law Project, Local Living Wage Laws and Coverage, updated July 2011, http://www.nelp.org/content/uploads/2015/03/LocalLWLaws CoverageFINAL.pdf (accessed Oct. 9, 2017); UC Berkeley Labor Center, Inventory of US City and County Minimum Wage Ordinances, updated Dec. 16, 2017, http://laborcenter.berkeley.edu/minimum-wage-living-wage-resources/inventory-of-us-city-and-county-minimum-wage-ordinances/ (accessed Oct. 9, 2017).

52. Ruth Milkman, *L.A. Story: Immigrant Workers and the Future of the U.S. Labor Movement* (New York: Russell Sage Foundation, 2006).

53. Stone and Cummings, note 6.

54. See George I. Lovell, Michael McCann, and Kirstine Taylor, "Covering Legal Mobilization: A Bottom-Up Analysis of *Wards Cove v. Atonio*," *Law & Social Inquiry* 41 (2016): 61; Reva B. Siegel, "Constitutional Culture, Social Movement Conflict and Constitutional Change: The Case of the De Facto ERA," *California Law Review* 94 (2006): 1323.

55. For a powerful argument for culture shifting, see Guinier and Torres, note 40.

56. See William E. Forbath, *Law and the Shaping of the American Labor Movement* (Cambridge, MA: Harvard University Press, 1991).

57. Richard Schragger, *City Power: Urban Governance in a Global Age* (New York: Oxford University Press, 2016), 16.

58. Gerald P. López, "Lay Lawyering," *UCLA Law Review* 32 (1984): 1.

59. Richard Kluger, *Simple Justice: The History of* Brown v. Board of Education *and Black America's Struggle for Equality* (New York: Vintage Books, 1977).

60. "Don't Waste LA Facts and Impacts," http://www.dontwastela.com/wp-content/uploads/2014/08/DWLAFactsandImpacts.pdf (accessed Dec. 18, 2017); see also Sabrina Bornstein, *Don't Waste L.A.: A Path to Green Jobs, Clean Air and Recycling for All* (Jan. 2011), http://www.dontwastela.com/wp-content/uploads/2013/06/DWLA_Report_Finalweb1.pdf (accessed Dec. 18, 2017).

61. Emily Alpert Reyes, "L.A. Council Overhauls Trash Collection for Business, Big Apartments," *Los Angeles Times*, Apr. 1, 2014, http://www.latimes.com/local/lanow/la-me-ln-trash-franchise-overhaul-20140401-story.html-ixzz2xgQePq3n (accessed Dec. 18, 2017); Soumya Karlamangla, "L.A. Opens Bidding for Firms to Serve New Trash Collection Zones," *Los Angeles Times*, June 11, 2014, http://www.latimes.com/local/lanow/la-me-ln-trash-collection-overhaul-20140611-story.html (accessed Dec. 18, 2017).

Index

Accountability in social movements, 4, 7, 15, 322, 325–333
Aerospace industry, 18, 32, 60, 67, 68
AFL-CIO, 11, 73, 106
AGENDA, 107
Agglomeration economy, 43
Air quality management districts,
  regulation of fixed sources of air pollution, 56
  South Coast Air Quality Management District, 58, 65, 77, 78, 90, 91, 92, 94, 95, 97, 100, 101, 110, 120, 134, 158, 280, 288, 325
Alameda Corridor Railroad. *See* Railroads
Albillo v. Intermodal Container Services, 71–72, 189, 304
Alhambra, 38
Allen, Donald, 69
Altshuler Berzon, 75
Amalgamated Textile Workers Union, 127
Amazon,
  class action against, 290
  warehouse, 257, 341
Ambulatory picketing, 221, 222, 226, 228, 229, 238, 255, 273, 310, 321, 340
American Federation of Government Employees Union, 128
American Federation of Labor, 28, 30
American Logistics, 221, 226, 228, 310
American Trucking Associations,
  Intermodal Carriers Conference, 123
  legal challenge to Clean Truck Program, 108, 130, 131–132, 147, 154–165, 168–178, 194, 301
  litigation strategy, 155–156, 179, 317–318
  lobbying, 182
  response to strikes, 250
American Trucking Associations v. City of Los Angeles,
  analysis of, 301, 316, 325
  campaign response to, 182, 194, 275, 306
  city attorney role in, 157, 169
  complaint, 154–155
  effect on Clean Truck Program launch, 164, 201, 337
  effect on port drivers, 164–165
  NRDC role,
    coordination with labor movement, 157
    coordination with Port of Los Angeles, 156–157
    joint defense agreement between NRDC and Port of Los Angeles, 156, 338
    motion to intervene, 156, 332
  permanent injunction,
    district court decision, 170–171
    Ninth Circuit decision, 171–173
    trial, 168–170

American Trucking Associations v. City of Los Angeles (cont.)
  port argument against preemption,
    market participation, 157
    safety exception, 158
    sovereign control, 157
    Tidelands Trust Act, 157
  Port of Long Beach settlement, 162
  preliminary injunction,
    amici, 160
    appeal, 159–161
    district court decision, 158, 162
    Ninth Circuit decision, 160–161, 163–164
    remand, 161
  protests in relation to, 159, 163
  Supreme Court,
    decision, 176–178
    oral argument, 175–176
    petition for certiorari, 174
America's Port, 1, 6, 15, 17, 33, 284, 285, 291, 293, 353
Amnesty program,
  approval, 253
  Assembly Bill 621, 253, 260, 262, 263, 272, 310
Annexation, 19
  Los Angeles attempt on Long Beach, 22
  of San Pedro and Wilmington by Los Angeles, 22–23
Antitrust law,
  impact on drayage trucking, 47
  labor exemption, 46, 208–209
  relation to independent contractors, 8, 55, 70, 208, 299, 315
  relation to shipping industry, 27, 36, 239
Appointment system, 287
Arambula, David, 227
Arian, Dave, 109, 128, 136
Arian, Sean, 130, 136
Asian American Center for Advancing Justice–Los Angeles, 199, 267
Asian Pacific American Legal Center, 192

Assembly Bill 2042, 90
Atkinson, Andelson, Loya, Ruud & Romo, 230
Audubon Society, 80–81

Backstop rule, 280
Baddeley, Kevin, 242, 255
Banares, Bernice, 170
Bankruptcy, 135, 159, 241, 248–249, 251, 261–262, 265, 270–272, 285, 316, 319–320
  Green Fleet, 248, 261, 320, 354
  Hanjin (*see* Hanjin)
  Pac 9, 262, 316, 320, 332
  QTS, 240, 249, 251, 320, 332
  Seacon Logix, 196, 224, 248, 261, 307, 320
  Total Transportation Services, Inc., 261, 320
  Tradelink Transport, 248
Banning, Phineas, 19
Banning's Landing, 64, 128, 241
Becker, Craig, 211
Beers, Roger, 82
Beezer, Robert, 160
Bell, Derrick, 326
Bensman, David, 191
Berk, Ed, 71
Berkshire Hathaway, 279
Berzon, Stephen, 75
Big Rig, The, 191–192, 194, 223, 243
Big Rig Overhaul, The, 243
Blackman, Jimmy, 129
Blake & Uhlig, 73
Bloomberg, Michael, 181
Blue Book Union, 29
Blue-Green Alliance, 6, 105, 126, 181, 284, 294, 324–325, 345, 353
Board of Harbor Commissioners,
  China Shipping, 81–84
  and Clean Trucks, 99–101, 122, 123, 125, 127, 129, 133, 135, 140, 141–153, 167, 277, 305, 327, 337

Index

Long Beach, 25, 42, 80, 90, 91, 125, 239
Los Angeles, 22–24, 37, 42, 53, 62, 64, 66, 95, 97, 99, 112, 113, 125, 343
   response to labor unrest, 227, 228, 232, 233, 237, 238, 255
   structure, 24, 54
Bond, Julian, 159
Bonds, 24–25, 34
Boston Consulting Group,
   analysis of port trucking, 139
   in relation to Campaign for Clean Trucks, 138, 180, 306
   report on Clean Truck Program, 148–149
Bramson, Plutzik & Birkhaeuser, 257
Break-bulk system, 33
Brennan, Joseph, 165, 167
Brewster, Rudi, 172
Bridges, Harry, 29, 45, 46
Brokerage, 324
Brown, Scott, 181, 188
Brown v. Board of Education, 311
Buffett, Warren, 279
Building and Trades Council, 109
Bunker fuel, 77
Burlington Northern Santa Fe. *See* Railroads
Burts, Ezunial, 61
Bush, George W., 47, 165, 167, 172, 211
Bush Gottlieb, 218, 220

California Air Resources Board,
   findings on port, 49, 97
   regulation of mobile sources of air pollution, 56
   regulation of ports, 77, 89, 93, 280, 283, 325
California Attorney General, 160, 190, 320
California Cartage,
   alliance with NRDC in lawsuit against SCIG, 279, 325
   class action against subsidiaries, 257

   DLSE claims against, 257
   Express, 268
   fines against, 257–258
   Teamsters organizing against, 257–258, 265, 267
   warehouse, 257
California Cleaner Freight Coalition, 280
California Coastal Act, 77
California Coastal Commission, 77, 80, 96
California Division of Occupational Safety and Health, 198, 227, 257
California Employment Development Department, 240, 245
California Environmental Quality Act, 55, 141, 257, 304
California State Assembly, 89, 113
California State University at Long Beach, 251
California Sustainable Freight Plan, 281
California Trucking Association,
   lawsuits against DLSE, 267, 321
   opposition to campaign, 184, 197, 231
Calko Transport, 215
Callahan, Colleen, 114, 125
Campaign, definition of, 297
Campaign for Clean Trucks, 2, 8, 13, 101
   aftermath, 288, 349
   assessment, 299–306, 311–318, 322–330
   attempt to amend FAAAA, 178–184
   background of, 19, 58, 76
   bottom-up approach, 193
   concession model, 107–108, 300, 314, 318, 322
   different outcomes for labor and environmental movements, 311–315
   formation, 13
   funding, 106

Campaign for Clean Trucks (cont.)
  goals, 10, 201, 209
  launch, 118
  lawyering for, 336–338
  legal theory, 110
  litigation phase, 154
  outcomes, 11, 15, 275, 284
  partnerships, 108
  passage of Clean Truck Program, 141–153
  planning, 105
  policy development, 126
  relation to labor law, 344–345
  termination of, 173
  top-down approach, 104
  transition to focus on misclassification, 185, 190, 201, 204
Campbell, Molly, 149
Canales, Humberto, 238
Cancer, 49, 79, 91, 97, 112
Canham-Clyne, John, 74, 129, 304
Cap-and-trade program, 282
Carder, Leonard, 219
Cardona, Amilcar, 222–223, 224, 236, 247
Carrix, Inc., 196
Carson, 37, 39, 67, 218, 220–221, 237, 241, 251, 256
Carter, Jimmy, 171
Carver, Ron, 71, 73, 144, 304
Castellanos, Patricia, 107, 114, 120, 127, 173, 193, 327
Castle v. Hayes Freight Lines, 168
Castro v. Pac 9 Transportation, Inc., 224, 225, 258, 259
Center for Food and Justice, 114
Cerritos, 230
Cervantes, Julio, 243, 260
Chang, Elaine, 170
Change to Win,
  formation of, 11, 73, 305
  National Ports Campaign, 74
  role in Campaign for Clean Trucks, 74, 75, 104–108, 112, 115, 118, 120, 121, 125, 126, 130, 146, 179, 181, 323, 324, 327, 328, 329
  role in Justice for Port Drivers campaign, 185, 186, 189, 191, 192, 193, 198, 202, 209, 223, 227, 244, 350
  strategic organizing center, 73
Chassis,
  in relation to cargo pickup, 49
  safety, 144
  shared pools, 237, 245
  shortage, 236, 237, 239, 252, 287
Chatten-Brown, Jan, 80
Cherin, Alex, 225, 227, 229
Chevron, 81, 84
China Ocean Shipping Company, 80
  planned expansion, 80–81, 90, 249
  relation to Intermodal Bridge Transport, 248, 254, 265
China Shipping,
  aesthetic impacts, 87
  community mitigation fund, 86–87
  community opposition to, 82
  enforcement failures, 280
  Environmental Impact Statement/ Impact Review, 82–83
  Gateway Cities Program, 87
  litigation against, 82–86
    federal court decision, 85
    settlement, 86
    state appellate court decision, 85
    state trial court decision, 84
    temporary restraining order, 85
  occupancy, 88
  terminal at Port of Los Angeles, 82
  U.S. Army Corp of Engineers evaluation, 82–83
  West Basin Transportation project, 82

Choi, Jean, 209, 255
Choke point, 300, 350
City of Industry Union Pacific Yard, 37
Class actions, 188, 195, 200, 224, 260, 262–263, 266, 269–271, 290, 304, 319, 332
Clean Air Act, 56
  exceptions to, 77
  litigation, 77–78
  preemption, 57, 58, 76, 94, 110, 158, 324
  state implementation plans, 56, 77
Clean Air Action Plan,
  approval, 101
  comments on, 118
  control measure HDV1, 100
  draft development, 99–100
  enforcement, 318
  Environmental Impact Report, 117
  final plan, 99
  financial incentives to trucking firms, 100, 283
  heavy-duty vehicles, 100
  near-zero emissions, 283
  provisions, 101
  relation to Clean Truck Program, 103–105, 111–112, 116, 117–122, 125, 128, 129, 133, 147, 165, 305, 323, 327, 335–336
  stakeholder input, 120
  technical report, 100
  update, 282–284, 325
Clean Truck Coalition, 197
Clean truck conversion,
  cost of, 134
  impact on environment, 284
  impact on port drivers, 15, 204, 328–330
  incentives, 123, 143, 148, 159
  as part of Clean Truck Program, 141, 162, 178, 312, 314, 324, 325
  replacement, 100, 101, 143, 151
  retrofit, 100, 101, 143, 151

Clean Truck Program, 10, 11, 301
  American Trucking Associations lawsuit against (see *American Trucking Associations v. City of Los Angeles*)
  approval of, 141–142
  attempt to amend FAAAA to permit (*see* Campaign for Clean Trucks, attempt to amend FAAAA)
  clean truck conversion (*see* Clean truck conversion)
  concession plan approval in Long Beach, 146
  concession plan approval in Los Angeles, 150–153
  differences between Los Angeles and Long Beach, 125, 146, 312
  different motivations of labor and environmental movements, 318
  economic analysis of, 134–135, 138, 148–149
  employee conversion (*see* Employee conversion)
  enforcement of, 150
  Federal Maritime Commission challenge to (*see* Federal Maritime Commission, Clean Truck Program litigation)
  fee, 122, 141, 143, 146, 147, 150, 166, 312
  industry resistance to, 130, 133, 135, 142, 144, 152, 194
  launch, 160
  leases, 243–244
  legal analysis of, 108, 130–132, 150, 151, 328, 336–337
  Los Angeles concession plan provisions,
    maintenance, 153, 154, 163, 168, 170–172, 174, 177, 280
    off-street parking, 139, 153, 154, 161, 168, 170–172, 177, 178, 301, 302

Clean Truck Program (cont.)
  placards, 124, 139, 150, 153, 163, 168, 170–174, 177, 178, 301, 302, 347
  truck routes, 121, 153, 162, 168, 170, 172, 327
  Los Angeles-Long Beach split, 146
  model year standards, 122
  organizing in support of, 124, 125, 126, 133, 143–144
  at other ports, 13, 288
  outcomes, 12, 15
    effect on air quality, 284
    effect on coalition, 275
    effect on drivers, 243
    impact on environmental movement, 275–277, 325
    impact on labor movement, 302, 306, 312, 325
    reasons for, 314, 317
  policy development, 100, 104, 112, 118, 121–124, 127–130, 305
    discussions with industry, 139–140
    at Port of Los Angeles, 136–141
  politics of, 132
  progressive ban on dirty trucks, 141
  reasons for passage, 314, 323–324, 327, 353
  relation to Justice for Port Drivers campaign, 185–189, 191, 209, 211, 212, 239, 243, 258, 273–274, 347–348
  relation to union organizing, 201, 215
  relation to waste haulers campaign, 350–351
  request for proposal, 121
  severability, 150–151, 337
  subsidies, 122
    Scrap Truck Buyback Program, 151
    Truck Funding Program, 151
    Truck Procurement Assistance Program, 151
  Tariff no. 4, 141, 149, 150
  trucking firm participation in, 201, 216
Clergy and Laity United for Economic Justice, 115, 193
Clinton, Bill, 158
Coal, 52, 65, 76, 349
Coalition for a Safe Environment, 67, 79, 112, 115, 120, 145, 278, 327, 352
Coalition for Clean Air, 78, 82, 89, 92, 114–115, 120, 144, 147, 156, 278–279
Coalition for Clean and Safe Ports,
  accountability to stakeholders, 328–330
  in American Trucking Associations litigation, 155, 156, 168, 173, 178
  coalition building, 10, 11, 104
  governance, 119, 327
  impact of Clean Truck Program on partners, 311–312
  industry responses to, 318
  LAANE's role, 105, 115–116, 126
  launch, 118
  members, 109–115, 119
  reasons for formation, 322–324
  relation to Justice for Port Drivers campaign, 193, 195, 196, 216, 224, 284, 351
  role in Campaign for Clean Trucks, 13, 119–126, 133–134, 144–153, 159, 160, 164, 305–306
  role in lobbying to amend FAAAA, 179–184, 349
  role of environmentalists, 110
  role of labor, 109
  split after Clean Truck Program, 14, 185–186, 275, 317
  strengths and weaknesses of partners, 3, 317
  tensions, 324–325
Coalition for Humane Immigrant Rights of Los Angeles, 115

Index

Coalitions,
  accountability in, 325
  cross-sectoral, 7, 293, 318
  divergence of interests, 324
  in local politics, 321
  social movements and, 2, 4–6, 14, 295
  sustaining, 29
Coast Bridge Logistics Inc., 227
Coast Guard, 137
Cold ironing, 77, 88–89, 92–94, 97, 99
Collective bargaining agreements, 269, 316, 324
Commerce, city of, 37, 67, 279, 347
Committee of Thirty, 20
Communication Workers of America, 127
Community benefits agreements, 96, 110
Compton, city of, 39, 81, 227, 237
Concession agreement,
  enforcement of, 243, 273
  as focus of ATA litigation, 154–165, 168–178, 301–302, 304–306, 312, 317, 328, 338, 347, 349, 350, 351
  as model for Campaign for Clean Trucks (see Campaign for Clean Trucks)
  in relation to Clean Truck Program, 2, 104, 105, 110, 112, 113, 115, 116, 118, 121–123, 130–132, 138–142, 146, 149–153, 182, 184, 301, 323, 324
  in relation to Federal Maritime Commission litigation, 167–168
  in relation to labor peace, 274–275
Congestion,
  impact on communities, 10, 216
  at ports, 10, 37, 43, 239, 242, 245, 246, 251, 252, 276, 277, 285, 286
  proposals to alleviate, 287
  traffic, 37, 76
Congress of Industrial Organizations, 30

Container Connection, 199–200
Container Freight, 268
Containerization,
  definition of, 33
  impact on labor, 41
  impact on neighborhoods, 43, 64
  impact on ports, 33–40
Contra Costa Superior Court, 279
Contreras, Miguel, 128
Cordero, Mario, 251
Corporate campaigns, 106
Costco, 228
Cotrell, Sharon, 113
Countermobilization, 7, 14, 316–318, 322
Courage, 15, 124, 205–206, 296, 346, 351–352
Craig Shipyard, 25
Crose, Joy, 130–131, 157
Culinary workers, 106, 352
Cunningham, Randy "Duke," 81

Daniels, Paula, 96
Davis, Cowell & Bowe, 75, 108, 191, 233
Davis, Gray, 96
Davis, Martin, 237
D.C. Circuit, 211
Delugach, Sharon, 96
Denver International Airport, 106
Department of Justice, 160
Department of Labor,
  campaign outreach to, 194
  investigations into misclassification, 192, 196, 200
  lawsuit against Shippers Transport Express, 234, 241, 242, 306, 331
  memorandum of understanding with DLSE, 196
  relation to Justice for Port Drivers campaign, 187, 190, 197, 198, 218, 340
  wage and hour division, 190

Department of Labor Standards Enforcement,
  amnesty program, 255, 260, 262
  Bureau of Field Enforcement, 190
  campaign outreach to, 193
  group wage enforcement actions, 190
  individual wage claims, 190, 196, 200, 347
  industry challenges to, 198, 263, 267, 272, 321
  leadership of, 192, 306
  memorandum of understanding with Department of Labor, 196
  misclassification claims filed with, 187, 190, 199, 201, 220, 302
  relation to Justice for Port Drivers campaign, 212, 218, 223, 225, 233, 238, 257, 261, 269, 302, 307–310, 315, 319, 340
  retaliation against drivers for filing claims, 222, 233
  rulings on port driver cases, 196, 218, 226, 227, 244, 255, 258, 259, 260, 332
  studies of outcomes, 223, 240, 269
  wage claim process, 190
Department of Transportation, 160, 182
Deregulation,
  impact of, 41, 42, 47, 48, 68, 299, 349
  of railroads, 36
  of shipping, 36, 56
  of trade, 35
  of trucking, 10, 36, 56, 57, 344
Destabilization rights, 335
Diesel fuel,
  particulates, 145, 284
  ships, 77, 280
  trucks, 10, 19, 49, 78, 159
Digges, Robert, 169
Dines, Rich, 228
Division of Advice,
  as part of NLRB (*see* NLRB)
  in relation to Justice for Port Drivers campaign, 255, 258
  mandatory submissions to, 264
  memorandum on misclassification, 258–259, 269, 309–310
Dominguez Channel, 63
Don't Waste L.A., 350–351
  waste hauling plan, 351
  zero waste goal, 351
Double conversion, 107, 301
Drayage trucking,
  concession plan, 150
  definition of, 37
  diesel trucks, 49
  employee firms, 202
  firm types, 100
  as focus of Campaign for Clean Trucks, 104, 107, 111, 118, 121, 122, 130, 139, 143, 148, 149, 150, 151, 161, 184, 300, 301, 323
  as focus of Justice for Port Drivers campaign, 231, 237, 238, 242, 243, 250, 253, 257, 260, 302
  frequent caller, 100
  immigrant drivers, 49
  impact of port growth on, 37, 38, 43, 64, 67, 285, 287
  independent contractors, 47, 48, 289, 299
  labor relations, 47, 49, 74
  at Los Angeles and Long Beach ports, 44, 49
  misclassification, 188, 196, 265
  numbers at ports, 101, 134, 201
  pollution, 49, 100, 283, 336
  in relation to American Trucking Associations litigation, 166, 172, 174, 175
  restructuring, 47, 48
  semi-frequent caller, 100
Dredging, 22, 25, 35, 39, 77, 80, 83
Durazo, Maria Elena, 105, 129

Earth Corps, 92
EarthJustice, 281

Index

East LA Yard, 37
East West Bank, 96
East Yard Communities for Environmental Justice, 68, 114–115, 278, 327, 347
Echoing Green Fellowship, 219
Eco Flow,
  collective bargaining agreement, 250, 253, 269, 316, 331
  creation of, 250
  driver vote to unionize, 250
  port solutions agreement, 249
  relation to Total Transportation Services, Inc., 249, 261, 320
Economics & Politics, Inc., 133
El Monte Thai Worker case, 192
Emission reduction proposals,
  cargo fees, 143
  carrier leases, 97
  Clean Air Action Plan, 94, 99, 101
  clean ship fuel, 97
  diesel truck replacement and retrofit, 100
  as goal of Clean Truck Program, 117, 118, 301
  state-level actions, 280
Employee,
  definition of, 55
  drivers, 121–123, 125, 150, 164, 166, 173, 185, 188, 190, 201, 205, 210, 213, 228, 229, 250, 262, 272, 304, 314, 353
  economic realities test, 189
  employer control, 189
  misclassification of, 9, 69
  recognition as goal of ports campaign, 2, 10, 11, 12, 55, 70–72, 117, 118, 188–189
  rights attached to, 28, 55
Employee conversion, 71, 107, 120–123, 132, 134–138, 140–143, 146–148, 150, 153–156, 159, 161–162, 164, 167–168, 171–173, 179, 191, 194, 204–205, 209, 211, 215, 255, 274–275, 299–302, 304–306, 310, 312–314, 316–318, 323–324, 327–328, 330, 336–338, 347, 349–352
End Oil/Communities for Clean Ports, 278
Engine Manufacturers Association v. SCAQMD, 110
Englert, Roy, 173
Environmental Defense Fund, 96, 110, 128
Environmental hazards, 44–45
Environmental Impact Report, 64, 80, 304
Environmental justice,
  advocacy around ports, 9, 11, 13, 15, 43, 79, 107, 108, 110, 111, 294, 325
  and freight, 288
  in relation to Campaign for Clean Trucks, 111, 114–116, 119, 126, 128, 327
  tensions with mainstream environmentalism, 6, 112, 333
Environmental movement, 6, 9, 11, 13–15, 76, 112, 178, 275, 293, 295, 297, 300–301, 311, 313, 317–318, 322, 324, 332, 334, 341
Environmental Protection Agency,
  emission standards, 88, 100, 116, 122, 137
  and freight industry, 288
  near-zero 2010 standard, 283
  regulatory jurisdiction, 56, 77, 90
  truck regulations, 88, 91
Estelano, Cecilia, 96
Experimentalism, 335
Expertise, 246–248
Exports,
  in relation to ports, 18, 23, 26, 32, 35, 65, 143
  in supply chain, 73
  volume, 35
ExxonMobil, 52, 62

Fair Labor Standards Act, 190, 196
Fair Pay & Safe Workplaces Executive Order, 268
Far East Wilmington Improvement Association. *See* Wilmington
Fargo, 243, 259
Farmer, Jeff, 213, 227
Federal Aviation Administration Authorization Act,
  effort to amend,
    briefing book, 180
    Clean Ports Act, 182
    driver survey, 183
    H.R. 572, 183
    H.R. 5967, 182
    impact of 2010 election on, 188
    lobbying, 179–184, 187, 349
    research for, 191
    S. 1011, 184
    sponsors, 182, 184
  preemption,
    attempts to change state employment law in response to, 191, 234, 244, 267, 320
    doctrine, 57–58, 324
    in relation to American Trucking Associations litigation, 154–165, 168–178
    in relation to Clean Truck Program, 12, 301, 315, 328, 336
    in relation to Don't Waste L.A. campaign, 351
  provisions, 57
Federal Maritime Commission,
  campaign analysis of, 108
  Clean Truck Program litigation, 165–167
    injunction, 166
    motion to dismiss, 167
  industry request to stop Clean Truck Program, 135
  permission for inter-port cooperation, 239, 245
  role of, 27, 56

Federal Motor Carrier Safety Administration,
  as amicus in ATA litigation, 160
  petition for waiver of driver rest period rule, 235–236
FedEx,
  class action against, 291
  misclassification, 290
  picketing, 228
  settlement in California, 291
Fernandez, Ferdinand, 160
Feuer, Gail Ruderman, 82, 87, 93, 94, 110, 305
Fitzsimmons, Frank, 41
Fletcher, Betty, 171, 173
Flores, Joan Milke, 62
Foster, Bob, 125
Framing theory, 7, 173, 235, 298, 322
Frank, Larry, 107, 127, 129, 219
Free Harbor Movement, 21, 22
Freeman, David, 96–97, 133, 135, 140, 170
Freeways, 37–38, 45, 84, 277
  Interstate 10, 38
  Interstate 110 (Harbor Freeway), 38, 44–45
  Interstate 405, 37
  Interstate 710 (Long Beach Freeway), 37–38, 45
  Terminal Island Freeway (State Highway 103), 37, 63
  traffic, 24, 35–41, 60, 62, 64, 67, 76, 82, 86, 89, 91, 139, 159, 170, 182–183, 252
Freight Advisory Committee, 282

Garcetti, Eric, 220, 232, 282
Garcia, Olivia, 254
Gateley, Jason, 202, 216
General Agreement on Tariffs and Trade, 35
General Motors, 135
Georgetown Law Center, 111

Gephardt, Richard, 180
Gerald Desmond Bridge, 133, 147, 252
Gig economy,
  and labor law, 9
  and misclassification, 289
  NELP report, 289
Gilbert & Sackman, 199, 308
Glick, Stephen, 196, 199, 308
Globalization, 341
  impact on ports, 8, 18, 32–43, 299
Gold Point Transportation,
  class action against, 200, 224, 332
  labor peace agreement, 267, 269, 316, 332
  relation to Harbor Express, 198, 200
  strikes against, 258
Gomez Law Group, 238, 260, 332
Good Jobs, Green Jobs Conference, 202
Great Depression, 18, 25–26, 28–29
Greater Long Beach Interfaith
  Community Organization, 278
Green, Elina, 114, 120, 125
Green Fleet,
  bankruptcy (see Bankruptcy)
  DLSE awards against, 218, 223–224
  Employment Development
    Department action against, 240
  labor peace agreement, 249, 253, 264, 269, 272, 316, 331
  NLRB actions,
    charges against, 220, 222
    complaint, 229–230, 348
    ruling against, 247–248, 269
    trial, 247
  rehiring terminated drivers, 235
  strikes against, 218, 220, 221, 228, 230–232, 248–249, 310, 352
  as target of Justice for Port Drivers
    campaign, 218, 226, 233–234, 309, 354
  10(j) injunction, 223, 234–235
Green growth, 95, 98–99, 128, 144, 152, 175, 234, 305, 323, 353

Green LA Port Working Group, 114, 327
Green Port Gateway, 279
Griffin, Richard, Jr., 264
Gurley, David, 196
Gutierrez, Philip, 234
Gutman Dickinson, Julie, 218, 227, 230, 236, 257, 265, 268, 275, 334

H&M Terminals Transport, Inc., 68, 69
Haberbush, Vanessa, 262
Hadsell & Stormer, 193, 199, 241, 308
Hahn, James, 83, 89, 93, 111, 128, 304
Hahn, Janice, 88, 93, 125–126, 147, 202, 220, 222, 232, 281, 285, 304
Hamilton, Calvin, 61
Hanjin,
  bankruptcy, 285
  impact of bankruptcy, 286–287
  stranded ships, 285
Hankla, James, 92
Harbor Belt Line Railroad. *See* Railroads
Harbor Express, Inc., 198
Harbor Rail Transport,
  change to XPO Port Services, 254
  litigation against, 238
  relation to XPO Logistics, 238
  strikes against, 237, 238, 248, 249, 265
Harbor Trucking Association, 197, 225, 227, 229, 235, 251, 255, 260, 263
Harbor-Watts Economic Development
  Corporation, 115
Harris & Ruble, 332
Harry Bridges Boulevard, 45, 61–62, 67
Harvard,
  law school, 326
  public policy school, 98
Hawkins, James, 200, 332
Hayes, Matthew, 200, 308, 331
Health care reform, 180, 182
Hermandad Mexicana, 115
Hernández, Roger, 253, 260
Hewlett Packard Foundation, 107
Hobart, 37, 67

Hoffa, James, 129, 194, 202, 258, 306
Holmes, John, 137, 140, 158, 170, 178, 276, 306
Horizon, 72, 252
Hotel Employees and Restaurant Employees Union, 74, 106, 181, 219
House Committee on Transportation and Infrastructure, 180, 182
Hricko, Andrea, 114
Hub Group Trucking, 240
Hundred Years' War, 59, 66, 76, 335
Hunter, Duncan, 80
Huntington, Collis, 21
Husing, John, 133, 138
Husing Report, 133–136, 138, 148–149

Ibarra, Cecilia, 142
Idling, 77, 89, 91, 113, 300
Immigration, 337, 344
  impact on unionization, 342
  and labor, 6
  trucking industry, 48
Impact Project, 114
Imports,
  from Asia, 35, 38, 60
  in relation to ports, 24, 35, 39, 252, 285, 286, 299
Independent contractors,
  as focus of Justice for Port Drivers campaign, 185, 307–311, 339
  in gig economy, 290
  misclassification, 7, 12, 184, 192–197, 209, 212, 233, 237, 247, 253–256, 258–259, 265
  in port trucking, 2, 8, 18, 47–48, 164, 214, 299, 329, 351–352
  as problem for labor movement, 9–10, 13, 47, 68–76, 208–212, 294, 313, 345, 347–349
  relation to Campaign for Clean Trucks, 10, 118, 121, 134, 139, 148, 161, 163, 169, 171, 178, 300–303, 327
  in relation to low-wage work, 55
  trucking industry efforts to maintain drivers as, 263–264, 267, 319, 321
Industrialization, 8, 18, 24–26, 35, 43
Inhofe, James, 81
Inland Empire, 217, 341
Integrated advocacy, 185, 187, 303
Interest convergence, 2, 9, 104, 110, 322, 330
Intermodal Bridge Transport,
  class action against, 332
  employee firm as campaign target, 309
  misclassification as unfair labor practice, 268, 269, 348
  NLRB actions,
    charges against, 257, 264
    complaint, 266
    trial, 268, 315
  relation to COSCO, 249, 254, 347
  strikes against, 248, 249, 256, 258, 265, 267–268, 321
Intermodal Container Transfer Facility,
  impact on ports, 39, 43, 46
  opposition to, 277
  original site, 37
  proposed expansion, 277, 278
Intermodalism, 18, 33, 36–37, 41–42
International Brotherhood of Boilermakers, 73
International Brotherhood of Electrical Workers, 106, 114
International Brotherhood of Teamsters,
  conflict with longshoremen, 29–31, 41, 46, 47, 69, 228
  coordination with NRDC, 126, 128, 144, 146, 175, 324
  efforts to address independent contractor problem, 70, 299
  first Southern California union contract, 31
  general counsel, 73
  history at ports, 28, 29, 31, 49

Index

impact of trucking deregulation on, 47, 49, 68
leadership of coalition, 327–329
litigation against, 268, 321
Local 63, 109
Local 469, 215
Local 848, 109
national ports campaign, 71, 72, 74–75
in relation to Campaign for Clean Trucks, 10, 108, 109, 112–115, 120, 129, 130, 143, 147, 181, 301, 303–306, 327, 337
relation to Change to Win, 28, 74
in relation to Justice for Port Drivers campaign, 185, 186, 189, 192–198, 205, 209, 269, 272, 273, 279–280, 289, 306–310, 339, 340, 347, 350
port division, 71
port strikes, 49
port trucking, 10, 72
union organizing, 164, 201–204, 316, 332
International Longshore and Warehouse Union,
conflict with Teamsters (see International Brotherhood of Teamsters)
contract with Pacific Maritime Association, 245
decision not to join campaign, 109, 115
employee status, 46
founding, 30
impact of containerization on, 33
Local 11, 163
Local 13, 109
negotiations with Pacific Maritime Association, 231, 244, 245
participation in green growth discussions, 97, 114, 128–129
port strikes, 41, 231
power at ports, 42

racism in, 30
representation on Los Angeles Board of Harbor Commissioners, 96
resistance to FAAAA amendment, 181
support for Justice for Port Drivers campaign, 228, 230
support for Long Beach Clean Truck Program, 163
tension with environmentalists, 109
International Longshoremen's Association,
Big Strike, 29
Bloody Thursday, 29
development, 28
founding, 28
in relation to Teamsters' national ports campaign, 72
split with ILWU, 30, 41
Inter-port competition, 27, 32, 34, 40, 276
Inter-port cooperation,
to address chassis shortage, 239
efforts to promote, 98
joint port meetings, 101, 119
joint public workshop on Clean Truck Program, 124
review by Federal Maritime Commission, 165
Interstate Commerce Act, 27
Interstate Commerce Commission, 27

Janavs, Dzintra, 86
Jobs with Peace, 127
Johnson, James, 279
Journal of Commerce, 250
Justice for Port Drivers campaign,
environmental groups, 205, 275, 325
focus on misclassification, 13, 186, 187, 273, 302, 303, 312, 315
goals, 2, 207, 350
hardship fund, 232
industry responses to, 319
launch, 3

Justice for Port Drivers campaign (cont.)
  leadership, 209, 215, 218, 326, 330
  litigation as a tool, 223
  outcomes, 240, 275
  policy, 310, 311
  relation to Campaign for Clean Trucks, 13, 185
  strategy, 12, 14, 209
  strikes, 226, 264, 310
  suing-and-striking model, 209, 210, 214, 218, 251, 255, 262, 269, 272–274, 291, 299, 302, 307, 316, 319, 320, 347, 348, 350
  use of DLSE, 269, 308
  use of NLRB, 209, 213, 268, 307, 309, 338–339, 344

K&R Transportation, 265, 267, 321
Kabateck, Brian, 200
Kabateck Brown Kellner LLP, 200
Kagan, Elena, 176
Kahn, Andrew, 108
Kahn, Simon, 157, 169
Kaiser International, 65
Kanter, Robert, 94
Kaoosji, Sheheryar, 209–210, 217, 227, 259, 272, 284, 347
Kaye Scholer, 131, 157, 167, 169, 337–338
Kearney, Barry, 258
Kell, Jackie, 92
Keller, Larry, 93
Kennedy, Ted, 181, 188
Kim, Candice, 114, 120
Kim, Kaylynn, 96
Kim, Peter, 198, 270–271
Klein, Jonathan, 181, 193, 202
Knatz, Geraldine, 92, 98, 126, 162, 169, 181, 276, 280, 305
Knight Transportation, 139, 159
Kramer, Robin, 129
Krause, Doug, 96
Kumetz, Fred, 70

Labor law,
  collective bargaining, 28, 29, 54, 55, 204, 206, 228, 231, 242, 244, 250, 252, 253, 260, 309, 316, 324, 331
  and deregulation, 18, 47, 299
  firms, 199, 200, 218, 219, 233, 304, 309
  at local level, 296, 344
  misclassification, 224
  National Labor Relations Act, 28, 55, 206
    section 7, 206, 207, 258, 259, 265, 266, 310, 339
    section 8, 207, 213, 254, 255, 258, 264, 269
    section 10(j) injunctions, 219
    unfair labor practices, 206, 207, 230, 307
  National Labor Relations Board, 206, 208
    administrative law judges, 207, 208, 247, 268
    appointment process, 211, 307
    Division of Advice, 254
    general counsel, 208, 264
    Office of Appeals, 254
    process for appeal, 207
    Region 21, 208, 212, 213, 218, 219, 220, 223, 225, 227, 229, 230, 233, 234, 246, 254, 257, 259, 264, 265, 266, 269, 309, 310, 315, 320, 339, 348
    regional director, 208
    regional offices, 208
    under Obama, 211, 307
  ossification of, 207
  preemption, 57–58, 234, 315, 324
  in relation to independent contractors, 8, 9, 13, 70, 187, 195, 209, 313
  in relation to ports campaign, 8, 186, 192, 204–205, 209, 211, 253, 302, 303, 314, 315, 325, 331, 338, 339

Index

in relation to port workers, 26
staffing, 208
striking under, 208–209
Taft-Hartley Act, 47
unfair labor practice strikes, 208
unfreezing, 206, 207
Wager Act, 206
weakness of, 54–55
Labor militancy, 28
Labor movement, 2, 9, 12–13, 19, 31–32, 46, 59, 76, 104, 106, 128, 178, 185–188, 198, 201, 204, 215, 217, 219, 243, 249, 283, 291, 293–294, 297, 300, 304–306, 313, 315, 321–323, 326, 328–330, 332, 342–344, 350–351, 354
Labor peace, 14, 41, 46, 118, 122, 186, 241–242, 249, 253, 261–262, 267–271, 273–274, 303, 305, 308, 316, 331–332, 341, 350–351, 354
Labor precarity, 243, 288, 293, 298, 300
LACA Express, 199, 237–238, 240, 251, 254, 267, 269, 315, 320
Landbridge, 36–37
Land use, 6, 17, 19, 46, 61, 63, 342
La Opinión, 192
La Rosa, Vic, 194, 230
Las Vegas, 74, 106, 202, 352
Latin American Truckers' Association, 69
LA Trailer and Intermodal Container Facility, 37
Law in action, 311
Law making from below, 296, 343, 345
Law Office of Jewels Jin, 268, 321
Law on the books, 311
Lay knowledge, 347
Leadership, 6–7, 21, 30, 68, 73–74, 95, 123, 180, 192, 205, 209–210, 212–215, 220, 276, 280, 282, 288, 306, 326, 328, 342–343, 352
Lebar, Weston, 251
Legal mobilization, 3–5, 7, 14, 295–296, 311, 321, 333–334, 338

Legal opinions, 132, 154, 269, 309, 336–337
Lenz, Thomas, 230, 255
Leon, Richard, 167
Less-than-truckload firms,
 employee companies, 289
 Teamsters organizing against, 238
Lewis & Clark Law School, 278
Libatique, David, 98, 130, 132, 135
Liberty Hill Foundation, 67, 114
Licensed motor carriers, 122, 150
Lien, 256, 260
Lin Perrella, Melissa, 101, 111, 120, 169, 176
Liquefied natural gas trucks, 139, 140, 148
Listserv, 259
Littler Mendelson, 71
Llanteros, 144
Local labor activism, 296, 314, 343
Local workplace regulation, 4, 342
Lochner v. New York, 344
Logan, Angelo, 68, 114, 120, 128, 288, 347
Logistics supply chain, 18, 300, 329
Long Beach,
 annexation of Terminal Island, 23
 aquarium, 43
 Chamber of Commerce, 90
 city attorney, 131, 156, 157
 City Council, 90–92, 126, 238, 260, 279
 city of, 205
 Clean Truck Center, 159
 formation of, 22
 mayor, 125, 142, 143
 oil, 24–26, 34
 Tidelands and Harbor Committee, 255
 unified school district, 279
Long Beach Alliance for Children with Asthma, 114
Long Beach Heritage, 80–81
Long Beach Press-Telegram, 147, 236, 238, 241, 245, 251–252, 256, 260, 277, 287
Long Beach Yacht Club, 113

Longshoremen,
  impact of technology on, 41
  legal distinction from truckers, 42, 46, 52, 317
  as port "labor aristocracy," 47
  solidarity and conflict with Teamsters (*see* International Brotherhood of Teamsters)
  union history (*see* International Longshore and Warehouse Union)
López, Gerald, 219
Lopez, Miguel, 109, 114, 120
Lopez, Steve, 229, 237
Los Angeles,
  big-box ordinance, 342
  business team, 98
  Chamber of Commerce, 21, 24, 28
  City Council, 22–24, 42, 53, 62, 64, 65, 69, 82–84, 88, 93, 98, 113, 125, 144, 145, 147, 152, 153, 304, 351
  city of, 21, 23, 44, 60, 65, 66, 304
  Community Redevelopment Agency, 66
  day labor sites, 342
  Department of Water and Power, 23, 96
  economy, 24, 63
  living wage, 6, 342
  mayor, 53, 66, 83, 84, 89, 90, 93–96, 98, 99, 101, 106, 107, 111, 114, 124–130, 132, 133, 136–138, 140, 144, 151–153, 162, 181, 202, 216, 219, 220, 223, 232, 233, 237, 239, 249, 250, 282, 304–306, 327, 336, 337
  Minimum Wage Enforcement Office, 342
  Neighborhood Services Office of the Mayor, 107
  Planning Commission, 61
  recession, 63
  secession from, 66, 67, 83, 89, 304
  unified school district, 132
Los Angeles & San Pedro Railroad. *See* Railroads
Los Angeles Alliance for a New Economy, 105
Los Angeles Container Company, 41
Los Angeles County Board of Supervisors, 251
Los Angeles County Federation of Labor, 105, 127–129, 214, 229
Los Angeles County Local Agency Formation Commission, 66
Los Angeles Economic Development Corporation, 133
Los Angeles International Airport, 96, 110, 113
Los Angeles River, 22, 37
Los Angeles Times, 21, 60, 63, 78, 86, 144, 146–147, 149, 162, 164–165, 173, 225, 229, 237, 243, 260, 267, 279–280, 318, 352
Lowenthal, Alan, 89, 90, 113, 143
Lowenthal, Suja, 255
Low-wage work, 3, 4, 15, 55, 75, 198, 199, 291, 314, 315, 322, 342, 344, 353
Luban, David, 326
Ludlow, Martin, 98
Luetto, Adam, 195

Mack, Chuck, 75, 193, 215
Maersk, 32, 40, 91, 129, 245
Mandatory arbitration agreements, 225
Manley, Michael, 72, 73, 75, 108, 117, 120, 123, 130, 154, 157, 175, 193, 194, 203, 209, 211–213, 218, 222, 226–230, 254, 255, 258, 304, 305, 334, 340, 347
Map of ports, 45
Mares, Mateo, 222, 234, 236, 247
Marine Workers Industrial Union, 29
Market participation, 57–58, 76, 110, 138, 151, 156, 160, 168, 170, 172, 174–178, 186, 274, 300, 302, 313–314, 316, 318, 323–324, 338, 350–351

Index

Marquez, Jesse, 67, 79, 112, 114, 120, 128, 352
Marron Lawyers, 263
Martinez, Adrian, 110–111, 117, 119, 120, 123, 126, 131, 132, 144, 146, 152, 155, 156, 229, 281, 305, 352
Martinez, Dennis, 229, 352
Marvy, Paul, 191
Mass actions, 238, 260, 308, 332, 340
Master Freight Agreement, 204
Masters, Julie, 82
Maynard, Barbara, 203, 227, 232, 246, 255–256, 265, 267
Mayor Logistics, 224
McCracken, Richard, 75, 108
McCutchen, Doyle, Brown & Enersen, 83
McKinsey & Company, 136
McNatt, Christopher, Jr., 169
Megaships, 252, 273, 276, 284, 286–287
  post-Panamax, 34, 36
Mejia, Leonardo, 242
Mendoza, Jerilyn López, 96, 98, 110, 125
Mendoza, Mark, 109
Mercedes-Benz/Daimler Truck Finance Company, 159
  collections department, 159
  protests against, 355
Mexican American Legal Defense and Education Fund, 132
Mexican-American War, 19
Milkman, Ruth, 342
Miller, John, 113
Millman, Robert, 71, 194
Minimum wage, 6, 165, 189–190, 195, 197–200, 209, 224, 290, 341–344
Mira Loma, 228
Miscikowski, Cindy, 113
Misclassification,
  bottom-up enforcement strategy, 193
  in DLSE, 196, 258, 267, 269, 307–308
  as focus of Justice for Port Drivers campaign, 3, 9, 12, 13, 14, 185, 302, 303, 306, 312, 315–316, 325, 330, 334, 350
  in gig economy, 289–291
  illegal business deductions, 194–195, 197, 199–200, 223, 226, 308
  industry responses to, 194
  trainings, 197, 263, 319
  labor law section 2802, 195, 200, 225
  legal clinic, 198
  litigation, 70, 187, 188, 191, 199, 200, 224, 263, 267, 269, 340
  plaintiff law firms bringing misclassification suits, 195, 196, 199, 200
  port policy to address, 239, 242, 260, 272
  public enforcement, 184, 190–191, 194, 196
  in Region 21, 212, 213, 220, 235, 246, 254, 264, 266, 309–310, 338–339
  relation to bankruptcy, 263, 320
  relation to truck leases, 233–234, 319
  relation to union organizing, 205
  state laws targeting, 260
  study of, 191–192, 223, 289
  as unfair labor practice (*see* Misclassification as unfair labor practice)
  willful, 195, 241
Misclassification 3.0 training, 263
Misclassification as unfair labor practice,
  at NLRB, 254, 259, 265, 269, 310
  test case approach, 213, 214, 220, 309
Molina v. Pacer Cartage, 332
Montemayor, Dickie, 268
Mooney, Gary, 222, 249
Morello, Tom, 232
Moreno Valley, 228
Morrison & Foerster, 83

Morrow, Margaret, 85
Moscowitz, Ellyn, 195
Motor Carrier Act
　1935, 27, 31
　1980, 36, 57
　in relation to *Castle v. Hayes Freight Lines*, 168
Movement lawyering, 5, 296, 326, 334
Moving Forward Network, 288
Multiple Air Toxics Exposure Study (MATES II), 79
Munger, Tolles & Olson, 96

NAACP, 159, 326
Nadler, Jerrold, 181
National Association of Waterfront Employers, 160
National Employment Law Project, 191
National Environmental Policy Act, 55
National Industrial Transportation League, 135
Natural Resources Defense Council,
　advocacy after Campaign for Clean Trucks, 249, 277, 288
　challenge to Pier J expansion, 91–92, 249
　in China Shipping case (*see* China Shipping)
　Clean Air Action Plan update, 282
　climate and clean air unit, 278
　interest in Clean Truck Program, 110–111, 323
　legal analysis of Clean Truck Program, 130–131, 336
　legal challenges to ports, 145, 152, 257, 306, 312, 313
　litigation as a tool, 313–314, 325
　low-emission proposal, 89, 325
　near-dock rail, 278–280
　no net increase, 90, 94
　organization as client, 332
　reports, 49, 89
　role in American Trucking Associations litigation (see *American Trucking Associations v. City of Los Angeles*)
　role in Campaign for Clean Trucks, 10, 101, 110–111, 117, 120, 122, 124, 126, 135, 144, 146, 147, 304–305, 324, 327
　role in effort to amend FAAAA, 180–182
Navy, 81
New England Journal of Medicine, 91
Nitrogen oxide, 77–78, 88
No net increase,
　concept, 89
　controversy, 93
　as foundation for Clean Air Action Plan, 95, 97, 305
　plan to achieve, 90–93
　report, 94–95
　task force, 93, 111, 305, 323
Nuclear option to eliminate Senate filibuster, 211

Obama, Barack, 166, 180, 211
Oberstar, James, 180
O'Brien, Robert, 81
Ocean carriers, 24, 27, 32, 34, 36, 43, 46, 52, 299
O'Connell, Beverly Reid, 234
Oil companies, 26
　ConocoPhillips, 44–45
　ExxonMobil (*see* ExxonMobil)
　Tesoro Los Angeles, 45
Orrick, Herrington & Sutcliffe, 111
Osoy, Eddie, 266
Otis, Harrison Gray, 21
Ovrom, Bud, 96

Pac Anchor Transportation,
　preemption lawsuit against attorney general, 191, 244, 267, 320
　unfair business practices, 191

Index

Pace Freight Systems, 237
Pacer Cartage,
  class action against, 224, 238, 266, 269, 332
  DLSE claims against, 244, 259
  relation to XPO Logistics, 238
  settlement, 266, 269
  strikes against, 237, 238, 248, 249, 265
  as XPO Cartage, 254
Pacific 9 Transportation,
  bankruptcy, 261, 262, 264, 320
  DLSE claims against, 220, 258
  effort to ban from ports, 243
  Employment Development Department action against, 240
  independent contractor firm, 220, 309
  labor peace agreement, 262, 316
  lawsuits against, 224, 225, 269, 332
  misclassification as unfair labor practice, 220, 233, 265, 310, 339
  NLRB actions,
    charges against, 221, 348
    complaint, 225, 233, 259
    dismissal, 262
    Division of Advice opinion, 254, 255, 258, 269
    settlement agreement, 225, 226
    strikes against, 220, 221, 226, 228, 230, 231, 234, 235, 237, 248, 249, 253–256, 258, 259, 310
Pacific Coast District, 28, 30
Pacific Freight Lines, 31
Pacific Green Trucking, 195
Pacific Maritime Association, 46, 69, 228, 231
Pacific Merchant Shipping Association, 90
Pacific Unitarian Church, 234
Paez, Richard, 160
Palos Verdes, 17, 44
Panama Canal, 3, 22, 24, 34, 36, 252, 286, 313
Park, Noel, 82, 89, 93, 112

Pasha Green Omni Terminal Demonstration Project, 282
Paul, Sanjukta, 193, 199, 267, 340
Paz, Alex, 238
Pearce, Jeannine, 260
Pelosi, Nancy, 181
Perez, Thomas, 245
Performance Team Freight System, 259
Petroleum coke, 65, 349
Pettit, David, 111, 124, 135, 169, 176
Picketing between the headlights, 226–227
PierPass, 109, 287
Pier S,
  Environmental Impact Review, 278
  proposed cargo terminal, 277
  temporary container facility, 239
Pier T, 80, 81, 229
Politeo, Tom, 112, 133, 300
Political opportunity theory, 11, 103–104, 273, 295, 298
Pope, Carl, 105, 323
Port commerce, 20, 24
Port devolution, 52
Port governance, 51–54
  Board of Harbor Commissioners, 53
  community participation in, 19, 59, 103, 252, 294
  executive director, 53
  regulatory capture, 10
Port growth, 9, 18, 19, 20, 32, 33, 35, 38, 39, 40, 43, 64, 80, 277, 286, 299
Port Innovations, 259, 265
Port of Long Beach,
  Board of Harbor Commissioners, 80, 90, 91, 99, 123, 125, 127, 142, 146, 167, 228, 238, 239
  cargo volume, 25, 27, 35, 38, 40, 276
  Carnival Cruise Terminal, 66
  founding, 25
  management, 92, 94, 133, 162, 243, 249
  Pier J, 90–92, 94, 249

Port of Long Beach (cont.)
  terminals,
    Long Beach Container, 228, 229, 232, 236
    Pacific Container, 249
    Total Terminals International, 229
  threatened litigation as hazardous waste site, 145
Port of Los Angeles,
  Board of Harbor Commissioners, 22, 37, 42, 53, 81, 125, 133, 135, 141, 143, 144, 227, 232, 237, 277
  cargo volume, 25, 27, 35, 38, 40, 276
  founding, 23
  management, 61, 87, 93, 95, 96, 98, 126, 137, 142, 149, 152, 162, 169, 170, 181, 276, 280, 281, 286, 305, 306
  Port Community Advisory Committee, 84
  terminals,
    APL, 133, 229, 231
    China Shipping, 83, 304
    Evergreen, 229, 231, 232
    TraPac, 45, 52, 67, 133, 135, 144, 152, 252, 281, 306
    Yusen, 229, 232, 260
Port of New York-New Jersey, 75, 192, 251, 343
  Clean Truck Program, 288
Port of Oakland, 3, 75, 192, 288
Port of Seattle-Tacoma, 75, 288
Port revenues,
  amount, 26
  cargo fees, 54
  permit fees, 54
  pressure to generate, 27
  rents, 54
  shipping income, 53
  uses of, 42
Ports,
  as agglomeration economy, 43
  in city building, 19
  definition of, 51
  economic benefits of, 133
  environmental challenges, 49
  expansion (*see* Port growth)
  externalities, 43–49
  geography, 17
  history of, 17
  as opportunity for labor movement, 10
  regulatory jurisdiction over, 53–57
  in relation to globalization, 32–43
  structure, 52
Potter, Fred, 209, 215, 225, 227, 240, 248, 261, 350
Preemption,
  ceiling, 56
  challenge to local labor law, 14, 315, 344
  FAAAA (*see* Federal Aviation Administration Authorization Act)
  federal law, 55–56
  floor, 56
  legal analysis of, 108, 110, 322, 324, 328, 336–337
  market participant exception to, 58, 144
  in relation to ATA litigation, 154–165, 168–178, 314, 338
  in relation to labor peace, 316, 325, 350
  role in Campaign for Clean Trucks, 3, 6, 8, 11–15, 94, 95, 104, 126, 131, 178–184, 300–301
  as a tool of industry opposition to ports campaign, 318
Private Attorney General Act, 224, 241, 270–271
Progressive cities, 6, 343, 345
Progressive era, 24, 343
Project labor agreement, 58, 324
Proposition 1B, 101, 134
Proposition 187, 107
Proposition S, 147

Proxies for labor rights, 13, 313
Public Citizen, 74, 175
Public Counsel Law Center, 158
Public trust doctrine, 52

QTS Inc.,
　affiliated firms, 238, 254
　bankruptcy, 240, 249, 251, 320, 332
　class action against, 199–200, 224,
　　238, 251, 267, 269, 331, 340
　Employment Development
　　Department action against, 240
　Erick Yoo, 238
　NLRB actions, 315
　strikes against, 237, 238

Radio frequency identification device,
　142
Radisich, Joe, 96
Railroads, 24, 28, 36–37, 39, 93, 97
　Alameda Corridor, 39, 43, 44, 67, 68,
　　82
　Burlington Northern Santa Fe, 37, 39,
　　67, 94, 114, 277, 279, 280
　electrical, 94
　Harbor Belt Line Railroad, 37
　Los Angeles & San Pedro Railroad, 20
　off-dock, 37
　on-dock, 37, 39, 52, 82, 91, 278
　Southern Pacific Railroad, 20–23, 37
　Union Pacific, 37, 260, 277
Rancho Dominguez, 226
Raymond, Brad, 75
Reagan, Ronald, 160
Registration and Agreement, 162, 243
Reid, Harry, 211, 307
Resource Conservation and Recovery
　Act, 145
Retaliation by trucking firms, 201, 221,
　222, 230, 237, 254, 265, 266
Ridley-Thomas, Mark, 251
Riordan, Richard, 66
Road to Shared Prosperity, The, 134

Roberts, John, 175
Rocha, Ernesto, 209
Rosenthal, Jonathan, 250
Rosenthal, Steven, 131, 157, 169, 175,
　338
Russell, Thomas, 130
Rutgers School of Management and
　Labor Relations, 191

Sabel, Charles, 335
Saenz, Tom, 132
Sailor's Union, 28
San Pedro,
　activism by residents, 59, 81–84, 89,
　　103, 112, 144, 294
　annexation by Los Angeles, 22–23
　Bay, 17, 21, 22, 90, 117
　Cabrillo Beach, 65
　contrast with Wilmington, 65
　Fort MacArthur, 65
　Harbor, 19, 21, 22, 26
　Harbor Administration Building, 97
　history, 20–23
　impact of port, 60, 62, 65
　labor movement, 29–30
　Ports O' Call Village, 65
　public library, 113
　recreation, 61, 65
　redevelopment, 66
　secession effort, 66, 304
　socioeconomic conditions, 44
　transportation to, 37–38
　zoning, 65
San Pedro and Peninsula Homeowners
　Coalition, 82
San Pedro Peninsula Homeowners
　United, 82
Santamaria, Carlos, 209, 215, 223, 227,
　249, 273, 330, 352
Santa Monica,
　Bay, 17
　planned port and railroad, 21
Santangelo, Jim, 129

Savannah, Georgia, 289
Sayas, Joe, 200, 234, 308, 331
Saybrook Capital, 234, 249, 261, 347
Scalia, Antonin, 175
Schlageter, Martin, 147, 160
Schneider, 139
Schragger, Richard, 345
Schwab, Gertrude, 64
Schwarzenegger, Arnold, 70, 92
Scopelitis, Garvin, Light, Hanson & Feary, 169
Scott, Judy, 75
Scrap metal processors, 76
Seacon Logix, 196, 224, 248, 261, 307, 320
Sea-Logix, 253
Sectoral strategies, 342
Senate Bill 588, 255, 259–260
Senate Committee on Commerce, Science, and Transportation, 180, 184
Seroka, Gene, 237, 281, 286
Service Employees International Union, 211
Sherman Act, 209
Shippers,
  and air pollution, 90, 99, 101, 109, 282
  and Clean Truck Program, 135, 139, 143, 148, 154, 288, 312
  effect of deregulation on, 18, 26, 27, 32, 33, 36, 285, 286
  relation to ports, 35, 42, 46, 49, 74, 252, 287, 313
  vertical integration, 52
Shippers Transport Express Inc.,
  class action against (see *Taylor v. Shippers Transport Express, Inc.*)
  collective bargaining agreement, 253, 316
  DLSE claims against, 225
  DOL lawsuit against, 196, 306
  labor peace agreement, 241
  lawsuits against, 241
  reclassification, 241, 253
  relation to Carrix-SSA Marine, 234
  settlement with DOL, 241, 331
  strike at storage yard, 221, 226, 310
Shipping Act, 27, 36, 56, 108, 166
  1916, 27
  1984, 36, 56, 62, 77, 127, 166
  reasonableness test, 108
Shipping industry,
  consolidation, 252
  effect of recession, 285
  oversupply, 286
Shoestring district, 22
Short-haul rail, 287
Sierra Club,
  Harbor Vision Task Force, 113
  litigation, 78
  Los Angeles-Orange County chapter, 113
  role in ATA litigation, 156
  role in Blue-Green Alliance, 105
  role in Clean Truck Campaign, 105, 113, 115, 133, 144, 181, 300, 323, 327
Signal Hill oil strike, 25
Simon, William, 335
Sirolli, Matt, 251
Skadden Fellowship, 192
Skechers Shoes, 222
Slangerup, Jon, 243, 249
Slip No. 5, 62–63
"Slip seat" practice, 139
Smart, William, 125
Smith, Gary, 113
Smith, Randy, 171
Smith, Rebecca, 191, 340
Snyder, Christina, 158
Social movements,
  adaptation, 185, 303
  and coalitions (see Coalitions)
  coordination between labor and environmental, 7, 296, 297

in local politics, 6, 343, 345
outcomes, 311–321
resilience, 14
role of law in, 4, 5, 298
role of lawyers in, 5, 303, 333–341
strategy, 297
"Sock on a stack" technology, 282
Solicitor general, 173–175
Solis, Hilda, 181, 194
Southern California Edison, 25, 125
Southern California International Gateway,
  Environmental Impact Review, 257, 279
  NRDC lawsuit, 257, 325
  opposition to, 114, 278–279, 286
  proposal for, 257, 277, 312
Southern California Logistics and Supply Chain Summit, 246
Southern Counties Express, 164, 201, 224
Southern Pacific Railroad. *See* Railroads
Southern Utah Wilderness Alliance, 278
Spiro Moore, 200
SSA Marine, 196, 221, 234, 241–242, 248, 270–271, 310
Stakeholder participation, 326, 335
Stand Up for Savannah, 289
Stanford, Leland, 20
Stanford Law School, 219
Starcrest Consulting Group, 93
State Admissions Act, 23
State laws on statutory employees,
  New Jersey, 289
  New York, 289
St. Cyprian Catholic Church, 234
Steel Workers Union, 181
Steinke, Richard, 162
Sterling Express Services, Inc., 251
St. Joseph Catholic Church, 234
Stop the SCIG!, 278–279
Strategy, definition of, 297

Strumwasser Woocher, 199
Su, Julie, 192, 263, 267, 306
Subsidence, 26, 34
Sun Pacific Trucking, 195
Superior Dispatch, 259
Supreme Court, 11, 58, 86, 108, 168, 173–176, 179, 191, 244, 275–277, 302, 312, 316, 319–320, 324, 328, 338, 344
Swift Transportation, 216, 352
Szymanski, Pat, 73

Tactics, definition of, 297
Talavera v. QTS,
  class action, 199, 200, 251, 331, 340
  consolidation with bankruptcy, 320
  settlement, 267, 332
Target, 273, 341
Tate, Eric, 203, 209, 218, 225, 227
Taylor v. Shippers Transport Express, Inc.,
  class action, 200, 234, 320
  relation to DOL suit, 234, 241, 331
  settlement, 242
Teamsters. *See* International Brotherhood of Teamsters
Teamsters Joint Council, 214
Tea Party, 182
Tenacity, 348–351
Tennessee Valley Power Authority, 96
"Terminal Impact," 105
Terminal Island, 22–23, 37, 44, 63, 65, 89, 277
Terminal Island Freeway (State Highway 103). *See* Freeways
Terminals,
  operators, 52, 54, 90, 93, 109, 221
  role in Clean Truck Program, 141–143, 150, 153, 177, 301
  striking, 229, 261, 272
  structure, 43, 52
  types, 52
Through rates, 27, 36
Tidelands Trust Act, 23, 27, 43, 53, 157

Todd Shipyard, 81
Toll Group, Inc.,
  collective bargaining agreement, 203–204, 209, 253, 269, 316
  employee model, 72, 202
  Hoffa meeting, 194
  relation to Justice for Port Drivers campaign, 212–214, 215–218, 246, 309, 347, 349
  Teamsters campaign, 188, 202–203
  union vote, 203
Total Distribution Service of Wilmington, 142
Total Transportation Services, Inc.,
  bankruptcy, 261, 320
  conversion to Eco Flow, 249
  DLSE claims against, 226, 233
  litigation against, 195, 224, 233, 331
  NLRB actions, 227
  retaliation against drivers, 233
  Saybrook Capital, 124, 347
  strikes against, 228, 230, 231, 235, 237, 310
Traber & Voorhees, 199, 308
Trade,
  barriers, 18, 35
  competition, 34
  composition, 39
  decline, 26
  deficit, 35
  growth, 27, 32, 35, 41
Tradelink Transport, 248, 259
Traffic,
  cargo, 24, 35, 37, 39, 89, 159, 252
  efforts to mitigate, 86
  impact on environment, 91
  impact on neighborhoods, 60, 64, 67
  ship, 40
  truck, 37, 38, 76, 82, 182
Transportation law, 8, 58, 178
Transportation Workers Identification Credential, 134, 172
Transport Maritime Association, 69

Transport Workers Union, 202
Transshipment, 43, 286
TraPac Container Terminal,
  enforcement failures, 281
  Environmental Impact Report, 144
  expansion, 67, 133, 144, 252, 306
  legal challenge to, 144, 352
  location, 45
  settlement of legal challenge,
    creation of Harbor Community Benefit Foundation, 152
    establishment of Mitigation Trust Fund, 152
Troutman Sanders, 167
Trucks,
  dirty, 49, 68, 103, 104, 121, 123, 149, 283, 293, 299, 324
  impact on neighborhoods, 64, 67
  maintenance, 10, 118, 162, 341
Turn times, 228, 237, 239, 252, 287
Twenty-foot equivalent units, 33
  fees on, 143
  growth, 35, 39, 40
2020 Plan, 39, 60, 64
Tynan, Roxana, 107

Uber,
  class actions against, 290
  and misclassification, 55
  unemployment benefit actions, California, 290
  New York, 290
Unfair business practices, 71, 190, 191
Unger, Roberto, 335
Union density, 6, 32, 47, 202, 205, 293, 342, 350, 354
Union Pacific. *See* Railroads
United Church of Christ, 234
United Farmworkers Union, 106
United Food and Commercial Workers Union, 217
United Steelworkers Union, 105
United Teachers Union, 106

Index

University of California at Los Angeles, 96, 128, 217
   labor center, 128
   public policy school, 217
   school of law, 96, 175
University of Colorado Law School, 110
University of Southern California, conference, 79
   Keck School of Medicine, 114
   research, 91

Valenzuela, Manny, 209
Venetian Hotel, 106
Verrilli, Donald, 174
Villages of Cabrillo, 279
Villaraigosa, Antonio,
   LA Unified School District takeover, 132
   progressive coalition, 95, 106
   role in Clean Air Action Plan, 94–99, 114, 305, 327
   role in Clean Truck Program, 127–130, 132, 133, 136, 142, 144, 146, 148, 152, 162, 181, 306, 337
   role in supporting Teamsters organizing, 202, 216
Vincent Thomas Bridge, 81, 89

Wage Justice Center, 199, 238, 267, 331
Wage theft, 192, 196, 238, 256, 260, 263, 272
Wall Street Journal, 250, 253, 263
Walmart, 32, 73, 135, 139, 140, 217, 273, 300, 341
Warehouses, 1, 30, 37, 43, 52, 73, 202, 217, 219, 221, 227–228, 231, 257–258, 265, 279, 290, 300, 310, 325, 341
Waste disposal facilities, 60
Waterfront Rail Truckers Union, 68
Waxman, Seth, 175
Wedekind, Jeffrey, 247

Weiner, Nick, 74, 105, 115, 179, 191, 209–210, 220, 227, 244, 251, 272, 304, 310, 323, 350
Western Freight, 224, 259
West Long Beach Association, 278
Whalen, Curtis, 148, 250
White, Stephen, 21
Willful Misclassification Law, 195
Wilmington,
   activism by residents, 61, 62, 64, 114, 294, 348, 349, 352
   annexation by Los Angeles, 23
   Banning Park, 62
   Cal Cartage warehouse lease, 257
   Coalition to Stop the Wall, 67
   community plan, 62
   contrast with San Pedro, 65
   East Wilmington, 61, 63
   environmental hazards, 38, 45, 60, 62
   Far East Wilmington Improvement Association, 63, 349
   Heinz Pet Food cannery, 63
   history, 20–22, 60
   impact of port, 38, 39, 64, 65, 67
   Industry and Economic Development Committee, 62
   oil field, 26
   proposed Southern California International Gateway Project in, 277
   redevelopment, 62, 64
   secession effort, 66
   socioeconomic conditions, 45, 60
   wall, 67
   Waterfront Park, 45, 231
   Wilmington Home Owners, 62, 65
   Wilmington Liquid Bulk Terminals, 62, 64
   zoning, 61
WinWin Logistics, 199, 237–238, 251, 254, 259, 267, 269, 315, 320
Wong, Kent, 128
Woodfield, Kathleen, 144, 152

Worker center, 330
Workers' compensation, 70–71, 189–190, 289, 304
World Cruise Center, 66
World War II, 26, 30, 32
Wyenn, Morgan, 176, 257, 277–278, 334

XPO Cartage,
  change from Pacer Cartage, 254
  class action against, 332
  DLSE claims against, 260
  NLRB actions, 266
  strikes against, 265
XPO Logistics,
  company, 227, 289, 347
  DLSE claims against, 244, 260
  litigation against, 260, 332
  NLRB actions, 309, 315, 320
  strikes against, 248, 256, 258, 260, 265, 348
  subsidiaries,
    Harbor Rail Transport, 238, 254
    Pacer Cartage, 227, 238, 254
XPO Port Services,
  change from Harbor Rail Transport, 254
  class action against, 260
  NLRB actions, 266
  strikes against, 265

Yeager, David, 261
Yoo, Erick, 199, 238

Zero-emissions port, 282
Zerolnick, Jon, 106, 134, 154, 164, 179, 200, 217, 305, 327, 347, 351
Zoning, 61–63, 65

Urban and Industrial Environments
Series editor: Robert Gottlieb, Henry R. Luce Professor of Urban and Environmental Policy, Occidental College

Maureen Smith, *The U.S. Paper Industry and Sustainable Production: An Argument for Restructuring*

Keith Pezzoli, *Human Settlements and Planning for Ecological Sustainability: The Case of Mexico City*

Sarah Hammond Creighton, *Greening the Ivory Tower: Improving the Environmental Track Record of Universities, Colleges, and Other Institutions*

Jan Mazurek, *Making Microchips: Policy, Globalization, and Economic Restructuring in the Semiconductor Industry*

William A. Shutkin, *The Land That Could Be: Environmentalism and Democracy in the Twenty-First Century*

Richard Hofrichter, ed., *Reclaiming the Environmental Debate: The Politics of Health in a Toxic Culture*

Robert Gottlieb, *Environmentalism Unbound: Exploring New Pathways for Change*

Kenneth Geiser, *Materials Matter: Toward a Sustainable Materials Policy*

Thomas D. Beamish, *Silent Spill: The Organization of an Industrial Crisis*

Matthew Gandy, *Concrete and Clay: Reworking Nature in New York City*

David Naguib Pellow, *Garbage Wars: The Struggle for Environmental Justice in Chicago*

Julian Agyeman, Robert D. Bullard, and Bob Evans, eds., *Just Sustainabilities: Development in an Unequal World*

Barbara L. Allen, *Uneasy Alchemy: Citizens and Experts in Louisiana's Chemical Corridor Disputes*

Dara O'Rourke, *Community-Driven Regulation: Balancing Development and the Environment in Vietnam*

Brian K. Obach, *Labor and the Environmental Movement: The Quest for Common Ground*

Peggy F. Barlett and Geoffrey W. Chase, eds., *Sustainability on Campus: Stories and Strategies for Change*

Steve Lerner, *Diamond: A Struggle for Environmental Justice in Louisiana's Chemical Corridor*

Jason Corburn, *Street Science: Community Knowledge and Environmental Health Justice*

Peggy F. Barlett, ed., *Urban Place: Reconnecting with the Natural World*

David Naguib Pellow and Robert J. Brulle, eds., *Power, Justice, and the Environment: A Critical Appraisal of the Environmental Justice Movement*

Eran Ben-Joseph, *The Code of the City: Standards and the Hidden Language of Place Making*

Nancy J. Myers and Carolyn Raffensperger, eds., *Precautionary Tools for Reshaping Environmental Policy*

Kelly Sims Gallagher, *China Shifts Gears: Automakers, Oil, Pollution, and Development*

Kerry H. Whiteside, *Precautionary Politics: Principle and Practice in Confronting Environmental Risk*

Ronald Sandler and Phaedra C. Pezzullo, eds., *Environmental Justice and Environmentalism: The Social Justice Challenge to the Environmental Movement*

Julie Sze, *Noxious New York: The Racial Politics of Urban Health and Environmental Justice*

Robert D. Bullard, ed., *Growing Smarter: Achieving Livable Communities, Environmental Justice, and Regional Equity*

Ann Rappaport and Sarah Hammond Creighton, *Degrees That Matter: Climate Change and the University*

Michael Egan, *Barry Commoner and the Science of Survival: The Remaking of American Environmentalism*

David J. Hess, *Alternative Pathways in Science and Industry: Activism, Innovation, and the Environment in an Era of Globalization*

Peter F. Cannavò, *The Working Landscape: Founding, Preservation, and the Politics of Place*

Paul Stanton Kibel, ed., *Rivertown: Rethinking Urban Rivers*

Kevin P. Gallagher and Lyuba Zarsky, *The Enclave Economy: Foreign Investment and Sustainable Development in Mexico's Silicon Valley*

David N. Pellow, *Resisting Global Toxics: Transnational Movements for Environmental Justice*

Robert Gottlieb, *Reinventing Los Angeles: Nature and Community in the Global City*

David V. Carruthers, ed., *Environmental Justice in Latin America: Problems, Promise, and Practice*

Tom Angotti, *New York for Sale: Community Planning Confronts Global Real Estate*

Paloma Pavel, ed., *Breakthrough Communities: Sustainability and Justice in the Next American Metropolis*

Anastasia Loukaitou-Sideris and Renia Ehrenfeucht, *Sidewalks: Conflict and Negotiation over Public Space*

David J. Hess, *Localist Movements in a Global Economy: Sustainability, Justice, and Urban Development in the United States*

Julian Agyeman and Yelena Ogneva-Himmelberger, eds., *Environmental Justice and Sustainability in the Former Soviet Union*

Jason Corburn, *Toward the Healthy City: People, Places, and the Politics of Urban Planning*

JoAnn Carmin and Julian Agyeman, eds., *Environmental Inequalities Beyond Borders: Local Perspectives on Global Injustices*

Louise Mozingo, *Pastoral Capitalism: A History of Suburban Corporate Landscapes*

Gwen Ottinger and Benjamin Cohen, eds., *Technoscience and Environmental Justice: Expert Cultures in a Grassroots Movement*

Samantha MacBride, *Recycling Reconsidered: The Present Failure and Future Promise of Environmental Action in the United States*

Andrew Karvonen, *Politics of Urban Runoff: Nature, Technology, and the Sustainable City*

Daniel Schneider, *Hybrid Nature: Sewage Treatment and the Contradictions of the Industrial Ecosystem*

Catherine Tumber, *Small, Gritty, and Green: The Promise of America's Smaller Industrial Cities in a Low-Carbon World*

Sam Bass Warner and Andrew H. Whittemore, *American Urban Form: A Representative History*

John Pucher and Ralph Buehler, eds., *City Cycling*

Stephanie Foote and Elizabeth Mazzolini, eds., *Histories of the Dustheap: Waste, Material Cultures, Social Justice*

David J. Hess, *Good Green Jobs in a Global Economy: Making and Keeping New Industries in the United States*

Joseph F. C. DiMento and Clifford Ellis, *Changing Lanes: Visions and Histories of Urban Freeways*

Joanna Robinson, *Contested Water: The Struggle Against Water Privatization in the United States and Canada*

William B. Meyer, *The Environmental Advantages of Cities: Countering Commonsense Antiurbanism*

Rebecca L. Henn and Andrew J. Hoffman, eds., *Constructing Green: The Social Structures of Sustainability*

Peggy F. Barlett and Geoffrey W. Chase, eds., *Sustainability in Higher Education: Stories and Strategies for Transformation*

Isabelle Anguelovski, *Neighborhood as Refuge: Community Reconstruction, Place-Remaking, and Environmental Justice in the City*

Kelly Sims Gallagher, *The Global Diffusion of Clean Energy Technology: Lessons from China*

Vinit Mukhija and Anastasia Loukaitou-Sideris, eds., *The Informal American City: Beyond Taco Trucks and Day Labor*

Roxanne Warren, *Rail and the City: Shrinking Our Carbon Footprint and Reimagining Urban Space*

Marianne Krasny and Keith Tidball, *Civic Ecology: Adaptation and Transformation from the Ground Up*

Erik Swyngedouw, *Liquid Power: Contested Hydro-Modernities in Twentieth-Century Spain*

Julian Agyeman and Duncan McLaren, *Sharing Cities: Enhancing Equity, Rebuilding Community, and Cutting Resource Use*

Jessica Smartt Gullion, *Fracking the Neighborhood: Reluctant Activists and Natural Gas Drilling*

Nicholas A. Phelps, *Sequel to Suburbia: Glimpses of America's Post-Suburban Future*

Shannon Elizabeth Bell, *Fighting King Coal: The Barriers to Grassroots Environmental Justice Movement Participation in Central Appalachia*

Theresa Enright, *The Making of Grand Paris: Metropolitan Urbanism in the Twenty-first Century*

Robert Gottlieb and Simon Ng, *Global Cities: Urban Environments in Los Angeles, Hong Kong, and China*

Anna Lora-Wainwright, *Resigned Activism: Living with Pollution in Rural China*

David Bissell, *Transit Life: How Commuting Is Transforming Our Cities*

Javiera Barandiarán, *Science and Environment in Chile: The Politics of Expert Advice in a Neoliberal Democracy*

Scott Cummings, *Blue and Green: The Drive for Justice at America's Port*

www.ingramcontent.com/pod-product-compliance
Lightning Source LLC
Chambersburg PA
CBHW071353300426
44114CB00016B/2047